A history of the study of human growth

A history of the study of human growth

J. M. TANNER

Professor of Child Health and Growth, Institute of Child Health
University of London

CAMBRIDGE UNIVERSITY PRESS
Cambridge
London New York New Rochelle
Melbourne Sydney

Published by the Press Syndicate of the University of Cambridge
The Pitt Building, Trumpington Street, Cambridge CB2 1RP
32 East 57th Street, New York, NY 10022, USA
296 Beaconsfield Parade, Middle Park, Melbourne 3206, Australia

First published 1981

Printed in Malta by Interprint Limited

British Library Cataloguing in Publication Data

Tanner, James Mourilyan
A history of the study of human growth.
1. Human growth
2. Human biology–History
3. Developmental biology–History
I. Title

612.6'009 QP84 80-40485

ISBN 0 521 22488 8

To my wife and colleague, Dr B. A. Tanner, whose sustained enthusiasm for ideas and uncompromising devotion to scholarship have been an inspiration and support through many years

CONTENTS

PREFACE

For the last thirty years, most of my professional life has been concerned with the physical growth of children: individual children growing up normally or coming to a clinic for growth disorders, and populations of children struggling to grow up in the forests of New Guinea, the slums of Caracas and the social-security-benefit areas of British industrial towns. I am by profession a human biologist and a medical man, not a historian. The fact brings difficulties from two directions, corresponding to the two categories of people, human biologists and historians, that I am endeavouring simultaneously to address. For historians, I have tried to describe in straightforward language, at appropriate junctures, the processes of growth, so that they will not sink in a morass of technology: to aid me I have the experience of writing one short book describing growth in terms comprehensible to the non-biologist reader (*Fetus into Man,* 1978) and I hope that book may be a suitable and certainly a sufficient background for historians who encounter any difficulty with the present one, including its later chapters. For the human biologists, I have tried to sketch at least the broadest outlines of the historical scene, and historians must bear with me when I pause to explain the family tree of Frederick the Great or the relation of Chadwick to the Factory Acts of nineteenth-century Britain.

I very much hope that the professional historian will forgive a biologist for trespassing on his domain. Though I can justly claim to know more about growth than any historian, any historian, with equal justice, may claim that I know less about history than he. For this I can but apologize, and hope the professionals may at least think it worth their while to write and tell me about mistakes and misconceptions, so that if a second edition should be called for, it can be an improved one.

Even the title, I fear, is somewhat pretentious, for this book is scarcely a history, more a simple chronicle. To turn one into the other, the movements of men's minds, even the elaborate driftings of the *Zeitgeist,* need to be traced. *The Origins of Human Biology* is a book not yet written; but whoever writes it will need this more modest list of events close to hand. The attitude of society towards the study of growth reflects the way in which people regard not only children but themselves, their origins and their destination. Thus I hope the historian of ideas will find something in this book to claim a portion of his attention.

I leave the worst difficulty till last. I found it impossible (by which I mean I refused) to choose between writing a book on the history of ideas concerning growth, and a book on how children grew in former times. I wanted to do both at once, and I have endeavoured to do so. I can only hope that historians and auxologists alike will be tolerant of and even interested in the reverse of their side of the coin.

For the references I have used the system familiar to scientists, which seems to me to have a substantial advantage over the various methods used in other disciplines. I make no apology for the appearance of the name of the author in the main text; it adds interest and piquancy, it is itself a bit of history. Especially in historical studies, where it is least used, the Harvard system seems the most appropriate.

The endnotes unashamedly serve two purposes. Some supply the familiar scholarly apparatus; others mount private hobby-horses to hunt down, or at least after, particular points I find of interest. The reader whose mind works like mine will be amused by them; as for the others, endnotes, after all, are tucked away at the end of the book.

I have pursued my subject (perhaps unwisely) practically up to yesterday: this sets obvious traps that I have tried not to fall into in too inelegant a posture. I have usually used the past tense in referring to the works and interests of contemporaries, for example, and the unwary reader should not imagine they are all dead.

Lastly I have many people to thank. First and foremost my wife, Dr B. A. Tanner, who did much to initiate and sustain the whole enterprise. Second, librarians; particularly Mr H. R. Denham of the Wellcome Institute of the History of Medicine, whose immense knowledge of early works and the art of bibliography saved me untold amounts of time and supplied material which I could never have hoped to find on my own; Mr Richard Wolffe of the Countway Library of Medicine in Boston, who was so cheerfully generous of photocopies of Stöller and Bergmüller, and showed me Bowditch's correspondence; Miss Sandra Colville-Stewart of the Biomedical Library of the University of California at Los Angeles, who miraculously located practically all the items I was preparing to give up on; Jan-Olof Friström of the University Library of Lund; the staffs of the Royal Society of Medicine, the British Library and the US National Library of Medicine; and, last strictly for reasons of politeness, Miss E. Brooke and her staff of the library of my own institution, who performed prodigies of tracking and themselves carried great volumes of nineteenth-century literature across London to save me time necessarily whittled out of a busy professional existence. Third, all those who have helped with data, sources, memories and criticism: especially Professor Roderick Floud of the department of history, Birkbeck College, for letting me use prior to publication his magnificent archive of eighteenth-century data on the boys of the Marine Society, and his collaborators Professor Robert Fogel of Harvard University and Dr James Trussell of Princeton for a similar access to their data on the heights of North American slaves; Dr R. W. Sharples of University College, London, for help with Greek syntax and bibliography; Herr Ewald Mertins of Berlin and Frau Liesel Mewes of Hanover for helpful and detailed correspondence concerning Berlin and Potsdam in the eighteenth century; Dr Wolfram Kaiser for similar aid in respect of his home town, Halle; Professor Wilhelm Theopold of Stuttgart for introducing me, some years ago, to the history of the Carlschule; Dr F. Würst of Vienna for first acquainting me with the works of Hippolyt Guarinoni; Miss T. M. Swann of Northampton Central Museum and Miss E. Vigeon of the Salford Art Gallery for information on design of clogs in the nineteenth century; Dr Gunilla Lindgren of Stockholm for successfully dragging the cellars of the Institute of Educational Research for a copy of Key's report; and Heinz Schubert of Basle, my long-time mentor in the arts of translation and film communication, for confirming my suspicion that what turned out to be one of the best-known quotations from *Faust* must have been written by someone more literate than the author who had lifted it. Finally I wish to thank my friends, Frank Falkner, Nancy Bayley, Howard Meredith, Bill Krogman, Lois Stolz, Frank Johnston, Arto Demerjian, Peter Fox, Hans van Wieringen and Bryan Senior, for reading parts of the manuscript and for answering questions as to their own and other's lives, their work and parts of the manuscript; and Reg Whitehouse, collaborator of thirty years, for drawing, impeccably as usual, the charts. My secretarial staff exercised unbelievable patience as well as energy in reducing continually altering manuscripts to forms again to be altered, and to all of them, especially Janet Baines Preece and Susan Barrett, I wish to extend a heartfelt thanks.

London and Stentwood J. M. T.
December 1979

1

Human growth and the Ancient World

Solon

The earliest surviving statement about human growth appears in a Greek elegy of the sixth century B.C. The poet is Solon the Athenian, better known to history as statesman and law-giver. It was Solon who, during his archonship in 594–593 B.C, began the transformation of Athens from a Homeric feudal system to a democratic *polis*, or city-state (see Ehrenberg, 1973). He reformed the constitution, abolished the prevalent custom of enslaving citizens for debt, and when his term as archon ended, rejected the request that he continue as permanent tyrant, insisting that archons must be replaced by election every year. The Greeks themselves, no less than later historians, regarded him as one of the chief 'midwives of the *polis* ideal' (Green, 1973) and numbered him among the Seven Sages of the Ancient World.

Solon's surviving poems are few and not very highly regarded by modern classicists (though Plato had one of his characters declare that if Solon had put his mind to poetry instead of worrying about statecraft, he would have rivalled Homer or Hesiod: *Timaeus*, p. 34 in Lee's 1971 translation). The poems cover a number of subjects, for in Ancient Greece poetry was the vehicle for philosophical and scientific speculation as well as for lyrical passion and political complaint. The poem concerning growth deals with the division of the human life-span into *hebdomads*, that is, successive periods each of seven years' duration. There are considerable difficulties over the translation of what may be a partly corrupt version of the original, long since lost (see note 1.1). A version that combines scientific sense with philological probability (if not, regrettably, with poetic elegance) runs as follows:

> A young boy acquires his first ring of teeth as an infant [literally, while unable to speak] and sheds them before he reaches the age of 7 years. When the god brings to an end the next seven-year period, the boy shows the signs of beginning puberty [or: of beginning pubic hair]. In the third hebdomad, the body enlarges, the chin becomes bearded and the bloom of the boy's complexion is lost. In the fourth hebdomad physical strength is at its peak and is regarded as the

> criterion of manliness; in the fifth hebdomad a man should take thought of marriage and seek sons to succeed him. In the sixth hebdomad a man's mind is in all things disciplined by experience and he no longer feels the impulse to uncontrolled behaviour. In the seventh he is at his prime in mind and tongue, as also in the eighth, the two together making fourteen years. In the ninth hebdomad, though he still retains some strength, he is too feeble in mind and speech for the greatest excellence. If a man continues to the end of the tenth hebdomad, he has not encountered death before due time.

The poem epitomizes a tradition current in pre-classical Greece but coming, certainly, from an earlier time and a more Eastern source (Roscher, 1919). The human life-span was only one of an enormous range of natural and supernatural phenomena which tradition parcelled into sevens. The number seven was celestial, literally at first, metaphorically later. There were seven moving bodies: the Sun, Moon and the five planets visible to the naked eye —Mercury, Venus, Mars, Jupiter and Saturn (see Aaboe, 1974). Correspondingly, the Babylonian temple or ziggurat was constructed with seven stages; later the seven vowels of the Greek alphabet became the planetary signs. The number persists in our seven-day week and appeared 'in the seven sleepers, the seven wonders of the world, and the seven ages of man . . . the seven states of Isis, the seven-stepped ladder of Mithras, the seven joys of the righteous in the Salethiel apocalypse, the seven angels and vials of the *Revelation*, the seven gates of hell, and the seventh heaven' (Tarn and Griffith 1966).

Indeed Solon's poem itself only reached us as part of a collection of seventh-day and seventh-fold allusions and practices in Jewish, Greek and Roman religion and tradition. The collection, *De septenario*, was assembled by Philo (*c.* 20 B.C.—A.D. 45), a Jewish philosopher living in Alexandria some 500 years after Solon's death. The notion of a seven-related division of the life-span has persisted right into modern times. It was responsible for the Roman law which fixed age 14 as the legal beginning of puberty. It found an economic embodiment in the Elizabethan apprenticeship, which combined an obligatory seven-year duration with a termination at an age not less than 21 (see Emison, 1976, p. 168). Until recently it governed our own laws on coming-of-age and the right to marry.

Solon's division of the life-span into ten successive seven-year periods, summing to the biblical seventy, was modified later (or perhaps had a contemporary, probably more Eastern, rival). The number seven persisted, but now as the divisor of the total life-span; life was partitioned into seven periods, not all of equal length. This system seems to have arisen from a desire to place each successive age under the aegis of one of the seven planets.

References to it are found in Greek writings throughout the classical and Hellenistic period (see Mansfeld, 1971), in particular in the pseudo-Hippocratic treatise entitled *Peri hebdomadôn* ('On Sevens'). The manuscripts of this work were unearthed by Littré, who presented it as *Des semaines* in his *Oeuvres complètes d'Hippocrate* (1853, vol. 8, pp. 634–73). No English translation has been made. Littré says he thinks the last original Greek text of the work perished in a fire in the Escorial in Madrid in 1671; only two Latin texts 'horribly barbarous and in places unintelligible' remained for him to translate. Probably the book dates from the first century B.C., well after Hippocrates and Aristotle (see note 1.2).

Des semaines deals with all manner of meanings of the number seven. There are seven winds, and seven parts of the year. The body is divided into seven parts (oddly, the head, hands, viscera, genitalia, urine plus sperm, intestines and limbs). The head has seven functions, the body seven elements. 'Thus also there are seven periods in nature called "stages": *puerulus, puer, adolescens, juvenis, junior, vir, senex.*' The first four stages are still each of seven years' duration. *Puerulus* lasts until the shedding of the teeth at age 7; *puer* until 'the emission of semen' at the age of 14 (a confusion, for normally this would be later than the first appearance of pubic hair, used as the transition point by Solon). *Adolescens* continues to 21 and the transition is marked by the beard having grown. The transition at 28 (from *junior* to *vir*) is marked by '*crementus corporis*' presumably meaning full growth of the body. (If this indeed is what the author of *Peri hebdomadôn* meant it perhaps throws some light on the question of the age of cessation of growth in classical Greece. The mystique attached to the seven-fold system makes reliance on the truth of any actual age impossible however; both at the time and for centuries later there is a tendency for all ages to be rounded to the nearest hebdomadal number. Nevertheless, the phrase may imply that cessation of growth occurred nearer age 28 than 21.) The hebdomadal tradition is a strait-jacket through the ages. Even in Shakespeare's *Romeo and Juliet* Juliet and her mother are both given age 14 as the age at which they become women. Perhaps the daughters of Shakespeare's acquaintances did indeed have their first menstrual periods at about 14; but more probably the age is simply a reflection of tradition.

In *Peri hebdomadôn* there is no mention of the planetary rule. This first appears in the writings of Proclus Diadochus (A.D. 412–485), the last major teacher in the tradition of Plato's Academy. Proclus was as much influenced by Eastern sources such as the *Chaldean Oracles* as by the original ideas of Plato, and in his commentary on Plato's *First Alcibiades* he specifies a sevenfold division of the life-span together with the appropriate planets for each age. Moon reigns over the infant, Mercury over the child, Venus over the adolescent, then follow Sun, Mars, Jupiter and Saturn. Proclus sees this as one of the many ways in which the celestial macrocosm interpenetrates the microcosm of man. That notion, together with the division of the life-span, persisted through mediaeval times and into the Renaissance. It appears, almost unchanged from Proclus, in Sir Walter Raleigh's *History of the World* (1614), and, with the planets dropped, finds its most lasting echo in Shakespeare's *Seven Ages of Man* (see note 1.3).

Hippocrates

In the strictly medical works of the Hippocratic corpus there is little said about growth. In *Airs, Waters and Places* Hippocrates (b. 460 B.C.) writes that puberty is attained late in districts with a cold wind coming from the north-west or north-east, with a water supply which is hard, cold and brackish (Chadwick and Mann's translation, p. 92). Concerning the final effects of growth as manifested in the stature and strength of the adult, 'the chief controlling factors are the variability of the weather, the type of country and the sort of water which is drunk' (p. 110). Echoes of this statement are heard all through the Middle Ages down to the nineteenth century. The great emphasis that Villermé, early in the nineteenth century, placed on poverty and undernutrition as the main determinants of growth has to be seen in the light of the current opinions he was arguing against, and those opinions were directly descended from *Airs, Waters and Places.*

Aristotle

Aristotle (384–322 B.C.), the first great mentor of European science, had something to say about the mechanism of growth and a great deal to say about animal embryology. The phenomenology of human growth, however, concerned him less. As D'Arcy Thompson (1913), the only president of a national Classical Association to be simultaneously a professor of natural history insists, Aristotle was first and foremost a biologist. The observations and dissections he reports were mostly made on just those plants and animals which occur in the areas where he passed his early boyhood and mature middle age (Chalcide in north-eastern Greece and the Ionian islands round Lesbos). Classicists disagree as to whether the whole of the extraordinarily voluminous writings we traditionally ascribe to Aristotle were actually written by him, or represent the combined efforts over many years of successive members of the Peripatetic School; but there is no dispute about the authorship of the zoological books *Historia animalium, De partibus animalium, De generatione animalium, De incessu animali* and *De generatione et corruptione.*

Aristotle's father was a physician – latterly to Amyntas II, King of Macedonia – and he himself must have received much of the regular Asclepian's training, which included an extensive study of the varieties and uses of plants. However his father died when he was 10, and he never completed his medical studies; indeed the Aristotelian corpus of writing contains less about medicine than about almost anything else (see Sarton, 1952–9, vol. 1, p. 537). In 367 B.C. when aged 17, Aristotle was sent from Macedonia to Athens to enter Plato's Academy. He became the Master's most celebrated, and some said most critical, pupil. Plato was forty-four years older than Aristotle, and when he died (in 347 B.C.) the headship of the Academy passed to his nephew Speusippus, to whose philosophical views Aristotle was opposed. At the same time life in Athens was made difficult for a Macedonian-connected metic by the war against Athens waged by Philip, the new King of Macedonia, Aristotle's exact contemporary and probably boyhood friend. Accordingly Aristotle withdrew to Assos on the Ionian coast, accompanied by his colleague Xenocrates. Here and on the island of Lesbos just opposite, Aristotle undertook his most concentrated period of biological research. After four years he joined the court in Macedonia. Though the legend that he was Alexander the Great's particular tutor is without foundation (see Grayeff, 1974), he did become a trusted member of Alexander's faction in the struggle which culminated in Philip's murder and Alexander's assumption of power (334 B.C.).

The following year Aristotle was sent back to Athens loaded with books and with teaching material of all kinds. Xenocrates had meanwhile become head of the Academy, so Aristotle founded a school of his own, located on the opposite side of the city in the Lyceum, a celebrated gymnasium established by Peisistratus. As in all Athenian gymnasia there were gardens, theatres and places for wrestling and running. The Lyceum was distinguished by having a particular covered walk called the Peripatos, and because Aristotle usually taught there his students became known as the Peripatetics. (In the same way, later, the students of Zeno were named Stoics because they met at the Stoa Poikile, or Painted Stoa, a building in the form of a long open colonnade.) Aristotle's school soon far out-ranged the Academy in the breadth of its courses, which included science and rhetoric as well as philosophy. 'One sees the Lyceum', writes Grayeff (1974), 'as the first university almost in the full, modern sense of the word ... the school housed the first comprehensive collection of books in history'.

By 200 B.C., however, its course was run, and when Cicero visited Athens in 79 B.C. he found no trace of either Lyceum or Academy; the places where Plato and Aristotle had taught were deserted ruins. Copies of the lecture notes of Aristotle and his colleagues (which is all we possess, for the writings actually issued in his life-time have disappeared) went to Pergamon and Alexandria and then found their way back to Athens, whence in 86 B.C. Sulla, the Roman plunderer of Athens, brought them to Rome. Here Cicero saw them and commenced the reappraisal of Aristotle as the greatest philosopher after Plato. Finally Andronicus, a scholar of Rhodes, collected the works and brought out in about 30 B.C. the definitive edition of the Aristotelian corpus.

'None of the other philosophers except Democritus made any definite statement about growth, except such as any amateur might have made. They said that things grew by accession of like to like, but they did not proceed to explain the manner of the accession.' Thus Aristotle opens the section of *De generatione et corruptione* which deals with growth. The tone is modern and Aristotle's explanation is modern too: 'We must think of tissues after the image of flowing water ... particle after particle comes-to-be (*ginomenon*) and each successive particle is different. It is in this sense that the matter of flesh grows (*auxanetai*), some flowing out and some flowing in fresh, not in the sense that fresh matter accedes to every particle of it. There is however, an accession to every part of its figure (*schêmatos*) or form (*eidos*)' (321b, Joachim's translation). Joachim summarizes Aristotle's meaning: 'Matter is a flux of different particles ... flowing in and out of the structural plan which is the form' (see note 1.4).

This passage was the subject of countless commentaries and counter-commentaries in mediaeval times. Only very recently have we made any real progress in understanding the mechanism whereby varying rates of growth in different directions generate and sustain form. As to the organizing forces of which form is the tangible outcome, of them we know scarcely more than Aristotle. Aristotle distinguished 'growth' from 'alteration' and from 'coming-to-be' or 'formation'. Growth concerns a quantitative increase without change of quality; alteration (*alloiôsis*) represents a change of quality in something with persistent identity (as in assimilation of food); coming-to-be is the appearance of a new identity.

As a descriptive zoologist, particularly of the marine species known to him in his childhood, Aristotle was superlative. His descriptions of the species of cuttle-fish and octopus were not equalled, let alone surpassed, till Cuvier completed Buffon's *Natural History* in the nineteenth century. Even then, says Thompson (1913), 'Cuvier lacked knowledge that Aristotle possessed'. George Henry Lewes (1817–1878), a naturalist and pioneer of the history of science (and the devoted companion of George Eliot), wrote in 1864 'I know of no better eulogy to pass on Aristotle than to compare his work with the "*Exercitations concerning Generation*" of our immortal Harvey. The founder of modern physiology was a man of keen insight, of patient research, of eminently scientific mind. His work is superior to that of Aristotle in some few anatomical details; but it is so far inferior to it in philosophy, that at the present day it is much more antiquated, much less accordant with our views.' Charles Darwin was equally impressed. When William Ogle published his translation of Aristotle's *The Parts of Animals* in 1882, Darwin wrote to him: 'From quotations I had seen, I had a high notion of Aristotle's merits, but I had not the most remote notion what a wonderful man he was. Linnaeus and Cuvier have been my two gods, though in very different ways, but they were schoolboys compared with old Aristotle' (*Parts of Animals*, Peck's translation, p. vi).

Aristotle gave a lengthy and detailed account of the development of the chick. On the third day he saw the heart beating, and a little later he made out the outlines of the body, with the big head and eyes and the blood vessels leading to the yolk sac and allantois. It was the early appearance of the heart that prompted him to make one of his few bad mistakes, in transferring to it the primary sense, or central seat of the soul, which Hippocrates and Plato located in the brain. Aristotle's embryology was the direct father of Harvey's study of reproduction and Haller's study of chick growth, 2,000 years later.

About human growth, however, Aristotle has not a great deal to say. He noted that 'Man, when young, has his upper part [i.e. above the pubes] larger than the lower, but in the course of time comes to reverse this condition and it is owing to this circumstance – an exceptional instance amongst animals by the way – that he does not progress in early life as he does at maturity, but in infancy creeps on all fours' (*Historia animalium*, 500b, Thompson's translation). He describes puberty: 'In man, maturity is indicated by a change in the tone of voice, by the increase in size and alteration in appearance of the sexual organs, as also by an increase in size and alteration in appearance of breasts; and above all in the hair growth at the pubes' (544b, Thompson's translation). 'When twice seven years old in the most of cases the male begins to engender seed, and at the same time hair appears on the pubes. At the same time in the female the breasts swell and the so-called catamenia [i.e. menstrual discharges] commence to flow . . . in the majority of cases catamenia are noticed by the time the breasts have grown to the height of two finger breadths' (581). Aristotle thought, incorrectly, that females developed more slowly than males in early fetal life, but, correctly, that 'after birth everything is perfected more quickly in females than in males', as shown by the earlier puberty of girls (*De generatione animalium*, 775a).

In discussing education he says that 'till the age of puberty [physical] exercises ... should be light [since] violent exertion may hinder the proper growth of the body. The bad effects of excessive early training are strikingly evident. In the lists of Olympic victors there are only two or three cases of the same person having won in the men's events who had previously won in the boys' and the reason is that early training and the compulsory exercises involved, has resulted in loss of energy' (*Politics*, 8, 398). Aristotle also says, though with less assurance, 'the physique of men is also supposed to be stunted in its growth when intercourse is begun before the seed has finished *its* growth' (*Politics*, 8, 383). Both opinions, like so much of Aristotle, survive till the present day, half-conscious and half-believed. (See note 4.5 regarding Tacitus' emphasis on the latter view, and its subsequent history.)

It seems very likely that some form of measurement was practised on children and youths in Ancient Greece, but no records have come down to us. Polyclitus' search for the canon of physical beauty apparently involved measurements of youths and adults (see Chapter 3). And, as regards younger children, Aristotle states that 'in about five years, in the case of human beings ... the body seems to gain about half the height that is gained in all the rest of life' (*De generatione animalium*, 725b). If we take this to mean that the gain in height from birth to age 5 is about the same as the gain from 5 to maturity, the statement corresponds fairly well to present-day reality. Indeed nowadays the present half-way mark between birth and maturity is reached in girls at about 4.9 years and in boys at about 5.6. Thus one might read into this statement of Aristotle's the implication that in 350 B.C. children were maturing at approximately the same rate as now. In contrast boys working in factories in England in the nineteenth century only reached the half-way mark at 6.6 years.

Soranus

During the period of the Roman Empire two physicians, Soranus and Galen, established reputations above all others. Both were Greeks by birth, but practised in Rome. Soranus was the older; he was born in Ephesus in the second half of the first century A.D., was in Rome during the reigns of Trajan (98–117) and Hadrian (117–138) and died about the time Galen was born (129). He was a member of the school of medicine known as Methodist, founded by Greek physicians in Rome in the period stretching from the first century B.C. to the time of Galen – who attacked them mercilessly. The Methodists based their physiology, pathology and therapeutics on a consideration of the state of the pores supposed to exist between the atoms of which the tissues of the body were composed. An excessive opening of the pores permitted excess passage of fluid (a flux or, in modern terms, a discharge) and should be countered by a styptic agent; an excessive closure represented a *status strictus*, whence comes our word stricture. Galen, on the other hand, based his pathology on humoral theory; four humours penetrated all parts of the body and disease resulted from their imbalance. Echoes of each of the two views are heard all through mediaeval and Renaissance medicine, and discussions of the relative importance of cellular and systemic factors are still in vogue today. Galen's views rapidly displaced Soranus' in the Greek and Arabic East, but in the Roman West Methodism maintained its predominance (see Temkin, 1956). Only with the spread of Arab influence in the eleventh century did Galen become the over-riding authority of mediaeval medicine.

Soranus himself probably departed somewhat from the strict practice of the Methodists, and, perhaps for this reason, he alone amongst his colleagues was spared the considerable force of Galen's ridicule. His best-known surviving work is the *Gynaecology* (translated by Temkin, 1956). In it he deals with the anatomy of the female reproductive tract, the hygiene of menstruation and pregnancy, the conduct of labour and the puerperium, and the care of the infant. He also wrote a book on fertilization and embryology, but it has not survived. There are minor works on surgery surviving in their original Greek and a major work *On Acute and Chronic Diseases*, transmitted to us in the Latin paraphrase of Caelius Aurelianus, of the fifth century (who also translated the *Gynaecology*). 'Menstruation', wrote Soranus, 'in most cases first appears around the fourteenth year' (Temkin translation, p. 17). Further on, in discussing the signs of approaching menarche (first menstrual period), he added: 'Menstruation which is about to appear for the first time must be inferred . . . from the growth of the breasts and from the heaviness, irritability and pubescence in the region of the lower abdomen' (p. 20). The context makes it fairly clear that 'pubescence' (*to hêbân*) refers to the appearance of pubic hair. Caelius Aurelianus' Latin translation (Drabkin and Drabkin, 1951) makes that interpretation certain, for the phrase he uses is '*pubis erumpentibus capillis pubertate surgente*'.

Since in modern data the most probable time for menarche is when the breasts and pubic hair are both at stage 4 (out of 5 stages, hence shortly before full development) Soranus, when telling physicians when to look out for the imminent occurrence of menarche, is saying more or less what we would say today. As to 'the fourteenth year' for the occurrence of menarche (which should represent the age 13.0 to 13.99 years but at least in later usage frequently meant 14.0 to 14.99), that would be very reasonable also, although we have always to remain suspicious of the seven-fold number. Soranus underlines the date, however, when he writes 'It is good to preserve the state of virginity until menstruation begins by itself . . . as a matter of fact in most instances the first appearance of menstruation takes place around the fourteenth year. This age, then, is the natural one indicating the time of defloration' (Temkin, p. 31). Elsewhere he recognizes that vigorous gymnastic training 'or vigorous vocal exercise' may stop menstruation; he says such a thing is natural and advises against giving any form of treatment (p. 135). Indeed we find some opinions in this respect that reverberate down the ages: 'all excretion of seed is harmful in females as well as in males . . . men who remain chaste are stronger and bigger than the others' (p. 27). 'Men rid themselves of surplus matter through athletics' (p.23). As to the age of menarche, the fifth-century Latin version of Caelius, as also a sixth-century Latin translation by Mustio (Muscio) which was popular throughout the Middle Ages, inserts after the passage indicating that girls have menarche in the fourteenth year the words 'however, some earlier and some later' (see Drabkin and Drabkin, 1951). In this way they echo, presumably, authors writing in the intervening period, notably Rufus and Oribasius (see below). Gradually empirical experience was allowed to modify the tenets of tradition wisdom.

Galen

Galen (129–199), physician to Marcus Aurelius and the most prolific writer on medical matters up to if not beyond Haller, has some general remarks about growth in *On the Natural Faculties* (Brock's translation, 1916). 'Erasistratus', he writes, 'imagines that animals grow like webs, ropes, sacks or baskets, each of which has woven on to its end or margin other material similar to that of which it is originally composed. But this . . . is not growth, but genesis [coming-to-be] . . . Growth belongs to that which has already been completed in respect to its form, whereas the process by which that which is still *becoming* attains its form is termed not growth but genesis. That which *is*, grows, while that which is *not*, becomes' (*auxanetai men gar to on, gignetai to ouk on*: p. 137). This is a rather more restricted definition of growth than Aristotle's, with its Heraclitean image of continuous flux. Galen limits the term to designate simple enlargement, citing as a model the change in a pig's bladder when a child blows it up. Only, he says, in nature substance is not thinned by distension, as is the pig's bladder; the 'nutritive faculty' increases the substance as it stretches (p. 27).

Galen, at least in the books which exist in modern translation, has not much to say about the growth of children. However, he does remark that it is not possible to put age limits on the various stages of life, except broadly. 'Some begin puberty (*hêbaskein*) at once on the completion of the fourteenth year but some begin a year or more after that' (*De sanitate tuenda*: Corpus Medicorum Graecorum; see also Green's translation, p. 288). Since Galen is talking of boys and *hêbaskein* certainly refers to the pubic hair, this would indicate an age of puberty a year to eighteen months later than in Europe at present.

Oribasius

Amongst Byzantine physicians, Oribasius and Aetios are the most remembered. Oribasius (325–*c*.400), was born at Pergamon and became physician to Julian the Apostate. He remarks that in most women menstruation begins about 14 'but in a few as early as 12 or 13 and in quite a number after 14' (collected works translated into French by Bussemaker and Daremberg (1851–76, vol. 4, p. 650). He seems therefore to indicate a generally later average than 14.0; and his text has an air of empirical observation rather than of simple tradition. He adds a further comment, quoting as authority Rufus, a contemporary of Soranus. Humoural pathology and the theory of temperaments (see note 1.5) were well established in Byzantium after Galen, and Oribasius links rapid or slow development with physical constitution, at least in girls. Fetuses and children were regarded as characteristically hot and fluid, from the amount of concentrated semen and blood that went into them; indeed the movement of growth was due to their excessive heat. It is thus no surprise to find that girls of a hot fluid constitution are advanced in growth, while those of a cold constitution, 'even if fluid', are delayed. Oribasius already associates physical appearance with temperament; thus girls with marked veins, a ruddy complexion, lots of flesh and well-developed thighs mature earlier; those with the reverse characteristics mature later. Give or take a few differences in our manner of describing physique we find the same in modern data: the endomorphic girls are early maturers, the ectomorphs late.

Aetios of Amida, physician to Justinian I, emperor in Byzantium from 527 to 565, subsumed all current medical knowledge in a work known as the *Tetrabiblon*. A Latin version was published by Cornarius in 1542, and translated into English by Ricci in 1950. The old authorities and opinions are liberally quoted but nothing new concerning growth appears. The same can be said of the works of Paulus of Aeginata who wrote in the sixth or seventh century (see Amundsen and Diers, 1973).

Roman law

Finally, a warning concerning ages quoted for puberty in Roman and later times should be given. As Amundsen and Diers (1969) have convincingly argued, Justinian's codification of Roman law equated the age of pubescence with the legal age for the ending of childish tutelage, and this was placed at 14 (i.e. 14.0). At first this law applied to boys only, but was later extended to girls, and for them the age of termination of tutelage was fixed at 12 years, probably (Plutarch argues) to allow early marriage and a better chance of virginity at betrothal (see Amundsen and Diers, 1969). However that may be, this age of tutelary termination, though loosely and misleadingly called the age of puberty in some accounts, cannot be taken as representing anything biological. Macrobius indeed states that 'After fourteen years, by the very necessity of age, one becomes pubescent. Accordingly then the generative power in males and the menses in females begin to be aroused. Therefore a male is released from childish tutelage as if already a grown man; nevertheless girls are freed from this by law two years earlier for the purpose of hastening the wedding vows' (Amundsen and Diers, 1969, p. 130). It is not clear what part this tutelary age of 12 for girls played in the writings about menarche in mediaeval times; what does seem to be clear is that menarche was much later than age 12. Zacchias, a seventeenth-century medico-legal encyclopaedist, reviewing Roman and mediaeval literature declared that 'hardly one in a dozen [girls] was menstruating at that age [12.0] and many not till after fourteen' (*Quaestiones medicolegales*, Book 1, 1.6.34: see Ploss, Bartels and Bartels, 1935).

2

The Middle Ages and the Renaissance

Mediaeval medicine was neither inventive nor exploratory, and it is vain to search there for any advance in the understanding of growth. Even in the Renaissance the old doctrines held firm; indeed sometimes, as their sources increased, they held still firmer. Thus between Galen and Soranus in the second century and Buffon and Haller in the eighteenth, growth was scarcely investigated at all. Such attention as was given to human measurements in the Renaissance came from artists rather than doctors, and will be discussed in the next chapter. But the present book has a dual purpose: besides describing the studies of human growth that have been made and the climates of scientific and philosophical theory that gave rise to them, it aims also to chart the actual changes in growth and development of children which have occurred from one period of history to another.

We can pursue the second purpose of the book by describing the tempo of children's growth in mediaeval and Renaissance times. Or rather, we could do so, if only we could give to the words of mediaeval writers the same credence that we assign to reports of nineteenth-century biographers, let alone contemporary scientific surveys. This is unfortunately impossible; for the mediaeval writers wrote above all to transmit and support authority, not to make known new facts or their own observations. Remembering this, we should still briefly survey what the more prominent mediaeval and Renaissance writers said of human physical development in their time. This usually means discussing one thing only: the age of menarche. Before measurements were made, this was the one landmark that was noted; indeed particularly so, since delay in the onset of menstruation might not only betoken disease but, in royal and aristocratic families, also put severe strains on projected dynastic arrangements.

Isidore of Seville

One writer stands at the transition of the Roman into the mediaeval world, linking the two and carrying forward an authority that lasted until the

Renaissance. This is Isidore (*c.* 560–636), Bishop of Seville and Primate of Visigoth Spain. Viewed from the past, he was the last of the long line of Roman encyclopaedists that started with Varro in the first century B.C. Viewed from the future, he was the first of the great mediaeval compilers, the forerunner of Vincent of Beauvais, Albert the Great (Albertus Magnus) and Bartholomew the Englishman. It was the Latin authors that Isidore compiled and explained, for his knowledge of Greek, like that of nearly all West Europeans of his time, scarcely extended beyond the comprehension of the occasional phrase. He wrote many books, the best known being the *Etymologiae*, an encyclopaedic dictionary that he worked on up to his death. Throughout the Middle Ages and beyond he was quoted and requoted incessantly. Sharpe (1964), whose scholarly review of Isidore and his forerunners is invaluable, thinks it fair to suggest that during the Middle Ages 'so far as non-liturgical works were concerned, manuscripts of the *Etymologiae* were exceeded in number only by those of the Vulgate text of the Bible'. Isidore laid no claim to originality. 'The reader reads not my work', he said, 'but rereads that of the ancients' (Sharpe, p. 7). Not much critical acumen went into his compiling, however, and a modern reader can scarcely disagree with Reynolds and Wilson (1974) when they say that the *Etymologiae* 'is packed with information and misinformation on every topic from angels to parts of the saddle, [and] descends so often into false etymologizing and the uncritical parade of absurd bric-a-brac that it cannot be read without a smile. But Isidore wins our respect and even affection, by his obvious appreciation of knowledge for its own sake.'

Isidore's polymathic interests naturally embraced medicine. Book XI of the *Etymologiae* discusses 'man and his parts' and Book IV lists 'man's diseases'. Unfortunately, there is no information given on the contemporary age of puberty in either sex; but Isidore does react against the Roman legal custom of assigning puberty to a fixed age. 'There are some', he says, 'who calculate puberty from age, reckoning that individual to be pubescent who has completed fourteen years, even though he may reach puberty very late; but most certainly anyone who shows its signs on his body [an earlier passage indicates Isidore means pubic hair] and can beget children has reached puberty' (XI.2.13; Sharpe's translation, here and later).

Isidore has a good deal to say about the ages of man (XI.2.1). There are, he writes, six. Although beyond the sixth there extends an extra age of senility, 'the terminus of the sixth age', this does not bring correspondence to Solon's and Proclus' seven ages (see note 1.3) any closer. Senility is not the age after 70, as one might expect, for the sixth age, *senectus*, fills that role, extending from 70 onwards 'for no definite span of years'. *Infantia* is the first age and extends from birth to 7 years. It is so named because the 'infant is still unable to speak, that is, he is unable to talk. His teeth are not yet properly arranged, and, as a result, his clarity of speech is limited'. The second age, *pueritia*, extends from 7 to 14. The third age is *adolescentia* and lasts till 28 (giving a double hebdomad). It is so named 'because [the young man] is sufficiently grown up to beget or because he is maturing and increasing in size'. The fourth age, *juventus*, extends from 28 to 50 (that is three hebdomads approximately). This is the strongest age. The fifth age is *gravitas*, the age of mature judgement (a long-range echo of Solon perhaps), and extends from 50 to 70. The sixth age is *senectus*. On the division of ages, therefore, Isidore seems to have struck out on something of a line of his own, neither Solonic nor Proclan. Subsequent literature tended to echo a mixture of all three systems (see note 1.3).

Michael the Scotsman

The gradual revival of learning in the West was at first heavily dependent on the translations from the Greek that the conquering Arabs brought to southern Europe. Much before the revival of Greek learning, Latin translations were made of Arabic authors such as Avicenna and Averroes and schools of translators grew up in Toledo and southern Italy where Europeans, Jews and Arabs together prepared Latin editions of all the major works. One of the scholars chiefly responsible for the activities at Toledo was Michael the Scotsman (Michael Scotus) (died *c.* 1235: Thorndike, 1965). Michael, though a cleric, was not a member of an Order like Vincent of Beauvais or Bartholomew the Englishman (see below). Born in Scotland, as his name implies, he was in Toledo from around 1210 to 1220, and translated many Arabic manuscripts including those of Aristotle's *Historia animalium*, *De partibus animalium* and *De generatione animalium*. He was clearly an Arabic scholar of substance, but Thorndike thinks he probably knew little Greek.

In 1200 he became astrological, medical and philosophical advisor to Frederick II, Holy Roman Emperor and King of Sicily, whose artistic and intellectual curiosity and sophistication earned him the appellation of *Stupor mundi*, the 'Wonder of the World'. Thereafter Michael lived at the great Emperor's court until his death about 1235. He left a number of books and in one, called originally *Physionomia*, but *De secretis naturae* later, he discussed menstruation. The first edition was published about 1230; some twenty printed editions are known. Regarding menarcheal age Michael wrote 'This purge, which the force of nature throws out each month from 12 up to 40–50 years' (1615 edn, Strassburg). Why exactly he chose the earliest of all suggested dates for menarche without even giving an alternative range of ages is of course unknown, but it might perhaps have been due to his contact with vernacular Arabic literature, and to experience with Arab and southern Spanish and southern Italian women. Data of the 1970s make it clear that, as explained further below, girls nowadays in well-off circumstances in the Mediterranean area have an average age of menarche six months to a year earlier than the averages recorded in most other parts of Europe (Eveleth and Tanner, 1976, p. 214). Michael's contacts amongst the Arabs, and amongst the Sicilians and Neapolitans too, must have been predominantly with aristocrats. It is very possible that children of the Mediterranean area grew up faster than children in the more northerly parts of Europe, then as now.

Michael's book also says that 'the beard in men appears (*nascitur*) at about 14 years, little by little, from one day to another' (1615 edn, p. 365). This also is very early, but not wholly out of line with a menarcheal age of 12.

Vincent of Beauvais

Amongst the many compilations of the Middle Ages, the *Speculum majus*, or *Great Mirror*, stands out, and not only by reason of its enormous size. The compiler was Vincent of Beauvais (Vincentius Bellovacensis), a man probably born in Beauvais about 1190 and educated in Paris, who was one of the earliest members of the Dominican Order. In about 1233 he entered the priory of the Order of Beauvais and began his immense work. News of it reached the King, Louis IX (1215–1270), who thereafter supported it and became a life-long friend. Vincent was installed in the Cistercian Abbey at Royaumont and played a large part in educating the royal children, writing a book on education, *De eruditione*, at the specific request, it is said, of Queen Margaret (McCarthy, 1976). The first version of the *Speculum majus* was issued in 1244, a second in 1247 and a third about 1250. Traditionally, it consists of four parts: the *Specula Naturale, Historiale, Doctrinale* and *Morale*. However, only the first three are genuine. The fourth was written about 1310–25 and contains quotations from Albert the Great (1193–1280) and St Thomas Aquinas (1225–1274) as well as other post-Vincentian writers. Vincent himself died in 1264. The first printed edition appeared in 1473–6, and a modern reprint was issued in 1964–5 by the Austrian Academy of Sciences, based on the emended, definitive edition of 1591.

The *Speculum majus* is an astounding compilation, an inexhaustible source of writers from Hippocrates and Aristotle onwards. Like Isidore, Vincent was probably ignorant of Greek, but by the mid-thirteenth century the translators in Toledo and elsewhere had produced Latin editions of many Greek works, though mostly via their Arabic versions. Creutz (1938) has analysed the medical quotations in the *Speculum*. In the *Naturale* Hippocrates is quoted 11 times, Aristotle 21 times, Galen 5 times and Isidore 56 times. Constantinus Africanus (eleventh century), a monk of Monte Cassino, born in Carthage, is quoted 64 times, and the Arabic writers Rhases, Haly Abbas and Avicenna 57, 60 and 38 times respectively. In the *Doctrinale* Hippocrates gets 16 quotes, Galen 12 and Rufus 1; Isidore has 52, Constantinus 39 and the three Arabic writers 81, 118 and 92. In the whole book 450 authors are represented.

Though the *Speculum* contains much on how to bring up children (1964–5 edn, pp. 1089–179) there is little on physical growth. The ages of man are described, Isidore being quoted, together with Avicenna (p. 2348). The humours are, of course, discussed, and a predominance of each is assigned to a particular age. 'Thus the ages are divided into four, *phlegma* in childhood from the beginning until 14 years. Then *cholera* in the youthful until 28 years. Then in subsequent years, till 60, *sanguis*. And then in old age *melancholia* (p. 2345). This division is ascribed to Hippocrates, though wrongly. It must originate in Galen or a later writer; since Isidore does not have it (at least not in Sharpe's extracts) it seems to be a mediaeval implant. The idea of linking different humoural predominances with successive ages gained wide acceptance during the Middle Ages and the Renaissance and was still current into the seventeenth century (see note 2.1).

Vincent writes of age at menarche: 'Thus it is proper [i.e. normal] that the flow of menses should be discerned after twelve years have been completed, and [it occurs] in the majority if this period is extended to fourteen years' (*Menstruorum exitum . . . oportet intellegi, itaque cum 12 annum expeleverunt et plurimum in quartodecimo protenduntor*). The phraseology probably implies that menarche occurs within a range from 12.0 to 14.0. Vincent attributes this sentence to Constantinus Africanus. The use of *oportet* (meaning it is proper) may simply imply a reference to Roman Canon law, which gives 12 as the age of puberty in girls, so leaving 14 as the usual real age of menarche, a figure consonant with that given earlier by Soranus. Alternatively, the explanation advanced above for Michael's figure may hold here too.

It is this continual echoing of Ancient authorities that frustrates any attempt to argue from the scholastic encyclopaedias to the actual age of menarche in mediaeval times. Post (1971), in a valuable paper, suggests that later-written books, and particularly those in vernacular languages, are more likely than those written earlier to represent the views and even the experience of the authors. In the fifteenth and sixteenth centuries English-language versions, he thinks, were specifically written so they could be understood by common folk and not just by the schoolmen. He thus argues that in mediaeval times menarche occurred at much the same age as now: that is, in most girls between 12.0 and 14.0. The argument is not wholly convincing. The mediaeval mind accepted authority – a form, after all, of revelation – as superior to experience. (Similarly, Sarton (1957, p. 6), has said of the Renaissance that 'more weight was given to Strabo's words than to those of Columbus or Vasco de Gama, and the expedition of the Argonauts was considered more interesting than Magellan's'.) There is little reason to think that an attempt to transmit knowledge to a wider range of people would be the excuse to dilute the hallowed phrases with a mere translator's personal observations. To a modern auxologist, the absence of any mention of differences between the well-to-do and the peasants or between the urban rich and urban poor makes the whole account suspect. We would surely expect such differences, but it is not till Guarinoni in the early 1600s that, at last, we find them.

Bartholomew the Englishman

The *Great Mirror* is said to have been the source of much of Chaucer's scientific and medical knowledge (see McCarthy, 1976, p. 11), and the same is probably true of a similar, though smaller, compilation made by an English Franciscan named Bartholomew. Bartholomew the Englishman (Bartholomeus Anglicus) taught theology in Paris shortly before Albert the Great, and so was contemporary with Vincent. His compilation was entitled *De proprietatis rerum.* Bartholomew quotes Aristotle as saying 14 is the usual age for menarche, but gives no age himself. He describes the ages of man in terms derived closely from Isidore. The book was translated into English by Stephen Batman, professor of divinity at Oxford, and published in 1582, in a version known as *Batman uppon Bartholome.* (Vincent was never translated, save into French in 1495; one look at the book's bulk would have deterred the most hardened translator.)

Stephen Batman exerted a considerable influence on English thought in both Elizabethan and Stuart times, especially in relation to the ages of man (see note 1.3). The first age, *infantia,* is 'the childhood without teeth, and dureth seaven months ... and childhood that breedeth teeth endureth and stretcheth seaven yeares. And such a child is called Infans, that is to understand, not speaking ... for his teeth be not yet perfectly grown and set in order, as sayeth Isidore'. *Puericia* goes to 14; *adolescentia* to 21, 'but Isidore sayeth that it endureth to the fourth seaven yeares that is to the end of eight and twentie yeares. But Phisitions accept this age to the end of thirtie or five and thirtie years'. *Juventus* follows to 45 or 50 years, then *senectus*, which some say ends at 70 and others at an indefinite age. *Senectus* brings 'passing and failing of wit'. The last part of this age is called *senium*. A fifteenth-century illustration of the seven ages, from a French translator of Bartholomew, is shown in Fig. 2.1.

Trotula

Post (1971) has also pointed out a technical problem in the transcription of Roman numbers: it is very easy to omit a digit and write xiij instead of xiiij. He examined the eighteen manuscripts available in Oxford of the most popular book on gynaecology of the Middle Ages. This is by Trotula, a woman physician (or perhaps patroness) of the famous medical school of Salerno, near Naples. This was the first medical school established in Europe, and is best remembered for producing – or having ascribed to it – the *Regimen*

Fig. 2.1. A fifteenth-century illustration of the seven ages of man from a French translation of Bartholomew's *De proprietatis rerum* (Lyon, 1491). Reproduced from Garnier (1973).

sanitatis salernitanum, a book which passed through 240 editions (Gordon, 1960, p. 317). There are two books by Trotula, *De egretudine mulierum* ('Trotula Major') and *De ornamentia mulierum* ('Trotula Minor'). They were probably written in the eleventh century, and the extant manuscripts date from the thirteenth to the fifteenth centuries. There is a passage on menarche, which is present, however, in only twelve of the manuscripts in Oxford. The translations read, more or less consistently: 'This purge occurs in women around the fourteenth year, or a little earlier or later, according to whether they have more or less heat or cold'. However, whereas seven of the manuscripts read 14 (xiiij), four read 13 (xiij) and one 15 (xiiiij), presumably due to copying errors rather than the personal experiences of the scribes. There are ambiguities in language too, then as later. 'Around the fourteenth year' is written 'circitur xiiij annum' and presumably means around the fourteenth birthday (14.0), hence say 13.5–14.5. But it might mean around the fourteenth year itself, which is the year ending at 14.0 and having its centre at 13.5, whence the range might be 13.0–14.0 (see discussion in note 11.8).

Another mediaeval book, more or less about gynaecology, and very popular, was the *De secretis mulierum,* attributed to Albert the Great but probably by one of his pupils. It was written about 1260, and first printed in 1478. In it menstruation is said to happen to women every month 'when they are of due age, namely 13 years' (Post's translation); but the 1615 edition, undoubtedly emended by subsequent hands, reads '12, 13, or 14 and frequently 14'. Gilbertus Anglicus, a thirteenth-century English physician, wrote in his *Compendium medicinae* (*c.* 1230: see Talbot, 1967, p. 72) 'menses are withheld naturally below twelve and above fifty years' (Post's translation). However, Gilbert, who probably studied at Salerno, drew most of his work from Constantinus Africanus and hence, ultimately, from the Romans and Greeks. We have again the difficulty that in Roman custom 12 was the 'correct' age for menarche, just as 14 was the 'correct' age for puberty in boys. We have seen that even in Rome the correct legal age was an incorrect biological one. What we cannot determine is the precise extent of this legal influence. It may be significant, however, that 12 and 14 remain to this day the legal ages of puberty in Canon law (Amundsen and Diers, 1973). Alternatively, the figure may reflect an actual earlier age of menarche in Mediterranean children, as described above.

To sum up: both Post (1971) and Amundsen and Diers (1973) have argued that in classical and mediaeval times menarche took place at an average age that was broadly comparable to the average at which it occurs nowadays. In the Mediterranean area, to which most of the old data relate, the mean age at present is a little under 13, say 12.7 years. It seems likely that in mediaeval times the average age was really rather later than this. On the other hand the girls the mediaeval writers are considering certainly had menarche much earlier than poorly-off working girls of the nineteenth century who had an average between 15.5 and 16.0. However, this may merely mean that the older documents relate purely to the more or less well-off and ignore the weavers' daughters and the daughters of country labourers.

The Renaissance

In the early stages of the Renaissance the recovery and translation of the Greek originals of Hippocrates, Aristotle, Galen and the rest, and their widespread dissemination by printing, served to reinforce ancient authority even further. Thus a dissertation with the promising title of *De incremento* (Landus, 1556) is scarcely more than an elegant commentary on Aristotle's views on growth, alteration and generation, fluffed out with some travellers' tales. Pavisi's *De accretione* (1559), also a graduation thesis, covers much the same ground at greater length. Pavisi (Pavisius) was a Calabrian who studied at Padua, but in the generation before Galileo and Harvey; Fallopius would probably have been his teacher. The tone of these two theses is quite different from writing on the same subject a century later. Pavisi follows custom in quoting Aristotle, Galen, Avicenna, Averroes and Michael the Scotsman, and in attributing the slowing of growth to the increasing dryness of developing tissues. At one point in the thesis, though, he is quite unusual in making what is perhaps the first apparently quantitative statement about children's rate of growth since the Greeks. However, it is scarcely auspicious. 'The growth of infants and children is quite swift and often in two or three years they add two or three cubits' he wrote (*et infantium et puerorum, velox admodum sit incrementum et cum in totis saepe duobus, aut tribus annis, duorum et triorum cubitorum adiectionem ceperint* (pp. 28–9). A cubit is traditionally one and a half feet, and though it is possible, and indeed likely, for a child to grow two cubits, or three feet, in the first two years, no child could add a further cubit in the year following. Perhaps Pavisi has confused feet and cubitis.

Such information as exists on age at menarche in the sixteenth century points to a figure in close accord with mediaeval values. Marinello (d. *c.* 1576) in Italy said of the menses 'I am not able to be very definite (*non sappiamo troppo bene*) about what time in childhood they appear, because some [girls] have not completed their twelfth year when they appear; others, the majority, are in their thirteenth year' (Marinello, 1574, p. 96). These ages are really quite early, even for modern times, and as if to underline the fact Marinello continues 'A doctor told me he had seen a pregnant woman in Pavia who was no more than nine years old, but she was a rarity . . . The cause of the variation is [differences in] the natural composition of the body, or complexion or habits; thin and long girls [menstruate] later fat and strong ones earlier'. Houiller (Hollerius: d. 1562), writing of France about 1550, gave as the average 14 years, 'when the young girl has grown to the proper size: because nature does not wish to be uncircumscribed the time for growing is earlier than this (*est praestitutum incrementi tempus*) (Hollerius, 1611, p. 610). Houiller added that in some circumstances menstruation might be delayed till 17 (Hollerius, 1572, p. 253: cited in Müller-Hess, 1938, who gives a thorough review of theories of the causation and significance of the menses from the sixteenth century onwards).

The distribution of age at marriage in Tuscany a century earlier is known, thanks to the *catasti*, or censuses, taken there (Klapisch, 1973). The mean age of girls at marriage in 1427–30 was 17.5 years, with the lower limit (excluding the 2.5 per cent lowest) at about 13.5. Girls were allowed to marry six months before the canonical age of 12 'provided they were *pubère*', and this argues that marriage did not precede menarche, though it may often have followed it rather closely. These figures give us an upper limit to the mean age of menarche of about 16 years, but the marriage figure is by no means incompatible with menarche at an average age of 14 or 15.

It might be expected that data on the ages at menarche of their daughters would be available in diaries of women of quality of Renaissance and later times. But these sources, if indeed they exist, seem never to have been tapped. There is one piece of information concerning the male. In the famous *Journal* of the physician Jean Héroard on the childhood of Louis XIII, the entry for 1 August 1624 says that Louis then aged 23, '*se fait raser la barbe pour la première fois; il ne y avoit que du poil imperceptiblé* (Soulié and BarthOlémy, 1868, vol. 2, p. 297). The lateness of his puberty is, however, rendered less significant by some indications that he may have been suffering from an illness.

Fernel and the physiology of growth

An idea of Renaissance conceptions of the physiology underlying the growth process can perhaps best be obtained through the works of Jean Fernel (1497–1558). Fernel was himself the first to use the word 'physiology', though he is careful to say it is a study of which Aristotle was the founder. Fernel was the oustanding physician of his generation, doctor to Henri II, Catherine de Medici and Diane de Poitiers, and professor at the University of Paris. Sir Charles Sherrington's profound and sensitive biography (1946) is the tribute of one great scholar to another. In 1542 Fernel wrote a treatise entitled *On the Natural Part of Medicine.* The adjective refers to the medical faculty's classification of studies into 'things natural, things non-natural and things contra-natural'. A few years later, on the issuance of a new edition (1554) as part of a whole textbook entitled *Medicina,* the 'natural part' was labelled *Physiologia.*

Fernel divides his book not into accounts of respiration, digestion, circulation, and so forth – that was to wait almost till Haller two centuries later – but into the Elements, the Temperaments, the Spirits, the Innate Heat, the Faculties, the Humours and the Procreation of Man (see Sherrington, 1946, pp. 64ff.). The Elements were earth, air, fire and water, and the Temperaments were the combinations of hot, cold, solid and fluid (see note 1.5). All things, inanimate as well as animate, had a temperament which resulted from a particular balance of substances. The physician's first task was to judge correctly the temperament of his patient, for the same disease acted differently in persons of differing temperaments (see note 2.2). The humours were the 'blood', equivalent to air, hot and fluid; the 'yellow bile' or choler, equivalent to fire, hot and solid; the 'phlegm' or pituita, equivalent to water, cold and fluid; and the 'black bile' or melancholer, equivalent to earth, and cold and solid. Fernel objected to the common practice of using the humours to designate temperaments – for example talking of a 'sanguine' temperament – on the grounds that it was not the blood that caused the characteristic temperament; the temperament was a property of the material *composing* the body, whereas the blood, like the other humours, was only a substance contained *in* the body.

'Faculties', for Fernel, represented 'efficient causes' (that is to say 'causes' in our modern sense of the word, as used in 'cause-and-effect'). To explain anything, one had but to postulate a faculty. The primary faculty was the nutritive, which worked the first coction in the stomach and drew from the veins material which renewed the flesh. But Fernel remarked that a child while wasting from fever could yet continue to grow in height; therefore, he said, there was also a 'growth faculty' (Sherrington, p. 94). Faculties were the means by which the immaterial tenant of the body, the life-soul, operated the body. It did so through the agency of spirits, which were material and divided into natural spirit *(pneuma physikon)*, vital spirit *(pneuma zotikon)* and animal spirit (*pneuma psychikon*). It was through the natural spirit that the life-soul activated the faculty of growth to make a child grow.

The life-soul of the parent, says Fernel, is not the source of the life-soul of the offspring. The reservoir of souls (or 'forms' in the Platonic sense) is in the heavens, in the eighth sphere, that of the stars. Every new life comes from there, carrying a portion of the divine with it (in an earlier work Fernel had specified that the soul came by way of the planet Sol). At the beginning of the fourth month of pre-natal life the soul enters, quite suddenly. By this time the heart and brain are sufficiently advanced to receive it but 'owing to the plethora of the humours at that time, it remains in a state resembling drunkenness or torpor, unable to exercise its proper functions' (Sherrington, p. 91). During this time the soul differentiates, for at first it is animal, later human, and lastly individual. Thus at first growth resembles that of other animals, but later becomes characteristic of the individual.

William Clever

Thus Renaissance physiology: a curious but not inconsistent mixture of pre-Christian mystery religion, Greek science and Christian dogma. As regards the physiology of growth it was the question of heat which gave most trouble. Aristotle had thought that the elemental heat entered at conception but that the cooling power of the infant's relatively large brain made infants cooler than children. There was argument as to whether children were cooler or hotter than adolescents. Indeed though some authors thought that children grew by virtue of their superior heat, others denied that children were hotter than adults, let alone adolescents (see Sherrington, p. 96; and Paparella, 1573, chaps. 14 and 15).

This issue was discussed at length by William Clever (dates unknown), personal physician to Sir John Rooper (Roper) of Lynstead Park in Kent (later Baron Teynham, the great-nephew of the William Roper who was Sir Thomas More's son-in-law and biographer), in his book *The Flower of Phisicke* (1590). Galen thinks that children are more hot and moist than adolescents, said Clever, because they have more blood and a larger appetite. However, he himself thinks that infants are at first cold and hence given to sleep 'but growing up to children, are every day more sanguine and therefore more hote and moyst: for as heat provoketh appetite, so moystness is the cause efficient, as well to nourish great sleepe in the body as to advance therewith the office of good nourishment' (p. 73). As children develop so their moistness deserts them; adolescents and young men are drier (i.e. more solid). Though young men may not actually be hotter their heat is sharper, whereas the heat in children is sweeter (this apparent quibble had a respectable past and future history: see Paparella, chapt. 15, and note 2.3). Thus medicines need to be stronger for young men, otherwise they are unable to affect the constitution of adolescence.

This drying-out or solidifying process is very important. In the years from adolescence to 20 or beyond 'driness hath his best place and chiefest felicity, although heat and moystnesse in moste part of them continueth until thirtie and three, as the last and farthest drift of adolescencie. Arnaldus of Nova Villa saith that the most part, especiallie of women, beginning their adolescencie before ripenesse of age hath given them libertie therunto, doo afterwards live like untimely fruite: as peares, plummes, or apples gathered before seasonablenesse and ripeness hath perfected them, doo most speedilie drie, rot, decay and utterly perish' (p.70) (see note 2.4). Guarinoni (see below), using a similar comparison, adds that it is an abundance of moist and succulent food which brings on the fruit-too-early syndrome.

Lemnius

Levinus Lemnius (1505–1568) is more concerned with insufficiency of food. He was a doctor of Ziricsee in Holland, who studied medicine in Ghent and Louvain, practised for a while, then entered the priesthood. The first edition of his book *De habitu et constitutione corporis* ... was published in Antwerp in 1561, a second edition in Erfurt in 1582, and a third in Frankfurt in 1619. An English translation was issued in 1633 entitled *The Touchstone of Complexions.* 'Comely tallnesse and length of personage', said Lemnius, 'cometh and is caused of the abundance of heat and moisture, where the spirit is thoroughly and fully perfused' (p. 42). 'Schoolmasters and others that take the chance upon them to teach and boord young boyes ... pinch their poore Pupils and Boorders by the belly, and allow them meate neither sufficient nor yet wholesome ... whereby it cometh to passe, that in growth they seldome come to any personable stature, to the use of their full powers, to perfect strength and firmity of their members, or to any handsome feature or composition of bodily proportion: and the cause is for that in their tender and growing age, being kept under by famine and skanted of common meate and drinke, their natural moisture which requireth continuall cherishing and maintenance, was skanted and bebarred of his due nourishment and competent allowance' (p. 43). Evidently Lemnius' acquaintance, at least, was not only with the wealthy. Thus his opinion on menarche is especially interesting: he says it begins at 14 or 15 years old, in some later, in some earlier.

Bacon

Francis Bacon (1561–1626) subscribed to the theory that the natural heat of the child was one of the influences which made him grow. In his *Catalogue of Particular Histories* (in *Novum organum,* 1620) he listed the 'History of the

Growth of the Body, in the whole and in its parts' as one of the 130 subjects to be investigated, and in the posthumously published *Sylva sylvarum* (1626) he comments on how to accelerate growth or stature. 'It must proceed either from the Plenty of the Nourishment; or from the Nature of the Nourishment; or from the Quickning and Exciting of the Natural Heat. For the first, Excess of Nourishment is hurtfull; for it maketh the Childe Corpulent; And growing in Breadth, rather than in Height . . . As for the Nature of the Nourishment; First, it may not be too Drie; And therefore Children in Dayrie Countries do wax more tall, than where they feed more upon Bread and Flesh . . . Overdrie Nourishment in Childhood putteth backe Stature . . . As for the Quickning of Natural Heat, it must be done chiefly with Exercise. And therefore (no doubt) much Going to Schoole, where they sit so much, hindereth the Growth of Children, whereas Countrey-People, that go not to Schoole, are commonly of better Stature' (2nd edn, 1629, p. 92. section 354).

In thinking that 'natural heat' could be stimulated by exercise Bacon exemplifies perfectly the transitional man of the late Renaissance. Evidently he is aware of the very recent physiological approach of Harvey, and perhaps of the experiments of Sanctorius (see Chapter 4), but he confuses heat of this sort with the innate, or even elemental, heat of Aristotle. His remarks on 'overdry nourishment' exemplify also the theory of the time, though this was a theory that remained current for a further 200 years. The idea that dry foods delay growth appears in Hamberger's 1751 remarks on menarche (p. 94) and later still, and very precisely, in Virey's 1816 encyclopaedia article on giants (see p. 119).

What Bacon has to say about the country children is interesting. It seems to go against Guarinoni's view in Austria (see below), though perhaps it could be harmonized with Lemnius' in Holland. In the late nineteenth and early twentieth centuries country people were smaller, both as children and adults, than the inhabitants of the towns in all countries of Europe except, significantly, England. Here the earliest available statistics, for 1911, actually show country children as taller. Perhaps we may accept Bacon's observation while rejecting his reasons.

After Bacon there are a number of authors who discuss the relation of growth to innate heat, without, however, throwing any fresh light on the supposed connection. Sinibaldi (Sinibaldus), writing in Rome, is one of them, whose book *Geneanthropeiae* (1642) is well known, but so far as growth is concerned rehearses only the classical opinions. Another is Wechtler, physician to the Emperor Leopold, who wrote an immense textbook entitled *Homo oriens et occidens* (1659). Wechtler gives a lengthy discussion on the relation of heat to growth and on whether growing boys have more innate heat than adults (pp. 329–43, 375–80). The great Haller (see p. 85), perhaps a little uncharitably, described Wechtler's book as 'A vast tome worthy of oblivion . . . the epitome of the scholastic method . . . full of Arabistic and useless speculation' (Haller, 1776–9, vol. 3 (1779), p. 79). In England, Walter Charleton's *Enquiries into Human Nature* (1680) mercifully escaped Haller's notice. Charleton (1619–1701) was a well-known physician in Stuart and Commonwealth days; what he has to say (or to omit saying) is important in that Webster (1975), the recent historian of science and medicine in England under the Commonwealth, calls him 'the intellectual barometer of the age' whose 'numerous works are a valuable index to contemporary fashions' (p. 278). Charleton's book consists of six lectures on anatomy and physiology; his account of the physiology of digestion, and, to the extent that he deals with it, of growth and accretion, departs hardly at all from Fernel's a hundred years before.

Guarinoni

Only as the Renaissance was giving way to the Wars of Religion do we begin to hear the authentic voice of personal observation. With Hippolyt Guarinoni (1571–1654), an exact contemporary of Francis Bacon, we are in a world at once more innocent and more reliable. Hippolyt was the son of Bartholomew Guarinoni, a man of Milanese origin who was *Leibesarzt* (personal physician) in Prague to the Emperor Rudolf II, King of Bohemia and Hungary (see Evans, 1973). Hippolyt was born in Trient (Trento) in 1571. As a child he served as a page in the household of the Cardinal Archbishop Karl Borromäus in Milan and he then attended the Jesuit school in Prague for eleven years. In 1593–7 he studied medicine in Padua, following his father. These were the years of Padua's ascendancy (see p. 66). Guarinoni's teachers would have included Sanctorius and Fabricius; Galileo was professor of mathematics and William Harvey arrived as a student the year after Guarinoni left. After graduation, Guarinoni was appointed physician to the Royal Convent in Hall, near Innsbrück (founded by three daughters of the Emperor in 1567), and he lived in Hall from 1598 onwards. In 1607, when the Grand Duchesses Eleanor and Christina, daughters of Karl of Steiermark, entered the convent, he became their physician and after their deaths he continued as doctor to the convent and the town.

It is in this last capacity that Guarinoni is still, and increasingly, remembered (Grass, 1954). The entry in the *Allgemeine Deutsche Biographie* of 1879 is unable to record even his dates of birth and death. In 1903 he was remembered in a short but illuminating article published in Brixen, a small town in Tirol (Rapp, 1903). In 1927 he was hailed as the forgotten forerunner of preventive and social medicine in Germany (Fischer, 1927). In 1954 the tercentenary of his death was the occasion of a full-scale memorial symposium published by the University of Innsbrück (Dörrer *et al.*, 1954) and by 1968 he had become the subject of several PhD theses (e.g. Bücking, 1968). He emerges, particularly from Rapp's account, as a straightforward and good man of immense energy and enthusiasm, a people's doctor 'ready even in his old age to visit the lowliest hut at any hour of the night' (Rapp, 1903) and a stauch campaigner for clean air, pure water and belief in an old-fashioned God. More zealous than his masters the Jesuit reformers (who on occasion told him, as the Pope had told Arnald of Villanova, to stay out of theology) he taught an active if slightly eccentric religion to the youth of his province and told his patients that belief in God was as therapeutically powerful as the six *res non naturales* of Hippocrates.

In 1613 he made a pilgrimage to Rome, with four companions, to solicit on behalf of the Grand Duchesses Eleanor and Christina the gift of the bodies of the two virgins and martyrs, Ste Vincentia and Lea. Parts of his diary of the pilgrimage were published in the nineteenth century (Stampfer, 1879). Negotiations for the relics were made somewhat difficult by the fact that the general of the Jesuit Order, Klaudius Aquaviva (in whose garden the bodies had been found), had promised to send them to the Jesuit mission in China; however, Cardinal Bellarmin persuaded him to redirect them because of the great favours the Grand Duchesses had shown to the Society. Guarinoni's joy at the success of his mission still rises like incense from his diary's extraordinary prose. Later in life he founded a church and cloister at Volders, a bridge over the Inn about an hour's journey from Hall. He laid the foundation stone with his own hands in 1620 (the year Elizabeth of Bohemia was being chased out of Prague by the forces of the Counter-Reformation, leaving her copy of Walter Raleigh's *History of the World* behind) but died six weeks before the consecration, in 1654. It is, say the guide-books, a nice baroque church just off the road to the Brenner Pass, painted by Martin Kroller in 1764 (Riehl, 1898).

Guarinoni loved the countryside and the peasants. He extolled the value of a simple life, and thought the *Bauerschaft* (farmers and peasants) both healthier and nearer to God (not that he distinguished those two things very clearly) than the nobles and merchants. When rebuked for using language unsuitably coarse for a Paduan graduate he cheerfully replied that he talked the language that his patients understood (Rapp, 1903, p. 16). His portraits (Dörrer *et al.*, 1954) show, in the conventional pose of piety, a thick-set, muscled man, well suited to the plough. Guarinoni's reputation as the forerunner of public health medicine in Germany rests not only on his environmentalist concerns but also on his insistence on a healthy diet, in which bread and sauerkraut figured prominently. Many of his patients, he said, overate and would do better on the soup served in the Hall hospital for the poor. Not only had several dozen of his women patients died of excessive eating (Bücking, 1968, p. 151) but many children too in his region had died of being given too rich food (Rapp, p. 21).

He deplored the excessive use of alcohol and his largest book, *Die Grewel der Verwüstung menschlichen Geschlechts* (1610; *The Abomination of Desolation of Humankind*: see note 2.5) contains hair-raising stories of alcoholic quarrels and violence. *Die Grewel*, dedicated to Rudolf II (and also the Virgin Mary) ironically as a *buchle*, or little book, boasts 1,330 folio pages and is bound and locked like a family Bible. It is, if such a thing is possible, a book of devotional hygiene. It contains seven sections, respectively on God, the soul or spirit (*Gemüt*), air, nutrition, care of the body, exercise, and sleep. It gives sound commonsense advice on, for example, ventilation and clean water, followed by faults, or what not to do (see note 2.5). One in particular concerns us: Guarinoni, it seems, was the first to describe retardation of growth caused by emotional stress in school. 'Many children do not grow properly despite good food', he wrote, 'because coming home from school they feel still the pain of rough blows [i.e. given there] and look forward to their renewal with anxiety and fear, so that they are never happy or light-hearted' (p. 246: *Viel Kinder bei guter Kost nicht wachsen, weil sie von Schul auss der Schmertzen von grossen Streichen daheim stets empfinden, und auff kunfftige widerumb Sorg und Forcht haben, also niemals frölich, noch sich von Hertzen ergötzen mögen*).

Die Grewel is also full of stories, personal anecdotes illustrating the points at issue, and devotional exclamations and harangues. It is an immense panoramic scrapbook of the Tirol and nowadays recognized as a major source of knowledge on peasant life in seventeenth-century Germany (see, e.g., Bücking, 1968) as well as a prime document in the history of public health. Guarinoni, living amongst the peasants of a mountainous country, had this to say about menarche: 'The peasant girls of this *Landschaft* in general menstruate much later than the daughters of the townsfolk or the aristocracy, and seldom before the seventeenth, eighteenth, or even twentieth year. For this reason they also live much longer than the townsfolk and aristocratic children and do not become old so early. The townsfolk have usually borne several children before the peasant girls have yet menstruated. The cause seems to be that the inhabitants of the town consume more fat (moist) foods and drink and so their bodies become soft, weak, and fat and come early to menstruation, in the same way as a tree which one waters too early produces earlier but less well-formed fruit than another' (The translation is by Dr F. Würst: see Würst, Wassertheurer and Kimeswenger, 1961; Würst, 1964).

This has the real ring of observational truth. The figures for peasant girls accord well with twentieth-century agricultural communities in the highland districts of under-developed countries such as New Guinea (see Eveleth and Tanner, 1976). Even in Europe in the 1960s there remained a difference of nearly a year between the age of menarche in town girls and girls in mountain villages in Austria and Slovenia. (Würst *et al.*, 1961; Würst, 1964; Valsik *et al.*, 1963). As for the ruro-humoral prejudices and the attribution of long life to slow maturers, those were fairly general in Guarinoni's time and are perhaps gradually creeping back in our own century. Guarinoni does not quote Arnald as his source, as does Clever twenty years earlier, but his analogy is exactly Arnald's. In another passage (*Die Grewel*, p. 1124) Guarinoni says that girls in German lands do not menstruate as early as southerners and 'seldom have their Flowers before 18 years, or at the very earliest 16, while the southern (*Mittäglichen*) girls all show at 11, 12 or 13'. Not, he goes on, that the Flowers are a true sign of being a woman; the real change is four or seven years after the first period, thus at 24 or 27: that is quite early enough to enter into the powerful immodesty of the married state.

Another writer whose work has the truthful ring of observation is Nicholas Venette (1633–1698), professor of anatomy and surgery at Rochelle. Venette was the author (pseudonymous in the first edition) of the most successful sex manual of the seventeenth century, at least to judge by the number of its editions and translations. Regarding puberty, he wrote 'The most assured sign of being in a condition of Engendering is, according to the Sentiments of Physicians, when a Boy can ejaculate Seed and the Terms appear in a Girl . . . Those flowings of Humours appear very seldom at 9 or 10 years, one shall hardly see Girls of 12 and Boys of 14 capable of obeying Love, and to produce such matter as forms Men. A young Woman would be very slow, if she was not capable of perpetuating herself by the production of a Child at Age 16 and a young Man of 18 would be esteemed very cold if lying with such a Woman he should find it impossible to partake of the pleasures of Love' (Venette, 1696; translation of 1712). Venette is, of course, writing for Guarinoni's 'townspeople and aristocrats' but, in this passage, it seems from his experience. According to him, menarche in such girls would average about 14.0 to 15.0 years, which is entirely compatible with what Guarinoni was saying fifty years before. It is also comparable with what Nicholas Culpeper (1616–1654), a popular English physician of the seventeenth century, wrote in his *Directory for Midwives*, a book clearly aimed at what the next two centuries would have called 'mechanic women', not scholars. 'The menstruis', he wrote, 'stop in a woman, 1, naturally; 2 against Nature. To know the difference between these, you must have regard to the age of the party; in many the menstruis appear not until after the 14th year, in a few before, in none till after 12' (1675 edn, p. 75). This would seem to indicate a mean of 14.5–15.0 amongst the generality of girls. Mrs Jane Sharp (1671), a nearly contemporary midwife writing for midwives, essentially agrees: 'Generally maids have their terms at about fourteen years old . . . fullness of blood and plenty of nutrient brings them down sometimes at twelve years old'.

Venette says it is clear that girls mature earlier than boys, but he doubts whether this is due to their innate heat being greater, as most physicians allege. We have penetrated into the century of the iatrophysicists, when very gradually it was recognized that heat was something measured by a thermometer. To that century we shall turn in Chapter 4; but first we shall step aside to examine a tradition that in our time, but not then, would seem very different. This is the tradition of the artists and sculptors.

3

Human proportion and the canons of beauty: the artistic and philosophic tradition

Anthropometry was born not of medicine or science, but of the arts, impregnated by the spirit of Pythagorean philosophy. Painters and sculptors needed instructions about the relative proportions of legs and trunk, shoulders and hips, eyes and forehead, so they could more easily go about what we might nowadays consider the mundane occupation of making life-like images. In Ancient, mediaeval, even Renaissance times, however, such image-making was infused with a deeper significance. Man was made in the image of God. And God, according to the Pythagorean–Platonic tradition, was most immanent in number, proportion and harmony. Thus the numbers which expressed the bodily proportions of the ideal human representation were in some sense divine (to the extent that some later, more magically degenerate traditions held that if the proportions were exactly right, it took but the conjunction of a name to make the image move). So artists searched for ideal, Platonic, asymptotical beauty and believed that a scale of proportions of a relatively simple sort, such as governed the positions of the planets or the harmonies of musical composition, would procure it. Man was the microcosm set in the macrocosm of the universe, subject to and manifesting the same laws of form. Such a view was by no means exclusive to the West; it was shared by Hindu and Buddhist artists (Tagore, 1921) and probably by the Ancient Egyptians also (Iversen, 1955; Müller, 1973).

In Western art the tradition was powerfully revived at the Renaissance; continued by those neo-Platonists of the sixteenth and seventeenth centuries who were so instrumental in generating modern science; had a nineteenth-century flowering in no less a person than Adolphe Quetelet; and then split, on the one hand into Henry Wampen's (1864) instructions to tailors and the dry rules of first-year art students, and, on the other, into the continuing search for the mathematical formula of the ideal growth curve. No work on the proportions of adults or the growth of children has come down to us from Greek times, although in Galen and Plutarch there are a number of statements attributed to Polyclitus, the fifth-century B.C. founder of the

Greek rule or canon of proportion (Hiller, 1973). Polyclitus was the sculptor of the Doryphorus, a statue known to us only through a Roman copy, which enshrines the Greek canon and hence beauty. Polyclitus and his school probably made actual measurements on subjects, since they are said to have taught that only by taking the means of the measurements of many persons could a harmonious proportion be embodied (cf. discussion of Quetelet, Chapter 6). The first surviving book that deals with human proportions, however, is that of the Roman Vitruvius, who lived in the reign of the Emperor Augustus at the time of Christ, a little more than a century before Galen.

Vitruvius

Vitruvius' work was never entirely forgotten during the Middle Ages (see the very informative account of Braunfels, 1973); it is cited, for example, in Vincent of Beauvais' thirteenth-century encyclopaedia. However, the rediscovery and study of his book *De architectura* had a profound effect on many aspects of Renaissance thinking and spread far beyond architecture itself, since amongst the architectural subjects Vitruvius included geometry, arithmetic, painting, music, astronomy, military fortification and the construction of all machinery such as pumps and pulleys. *De architectura* was, as John Dee said, 'The storehouse of all workmanship' (See Dee's preface to the English Euclid, reprinted in Yates, 1969, appendix A). The *editio princeps* of Vitruvius is dated 1486, and was issued at Rome; many editions and translations followed, the best known of which was Cesariano's, published in Como in 1521. The third book (of ten) deals with human proportion. It does so in the context of the design of temples, since, says Vitruvius, the symmetry and proportion of a temple should truthfully reflect the symmetry and proportion of the human body.

'The navel', he writes, 'is naturally the exact centre of the body, for if a man lies on his back with hands and feet outspread, and the centre of a circle is placed on his navel, his fingers and toes will be touched by the circumference. Also a square will be found described within the figure, and in the same way as the round figure is produced. For if one measures from the sole of the foot to the top of the head, and applies the measure to the outstretched hands, the breadth will be found equal to the height' (Granger, 1931–4 translation, p. 161). This statement is the origin of many Renaissance drawings (see Fig. 3.1), including the famous one by Leonardo da Vinci. It also became the guiding principle in the construction of Renaissance round churches (Wittkower, 1949), and was for centuries the basis on which physicians diagnosed bodily disproportion, especially that ascribed to lack of male sex hormone. Ghiberti (1378–1455) proposed that the pubes ideally was just half-way up from the soles of the feet to the top of the head, and the drawings incorporate this rule also. Vitruvius continues: 'The ancients collected from the human body the proportionate dimensions which appear necessary in all building operations; the finger or inch, the palm, the foot, the cubit ... [They] determined as perfect the number which is called ten ... for the reason that from individual things which are called monads ... the decade is perfected (i.e. 1 + 2 + 3 + 4 = 10) ... but mathematicians, disputing on the other side, have said that the number six is perfect for the

Fig. 3.1. Illustration of Vitruvius' rule of human proportion. From *De architectura,* Cesariano edn of 1521, by courtesy of the Warburg Institute, London.

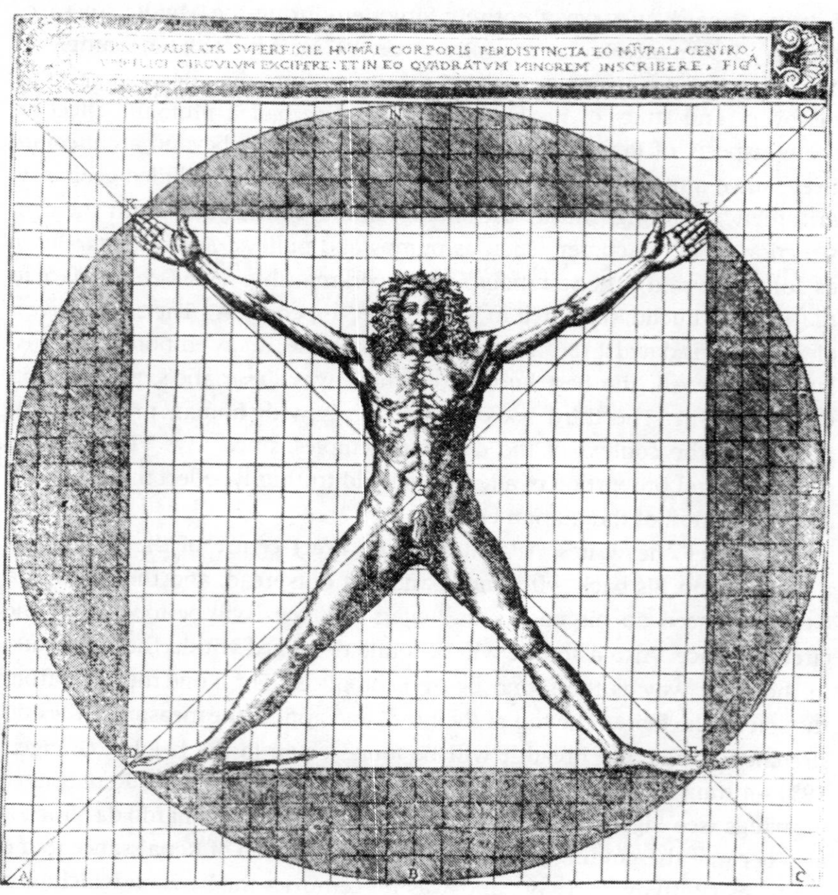

reason that this number has divisions which agree by their proportions with the number six [he means 6 is the lowest common multiple of 1, 2, and 3]... not less also because the foot has the sixth part of a man's height' (Granger, pp. 161–3). Vitruvius found confirmation of the decimal perfection in the relation of the face height to stature: from the lowest point of the chin (menton) to the root of the hair on the forehead (the length the Italians call *il volto*) was one-tenth of stature. The length from chin to vertex was one-eighth of stature. The cubit was one-quarter of stature, and the length of the chest the same. Vitruvius divided the face into three equal parts: from chin to bottom of nostrils; thence to the junction of the two eyebrows; and thence to the hair line (from which Dürer derived some of his illustrations: see below).

Alberti

Leon Battista Alberti (1404–1472) seems to have been the first person to have constructed an instrument with which to measure people, at least since Ancient times. An excellent account of his work and outlook can be found in Gadol (1969). Alberti was one of the extraordinary polymathic figures of the Renaissance, revered by all his contemporaries and occupying much the same niche in his time as did Leonardo da Vinci a generation later. Though he wrote works of general literature, and indeed was one of the founders of the Italian written language, his chief fame, then as now, rested on his application of geometrical and mathematical techniques to the arts of architecture, painting and sculpture. His great book *De re aedificatoria* (written about 1452) became the bible of architects right down to modern times; furthermore, it was written when he was serving in Rome as advisor to Pope Nicolo V and actually engaged in 'drawing the plans which ultimately restored the visual majesty of the Eternal City' (Gadol, p. 94).

Before that, from 1434 to 1436, he was in Florence and friendly with the painters and sculptors Ghiberti, Donatello, Della Robbia and Brunelleschi. In 1425 Brunelleschi had demonstrated for the first time how the laws of perspective work in the formation of images of three-dimensional objects on a two-dimensional surface. He discovered what we now call the vanishing point, and showed how the artist could place it in the centre of his drawing space to act as the single locus for all converging architectural lines (Edgerton, 1975). In the same year Masaccio and Donatello produced the first works employing linear perspective. Brunelleschi and his friends were not scholars, however, but artisan-engineers, educated in the *abacco* to understand arithmetic (for mercantile Florence led the world in such education) but not versed in science or the humanities. Alberti, on the contrary, was a patrician and a Florentine one at that, though born during his family's exile in Genoa. He had studied science, including optics, at Bologna, and knew how to write. In consequence he set his hand to composing a book in which he set forth the geometry and optics of painting in a manner comprehensible to his artisan readers, and gave directions for the construction of linear perspective. This was in the service of *istoria* or history-painting, which Alberti conceived of not primarly as a record of notable events, but as a powerful influence in moulding society by placing before its members examples of classical ideals expressed in geometric harmony. The book was issued in both Latin and Italian versions (*De pictura* and *Della pittura*), the latter dedicated to Brunelleschi. Alberti urged his readers to pay special attention to proportion, using measured geometrical relationships between large and small, near and far objects. All was to be in harmony, and thus exemplify the divine masterplan of the Harmonious World.

A few years later, when he had returned to Rome, Alberti wrote a companion, though shorter, volume on sculpture, *De statua*. Definitive English translations of both books are given by Grayson (1972). In *De statua* Alberti was concerned with the production and reproduction of figures in due proportion, and it was to this end that he adapted an instrument he had first used, it seems, in the mapping of Rome. Cartography was at that time a suddenly expanding field. In 1400 Ptolemy's world atlas, the *Geographia*, had been brought back to Florence. In 1406 a Latin translation of the text, with its instructions for constructing hitherto quite unknown map projections as a grid system, was completed, and ten years later copies of Ptolemy's maps annotated in Latin began to circulate. Edgerton (1975) in particular has stressed the impact this new representational system had on Florentine artists, who in making frescos abandoned the old method and began to use a grid technique to transfer details from preliminary drawings to the full scale of the wall space. Alberti combined his architectural and geometrical skills and worked out a mapping system based on Ptolemy, which he used in making the new city plan of Rome. The details were published about 1450 in *Descriptia urbis Romae*. Alberti started with a circular disc called a *horizone* held in the horizontal plane. The outer edge was divided into equal parts 'like those astronomers inscribe on astrolabes'. These parts were called degrees, and each was divided into lesser parts called minutes. A ruler, called the *radius*, had its end fixed at the centre of the *horizone* and could be moved around it. Alberti set up the instrument on the Capitoline Hill, placed zero degrees towards north, and took bearings on the various buildings. By pacing out the distances to each he constructed a simple system of coordinates for locating precisely each building, a system still used in the tourist maps displayed in the squares of many European cities today.

In *De statua* Alberti used the same instrument. He divided the task of measuring the body into *dimensio*, measurements in a static upright position, and *finitio*, measurements of angles and distances enabling the pose of a body to be exactly copied. '*Dimensio*', he wrote,

> is the accurate . . . recording of measurements, whereby the state and correspondence of the parts is observed and numerically represented, one in relation to another and each to the whole length of the body. The observation is carried out with two instruments, the *exempeda* and the moveable squares. With the *exempeda* we measure the lengths and with the squares the other diameters of the limbs. The *exempeda* is a thin wooden ruler as long as the length of the body you wish to

Fig. 3.2. Alberti's squares. From *De statua* (1961), by courtesy of University of Catania Press.

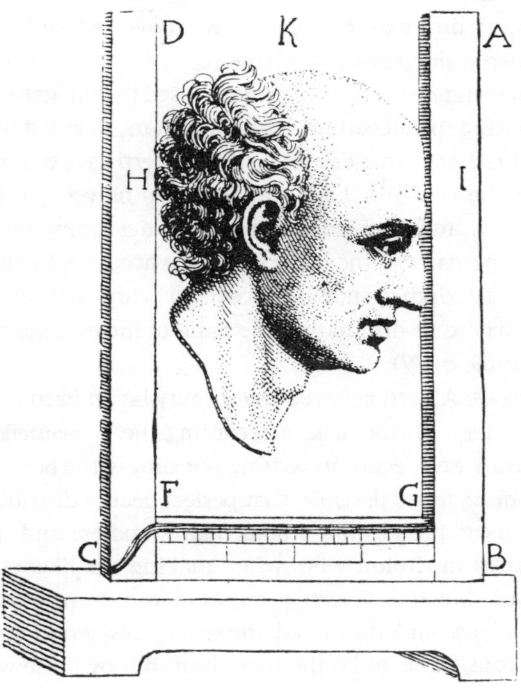

> measure, from the top of the head to the soles of the feet. From this it follows that the *exempeda* for a small man will be short, and for a tall man longer. Whatever the size of the chosen figure, we divide it into six equal parts which we call feet, and this is why we give the ruler its name *exempeda*, from the number of feet; then we divide each of these feet into ten equal parts, which we call inches ... then I divide each inch into ten small equal parts called minutes ... If we wish to measure a standing man, we set up [the *exempeda*] beside him and note for each of the limbs, the heights from the sole of the foot, ... say up to the knee, the navel, the collar-bone and so on' (Grayson, pp. 126, 127).

The squares are illustrated in Fig. 3.2. The base is 'not less than 15 inches ... of the *exempeda* of the body you are to measure: measurement of diameter and breadths proceeds in the way shown'.

'*Finitio* is the means whereby we record with sure and accurate method ... the variations in limbs from time to time caused by movements and new dispositions of the parts'. For this Alberti used the horizon and radius combined with a plumb-line (see Fig. 3.3). The horizon was made by inscribing a circle of diameter three feet of the *exempeda* used. The radius was a thin ruler three feet of the *exempeda* long. Inch and minute markings were made upon it. A plumb-line was attached to its moveable end. 'The whole instrument I called a *finitorium*', he wrote (Grayson, p. 131). The instrument was directly descended from the astrolabe, used by navigators and astronomers for measuring the elevations of sun and stars, adapted by being placed in a horizontal rather than a vertical plane. Alberti says that by use of *exempeda* and *finitio* he could make a statue in two halves and they would fit together perfectly, and claimed that he could determine the contours and proportions of the parts of the body with such accuracy by these two 'rules' that if a statue by Phidias should be covered with earth or clay he would know exactly where to drill to find the pupil of the eye, the thumb or other parts (Gadol, 1969, p. 79).

Like the Greeks, Alberti believed that beauty lay in harmony and proportion. '[When] I took on the task of recording the *Dimensione* of man I proceeded to measure and record in writing not simply the beauty found in this or that body but, as far as possible, that perfect beauty distributed by Nature, as it were in fixed proportion, among many bodies; and in doing this I imitated the artist at Croton who, when making the likeness of a goddess, chose all remarkable and elegant beauties of form from several of the most handsome maidens, and translated them into his work. So we too chose many bodies, considered to be the most beautiful by those who know, and took from each and all their dimensions which are then compared one with another, and leaving out of account the extremes on both sides, we took the mean figures validated by the *exempeda*' (Grayson, p. 135). There follows a long list of proportions representing the ideal. If the man should be 6 feet tall, then from soles of feet to 'the bone above the base of the penis' is 3 feet, to the point of the shoulder $5\frac{1}{10}$ feet, to the 'hole of the ears' $5\frac{5}{10}$ feet. The distance between the hip joints is $1\frac{1}{10}$ feet; and many other transverse and antero-posterior widths are listed.

Thus Alberti revived the old Graeco-Roman anthropometric sculptural practice and theory of proportions (see Panofsky, 1968, a fundamental paper in this field). The tradition had been eclipsed in mediaeval times, when

Fig. 3.3. Alberti's *finitorium*. From *De statua* (1961), by courtesy of University of Catania Press.

symbolic considerations outweighed naturalistic ones. The Byzantine artists, using the perfect figure of the circle the perfect number of three times to symbolize the perfection of the Trinity (a sort of belt and braces procedure), took the root of the nose each time as centre and drew with the first circle the outline of the face, with the second the outline of the head, with the third the outline of the halo. The height of the body was nine (3 × 3) times the face module. The Byzantines took the head as module because it was the seat of mind or spirit, and Alberti was apologetic and evidently a little disturbed to have to choose, for functional reasons, the 'less worthy' foot. Alberti was the link between mediaeval and Renaissance attitudes, just as Newton was the link between Renaissance and modern ones. He was, after all, an older contemporary and occasional pupil of the Florentine neo-Platonist Marsilio Ficino, the translator of Plato and of Hermes Trismegistus, who resurrected Plato's Academy in Florentine form, revived the Orphic mystical religion of pre-Christian times, and even, such was his power, ensconced its prophets as mosaics in the hallowed precincts of Siena cathedral. Ficino was the man who designed Botticelli's *Primavera*, and he designed it, in all probability, 'as a complex talisman, an image of the world arranged so as to transmit only healthful, rejuvenating, anti-Saturnian influences to the beholder', an unstable adolescent cousin of the ruling Medici (Yates, 1964, p. 77; Gombrich, 1972, p. 33ff.). The Hermetic tradition, that perfectly proportioned statures or perfectly harmonized music would exert a magical effect on the subject blessed by them, was never far away.

Paccioli

Leonardo da Vinci (1452–1519) incomparably illustrated his predecessor's ideas, and incorporated a good deal of Alberti, without acknowledgement, after the manner of the times, into his own treatise on painting. So far as we know he made no systematic measurements. It was he who gave the name of the 'Golden Section' (an alchemical reference) to that division of a line in two so that the ratio of the smaller to the larger equals the ratio of the larger to the whole. His friend Luca Paccioli (1445–1510) had rediscovered the Section in Greek texts and described it in a work called *De divina proportione* published in Venice in 1509 (see Ghyka, 1927, especially the charming preface). Leonardo da Vinci also did the magnificent drawings which illustrate the series of solids described in the book (see note 3.1). *De divina proportione* starts with an encomium to mathematics and an argument that it is essential to science, it underlies the arts of architecture, sculpture and painting, and besides, it is necessary for making engines of war; its cultivation plays a vital part in the defence of the Republic. Paccioli published one of the first Renaissance books on mathematics, in 1494, and played the same part in initiating mathematical studies in Italy as did John Dee in Elizabethan England a century later.

Dürer

In 1506, not long before his death, Paccioli was visited, it is believed, by a young artist from north of the Alps who rode over from Venice to Bologna 'to take lessons in secret perspective'. The artist was Albrecht Dürer (1471–1528), who wrote to a friend of his intention (Benesch, 1966, p. 35). Though he did not specifically mention Paccioli, Paccioli's pre-eminence as a geometer must necessarily have directed Dürer to him and his school. Dürer clearly felt he had joined the fraternity, for he wrote a book on the geometry of measuring figures (1525, German; 1532, Latin) and afterwards, on the title page of his most celebrated book, usually referred to as the *Proportionslehre* (1528), he styled himself 'geometer and painter'. *Proportionslehre* was first published, in German, in Dürer's home town of Nuremberg, in the year of his death. Translations into Latin and French were issued in 1557 and 1613 and into English in 1666. It is nowadays the best-known book on human proportions of its time, especially amongst scientists, because of D'Arcy Thompson's (1917) reproductions of the enormously forceful drawings of faces (Fig. 3.4).

Adolphe Quetelet (1870) in his review of the study of human proportions says he thinks Dürer measured few if any subjects, and indeed there is little evidence in the book of any intention actually to measure. Dürer was providing a scheme – in fact a library of schemes, most thoroughly analysed in Winterberg (1903) – for artists to follow in practice when drawing figures. His aim was apparently severely practical, and he, almost alone of his contemporaries in the field, was seemingly unconcerned with coralling beauty. Geometric construction predominated (see Fig. 3.5). After giving a system for drawing the human figure he goes on to provide a whole group of transformations by which elongated or compressed bodies can be drawn, for example by bending lines as though refracted through a lens. For faces (see Fig. 3.4) he gives more transformations still, including curvilinear ones. Clearly he was fascinated by the many different ways in which such transformations could be made. His text is a geometrical exercise of immense ingenuity and detail and though he starts the book with the disarmingly simple phrase – 'If you want to draw a man . . .' – he ends by himself acknowledging it is not an easy work. But he adds 'If anyone likes to use our principles . . . he will see what marvellous transformations and variations in the figures he can make'. (Nearly 400 years later D'Arcy Thompson did just that, changing one species of fish into another in a way that would surely have delighted Dürer, and drawn from him appropriately symbolic appellations for *Diodon* and *Orthagoriscus*.)

Dürer's main interest seems to have been in how to construct different

Fig. 3.4. Dürer's drawings of faces, with geometrical transformations of proportion. From *Proportionslehre* (1528), by courtesy of the Victoria and Albert Museum, London.

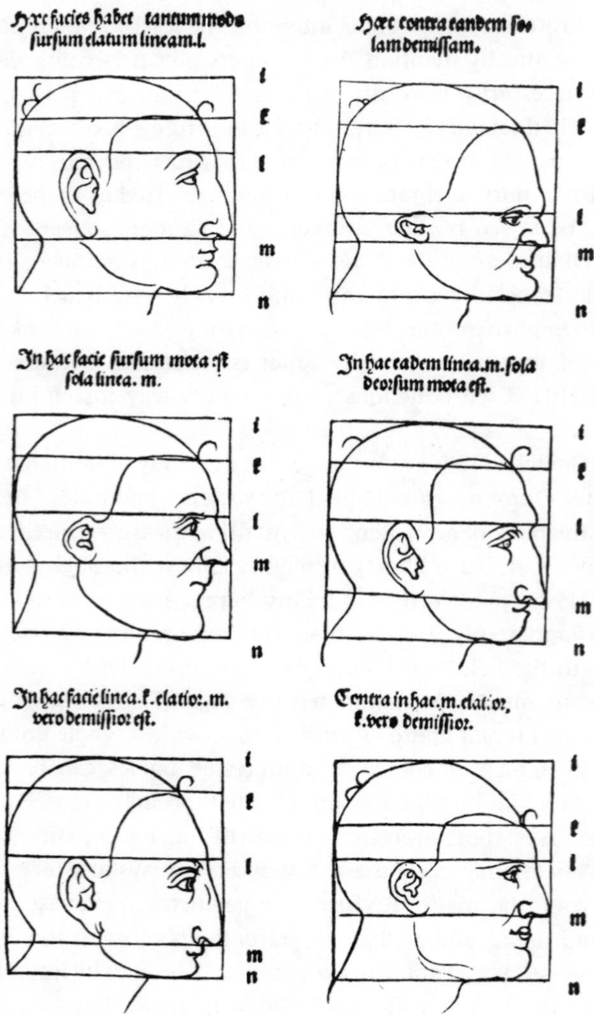

Fig. 3.5. Dürer's illustrations of a slender and robust form. From the Latin edition of *Proportionslehre* (1557), by courtesy of the Wellcome Trustees.

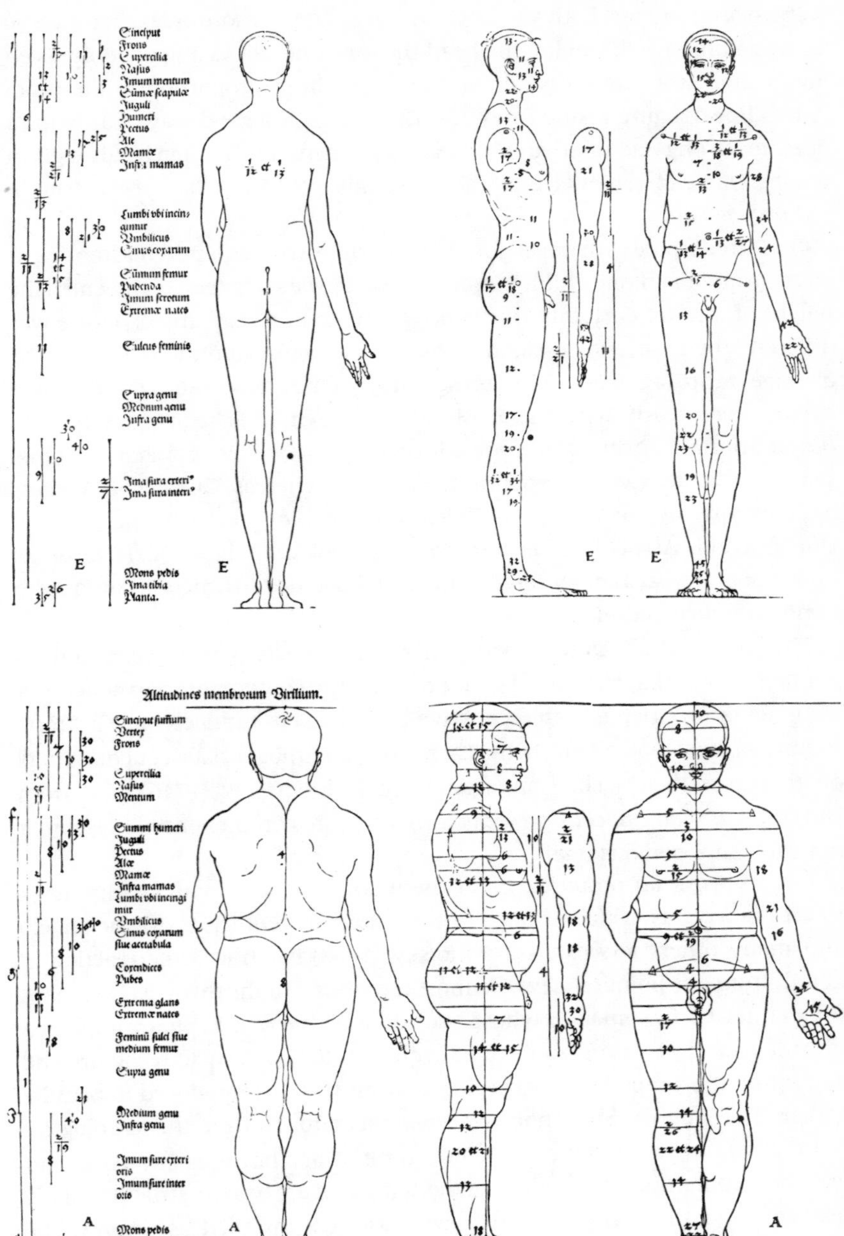

sorts of people. He begins, as did Vitruvius, by taking one-eighth of stature as the head (chin–vertex) height and giving proportionate values for other body measurements, both vertical and transverse. A more or less naturalistic figure results. Then, leaving the head the same, he draws a figure only seven heads high; the transverse dimensions remain in proportion to the head. A less linear figure results; and this, Dürer says, is a good way of drawing a peasant, strong and gnarled. He then draws men nine and ten heads high, of whom Quetelet remarks that they certainly could not have been seen in nature.

In Dürer's book we meet for the first time women and children; or at least, women and one child. Women are said to be seventeen-eighteenths the height of a man. Certainly this is not taken from nature unless nature was different then; seventeen-eighteenths is 0.943 and Galton (1889) gives the average ratio, agreeing with present-day values, as 0.926. Dürer's value agrees more nearly with the artistic canon given by Schadow (see below). (Come to that, Vitruvius' feet are a little long, as Quetelet pointed out. One-sixth of stature gives a proportion of 0.167, whereas Quetelet's Belgians had actually a value of 0.154.) There is no mention of the intended age of Dürer's child. As his head is about a quarter of his height, Quetelet thinks he is about 1 year old. Dürer give the usual detailed instructions on how to draw both face and body.

The influence of Dürer's book on art theoreticians was great, and on painters too perhaps not negligible: at least the contemporary woodcuts of Hans Baldung Grien and the seductive Eves of Lucas Cranach have a strong *Proportionslehre* look about them. There were a number of subsequent writers on the same theme, chief amongst them Lichtensteger (1746), also from Nuremberg, who, indeed, presents a much more comprehensible system, if not such an ingenious one.

Dürer's work on proportion and the whole idea of a mathematical basis to art theory was criticized later in the sixteenth century, by Michelangelo and many others less exalted. Panofsky (1968) has traced the reaction in detail: from our point of view the most interesting of the theorists is perhaps the painter G. P. Lomazzo, whose book *Idea del tempio della pittura* appeared in Milan in 1591. Lomazzo devoted a chapter to 'The way to recognize and determine the proportions according to beauty': it is reproduced in parallel Italian and English by Panofsky (1968, appendix I). Here the astrological trains of thought traceable to Ficino's translations have penetrated art no less than contemporary medicine; the temple of art has the structure of the heavens, and there are naturally seven painters installed as regents. The essence or spirit of beauty radiates from the heavens, but is differentiated by passing through the consciousness of the angels associated with the heavenly bodies. Thus there arrive on earth ideas of, for example, Jovial, Saturnine and Martial beauty. Martian types, being hot and dry, should be represented with large limbs, robust and hairy. Jovians, hot and moist, are more harmonious, 'delicate to the touch' and full-bodied. Lunar types, on the other hand, are smaller than Jovians, disproportionate, brittle and weak. We are back in the world of Fernel, Clever and Chaucer's physicians.

Elsholtz

The inventor of the term *'anthropometry'* was a German physician, Johann Sigismund Elsholtz (1623–1688). He seems to have been the first medical man to concern himself with measuring the human body. Born in Frankfurt-on-Oder, he studied at Wittenberg and Königsberg and travelled in Holland, France and Italy in the usual manner of the time, graduating eventually in 1653 at Padua (Munster, 1966). He returned to become personal physician to Frederick-William, the Great Elector of Brandenburg (see note 3.2) (whose youngest son, Christian Ludwig, was the man for whom eighty years later J. S. Bach wrote the Brandenburg concertos and whose great-grandfather Joachim Frederick had tried eighty years earlier to breed a race of dwarfs: see p. 121). Like Buffon in France in the following century, Elsholtz was appointed director of the state botanical garden, in this case newly founded. In 1666 he published a six-volume treatise on gardening, in 1674 a book on distillation and 1682 six books on foods and their preparation. His name is perpetuated in the small and not very spectacular genus of Chinese herbal shrubs, *Elsholtzia*.

Elsholtz's graduation thesis was entitled *Anthropometria* in the first edition, issued in Padua in 1654, and more explicitly *Anthropometria sive de mutua membrorum corporis proportione, item de naevorum harmonia libellus* in subsequent editions (1663 in Frankfurt-on-Oder, 1667 again in Padua, and 1672 in Stade, a town in Brunswick near Hamburg). The frontispiece and title page of the Stade edition are reproduced in Figs. 3.6 and 3.7. Elsholtz's anthropometer is illustrated in Fig. 3.8. Quetelet (1870) remarks of Elsholtz that 'A glance at his illustrations suffices to show he was inspired to the idea by Alberti, although one searches the book in vain for his name'. Elsholtz divided his feet into twelve inches, however, and not ten. Elsholtz's instrument has a transverse rod (the *regula*) moving up and down the vertical rod, 'similar to the draughtsman's rule except that at one end it has a special wider portion composed of two projecting angular pieces made so that when held against any object having a rectangular section it will be kept in a straight line' (1672 edn, p. 28). To explore 'to what degree a person approaches universal dimensions', Elsholtz goes on, 'You place him standing erect with his feet together and the outside of his left ankle touching the bottom of the rod, with the transverse rod pressing lightly on the top of his head. The *regula* applied to the anthropometer ... placed in contact with [the root of] the head-and-neck and, held there, will now show its correct length measurement. Moved twenty inches it corresponds with the

Fig. 3.6. Frontispiece of Elsholtz's *Anthropometria* (1672 edn), by courtesy of the Wellcome Trustees. Note the plants, which refer to Elsholtz's position as director of the Brandenburg Botanical Garden.

precordium, observe. Twenty-eight inches, to the umbilicus, thirty-six to the pubes' (p. 28). Like Alberti, Elsholtz started by making a 'straight rod, of square section about a thumb's width, of the height of the man to be measured' (p. 25, wrongly marked 23 in the 1672 edn). His transverse rod was to be made about one-sixth the length of the vertical rod and appears to be also marked in inches, which certainly recalls Alberti's *finitorium*.

Elsholtz was concerned with the Galenical problem of the correct balance

Fig. 3.7. Title page of Elsholtz's *Anthropometria* (1672 edn), by courtesy of the Wellcome trustees.

JOANNIS SIGISMUNDI
ELSHOLTII
D. & Medici Electoral. Brandeburgici
ANTHROPOMETRIA,
sive
De mutua membrorum corporis humani proportione, & Nævorum harmonia Libellus.
Editio post Patavinam altera,
figuris æneis illustrata,

STADÆ,
Apud ERNESTUM GOHLIUM,
Bibliopol. Stad.
ANNO M. DC. LXXII.

of humours, but as reflected in the correct or perfect (well-tempered) physical proportions so much under discussion by artists and philosophers. Presumably he thought of linking departures from such proportion (distempers) with susceptibility to various diseases, but he seems not to have pursued the problem once his dissertation had been accepted. Perhaps the tradition lingered on in Padua, for it was here that the first serious effort to carry out such a programme was made, by Achille De Giovanni nearly 200 years later (De Giovanni, 1904–9). Whether Elsholtz really used his

Fig. 3.8. Elsholtz's anthropometer. From *Anthropometria* (1672, edn), by courtesy of the Wellcome Trustees.

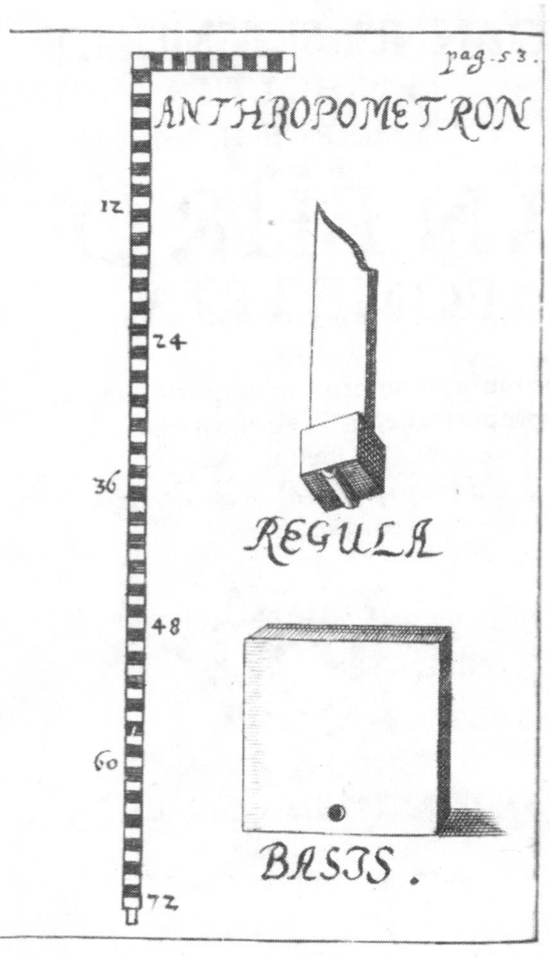

instrument much is open to doubt. He does remark (p. 61) that when the anthropometer is applied to the anterior and posterior surface of the neck it shows that turning the head upwards to the sky increases the length of the throat from two to four inches, with a corresponding decrease in the nape. And towards the end of the book there is a table of lengths and breadths of the parts of a 72-inch man, which shows, he says, all the species of numbers alluded to by Pythagoras, and exemplifies arithmetic, geometric and harmonic proportion. But the book mostly consists of references to Pliny and other classical authors on the uses of parts of the body, and it is permeated by a considerable odour of the then-popular physiognomy (citing Della Porta amongst others and even Henry Cornelius Agrippa). Medicine in the seventeenth century was a very backward art compared with painting and sculpture. Elsholtz dropped the study of proportion when he returned to Prussia but wrote many subsequent works on medicine, amongst them the posthumously published *Clysmatica nova* (1695) an early work on intravenous injections and the possibility of blood transfusion.

Bergmüller

At last, in 1723, there appeared a book specifically dealing with proportion in the growing child. Also called *Anthropometria,* it is by Johann Georg Bergmüller (1688–1762), 'oil- and fresco-painter' and director of the Academy in Augsberg (Kilian, 1765; Nagler, 1835). It is in German and consists of a frontispiece, two pages of text and eleven plates. There are few surviving copies of the book (one is still in Augsberg, another in Boston), and even Quetelet was unable to find a copy, though Schadow saw one (see below).

The Boston copy has two additional plates executed by Bergmüller's son Johann Baptista Bergmüller (1724–1785), also an artist (Nagler, 1835, pp. 439–40). They are quite different in flavour from the rest. One shows a man standing against a series of scales, the first representing a human scale of height, the second God's measurement of time, and others the time scales of the Old and New Testaments. The second plate shows a 120-year clock, with hours, days, months, seasons and signs of the zodiac with three triangles inscribed within it, and the words Macrocosmos and Microcosmos. It resembles the *Speculum naturae* of Robert Flood (see Yates, 1972) and harks back to Henry Cornelius Agrippa's man inscribed in an astral circle (see Yates, 1969, p. 19).

This is, of course, a side-issue and little to do with the drawings of Bergmüller the elder shown below. But it serves to remind us of the philosophical matrix from which the work on human proportion was chiselled, right down to the eighteenth century. Neo-Platonist imagery forms much of the chrysalis from which modern science slowly and fitfully emerged and even in the nineteenth century there are authors whose wings bear traces of their larval form.

Johann Georg Bergmüller gave, for the first time, a geometric law for the growth of a child. He illustrated the application of that rule in subjects aged zero, 1, 2, 3, 6, 9, 12, 15, 18, 21 and 24 years. In Fig. 3.9 his illustration of the rule is reproduced and in Figs. 3.10 to 3.16 a selection of the subsequent pictures. Bergmüller acknowledges Dürer as his predecessor but says that neither Dürer nor any others have previously published proportions from birth to maturity. He remarks that 'Because growth often is hindered or disturbed, by illness or accident, I have for the most part kept to the average (*Mittelmaass*) and have presented a neat and well-groomed (*angestalte*) proportion as the sensible way to represent healthy persons'. He has drawn his figures, he says, strictly according to the rules of proportion, and would not defend himself against criticisms levelled at them from a purely artistic viewpoint. There is no evidence that he actually measured children (indeed the falsity of his rule is evidence against it) but he is a careful and scrupulous man, for in a last note to the reader he warns 'Note that because the paper

Fig 3.9. Bergmüller's rule for human growth. From Bergmüller (1723), by courtesy of the Countway Library, Boston.

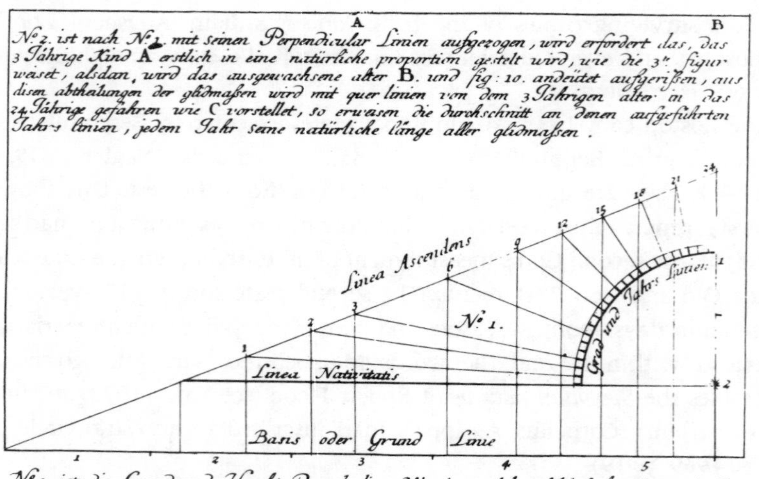

Fig. 3.10. Bergmüller's drawing of a child at birth, age 1 and age 2. From Bergmüller (1723), by courtesy of the Countway Library, Boston.

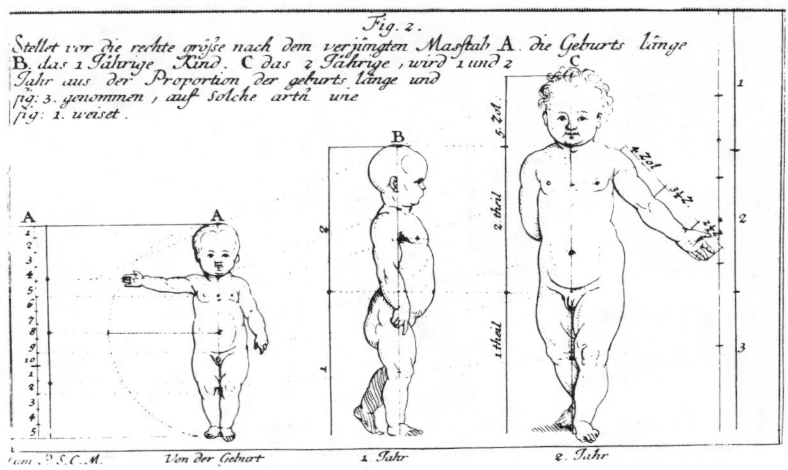

Fig. 3.11. Bergmüller's drawing of a 3-year-old boy. From Bergmüller (1723), by courtesy of the Countway Library, Boston.

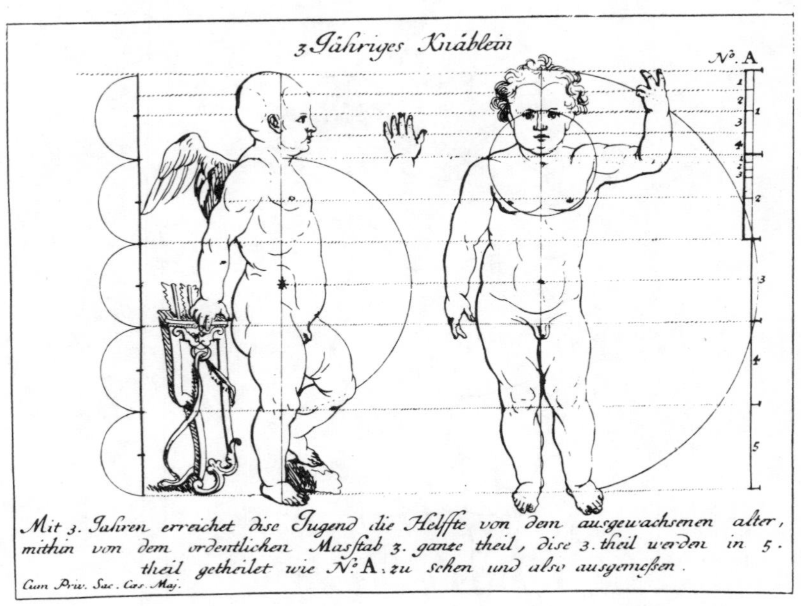

Fig. 3.12. Bergmüller's drawing of a 6-year-old child. From Bergmüller (1723), by courtesy of the Countway Library, Boston.

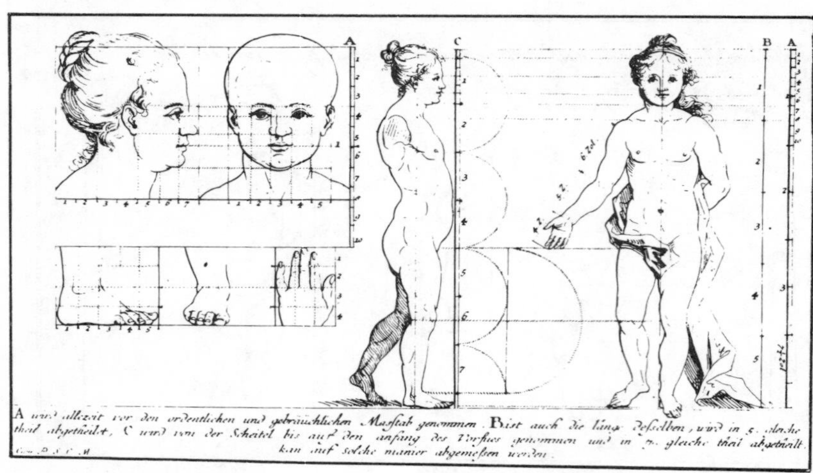

Fig. 3.13. Bergmüller's drawing of a 12-year-old boy. From Bergmüller (1723), by courtesy of the Countway Library, Boston.

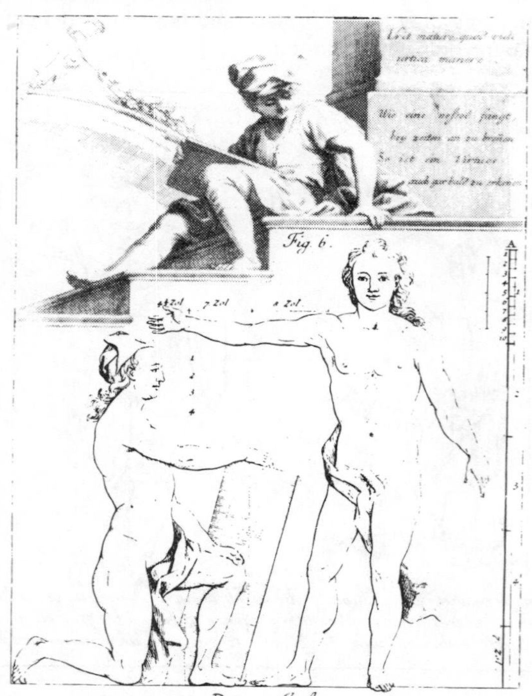

is sometimes unevenly pulled through the moistener, each figure and drawing must be taken individually according to its own scale (*Maass-Stab*)'.

Modernizing the terminology, the rule (Fig. 3.9) is made as follows: Draw a right-angled triangle, of height 2 units and base 5 units. Draw a line parallel to the base one-half unit up the vertical ('*linea nativitas*'). Using the intersection of this line and the vertical as centre draw a circular arc of radius 1 unit, rising from the line to the vertical. Divide the arc into 24 equal divisions after the manner of a compass. Draw lines from the centre of the circle through successive divisions. The points at which the lines meet the hypotenuse of the triangle (*linea ascendens*) represent the heights of children aged 1, 2, 3, etc. years. This makes height at age 3 exactly half adult height, as Bergmüller knows, for he shows it in a drawing not reproduced here. The physiologist Haller thought the same fifty years later (see p. 86). In fact the

Fig. 3.14. Bergmüller's drawing of a 15-year-old boy. From Bergmüller (1723), by courtesy of the Countway Library, Boston.

correct age is nearer 2, at least in modern times, so Bergmüller's children would be growing very slowly. It is just possible the rule was correct for the early eighteenth century, at least in this respect. But Bergmüller's rule also makes birth length a quarter of adult height, and this is too small by any standard, 0.30 being about the present-day value. In fact Bergmüller's rule does not define a very naturalistic curve of height growth; it represents a steadily decreasing growth velocity and omits all mention of the growth spurt at adolescence (and in this resembles Quetelet's curve, as we shall see later).

Bergmüller gives 21 years as the age at which 'the female sex reaches full height and proportion' and 24 years as the age at which a man reaches his full height. He ends the book proper with a little couplet reminiscent of the last words of Dürer in his *Proportionslehre*.

Bergmüller had no real successor for a hundred years, though there were

Fig. 3.15. Bergmüller's drawing of a young adult woman aged 21. From Bergmüller (1723), by courtesy of the Countway Library, Boston.

Fig. 9.
Das 21 Jährige alter.
In diesen Jahren wird das Weibliche Geschlecht in vollkommner gröse oder Proportion befunden, ist der ausgesezte Masstab um 3. Zol gegen dem nachfolgenden ausgewach„
senen alter fig: 10. verkleinert, alldieweilen auch selbe meistentheils kürzer als die Mans
Personen, wird solcher gestalten ausgemessen.

a number of imitators, such as Martinet (1777, vol. 1, p. 326). One or two anatomists in the eighteenth century became involved in devising rules for artists. The most distinguished was Peter Camper (1722–1789), professor of medicine in Gröningen. Camper's studies of the heads of apes and men laid the foundations of comparative anatomy and paved the way for evolutionary theory, and his comparisons of negro and European faces and skulls did the same for physical anthropology. Stimulated by Winckelmann's famous *Reflections on the Imitation of the Painting and Sculpture of the Greeks* (1755; English edn, 1765: see note 3.3) Camper wrote a treatise entitled *On the Connexion between the Science of Anatomy and the Arts of Drawing, Painting and Statuary* (1794) in which, though criticizing Dürer, he nevertheless also used a linear coordinate technique to show the difference in the proportions of the heads of newborns, infants and adults.

Fig. 3.16. Bergmüller's drawing of a young adult man aged 24. From Bergmüller (1723), by courtesy of the Countway Library, Boston.

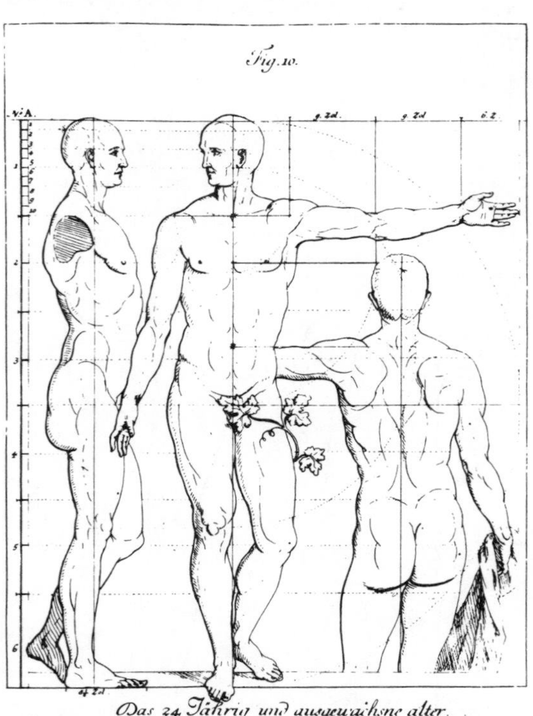

Das 24 Jährig und ausgewachsne alter.
N°. A. ist der volkomne Masstab, welches die eigentliche menschliche länge vorstellet so getheilet ein theil in 10. Zol, durch dises Mass können alle glidmassen so wohl der länge als dicke halber abgemessen werden.

Cum Pr. S. Cæs. Maj.

The statue-measurers

Winckelmann represents a school whose approach to the study of proportions was by a somewhat different, though parallel, route. The whole Renaissance study of proportion was part of an attempt to rediscover Ancient wisdom and thereby usher in a new Golden Age. *Jam redit et Virgo redeunt Saturnia regna* (Now Virgo returns, and the rule of Saturn) was the motto surrounding Vitruvian Man. Paccioli and Elsholtz worked through the secret of numbers, hoping some combination would generate Perfect Form. But the empirical approach was possible too; and in this case, it meant measuring the statues of Antiquity to see if laws could be deduced from the measurements observed. Some of this work resulted in beautiful books, most notably

Fig. 3.17. Audran's measurements of the Medici statue of Venus. From Audran (1683).

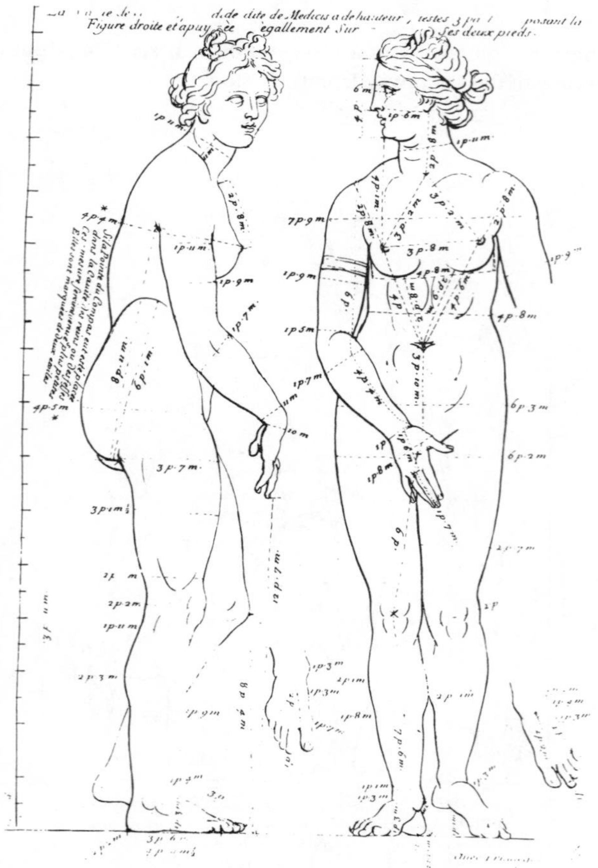

that published in Paris in 1683 by the royal engraver, Gérard Audran (1640–1703). It was entitled *Les proportions du corps humain mesurées sur les plus belles figures de l'antiquité*; some doubt surrounds its authorship (see note 3.4). There are thirty beautifully executed plates, two of which are reproduced here as Figs. 3.17 and 3.18. As usual, Audran says, he gives all the measurements as proportions of height, but he admits he has some problems determining the latter in, for example, the Laocoön.

This school became institutionalized in the art academies of the eighteenth and nineteenth centuries, where it figured increasingly as the quest for Ideal Beauty, in England in the hand of Joshua Reynolds, Flaxman, Walker and their successors. Amongst the various art academicians who concerned

Fig. 3.17. *contd.*

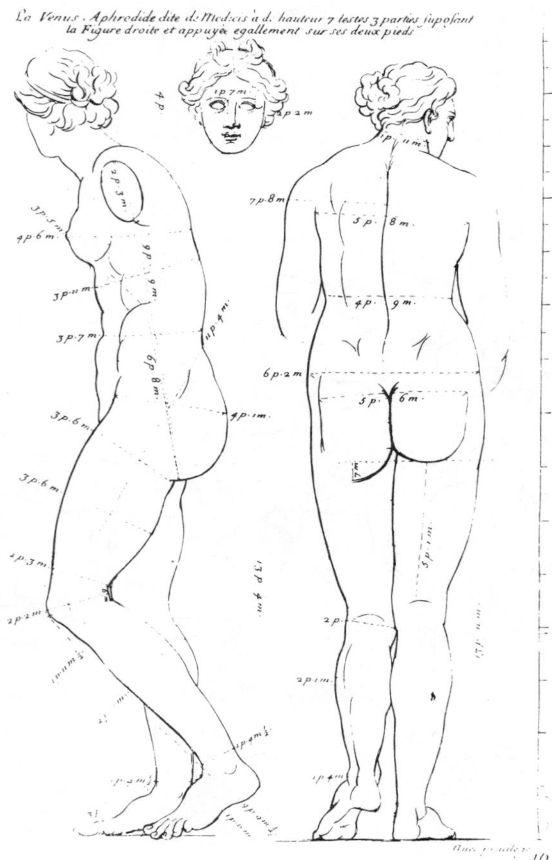

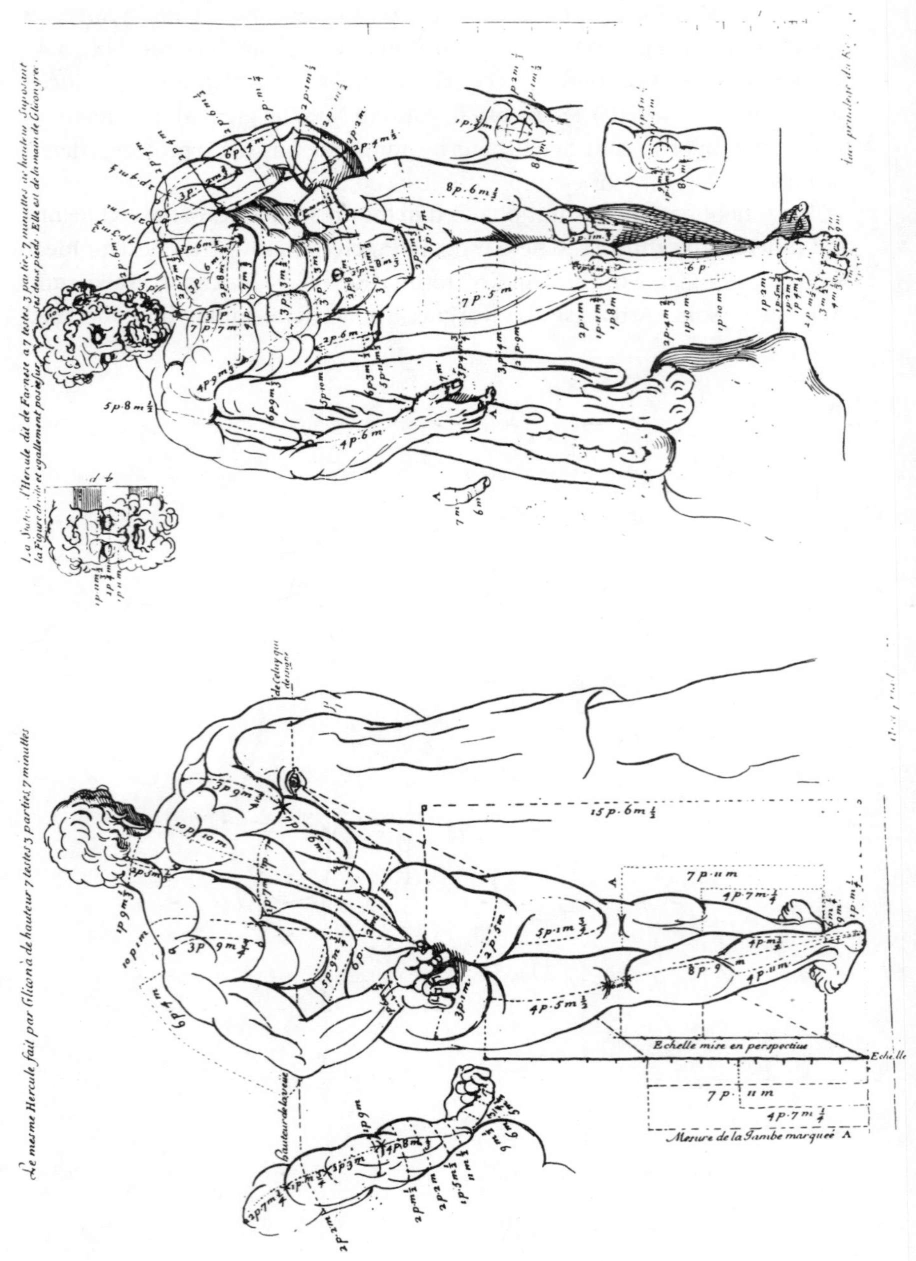
Le mesme Hercule fait par Glicon a de hauteur 7 testes 3 parties 7 minuttes
Echelle mise en perspectiue
Mesure de la Jambe marqueé A

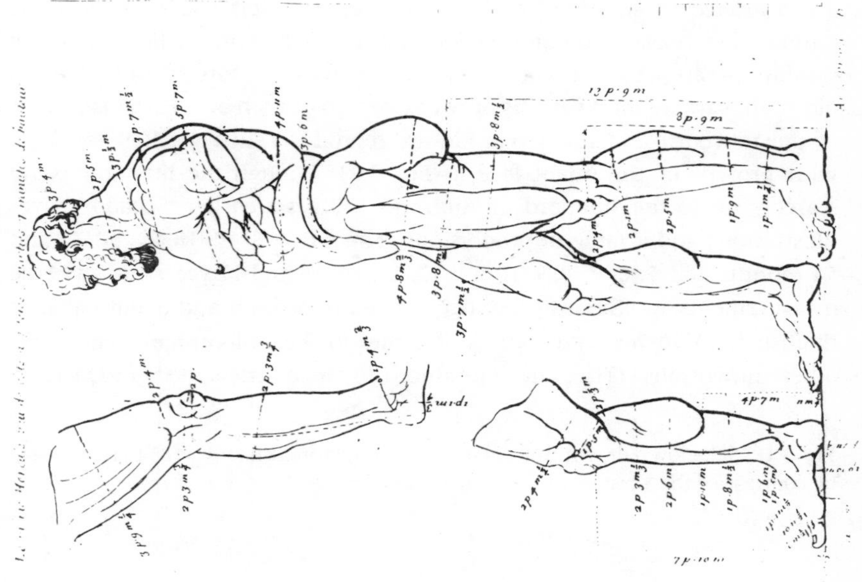

Fig. 3.18. Audran's measurements of the statue of Hercules. From Audran (1683).

themselves in this way, the nineteenth-century sculptor and director of the Berlin Royal Academy, Johann Gottfried Schadow (1764–1850), stands out much above the rest. His book *Polyclète, ou théorie des mésures de l'homme* was published in a parallel German and French text in Berlin in 1834, with a large folio atlas as a separate volume. A fourth edition appeared as late as 1882. Schadow actually measured babies and children, claiming, it seems correctly, that he was the first to do so (Jampert always excepted: see below). Before Raphael, he says, painters made the Infant Jesus with proportions which only betrayed their ignorance; 'Raphael was the first to realize that the eye of a child of 2 years was as large as that of an adolescent'.

Schadow follows only the male during childhood; his studies of women are restricted to adults. At each age he measured some subjects (with the sculptor's compasses) though he does not specify how many (for the idea of sampling was yet to be born). Quetelet, who knew him when he was an old man, says he thinks it was only one or two at each age. His illustrations of the newborn and the 1-year-old are reproduced in Figs. 3.19 and 3.20, with growth of the infant face in Fig. 3.21. Though the lithographs are scarcely up to the standard of Audran's 150 years earlier, Schadow gives illustrations and a table of proportions at ages 0, 4, 8, 12, 18, 24, 30, 48 and 54 months and 5, 6, 7, 8, 9, 10, 11, 12, 13, 14, 15 and 17 years. He gives the measurements in Rheinish feet (smaller than Parisian and about equal to English feet), inches and eighths of inches for head, foot, hand, cubit and other dimensions. (They are reproduced in the metric system in Quetelet,

Fig. 3.19. Schadow's drawing of a newborn infant. From Schadow (1834), reproduced by courtesy of the Trustees of the British Museum.

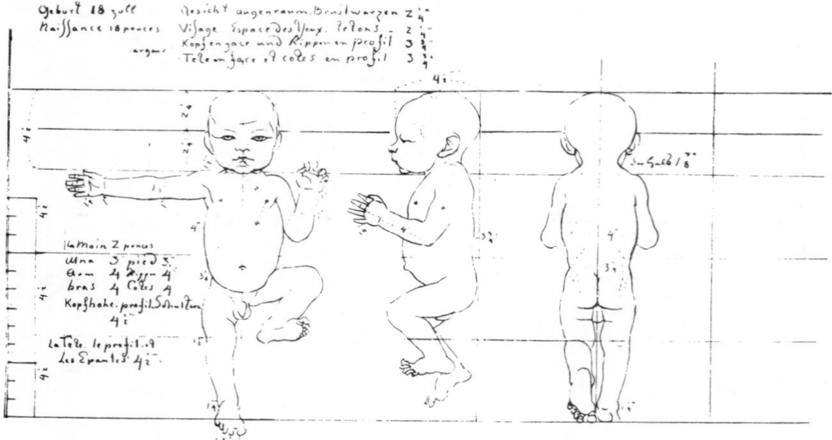

1870, p. 132.) Schadow's illustrations show a gradually falling velocity of height from birth till about 11, when from the very low figure of 1 inch per year it rises to 3 inches a year for each of the years 11 to 12, 12 to 13 and 13 to 14. Like everyone except Buffon before him, and like several after, he remained baffled by the supposed irregularity of growth at puberty. 'I have not succeeded', he wrote, 'in finding individuals in nature suitable for fixing the proportions from 15 to 19 years. The developments of puberty offer so many varieties that I leave to others the task of going into more detail on this subject' (cited in Quetelet, 1870, p. 130).

Schadow also measured a number of statues, noting some discrepancies between his measurements and those of Audran, who, he remarks, 'bent (*cédait*) sometimes to the rules of the school of his time'. Quetelet found the same when he, the latest and best of the statue-measurers, compared the Brussels copies of the Pythian Apollo, the Antinous and the Medici Venus with the Paris measurements of Audran (finding too little distance between the eyes, nostrils and breasts). Indeed it was Quetelet's interest in art and proportion that brought him into contact with the ageing Schadow, of whom he writes with warmth and admiration. Despite his advanced age

Fig. 3.20. Schadow's drawing of a 1-year-old child. From Schadow (1834), reproduced by courtesy of the Trustees of the British Museum.

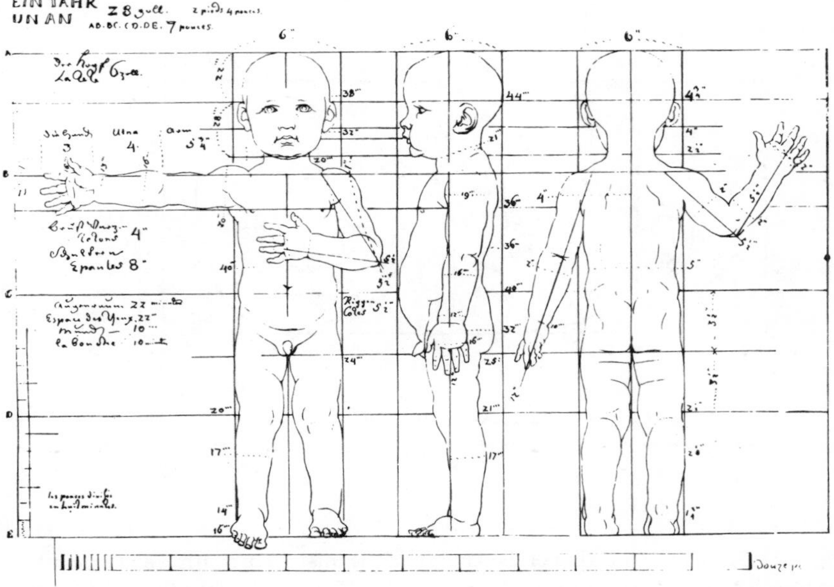

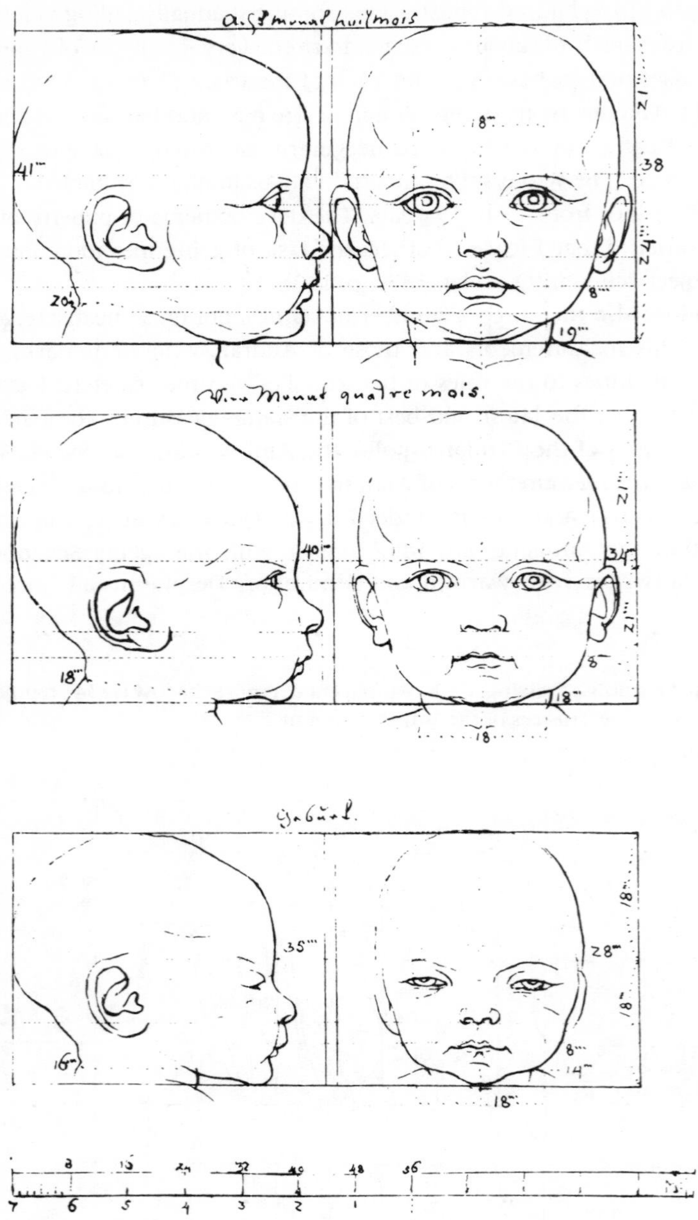

Fig. 3.21. Schadow's drawing of the growth of the infant face : birth, 4 and 8 months. From Schadow (1834), reproduced by courtesy of the Trustees of the British Museum.

and increasing blindness, Schadow helped Quetelet in many of the calculations of proportions and comparisons of statues given in Quetelet's *Anthropométrie* (1870).

Shadow's remarks about growth at puberty were taken to heart by Angerstein (1865), a Cologne teacher who measured his pupils successively over a period of five years, mostly in the age range 8 to 18. His data apparently relate to 1,000–1,200 boys and would constitute a highly interesting archive, filling in the gap in longitudinal studies between the Carlschule and the late-nineteenth-century studies (see below). They are presumably irretrievably lost. Angerstein gives only some representative values, obtained he does not say how, and finds, surprisingly, the maximum velocity of height growth to be at 13.0 years, almost certainly far too early. He is clear, however, that at puberty a great spurt in height occurs in most boys and quotes a few individuals' successive measurements, some of which must actually be regarded as products of an overenthusiastic technique. A faint echo of Guarinoni can be heard; none of his examples of very early or very late puberty, he says, occur in the sons of artisans or labourers. Angerstein does not quote Quetelet's 1835 *Physique sociale* (see below), thus demonstrating the lack of traffic between scientists and artists even at this time.

Adolf Zeising (1810–1876), another teacher and art theoretician, measured ten boys at each year of age in the range birth to 21 years (a cross-sectional study), and ten newborn girls (Zeising, 1858). Numerous measurements were taken, the object being to characterize the changes in body proportions that occur during growth. This was essentially Quetelet's recommended procedure for obtaining stable values (see Chapter 6); and Zeising criticized Schadow for using too few subjects and an inefficient measuring technique. Zeising's mean yearly increments, however, are further from modern values than are Schadow's. Presumably because of some problem of sampling, Zeising obtained an increment of only 0.6 cm from age 8 to 9; and later there are two equally large increments (7.7 and 7.5 cm) located at 12 to 13, and 15 to 16 (p. 806). Zeising's main preoccupation was the revival of the Golden Section mathematical series (Zeising, 1854). This he applied wholesale to explain not only the entire structure of vertical and transverse human dimensions, but also the rates of proportional growth in children. (A most sympathetic and mathematically interesting treatment of the series will be found in Cook, 1914.) Both Zeising and another seeker after beauty in mathematical series, Liharzik (1813–1866), achieved considerable fame in their life-times, and were quoted for several decades after their deaths. They were contemporaries but moved in very different worlds. Zeising was an art professor, Liharzik physician at the Children's Hospital in Vienna. Liharzik revived not only the Golden Section but a whole series of tetragons and magic squares. Starting as a pupil of the famous pathologist Rokitanski, he began (1858) by measuring chest and head circumferences, with the object of showing that persons with relatively small chests were particularly prone to tuberculosis, a thought very much in the minds of all medical men of the time. He progressed to a classification of human growth into twenty-four epochs, the first lasting 1 solar month, the second 2 months, the third 3 months and so on up to 300 months, or 25 years. Equal increments of growth were supposed to take place in each epoch. The curve of growth would thus be a steadily rising one like that of Bergmüller, with no adolescent spurt.

Liharzik gave values at each epoch for six measurements of length (vertex to chin; chin to top of sternum; top to bottom of sternum; bottom of sternum to pubes; pubes to top of foot; top of foot to floor), plus the length of the clavicle, to make seven measurements in all. By the use of a construction of squares and circles based on these landmarks and values, figures of the body for each epoch could be constructed. There is no evidence his values were based on measurements; indeed it is very clear they proceeded from what he considered the more solid ground of numerology (1862*a*, 1865). But the construction worked, at least sufficiently for statues enshrining it to be modelled in plaster, cast in bronze, and presented as a feature of the Austrian exhibit at the 1862 London Exhibition of Industry and Art (Liharzik, 1862*b*). They are pictured, together with Liharzik's portrait, in Levinson (1925). Liharzik eventually became totally absorbed in the manipulation of numbers and wrote a book proclaiming that the square was the basis of all natural proportion and the number seven the key to human structure (1865). Already by 1872 Hamy, an anatomist, had castigated Liharzik's work as 'a creation of the imagination'. But despite their mostly imagined provenance Liharzik's and Zeising's figures for children's heights found their way into several German textbooks of physiology and children's diseases (e.g. Vierordt, 1877, 1881) in the period 1860–90. Then the rapidly rising tide of scientific auxology cut between the artists and the doctors, leaving – with one exception – each increasingly isolated from the other.

The one exception was C. H. Stratz (1858–1924) (see p. 282), a German gynaecologist, prolific writer, connoisseur of photographs of the human body and inventor of a discredited classification of growth into two periods each of 'filling-out' and 'springing-up'. In *The Representation of the Human Body in Art* (1914*a*) Stratz analysed a large number of statues and paintings of all periods in terms of canons of proportion, in some instances using photographs of living models in the same poses as the works of classical antiquity or contemporary romanticism. Stratz was in the tradition of Schadow and, in a sense, of Quetelet; for him the mean of a large series gave only the average value and in no sense the 'normal' or standard one (*Normalwert*). If ten carefully selected persons were measured, that was different, and their average would give the type. In his best-selling book *The Beauty of the Female Form* (edns 1898 to 1941: see p. 282) he quotes Zeising (1854, 1858), Hay (1851) and Richer (1893). Stratz's love of art impelled him to take one very useful step in the analysis of growth patterns: he advocated, and practised, the recording of the body of the growing child by photography as well as by the traditional anthropometric measurements.

Absolute size and body proportion

What is perhaps surprising to the modern scientist is that through all the centuries of study of the human body scarcely any mention at all is made of absolute size. For painters and sculptors, of course, proportion was more important, since they had to reproduce figures in various sizes. Nevertheless, it might seem to us extraordinary that Alberti and Elsholtz began the process of measurement by making a rod the height of the subject and then dividing it up to form a scale. Such a procedure presumably precluded the measurement of substantial numbers of subjects. To us it seems so much easier to construct a fixed scale and calculate proportions by simple ratios. Scales of length, of course, varied from place to place and were only rigorously standardized with the introduction of the metre in 1795 (and even then Parisian, Rheinish and English feet continued to be used, with a Paris foot equal to 1.066 English ones: see Berriman, 1953). But this apart, proportion was regarded as much more interesting than absolute size. Proportion provided a pathway to the beautiful and perhaps to the healthy, too. Compared with that, what usefulness had mere size? It was the soldiers who answered that question, as we shall see in Chapter 5.

4

Iatromathematics and the introduction of measurement: the seventeenth and eighteenth centuries

We return to Hippolyt Guarinoni, with whom we ended Chapter 2. At the time that he was writing *Die Grewel der Verwüstung*, with its celebration of the simple, agricultural past, the wheels of the complex scientific future were beginning slowly to revolve. Galileo (1564–1642) is traditionally taken as the prime mover. Galileo made instruments to measure objects and observe events; he applied to his observations the rediscovered mathematics of the Greeks; and above all, if his results disagreed with accepted authority, he was nevertheless inclined to believe in them. It was the Aristotelian method; but it had been neglected in Europe for nineteen hundred years.

Galileo held the chair of mathematics at the University of Padua from 1592 to 1610. Padua at that time was an outstanding centre of learning, and its particular glory was its medical school. It was there that Vesalius (1514–1564) had reintroduced the dissection of the human body to Europe in the 1540s and, in his *De humani corporis fabrica* had laid the cornerstone of modern anatomy, and indeed of modern science, by extolling observation over authority. Vesalius was succeeded by his almost equally famous pupil Fallopius (1523–1562) and he by Galileo's older contemporary Fabricius of Aquapendente (1533–1619), who from 1598 to 1602 was the teacher of William Harvey (1578–1657). Galileo himself was a trained physician and he had as a contemporary in the medical school Santorio Santorio (1561–1635) (Sanctorius) who, like Galileo, introduced measurement into all phases of his work. Sanctorius compared the body to a machine and regarded the concoction of the food as similar to the burning of fuel in a furnace. Using an adaptation of the commercial balance to weigh himself, he showed that the body loses weight simply by exposure, and correctly attributed this to insensible perspiration. He even calculated, and with considerable accuracy, the amount of sweat lost in this way each twenty-four hours. He devised an instrument for measuring the force and rate of the pulse and adapted Galileo's thermometer to taking patients' temperatures. His book *De statica medicina* (1614) laid the basis for what became known as the iatrophysical or iatromathematical school, whose members sought to explain the workings of the body, in both health and disease, by the application of the methods and laws of the new physics (see note 4.1). To them was opposed, often enough, the iatrochemical school, who endeavoured to do the same using the still very unsatisfactory methods of chemistry. Michael Foster, in his lucid and charming *Lectures on the History of Physiology during the Sixteenth, Seventeenth and Eighteenth Centuries* (1901) points out that Harvey, though a product of Padua, belonged to neither of these schools but was a pure physiologist, describing only the biology of what he observed, and avoiding explanations either of a physical or of a chemical nature.

The iatrophysical school received a great, though somewhat left-handed impetus from Descartes (1596–1650). Though an eminent mathematician and philosopher Descartes was an amateur in physiology and anatomy; he made no personal investigations in these fields. But it was part of his philosophy to show that man consisted of an earthly machine inhabited and governed by a rational soul, and in 1662 he published a book *De homine liber*, which Michael Foster describes as a 'treatise of physiology, not . . . a contribution to physiological knowledge, but a popular exposition of the features of the earthly machine in illustration of its relations to the rational soul. The work thus stands out as the first Text-Book of Physiology written after the modern fashion, though in a popular way. We may perhaps speak of him as the Herbert Spencer of his age, in so far that his treatise on man bore somewhat the same relation to the physiological inquiries of his time as the *Principles of Biology* do to the biological researches of the present day' (Foster, 1901, p. 57).

It was Giovanni Alfonso Borelli (1608–1679), however, who really established iatrophysics, by applying Galileo's methods to elucidate human locomotion and the circulation of the blood. Borelli was born in Naples and spent much of his working life at the University of Pisa, then rapidly supplanting Padua as the outstanding academic centre in Italy. Malpighi (1628–1694), the great histologist who discovered the capillaries, was there and he and Borelli worked on many problems together. Borelli's major work, *De motu animalium*, was only published in 1680–1, shortly after his death, but it had been in manuscript and delivered as lectures to students at least since 1662. In it he describes the mechanics of walking and of muscular contraction in general, calculates the mechanical force of the heart at systole, and applies the principles of hydraulics to compute the flow of blood through the arteries in terms of the passage of liquid impelled by the piston of a pump. Borelli's methods were so successful in illuminating physiological problems that they established securely the application of physics, and particularly of techniques of measurement, to human biology and medicine. It was only a matter of time before measurement was applied to the study of growth.

In England the iatrophysical movement coalesced particularly around Isaac Newton (1643–1727). The most prominent medical man was Richard Mead (1673–1754), whom we shall meet in the next chapter as the physician who read to the Royal Society Wasse's paper on the diminution of stature with time of day.

With the eclipse (albeit a slow one) of the Galenical humours, new theories of growth were promulgated. No longer did children grow because of the innate heat implanted at conception by the incredible concentration of the semen. Heat became not the epitome of hot-substance but something to be measured with a thermometer. 'Heat is the effect only of the circulatory motion of the blood', wrote Archibald Pitcairn (1653–1714), a thoroughgoing and influential iatrophysicist from Edinburgh, who was briefly professor at Leiden (1692–3) where he had the distinction of having Boerhaave as a pupil. (Edinburgh University Medical School, which dominated medical education in the United Kingdom in the eighteenth century and beyond, was only founded in 1726.) Short persons are hotter than tall ones, Pitcairn went on, because their parts are less distant from the heart, so that the velocity of the blood when it reaches them is greater (Aristotle thought a similar thing, that organs distant from the heart were cooler than those near it, but considered that the heat was derived from inspired pneuma and attached to the blood once and for all in the heart). Pitcairn was a traditional man, still prepared to discuss the temperaments; but he thought they derived from differences in blood velocity and vessel size. There were three temperaments, not the Galenical four. In the Bilious the 'degree of fluxity of the blood' allowed a greater separation of bile in the liver than in others; in the Melancholic a greater separation of black bile in the spleen; and in the Pituitous a greater separation of pituita in the salivary glands (Pitcairn, 1718).

Boerhaave and van Swieten

Hermann Boerhaave (1668–1738) was a great deal more sophisticated. He was the dominant figure of his generation in European medicine, and his influence lasted for generations longer. He occupied the chair of medicine at the University of Leiden from 1701 until his death, and the chairs of botany and chemistry too from 1716 to 1729. He was an eclectic scholar who combined the iatrophysics that he learned from Pitcairn with an iatrochemistry by this time emerged from the background of alchemy and Paracelsian medicine to respectability in the hands of an earlier occupant of his chair, Franciscus Sylvius (1614–1672; chair from 1658 to 1672).

Boerhaave has some interesting things to say about growth. Adopting the iatrophysicists' view, directly derived from Borelli, he wrote: 'All appearances seem to teach us that the increase of stature depends on the elongation of the vessels by the fluid impelled through them; so that during those stages of life in which the vessels are most flexible and the action of the heart most quick and at the same time tolerably strong, growth is very rapid' (Swieten, 1744–73, vol. 12 (1765), p. 27, commenting on Boerhaave's *Of a Phthisis Pulmonalis*). Thus Boerhaave explained the fact that, as he put it, 'the body grows faster the nearer it is to its first origin'. Thus also he replaced Aristotle's concentrated innate heat by mechanical hydrodynamics. It was an advance, if in a somewhat circuitous direction. For Boerhaave was just as mistaken as Aristotle in his physiology of growth. Not that we moderns can afford to scoff, for we ourselves have no explanation for the diminution of growth velocity with age, that apparently most fundamental of all growth phenomena (see Tanner, 1963*a*). Nowadays we suppose that the slowing of growth represents either the cells progressively running out of capacity (at least as vague a word as heat) or else the blood reducing its content of growth-promoting humours, or both. The physiology of growth has advanced but slowly.

Boerhaave's two most seminal works were in origin scarcely more than lecture notes. The first, *Institutiones medicae in usus exercitationis annuae domesticos* was published in Leiden in 1700, and went through eight editions before Boerhaave's death in 1738, with a further seven in the next thirty years. The second, *Aphorismi de cognoscendis et curandis morbis in usum doctrinae domesticae*, first appeared in 1709 and had an equally enormous sale. It attracted a number of commentators, one of whom, Gerard van Swieten (1700–1772) was a close friend and student of Boerhaave and continued the lectures when Boerhaave died. Later he was dismissed on account of his religion and in 1745 became physician to the Empress Maria-Theresa in Vienna, where he not only developed the medical school, but rebuilt the whole university, adding physics and chemistry laboratories in which the spirit of his revered master could find a permanent home. He played a major part, too, in building and organizing the hospitals of Vienna and the rest of the Empire; after his death at Schönbrunn Maria-Theresa erected a statue of him in the university grounds. The first volume of van Swieten's *Commentarii in H. Boerhaavii aphorismis de cognoscendis et curandis morbis* was published in Leiden three years after Boerhaave's death, and subsequent volumes came out in 1745, 1753, 1764 and 1772. An English translation was made at once and published in eighteen volumes over the period 1744–73.

The bulk of the commentaries far outweighs that of the aphorisms; though van Swieten sticks fast to his teacher's principles, he adduces many examples, both ancient and contemporary, to illustrate them, and adds much extra information of which it is impossible to say whether it represents Boerhaave's actual lectures in their extended form or van Swieten's own observations. The comment above, regarding the mechanism of growth, closely follows Boerhaave's own writings, but the two further passages below, of considerable interest in a history of human growth, are not to be found in Boerhaave's original.

Concerning the events of puberty van Swieten commented 'In the human species the females as soon as they are fit for progagation, undergo a very remarkable and sudden change: they grow quickly taller; their breasts begin to swell up and ripen like two sister-twins; all the marks of puberty appear upon the pudenda: from that opening too, which is the particular mark of distinction of the sex, the blood now finds its way ... The menses, for the most part, begin to appear at 14 years of age, seldom before the thirteenth [year]: and they generally cease to flow after about the forty-fifth; in some however, they leave off sooner, in others later' (van Swieten, 1744–73, vol. 13 (1765), pp. 243–4; commenting on *Diseases Proper to Virgins*). Van Swieten, therefore, had a clear idea about the adolescent growth spurt. Furthermore, he went into detail:

> Those who deduce the cause of the menstrual flux from a plethora arising, for instance, in a girl's body when she has arrived at her full growth, don't seem to have considered how frequently it happens that girls grow remarkably taller and bigger after having had several regular periods ... after the first menstrual period there is for the most part a most remarkable additional increase to the body; though I have seen some, whose menses had begun to flow only at eighteen or later who never grew one bit taller afterwards. I am likewise of opinion, that the increase of the human body, according to what I have observed, does not always proceed in so regular and gradual a manner as is commonly imagined ... About the time of puberty, though in some later, there is frequently observed so sudden an increase of stature that even in a few months the body shall grow taller than it had done for two years together before that. It is very well known that young people, when seized with feverish disorders, especially of the acute sort, if they happen to escape, grow taller as they recover. This hath been explained from the force of the fever impelling the humours and by that means stretching out and lengthening all the vessels, whilst the bones at the same time, not having yet

> acquired their full hardness, easily yield to the same impressions. But when the disease is at an end this cause of the sudden growth ought of course to cease; and yet, from what I have seen, the principal increase of stature has generally happened some time after, rather than during the continuance of the disease (van Swieten, 1744–73, vol. 13 (1765), pp. 258–60).

This sounds like a description of catch-up growth (see Prader *et al.*, 1963), but van Swieten unfortunately went on: 'I recovered a young man of sixteen years of age, rather diminutive for his years, from the small pox which were of the discrete kind, though very numerous: the length of the body during his illness was increased somewhat indeed, though not very remarkably; yet the growth went on so fast, that when I saw him three years afterwards, I scarcely knew him, for he had not reached his twentieth year and yet was grown six foot high'. Van Swieten (or perhaps Boerhaave), like everyone else for centuries, thus confused catch-up growth with the adolescent growth spurt. But he felt something was wrong and ended: 'Hence there appear to be other causes of the body's increase, which cannot be so easily explained by anything which as yet we know of the human body' (p. 260).

Martin Weise

Boerhaave and van Swieten were not the first, or at least not the only ones in the early eighteenth century to describe catch-up growth. It is often said that the three great physicians of that century were Hermann Boerhaave, Friedrich Hoffman (1660–1742) and Georg Stahl (1660–1734). The last two were both professors at the newly founded (1693) University of Halle in Prussia, and Stahl, from 1713, was court physician to King Frederick-William I in Berlin. A number of doctoral theses written under their direction, or that of their immediate predecessor Wolfgang Wedel (1645–1721), dealt with growth or reproduction (e.g. Brebiss, 1705; Lindner, 1713).

One, by Martin Weise (1726) entitled *De proceritate corporis* ('Of tallness of body'), is especially interesting (Fig. 4.1). Supervised by Hoffman, the leading German iatrophysicist, it has all the hydraulistic explanations of growth that one would expect. Weise naturally quotes Keill, Boerhaave, Borelli and Bellini, Borelli's most famous pupil (see p. 67); but also, on several occasions, Elsholtz. Indeed Elsholtz was a colleague, at the court of Frederick-William I's grandfather the Great Elector of Brandenburg, of a Martin Weise the elder, a physician who was presumably the thesis-writer's father (see note 4.2). In his thesis Weise wrote: 'Most men grow in height until they are twenty-five years old . . . however, it sometimes, though rarely, happens that persons grow altogether too little in the early years, and then later for one or even two years are impelled in their height as though struck forward, those in whom the preceding ten years have held back something to be spent at another time' (p. 7). The context does not make it clear whether Weise is describing catch-up growth or the adolescent growth spurt in a marked form. Indeed he may not have distinguished the two, just as later Quetelet failed to distinguish them (see Chapter 6). But a few pages later (p. 11), in talking of rickets and saying that rachitic children seldom reach a proper height but remain puny, he remarks 'It sometimes happens, though, that the illness being overcome and with it the impediment to growth being removed, afterwards [the child] grows conspicuously, which gives rise to the German proverb *"Die Krankheit habe ihn ausgereckt"'* (literally 'Illness stretched him out'; but there is an additional nuance in that *ausgerecht* is a term used in hunting to mean fully-grown-to-maturity, of a deer's antlers).

Weise says Spigelius (1578–1625) and Antonius of Amsterdam (a historian) have already noted this phenomenon, but he is being perhaps overconscientious, for Spigelius confused it with gigantism (see note 4.3). Weise's thesis seems to represent the first published description, though surely not the first observation, of catch-up growth.

Fig. 4.1. Title page of Weise's thesis. Reproduced by courtesy of the Wellcome Trustees.

DISSERTATIO INAVGVRALIS MEDICA
DE
PROCERITATE
CORPORIS
EIVSQVE
CAVSIS ET EFFECTIBVS
QVAM
PROPITIO SVMMO NVMINE
PRÆSIDE
DN. FRIDERICO HOFFMANNO
FACVLTATIS MEDICÆ SENIORE ET h. t. DECANO
AD. D. JANVARII A. O.R. MDCCXXVI.
PRO GRADV DOCTORATVS
RITE AC LEGITIME IMPETRANDO
Solenni doctorum examini sistet
MARTINVS WEISE
BEROLINENS. MARCHICVS.

HALÆ MAGDEBVRGICÆ,
Typis, IOANNIS CHRISTIANI HILLIGERI, Acad.Typog.

J. A. Stöller

The notion of catch-up is carried further in a really remarkable work, hitherto ignored, by Johann Augustin Stöller. Stöller was born in the free Imperial city of Windsheim, a town some thirty miles west of Nuremberg, in 1703. His father, who at that time spelled the family name Stöhler, was cantor and organist at the church, and taught music at the *Gymnasium*. There is no extant account of Johann Augustin's life, but a brief description of him is given in Steyneger's excellent biography of his younger and more famous brother Georg Wilhelm (Steyneger, 1936, pp. 26–9). Georg Wilhelm Stöller (1709–46), later known at Steller, studied theology at Wittenberg but then changed his intended career and enrolled as a student in the medical faculty at Halle University, supporting himself there by teaching at the Francke'sche Waisenhaus. He rapidly made a career as a botanist, went to work in the developing Academy of Science in St Petersburg (see Kaiser and Krosch, 1967), served as surgeon-botanist on von Bering's expedition from Kamchatka to North America and became, in Stejneger's phrase, 'the pioneer of Alaskan natural history'. His letters from Kamchatka are a minor classic of the literature of exploration. He died young, of a fever, in the wilds of Siberia. His brother Augustin was prevailed on to write a short account of his life for the local literary magazine (Stöller, 1747).

Johann Augustin was six years the senior, and the oldest of ten siblings. He left school in Windsheim in 1722, and went to Leipzig to study medicine, but soon changed to the University of Halle, where in 1726 he presented his dissertation, with Michael Alberti as *praeses*. It was not on growth but on problems concerning venesection. Martin Weise, whom Stöller quotes as 'a learned colleague from Berlin', was a medical student at Halle at exactly the same time. Stöller's thesis was presented in January 1726 and Weise's *De proceritate corporis* a few months later (Kaiser and Krosch, 1966). Evidently Stöller found Weise's work more interesting than his own, for when he left Halle shortly after graduation to practise in the small Saxon town of Barby, he began work on a book on growth. He had two reasons for going to Barby: he had married the daughter of the pastor there and he had secured the position of personal physician to the reigning Duke. Perhaps he had two reasons, too, for writing his book on growth. Stöller said himself that he had a certain weakness for people in exalted positions, and his book was certainly aimed, with some precision and much force, at pleasing the King of Prussia himself. Whether it did so is unrecorded, but Stöller progressed to be physician to a dowager Princess in Köthen (presumably the widow of Prince Leopold of Anhalt-Köthen, J. S. Bach's employer from 1717 to 1723), and then, in 1738, went to Eisenach in Thuringia as physician to the reigning Duke of Sachsen-Eisenach. He was made *Rath*, or Counsellor, and after the death of the Duke and Duchess (and the absorption of Sachsen-Eisenach into Sachsen-Weimar) he remained in the town in retirement, dying in 1780. He had four children, one of whom, Friedrich Christian Stöller, born in Köthen in 1733, also qualified as a doctor at the University of Halle and achieved sufficient prominence to rate a short entry in the *Biographisches Lexicon der Hervorrägenden Aerzte aller Zeiter und Völker* (vol. 5, pp. 545–6), which is more than was accorded his father.

Johann Augustin and Georg Wilhem lived at the time when Prussia was ruled, and the whole of Europe disturbed, by Frederick-William I, Frederick the Great's irascible, porphyric and giant-loving father. Johann Augustin's book, dedicated to Frederick-William, has as its full title *A Medical Historical Investigation of Human Growth in Length, being what Doctors as well as all Growing and Grown-up People should know about its Natural Properties and Conditions, as also how Useful Diseases adjust the Growth in Length of the Body; and what Soldiers should know about their common Maladies and how to treat them. I. in a Theoretical Part, and II. in a Practical Part; including a Proposal for a Medical Review for the Betterment of the Many Distinguished Troops of the King of Prussia'* (see Fig. 4.2). It was published in Magdeburg in 1729 and runs to 224 pages. It is a rare book. No copy exists in Britain, but the Countway Library in Boston has one and so does the US National Library of Medicine. Haller refers to it several times in his *Elementa physiologiae* (1778, vol. 8, book 30, pp. 41, 45, 239) but from then till now it seems to have been passed over in silence. Yet it has much that is very sensible, especially about conditions governing growth. It presents no actual data on growth (the first were to come just twenty-five years later) nor does it have anything to say about the timing of puberty. It retails the still-current views about growth slowing down because of the decrease in elasticity of the body's fibres. Some of its animadversions on the Divinity put one in mind of Guarinoni. But it is, in a sense, the very first textbook of human growth.

In 1729 the Crown Prince, later Frederick the Great, was aged 17. He had been a small and puny boy, a fact utterly abhorrent to his father, whose chief joy was to collect and review his giant Grenadier Guards (see Chapter 5). With puberty, however, the Crown Prince had grown considerably and Stöller, in his dedication, claims that the proof of the usefulness and efficacy of his subject is that the Crown Prince 'in former years known to be delicate (*unpässlich*) has been cured (*hergeruhret*) by nothing else but a speedy growth in length. God grant him still further to grow, and, without check to his health, to bloom royally to the joy of his Most Serene Parents and the whole land' (see note 4.4).

After this preface Stöller launches into the first part of his book. The section headings, shown here slightly abbreviated, give an idea both of the scope of the work and the attitudes of its author.

> The might and wisdom of God everywhere to be seen; in this little work the marvel of tall people is specially considered. Nobility of soul accompanies tallness of body, but even the tallest, soldiers included, seek a life according to God's word. The exceptionality of giants. There really were giants, and there still are. Now begins a tract on growth, so we can see for ourselves properly what helps and hinders it. Who really should we designate as tall? Tall people seldom grow as proportionately as short, with examples. The causes and conditions to which we have to pay attention in our growth. The younger the person the faster he grows. Sex also influences growth, though very little in former times in Germany [cf. note 4.5 on Tacitus]. Also differences between regions, still seen nowadays, both in men and animals. Also manner of living. Hereditary disposition also belongs here. The effects air has. Late marriages and frugal concubinage help

Fig. 4.2. Title page of Stöller's book. Reproduced by courtesy of the Countway Library, Boston.

Historisch-Medicinische Untersuchung
des
Wachsthums
Der Menschen
In die Länge,
So wohl,
Was Medici, als alle wachsende und
grosse Leute von dessen natürlichen Eigen-
schafften und Umständen, wie auch von denen
mit dem Wachsthum des Leibes in die Länge
sich zutragenden Kranckheiten,
Ingleichen,
Was Soldaten von ihren gemeinsten
Maladien und deren Curen zu wissen
nöthig haben:
I. In einem theoretischen und
II. In einem practischen Theil,
Nebst einem Vorschlag
Von einer Medicinischen Revuë,
Denen vielen auserlesensten Königl. Preußis.
Trouppen zum Besten also abgefasset
von
Johann Augustin Stöller,
Medicinæ Licentiato & Practico in Barby.

MAGDEBURG,
Bey Christoph Seidels sel. Wittwe, und George
Ernst Scheidhauer. 1729.

very much the future growth of children. Children who are to grow briskly must remain the proper time in the uterus and not be born early. Also children must not be brought up over-tenderly. What advantages illness brings to growth. We now discuss material and efficient causes of increase in length. Growth in length stems principally from linear extension of the skeleton, and comes under the heading of nutrition (*ist ein species Nutritionis*). Why man cannot grow in length and breadth at the same time. It is certainly hard to be sure what causes growth, but probably there are two factors, the juices of the body and the condition (*Ansehen*) of the bones. The connection between all parts is the circulation of the blood. Above all, the growing person should neither be too weak nor too violent (*schwach; hefftig*). To what extent a determined size for a man is laid down in the egg. It is shown that the soul (*die Seele*) is the root of growth in both fetal and post-natal life. Whether one can see in quite young people whether they are going to be tall. Whether one should stimulate growth of the body. Whether one should intentionally make a dwarf.

The second part discusses illnesses and their treatment, including what doctors should do about dental disease. The book ends with the proposal for the founding of a Medical Review (never, so far as is known, accomplished). Some sections of the first part amply repay closer study. The section on giants is really just an introduction, presumably for the King's benefit. At p. 16 the real book begins. 'I have in mind', writes Stöller, 'to present human growth in length somewhat more rigorously than has yet been done in our country, using as my basis *Natur-Lehre* or physiology . . . Thus I take the opportunity presented by the differences in men's heights to write, from a medical point of view, a somewhat detailed account of growth and its bodily benefits and injuries, which so far as I can discover no one has done before. The book by the celebrated Conring cited above [*De habitus corporum germanicorum antiqui ac novi causis*: see note 4.5] and the inaugural dissertation *De proceritate corporis* of the learned Berlin physician Dr Weise, deserve to be reread, for the thoughts expressed in them were the beginning of which the present material is a further development'. Thus the line of contact and citation runs from Elsholtz to Weise to Stöller. As for tallness, anyone over six feet of the measuring rod (*sechs Schuhen der Mass-Stab*) should be called 'tall', although they may not be as tall as that huge Swede brought over by his majesty the King of Prussia in 1728 who was eight and a half feet (p. 18). In the same vein, later, he says that in 1718 twenty Swiss, all over seven feet tall and three feet wide at the shoulders, came to Berlin and volunteered for service in the Grenadier Guards (p. 22).

Stöller is quite clear that one grows more in fetal life than later. In discussing this he gives some of the very few figures appearing in his book. They are taken directly from Weise (1726). An embryo, he says, is the weight of a barley-corn; when it is nine months old it weighs 6 pounds, and a 25-year-old man weighs 200 pounds. (Weise's and his estimate of 6 pounds for birthweight, before Roederer, is much nearer the truth than all those that Roederer cites (see p. 96). This may reflect the work of the Gynaecological and Obstetrical Clinic in Halle, founded about 1720 by George Daniel Coschwitz (1679–1729) (see Kaiser and Piechocki, 1969), though there seems to be no record of any actual measurements.)

Stöller has clear and interesting views about hereditary disposition. Though in principle like begets like, this principle by no means always holds. The smallest and weakest parents are often seen to produce not merely medium-sized but even tall offspring: 'One frequently sees, too, that children may resemble their parents, grandparents, or this or that particular person in the family in character, temperament and facial features, while at the same time clearly surpassing them in height' (p. 42). Although growth may be disturbed, hereditary disposition restores the child to reach the height of his parents. But Stöller thinks this disposition is more of the mind than of the body: 'The main basis of the hereditary disposition towards a large growth rests in the soul (*Seele*) and its activity (*Bewegung*) at given times rather than in the body itself' (p. 43). Many persons, he remarks, bring with them at their entry into the world a disposition to be large but this may be frustrated by weak small and defective parents attaching infirmities and defects upon them right from the beginning of their lives.

'Amongst the things which hinder growth I think that an artificially soft treatment and regimen is the greatest. The more luxurious (*zärtlicher*) is the living and upbringing in youth, the more is the sickness and the smaller the growth' (p. 59). Wild animals are healthier and stronger than domestic ones. Children covered with warm clothes, taking warm baths and drinking tea or chocolate or coffee every day get fatter and softer and ruin their bodily constitution. Rich people's children have often a defective growth; whereas the children of the poor and of soldiers grow almost visibly (*gleichsam zusehens*) and they are for the most part poorly and lightly clothed and have poor meals chiefly of bread and water (p. 60).

Stöller then turns to catch-up growth. 'Lastly', he writes, 'I note that people grow really visibly (*recht zusehens*), and chiefly in length, when they have fully overcome a severe illness, provided they behaved appropriately during it, and the course of the illness was not disturbed by false treatment. Frequently illnesses stop people growing ... But if a feverish or not very long-standing malady is properly overcome then people grow very much;

so as a rule those persons shoot up in height, who particularly in their childhood have been held in check by hot or cold fevers. Spigelius . . . gives several examples . . . and experience teaches it daily. This is the basis of the well-known proverb: "Illness laid him low and stretched him out" (*Die kranckheit habe einen gestrekt, ausgereckt*). Children who have been properly sickened in their youth are the healthiest and largest' (p. 62).

As for the possibility and morality of stimulating growth in height, Stöller declares it is a problem he should discuss, especially as he is himself a man of short stature (p. 97). So far as healthy people are concerned he dismisses the idea with the usual quotation from Matthew (vi, 27). To tamper with God's work is impious and what might not follow if everybody by artificial means could compel the growth of his body? (p. 99). As for artificially creating dwarfs, which a beggar in Prague was alleged to have done by applications of rat-, bat- and mole-fat in early childhood, that was equivalent to murder.

Schurig and Rall: sex activity as a cause of early puberty

There are few other books of the time to compare with Stöller's. Martin Schurig (1656–1733), a Dresden physician, wrote an astonishing series of eight monographs, published between 1720 and 1744, which included books entitled *Spermatologia, Parthenologia* and *Embryologia* (see Needham, 1934, p. 165). In *Parthenologia* there is a long review of the various events of puberty and of the previous opinions concerning them. From it emerges clearly the confusion created by some authors in referring to the initial appearance of pubic hair as *pubertas*, following the Aristotelian and Roman tradition set forth for boys, while other authors refer to the menarche. Schurig says that the temperament, whether hot or cold, and not only the geographical region, effects the age of puberty, 'which is why puberty (*pubertas*) cannot be assigned to a definite time but varies (*errat*) from the tenth year to the twenty-second' (p.6). Schurig, in an otherwise rather pedestrian review, does however present one important viewpoint not mentioned by many previous writers. He quotes it from Rall (*De generatione animalium*, 1669, p. 164: see note 4.6). Bodily maturation in girls, he says, can be accelerated by indulgence in conversations with men, by kissing or by other sorts of sexual encounter; this is why prostitutes have an early menarche. This opinion, still held by many (mostly non-medical) authors today, crops up fairly regularly in the literature of the eighteenth and nineteenth centuries (see, for example, Maygrier, 1819, and Raciborski, 1868). Jean-Jacques Rousseau shared it, attributing also a powerful role to romantic literature and harmonious music: (see note 4.7). It seems most likely to have derived in the first place from the observation of farm animals and, very possibly, from farm practice. In pigs and sheep it is well known that introducing a male to a group of female animals not yet pubescent but past a certain critical size, will precipitate the changes of puberty. Contact is unnecessary; the stimulus seems largely to be carried by air-borne hormones, the pheromones –in other words, highly specific smells. Observations on the child prostitutes of Eastern countries may also have played a part in strengthening the idea. Girls in such places as Egypt did probably mature a little earlier than those in Europe and perhaps had a tendency to falsify their ages (when known) downwards, for reasons of trade. There was also the greater probability of prostitutes having vaginal discharge because of disease.

But the notion of sexual encounter accelerating puberty may also derive from theoretical considerations. Heat was still held in many quarters to be the engine of growth. Bacon, it will be recalled, thought exercise stimulated growth by increasing natural heat. Bacon's exercise was anything but sexual in content, but sexual exercise also warmed the body, and even, if carried far enough, produced or received that hottest of all substances, semen. It would have been against nature to suppose that such a thing was without effect.

Another variety of heat was also cited as stimulating development. In the thesis of Demol (1731), for example, a student at Montpelier, one reads 'It is established by my own and others' observations that women inhabiting France and the temperate zones have their first menstrual purge at age 12, 13 or 14 ... moreover in those who receive the first rays of the sun and are warmed by them longer, as in Spain, Turkey, Persia, Libya and the Eastern Empire, the catamenia begin at 9, 10, 11 or 12 ... Whereas those who breathe northern airs, in Holland, Belgium and Sweden have their first purge later, at age 15, 16 or 17' (p. 4). The origin of the belief that in hot climates growth is faster appears here clearly. The heat of the sun, no less than of Bacon's exercise, enhances the innate heat and hence speeds up growth.

Buffon

Though Boerhaave and his followers were certainly quantitatively minded there is no evidence that they measured fetuses or children. The first to do so, it seems, were Buffon and his collaborators, probably in the 1740s. Later, in 1759–77, Buffon influenced his closest friend, Montbeillard, to make the first longitudinal growth study from birth to maturity (see Chapter 5, p. 102; but in the 1740s he was concerned more with growth of the fetus and young child. The circumstances under which his measurements were taken are unknown. Both sex and numbers are unstated. But the excellence of the results, at least for the post-natal period, speaks for itself.

George LeClerc, Count of Buffon (1707–1788), was one of the most magisterial figures of eighteenth-century Europe. When Roger Heim (1952) introduces the volume on Buffon in the Great French Naturalist Series (which Buffon himself started) he searches for someone of the same dimensions whom he can place beside him. He finds only Goethe, a naturalist, scientific inquirer, practical man, writer of limpid prose. Buffon was all those too; his *Epoques de la nature* (1780) laid the foundations of modern geology and prehistory (and put the cat amongst the theologians of the Sorbonne: see Baudier, 1952, p. 44); and he also wrote a book on literary style which was a model for his time. With Rousseau, Diderot and Voltaire, Buffon formed the core of the Encyclopaedists, that most potent of all educational time-bombs.

Buffon lived all his life at Montbard, near Dijon. When he was 7 his mother died, and with the money that she left, Buffon's father, a country bourgeois, bought a large estate and the title of Count. Buffon was educated at the Jesuit College in Dijon and was first attracted to mathematics. When he became a junior member of the Academy of Sciences in Paris it was with a paper on the calculation of the probabilities of a coin touching one or more lines when thrown on a checkered floor (see Hanks, 1966). However, the care of his own forests soon occupied much of his interest and he became an expert silviculturist. In 1735 he translated *Vegetable Staticks*, the foremost work on the physiology of trees, by the celebrated English parson experimentalist Stephen Hales (1677–1761), and in 1739 he was appointed director of the Royal Medicinal Herb Garden in Paris, one of the most important posts for a naturalist in France. Here he was directed to prepare a catalogue of the Royal Collection, to be printed by the Royal Press. Never hesitant in search of glory, Buffon went one better; he proposed to make a catalogue of all nature, a work comparable with that of Aristotle, only larger. His capacity for work was indeed phenomenal; throughout each day he read and wrote in his library, emerging only in the evening for dinner. He revised continually, setting aside a manuscript till he had forgotten its contents, then having it read out loud by a person who had no technical knowledge of the subject; everything that person failed to understand was rewritten (Flourens, 1860). The prospectus for the catalogue of nature foresaw fifteen volumes covering all the animal and vegetable kingdoms, together with rocks, fossils and minerals. In the event the task proved even greater than it looked. Instead of fifteen volumes, Buffon had published thirty-five, with the thirty-sixth in press, by the time of his death, and covered only man, quadrupeds, birds, minerals and earth history. But the series continued, called *Suites à Buffon*, and under the editorship chiefly of Lamarck and Cuvier (a graduate of the Carlschule: see below) the definitive description of the whole animal kingdom was completed (see note 4.8).

The first volume of the *Histoire naturelle, générale et particulière* appeared in 1749, 3,000 copies being sold in six weeks. The fifteenth and last volume of this first series was published in 1767. The natural history of man occupies part of vols. 2 and 3; both were released in 1749 (see Fig. 4.3).

Buffon describes what are clearly his own and his colleagues' original observations as follows:

> I will not discuss the first period after conception, nor the growth which occurs immediately after the formation of the fetus; I take the fetus at 1 month, when all the parts are developed. It then is one *pouce* [2.7 cm] in length (*hauteur*) [see note 4.9] while at 2 months 2¼ *pouces* [6.1 cm], at 3 months 3½ *pouces* [9.5 cm], at 4 months 5 *pouces* [13.6 cm] or a bit over, at 5 months 6½ or 7 *pouces* [18.3 cm], at 6 months 8½ or 9 *pouces* [23.7 cm], at 7 months 11 *pouces* [29.8 cm] or more, at 8 months 14 *pouces* [37.9 cm] and at 9 months 18 *pouces* [48.8 cm]. All measurements vary a great deal between different subjects and it is only by taking the averages (*les termes moyens*) that I have determined these figures. For example, there are newborn infants of 14 and 22 *pouces* [38 and 60 cm] . . . but if an infant is 18 *pouces* long at birth, he will only grow 6 or 7 *pouces* in the following 12 months [i.e. 16 or 19 cm] so that by the end of the first year he will be 24 or 25 *pouces* long [65 or 68 cm]; at 2 years he will be 28 or 29 *pouces* [76 or 79 cm]; at 3, 30 or 32 *pouces* [81 or 87 cm] and after that he will scarcely grow 1½ or 2 *pouces* [4 or 5½ cm] each year up to the age of puberty. Thus the fetus grows more in one month at the end of his time in the uterus than the child grows in one year up to puberty (Buffon, 1749–1804, vol. 2, p. 472).

Evidently Buffon, his assistant Daubenton, or some other acquaintance had measured some aborted fetuses and some newborns. They were probably the first persons to do so. The newborn length (49 cm) is very close to modern figures (50 to 51 cm depending on sex and conditions of pregnancy). Buffon's limits however (38 and 60 cm) are far outside normal ones, in both directions, so presumably he was measuring pre-term infants and perhaps some who were infants of diabetic mothers or pathological in other ways Buffon (1749, vol. 2, p. 452) gives the newborn weight as 12 pounds (and sometimes 14) and in this he makes the same error as others of his time, presumably due

Fig. 4.3. Vignette from Buffon's *Histoire naturelle* (head of chapt. 1 in vol. 2). Buffon himself is sitting at the end of the table conversing with the Abbé John Turberville Needham (1713–1781). Buffon's assistant Louis Daubenton (1716–1799) is believed to be the person looking down the microscope. The dissector may be Guéneau de Montbeillard (1720–1785) or T. F. Dalibard (1703–1779), two other collaborators. The identification of the persons was made by Professor R. C. Punnett (see Needham, 1934, p. 194). The mammalian reproductive system is being studied. Reproduced by courtesy of the Trustees of the British Museum.

to inefficient scales. Before Roederer in 1753 (see below, p. 96) French authors gave figures that were much too high. Buffon may be measuring in Charbraque pounds which are equivalent to only 0.8 lb, but even so the value would be excessive. More likely he is following Mauriceau (see p. 96) who gives the 12 pound figure.

After birth Buffon's children grew very slowly by modern standards. The 1-, 2- and 3-year values are perfectly consistent however; and the larger of the alternative values he gives (i.e. 68, 79 and 87 cm) are respectively 3, 1½ and 0 cm below the present 3rd centile for British boys. That is to be expected if the children measured were from the classes using public hospitals. It was nearly a full century before any further figures for children were given. These were Quetelet's (see p. 130), obtained on Brussels children of the labouring class. The lengths (averaging the sexes and the two series of 1831 and 1832) were birth 49 cm, and 1, 2 and 3 years 69, 79 and 86 cm: thus almost identical to Buffon's.

Buffon's fetal lengths, however, are less accurate, perhaps because of errors in assigning ages to the abortuses. His successive calendar-month values from 2 to 9 months are 6, 10, 14, 18, 24, 30, 38 and 49 cm. These compare with present-day values of about 5, 12, 22, 31, 40, 45, 48 and 50 cm. Buffon concluded that the fetus grew ever faster until the moment of birth; it took a further hundred years to establish that growth in length slowed down during the last few months of pregnancy, so that the maximum velocity was reached at about 5½ months, and it took fifty years more to establish it securely (see Chapter 11, p. 262).

At this time Buffon had no personal data on growth at puberty, but he was sensible of what Boerhaave had written, and no doubt had his own informal observations. 'There is a quite remarkable thing about the growth of the human. The fetus grows more and more [rapidly] up to the moment of birth; in contrast the child grows less and less up to the age of puberty, when he grows, one might say, in a bound (*tout à coup*) and arrives in very little time at the height that he has for always' (vol. 2, p. 472). Thus Buffon was fully aware of the adolescent growth spurt before its beautiful demonstration from Montbeillard's data a quarter of a century later.

Buffon made other remarks on puberty:

> In the whole human species females arrive at puberty earlier than males, but the age of puberty differs in different people, and seems to depend partly on the temperature and climate and on the quality of the food; in the towns and amongst people who are well-off, children accustomed to succulent and abundant food arrive earlier at that state, while in the country and amongst poor people children

> take two or three years longer because they are nourished poorly and too little. In all the southern parts of Europe and in the towns the majority of girls have puberty (*sont pubères*) at 12 years and the boys at 14, but in the north and in the countryside the girls scarcely reach it by 14 or the boys by 16 (vol. 2, p. 489).

(Presumably by *pubère* Buffon means the traditional first appearance of pubic hair.) All this is pure Guarinoni, shorn of value-judgements.

Buffon's 1749 values, and indeed much of his text verbatim, appeared in the article entitled '*Accroissement*' in the first edition of Diderot's great *Encyclopédie* published in 1751. The article was written by Pierre Tarin (*c.* 1725–1761), a young Parisian anatomist who had recently established his reputation with a work on dissection entitled *L'anthropotomie, ou l'art de disséquer* (1750). Tarin was a great encyclopaedophil. He translated Haller's *Elementa physiologiae* into French in 1752 and issued a French edition of Boerhaave's *Elementa chemiae* in 1753. He himself made several judicious compilations of anatomical and surgical authors. In the second edition of Diderot's *Encyclopédie,* however, which appeared in 1777, sixteen years after Tarin's early death, the great Haller wrote an enlarged article, displacing entirely Buffon's observations on humans and substituting his own data, for the most part on the chick embryo. The editors, perhaps feeling that their gain had been accompanied by some loss, printed the full text of Tarin's first-edition article at the conclusion of Haller's contribution.

As to the physiology of growth, Buffon (or Tarin) followed Boerhaave directly. The fetus has a very large number of blood vessels and its tissues are very mucous. Hence the heart can distend the vessels more than in the harder and drier adult. 'Ossification occurs', wrote Tarin, 'when the gelatinous juice confined between two parallel vessels [suffers] the reiterated beating of these vessels. The bones enlarge because the vessels running along their fibres become stretched by the force of the heart; these vessels actually carry the fibres along with them, and push back the cartilages at the ends of the bones' (Diderot and D'Alembert, 1777, p. 372). Tarin attributes the slowing of the velocity in childhood to increasing solidity and lack of suppleness of the tissues. We are in the century of physiological hydraulics.

Indeed in his doctoral thesis the Englishman Freind (1675–1728), whose hero was Bellini, Borelli's pupil, skilfully combined the new hydraulics with the old ideas of plethora to explain menstruation. As children, he says, women have more blood than as adults, so they can nourish the growing parts. At puberty, which 'does not occur before the second Septenary' (i.e. 14.0), the fibres get more solid and hard (the old idea of drying up) so the pores constrict. Thus the plethora of blood, previously able to transpire through pores in the skin, breaks out as a vaginal discharge. Coition brings on menarche, in his view, because 'the blood upon admission of semen becomes more intensely rarified and circulates through the canals with a greater velocity, so that it may easier break through any impediments'. Menarche is retarded by cold, sorrow and fright (Freind, 1703, 1729).

Havers (d. 1702), of Haversian canal fame, had much the same view of bone growth as Tarin. In *Osteologia nova* (1st edn 1691, 2nd edn 1729) he discusses a question which perplexes us still: 'How Accretion is performed in young Animals till they grow to the convenient and ordinary magnitude of their own Kind' he says, 'and comes to cease, after the dimensions of the Animal are carried to the common and natural limits set to every Species, is not so easy to be explicated ' (2nd edn, p. 90). After much argument about this he concludes (echoing the Pneumatists) that there is no diminution in the *succus nutritius*; it is the fibres that are now so close together that the *succus* just cannot get into the tissue. Cell physiology was still 160 years away; but rephrased, Havers' explanation still seems to stand.

Haller

Albrecht von Haller (1708–1777), whose observations displaced, or at least took precedence over Buffon's in the *Encyclopédie* of 1777, was as formidable, and nearly as famous, as Buffon. He was born in Bern, and studied medicine first at Tübingen and then under Boerhaave, obtaining his doctorate at Leiden in 1727. He subsequently worked in Basel with the Bernoulli family of mathematicians before returning to Bern to practise. In 1736 the Elector of Hanover, also George II of the United Kingdom, founded a new university at Göttingen. A chair of anatomy, surgery and botany was created and offered first to G. E. Hamberger, the professor at Jena, of whom we shall have more to say in a moment. Hamberger's translation was blocked by his patron, the Duke of Sachsen-Weimar, and the second choice fell on Haller (Griese and Hagen, 1958).

Haller remained at Göttingen from 1736 to 1753, when increasing ill-health, over-weight and cantankerousness caused him to return to Bern in semi-retirement to complete his great achievement, the *Elementa physiologiae*. The first volume of this massive work appeared in 1757, the eighth and last, entitled *Life of the Fetus and Man*, in 1766. A second edition was issued in 1778. Haller died in 1777, the dominant, almost crushing figure in European medicine and physiology.

Haller was one of the first experimental embryologists. Boerhaave had himself made chemical experiments on the properties of chick eggs (*Elementa chemiae*, 1732: see Needham, 1934, p. 166) and a number of investigators, including Harvey and Malpighi, had followed Aristotle in watching embryonic chicks develop. But Haller measured the changes. He drew up a table (reproduced in Needham, p. 173) of the lengths and weights of the bones of the embryonic chick limb from the sixth to the twenty-second day after fertilization. He thus produced the first table of growth (apart, that is, from Jampert's: see p. 90) eleven years before the publication of Montbeillard's measurements on his son. Haller says that the chick grows fastest at first and more slowly later and thought the same to be true of man. Considering, as he did, gain of weight in proportion to weight already present, this is true in the human fetal period and up to adolescence.

Haller reviews the data available on human growth, but in this sphere he has no observations of his own to make (1778, vol. 8, part 1, pp. 378–80). He reproduces some of Buffon's (1749) measurements of fetal length and also the measurements given in the second edition of *L'art des accouchemens* (1761) by Levret (1703–1780), a celebrated Parisian obstetrician. Levret's measurements are in fact a considerable improvement on Buffon's from 4 months till term, being 22, 27, 32, 38, 43 and 49 cm. They are still too small from 5 to 8 months inclusive, but larger than Buffon's. As to growth after birth, Haller reports (1778, vol. 8, part 2, p. 27) only the very fallacious measurements of his rival Hamberger, already withdrawn by Hamberger himself. He quotes, without objection, the tradition that by age 3 'a boy's stature is half of the stature of a grown man, so that in the first three years growth is no less than in the following fifteen' (vol. 8, part 2, p. 26). He quite expected, therefore, a growth rate that would have been considerably lower than nowadays. About the adolescent spurt he has nothing much to say, but he notes that 'after puberty, especially if it is little marked (*utique exiguum est*), most boys and most girls also, grow, the latter at least till 15 and the former till 18' (vol. 8 part 2, p. 34). In another volume of his book he reviews age of menarche, but not very critically (vol. 7, part 2, p. 139).

Haller notices change in shape as well as in size. The head is one-half the length of the body at birth, he says, one-fifth at age 2, one-sixth at 4, and at maturity one part to seven and a half. (This was an important point, since in 1672 Kerckring had published in the *Philosophical Transactions of the Royal Society* 'drawings' of skeletons of human fetuses of 3, 4, and 6 weeks after conception, which were simply reduced copies of the skeleton of the newborn with no notion of change in shape.) The lower limb, according to the measurements made on skeletons by Sue (see note 4.10) was equal in length to the upper at birth, but 2 inches longer at age 25.

Haller has sections on tooth eruption and on age at walking and talking;

on the reasons (chiefly iatrophysically viewed) for causation of growth; on excessive growth and precocious puberty; on giants and on dwarfs. (The last two are by no means over-critical.) He deals also with *decrementum ('Tristis, sed copiosa haec est materies'*: vol. 8, part 2, p. 68), senility and death.

There could scarcely be a greater contrast between Haller's and Buffon's books. Haller's great book, with its quantitative tone and its massive list of references (however maddeningly presented), takes us suddenly into the world of modern science. As Michael Foster says, 'When we open the pages of Haller's *Elementa* we feel that we have passed into modern times ... We seem to be reading a modern text-book, of the most laborious and exhaustive kind. Haller passes in review all the phenomena of the body. In dealing with each division of physiology he carefully describes the anatomical basis, including the data of minute structure, physical properties and chemical composition so far as these are known ... He finally delivers a reasoned critical judgement, expanding the conclusions which may be arrived at, but not omitting to state plainly where necessary the limitations which the lack of adequate evidence places on forming a decided judgement' (Foster, 1901, p. 205). By contrast Buffon's huge work is an elegant and scholarly, but lighter, distillate.

G. E. Hamberger

Hamberger's shadow is a little less long than those of Haller and Buffon. He was born in Jena in 1697 and died there in 1755, twenty-two years before Haller's death and thirty-three before Buffon's. His father was professor of mathematics and physics at the University of Jena and he succeeded him in these chairs, adding the chair of medicine later. As his background would lead one to suppose, he was an iatrophysicist, a follower of the Italian school of Borelli and Bellini. Haller was the devoted pupil of the Dutchman, Boerhaave, and between Jena and Göttingen there was no love lost. The chief of their chronic disputes related to the mechanics of respiration, and the action of the muscles of the chest (Griese and Hagen, 1958, pp. 214–23).

Hamberger's chief claim to a place in the history of growth in his quantitative study of the water content of the embryo. This was published in his *Physiologia medica* (1751). He showed that 'The parts in the fetus are much less solid than in the adult. The cortical substance of the brain of an embryo loses 8,694 parts out of 10,000 on drying, but in the adult it only loses 8,096 ... the maxillary glands of the embryo lose 8,469, the liver 8,047, the pancreas 7,863, the arteries 8,278 and even the cartilages lose four-fifths of their weight. In the adult, the corresponding figure for the liver was only 7,192'.

Hamberger also published a small table of the lengths of children of 1½, 4½, 13½ and 18 years (1751, p. 319). He gives no hint as to how the table was obtained, and the result for 1½ years, as also for birth, is clearly erroneous. The 4½-year value is at the modern 5th centile, and the two subsequent values represent children who would average 10.0 and 13.0 years nowadays. There are crosses in the table which may indicate the children were dead (the 18-year-old being the one without a cross). In the second edition of the *Physiologia medica*, issued in 1757, the table is omitted, perhaps because in the meantime Buffon's work had become known and because Roederer, at the rival University of Göttingen, had published his famous observations on the lengths and weights of twenty-seven newborn infants (see p. 96).

C. F. Jampert

Presumably it was because of this uncertainty about the lengths of children that one of Hamberger's followers, Christian Friedrich Jampert (d. 1758), became the first person ever to publish a real table of measurements of the growth of the human. Jampert clearly indicated his debt to Hamberger throughout his thesis, but there is no evidence that he ever actually studied with Hamberger in Jena. His home town was Berlin and very probably he had studied at the Medico-Chirurgical College there, since the measurements of children which formed the subject of his thesis were made at a Berlin orphanage (see below). However he actually graduated (i.e. was 'promoted') in Halle, in October 1754. His thesis was entitled *De causas incrementum corporis animalis limitantes.* The promoter was H. L. Alberti, who thirty years before had supervised Stöller's thesis. (Stöller's son was a student with Jampert, graduating two years later.)

Jampert was later appointed to the position of *Privat-Dozent* in Halle, but had the misfortune to have as his professor ordinarius of anatomy one Philip Adolph Böhmer (1717–1789) who, according to Völker (1979), was more distinguished for political and financial pull than for pedagogical or scientific merit. Böhmer's father was a powerful Halle lawyer and he himself had the overwhelming advantage over his competitors for the Chair of being able to buy outright the new Anatomical Theatre built by Coschwitz only a few years earlier. Jampert soon fell out with Böhmer, and in 1775 applied for the position of ordinarius in anatomy in Frankfurt-on-Oder. Böhmer refused to back his application, having just quarrelled with him over the question of obtaining bodies from the town for dissection (Kaiser and Krosch, 1965). Jampert continued teaching in Halle, however, and must have had wide-ranging interests, to judge from the dissertations he promoted and the lectures he gave, which included a course on animal experiments as a basis for toxicology. He joined a Freemason lodge in Halle in 1757 and it is from their records that we know that he died in the following year. Jampert's younger brother came to Halle as a law student in 1757 and died of lung disease in 1763, only five years after Christian Friedrich. Perhaps Christian Friedrich Jampert also died of phthisis.

Jampert's thesis has been totally forgotten. Haller quoted it in a critical footnote to volume 8 of the *Elementa physiologiae* (1778 edn, part 2, p. 26), and in his *Bibliotheca anatomica* (1774–7 vol. 2, p. 506) he lists it with the dismissive summary '*Tabulae incrementi puerorum in pondere et statura non justae*' (Tables of the growth of boys in weight and height which are incorrect). Neither summary nor footnote are justified. Girls as well as boys were measured and many measurements beside height and weight were taken. The results, discussed below, were not that unreasonable, and far better than those of Hamberger that Haller quotes instead. As for the critical footnote, Haller simply takes Jampert's own very apposite criticism of his work, that is that his results were, as we should now say, cross-sectional whereas they should have been based on a longitudinal study (see below). Jampert had died too early to defend and perhaps elaborate his work.

Though Jampert's title 'On the causes which limit growth . . .' recalls, perhaps, Weise's preoccupation with tallness, in fact the Tacitan myth is mercifully absent. For the most part the thesis is strictly iatrophysical. However Jampert went far beyond all his predecessors in that he actually measured a series of children covering all ages from 1 to 25. He wrote:

> I was able to construct the attached table in the Royal Orphanage of Berlin, called Frederick's [see note 4.11], by kind permission of the Illustrious Director and Patron. I began the experiments in springtime, soon after Easter of this year, when the tissues were neither constricted by cold nor yet made turgid by heat. The experiments were made each day at the same time, two hours before meal-time, so that measurements of weight and the abdominal circumference would not be in error. In reckoning weight I always subtracted the weight of the clothes. So far as possible I chose subjects grown in healthy proportion. However it must be realized that I only measured a single individual of each sex [at each year of age] so that this table is not offered as a universal schema, from which, according to the principles of Arithmetic and Algebra, Universal quantities may be inferred. I used the duodecimal system because it divides length into more and smaller parts than the decimal system and thus makes the observations more precise. To measure the circumferences I took strong and finely-woven linen-paper with the divisions exactly marked (Jampert, 1794, p. 6).

Jampert goes on to describe the technique he used for the various measurements shown in his table (except, unfortunately, for height, evidently taken for granted). Arm length was from the acromion to the tip of the middle digit; head circumference was measured 'in a horizontal plane passing above the eyebrows' (exactly as nowadays). Chest and abdominal circumference were taken in horizontal planes cutting the nipples and umbilicus respectively, arm circumference by applying the measure in the depression under the axilla, going round the deltoid. The table is reproduced as Fig. 4.4.

Not only is Jampert sophisticated in his measuring techniques: he understands quite clearly the problems of variation and sampling. 'The amount of growth varies greatly in different subjects ... in the table given this variation is not as obvious as it might be because so far as possible the selected subjects were those whose growth was proportionate to their age; in those left out the deviation was greater. Thus measurements are not linked exactly to a year of age' (p.10). The meaning of this last sentence is a little obscure but presumably it implies that subjects were not measured necessarily on or near their birthdays. In the few hundred children available to Jampert in the Friedrichshospital it would have been impossible both to select fairly typical members of the 10-year-old age group, for example, *and* to have the selected person at or near the tenth birthday. Jampert seems to have resolved this dilemma in favour of typicality and against exact birthday age. 'The old saying', he writes (p. 13), 'that the child reaches half his growth by the end of his third year appears to be true. It is more valid for universal height growth, however, than for the growth of particular individuals (*de universali longitudine vero potius valet, quam de particulari*). For in the table given the measurements are made not of the same subject, but of different ones' (p. 13). Jampert, then, is the first to realize, or to state in print, the potential difference between the results of cross-sectional and longitudinal approaches. Actually his table accords exactly with the 3-year rule in the case of the girls represented, but only approximately in the case of the boys. For the latter the half-way mark would be reached in Jampert's data at about 3½ years (but his adult heights are surprisingly large).

Jampert's results have to be looked at with his method held in mind. Each age is represented by a single person, but the persons are evidently not chosen at random from their age groups, but as the most typical representatives that Jampert could see. We do not know the exact number of children in the Friedrichshospital (or Friedrichswaisenhaus) in the 1750s but judging from the situation a few decades later (see note 5.1) it is very unlikely there were more than 150 boys and 150 girls inclusive of all ages, and the number might have been only half this. Jampert made his selection probably in the

Fid. 4.4. Tables of growth of girls and boys, reproduced from Jampert (1754). The measurements are of single individuals of each sex at each year of age, selected from children in the Friedrich hospital (Orphanage), Berlin. Reproduced by courtesy of the Trustees of the British Museum.

Tabula Incrementum Virginum exhibens.

Aetas.	Pondus absolutum.	Longitudo.	Longitudo. ad genua.	Longitudo ad os Ileum.	Longitudo ad Acromium.	Longitudo Brachii.	Peripheria Capitis.	Peripheria Pectoris.	Peripheria Abdominis.	Peripheria lumborum	Peripheria brachii.
1 An.	13 ℔.	2′. 8/12″.	5 11/12″.	10 5/12″.	18″.	9 5/12″.	16 5/12″.	14 10/12″.	13″.	7 3/12″.	4 12/12″.
2.	17	2′. 1 1/4″.	6 5/12.	12 10/12.	20.	10 3/4.	17 1/4.	17 2/12.	18 4/12.	8 4/12.	5.
3.	21 ℔. ℥vi.	2′. 7 1/4″.	8 2/12.	15 7/12.	22 1/4.	12 1/3.	17 10/12.	19 1/3.	19 10/12.	9.	5.
4.	26 ℔. ℥i.	2′ 10 1/2″.	9 1/2.	18 1/4.	26″.	14.	18 1/4.	20 1/4.	21.	10 5/12.	5 2/12.
5.	28 ℔. ℥ii.	3′. 1. 1/2″	10 2/12.	21 1/4.	28 1/2.	14 8/12.	18 1/4.	19 3/4.	22 8/12.	9 7/12.	4 1/4.
6.	31 ℔. ℥xii.	3′. 2 1/4″	11 1/4.	22 1/2.	29″.	15 1/2.	19 2/12.	20 1/3.	21 2/12.	11 11/12.	6 1/12.
7.	37 ℔.	3′. 5 1/4″.	12 4/12.	24 5/12.	32 2/12.	18 11/12.	19 4/12.	21 5/12.	21 2/12	12 3/4.	5 8/12.
8.	39 ℔. ℥iv.	3′ 6 5/12″.	12 4/12.	24 10/12.	31 10/12.	19 8/12.	19 4/12.	21 11/12	20 1/4.	12.	6.
9.	45 ℔ ℥iv.	3′. 7 1/4″.	12 1/2.	25 4/12.	33 1/4.	20 1/12.	19 1/2.	22 7/12.	22 8/12.	13 5/12.	6 10/12.
10.	57 ℔.	3′. 10 2/12″.	13 11/12.	25 11/12.	37 5/12.	20 5/12.	20.	24 1/4.	22 9/12.	13 3/12.	7 1/12.
11.	61 ℔. ℥ii.	3′. 10 1/2″.	14 2/12.	26 2/12.	37 1/2.	22.	18 8/12.	24 8/12.	22 5/12.	14 1/4.	7 1/2.
12.	66 1/4 ℔.	4′. 1 1/2″.	14 5/12.	26 7/12.	40 1/4.	23.	19 10/12.	25 4/12.	23 10/12.	15.	7 7/12.
13.	69 ℔.	4′. 3 1/2.	15 8/12.	28.	42.	23 11/12.	20.	25 10/12.	23 4/12.	14 1/4.	7 2/12.
14.	71 ℔. ℥iv.	4′. 4 8/12″.	15 1/4.	29.	43 4/12.	25.	19 1/4.	25 11/12.	24 5/12.	14 1/4.	7 8/12.
15.	75 ℔.	4′. 5 1/2″.	16 10/12.	29 1/4.	44 10/12.	25 1/4.	19 7/12.	26 10/12.	26 10/12.	15 5/12.	7 11/12.
16.	79 ℔.	4′. 6 1/4″.	16 7/12.	30.	45.	25 1/4.	19 7/12.	26 7/12.	28 1/12.	15 1/2.	7 9/12.
17.	93 ℔. ℥vi.	4′. 10 2/12″.	18 4/12.	33.	47 5/12.	28 2/12.	20 10/12.	27 11/12.	30 8/12.	15 3/4.	8.
18.	107 ℔ ℥ii.	5′. 2 1/2″.	19.	33 10/12.	48.	30 2/12.	20 11/12.	27 8/12.	31.	15 7/12.	8 5/12.
19.	114 ℔. ℥i.	5′. 1 8/12″.	19 5/12.	34 2/12.	48 2/12.	30 10/12.	21 3/12.	28 2/12.	32 7/12.	16 4/12.	8 1/4.
20.	128 3/4 ℔.	5′. 2 7/12″.	20 8/12.	34 9/12.	50 9/12.	31 7/12.	21 1/3.	29 5/12.	33 2/12.	17 5/12.	8 10/12.
21.	129 ℔.	5′. 3″.	22 7/12.	34 8/12.	50 10/12.	31 3/12.	21 11/12.	30 7/12.	33 9/12.	18 3/4.	9 4/12.

Tabula Incrementum Juvenum exhibens.

Aetas.	Pondus.	Longitudo.	Longitudo ad genua.	Longitudo ad os Ileum.	Longitudo ad Acromium.	Longitudo Brachii.	Peripheria Capitis.	Peripheria Pectoris.	Peripheria Abdominis.	Peripheria lumborum.	Peripheria brachii.
1 Anni.	12 1/2 ℔.	1′. 9 7/12″.	6″.	11 1/2″.	18 7/12″.	9 8/12″.	16 5/12″.	14 4/12″.	17 5/12″.	6 3/4″.	3 11/12″.
2 An. 4 M.	26 ℔.	2′. 6 1/2″.	8 1/4.	15.	24 1/4.	12 7/12.	18 1/4.	19 2/12.	18 8/12.	11 7/12.	5 11/12.
3 Anni.	27 ℔.	2′. 7 7/12″.	7 10/12.	15 2/12.	24 1/12.	12 11/12.	18 11/12.	18 8/12.	18 8/12.	11 7/12.	5 11/12.
4.	29 ℔.	3. 1/12″.	9.	18.	28 10/12.	15.	18 7/12.	21 1/12.	18 2/12.	9 1/4.	6 1/2.
5.	30 ℔. ℥iii.	3′. 1/4″.	9 1/4.	21 1/4.	28 1/2.	15 11/12.	18 1/4.	20 1/2.	18 1/2.	10 1/2.	6.
6.	36.	3′. 1″.	9 1/4.	22.	30 1/2.	17.	19 8/12.	21 1/2.	21 1/2.	11 2/12.	6 1/4.
7.	39 1/2.	3′. 4″. 2‴.	11 1/4.	23 10/12.	33 1/4.	18 1/4.	19 1/4.	22 11/12.	22 4/12.	11 1/4.	6 2/12.
8.	41.	3′. 5″. 3‴.	11 1/4.	24 1/2.	34.	19 2/12.	19 10/12.	23 1/4	23 1/4.	12 1/2.	6 1/4.
9.	44	3′. 7 1/2″.	12 5/12.	25 1/2.	35 10/12.	19 2/12.	19 1/2.	24 5/12.	25 4/12.	13 1/4.	6 7/12.
10.	47 ℔. ℥iii.	3′. 9″.	13.	27 7/12.	36 5/12.	21 1/12.	19 1/4	24 8/12.	25.	13 5/12.	7 1/12.
11.	49 ℔. ℥v.	3. 11″.	15.	28 11/12.	38 1/2.	21 8/12.	20 5/12.	25 10/12.	22 4/12.	13 8/12.	7 1/12.
12.	56.	3′. 10″. 10‴.	15.	29 2/12.	39 1/2.	22 4/12.	19 10/12.	26 2/12.	24 1/12.	13.	6 11/12.
13.	60 ℔. ℥xiv.	3′. 11″. 1‴.	15 1/4.	25 7/12.	41 1/2.	23.	21 8/12.	27 1/4.	25 10/12.	14 7/12.	7 9/12.
14.	62.	4′. 4. 1/4″.	16 7/12.	29 10/12.	42 9/12.	23 1/4.	20 8/12.	26 11/12.	25 8/12.	16.	7 11/12.
15.	61 1/2.	4′. 5″.	17 1/2.	30 2/12.	42 1/2.	24 1/4.	21 1/4.	25 10/12.	25 1/4.	15 10/12.	8 1/4.
16.	92.	4′. 10″.	17 4/12.	30 7/12.	43 5/12.	26 10/12.	21.	27 1/4.	27.	17 1/4.	8 10/12.
17.	103 ℔. ℥iii ß.	5′. 1 7/12″.	17 5/12.	30 11/12.	44.	27 4/12.	21 5/12.	27 7/12.	27 9/12.	18 1/4.	9 2/12.
18.	115.	5′. 3 10/12″.	18 1/4.	31 2/12.	49 10/12.	29 5/12.	21 2/12.	29 1/2.	28 1/4.	18 1/2.	9 5/12.
19.	122 ℔. ℥x.	5′. 4 1/2″.	18 9/12.	31 10/12.	47 7/12.	29 4/12.	21 1/2.	28 10/12.	28 5/12.	18 8/12.	9 7/12.
20.	126 ℔. ℥xiv.	5′. 5 1/12″.	19 2/12.	32 4/12.	48 1/4.	30.	21 7/12.	29 3/12.	28 9/12.	19.	9 10/12.
21.	129 ℔. ℥i ß.	5′. 5 4/12″.	19 7/12.	32 7/12.	49 10/12.	30 5/12.	21 11/12.	29 7/12.	29 2/12.	18 11/12.	10 1/12.
22.	130 ℔. ℥ix.	5′. 7 1/12″.	20 1/12.	33 5/12.	50 2/12.	30 11/12.	22.	30 1/12.	29 11/12.	19 2/12.	10 4/12.
23.	134 ℔. ℥xii.	5′. 6 10/12″.	20 2/12.	35 1/4.	50 4/12.	31.	22 1/12.	30 5/12.	30 4/12.	19 10/12.	10 11/12.
24.	137 ℔. ℥i.	5′. 7 9/12″.	20 5/12.	34.	50 8/12.	31 7/12.	22 9/12.	30 9/12.	30 11/12.	20 5/12.	11 4/12.
25.	140 ℔. ℥xv.	5′. 8 5/12.	20 10/12.	34 2/12.	51 1/12.	32 1/12.	23 8/12.	31 2/12.	30 10/12.	20 2/12.	12 1/12.

same way that his older contemporary Bergmüller (see Chapter 3) chose his subjects: and he too reached the same conclusion about the validity of the Aristotelian 3-year rule. Jampert does not quote Bergmüller, however, nor even Elsholtz.

Jampert's height measurements are plotted on charts for modern British boys and girls in Fig. 4.5 and 4.6. Since only single children were measured at each age the points have not been joined up. It is noticeable that in both sexes the children are far below even the 3rd centile of modern data (though the girls less far below than the boys) up till age 16. Men over 22 and women over 18 are between the 25th and 50th modern centiles. However it should certainly not be assumed that this represents the true growth curve of the time (which would imply an enormous degree of growth delay, eventually compensated). It is all too likely that the principle of selection at older ages was radically different from that at younger ones and influenced, very likely,

Fig. 4.5. Jampert's measurements of boys (one at each year of age) plotted on modern British height standards (see text).

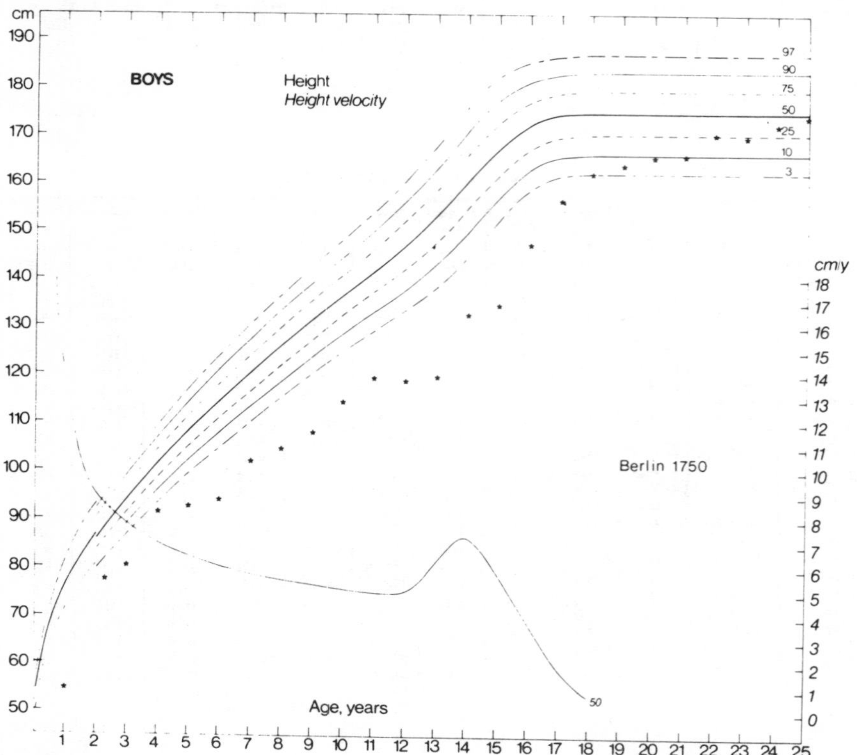

by what the heights of full-grown men and women were traditionally assumed to be.

As to the causes of growth and its limitations, the subject of his thesis, Jampert thinks it impossible to know these in the individual case. Truly to enumerate all the causes of limited growth in individuals, he says, 'is scarcely the work of a single man, much less the theme of a modest dissertation ... Growth is effected by the coming together simultaneously of many causes', whose sum gives the determinate cause (p. 12). So far as universals are concerned growth ceases 'when the resistance of the fibres is equal to the impetus of the fluids ... If the resistance is less, growth occurs, if it is greater, decrement ensues' (p. 53). The fibres of girls have less resistance and so can be more easily stretched. Also since women work less than men their copious humours are not dissipated (presumably in sweat) and fluid is retained. Hence women in general grow up faster and cease

Fig. 4.6. Jampert's measurements of girls (one at each year of age) plotted on modern British height standards (see text).

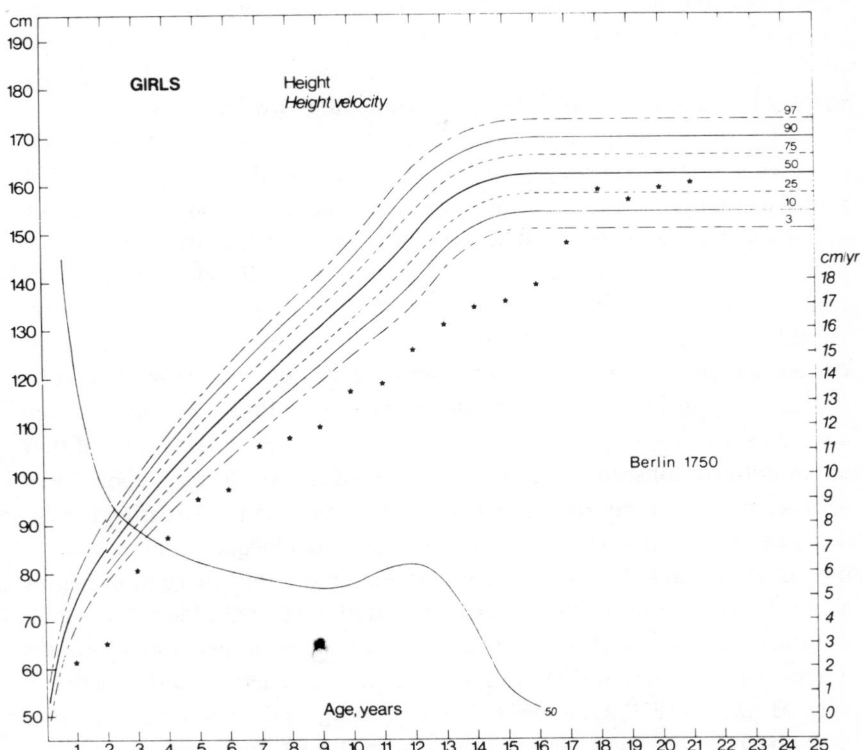

growing earlier than men. However 'working girls do not end their growth early, and rarely menstruate before 17 to 20 years, while girls who are less toughened menstruate at 14 to 16' (p. 62).

Age at menarche in the eighteenth century

Jampert's ages of menarche agree closely with those given by Guarinoni a century before. Other eighteenth-century views may be briefly summarized. Hamberger himself had some remarks to make on the age of cessation of growth. 'Up to the age of 16', he wrote, 'parts are still cartilaginous which after that age naturally become bony' (Hamberger, 1751, p. 320). Presumably he is referring to the closure of epiphyses, in which case what he says would imply a delay in growth of two or three years compared with nowadays. About age of menarche he writes 'The time when the first [menstrual] flow appears usually falls between 12 and 20 years; to be more precise it comes earlier in chubby girls (*succulentibus virginibus*) who lead a sedentary life with abundant food in sumptuous surroundings, later in lean (*siccus*) girls who do hard work and have scanty, dry and thick (*crassus*) food' (p. 698). He goes on to say that the early-maturing women 'only cease [menstruating] with difficulty before their forty-ninth year, whereas those who hold back [the menses] till age 20 are free of this incommoding flow by age 38'.

As for the obstetrics and gynaecology textbooks of the time, they vary somewhat in the age they give for the menarche. Berdot le Fils, in 1774, said that menarche occurred in temperate climates between the fourteenth and sixteenth years, but that in hot climates girls were 'more precocious' and in cold countries much more delayed, 'especially girls living in the countryside' (p. 26). Hamilton in Edinburgh in 1781 said menarche occurred in the 'thirteenth, fourteenth or fifteenth year; in this climate rarely earlier and seldom later' (p. 42). Denman (1788) in London wrote that 'in this country, girls begin to menstruate from the 14th to 18th year of their age and sometimes at a later period without any signs of disease, but if they are of delicate constitution and luxuriously educated, sleeping upon down beds, and sitting in hot rooms menstruation commences at an earlier period' (*Introduction to Midwifery*, p. 150). Though Baudelocque in France, writing in 1781, said that 'in our temperate climate, this evacuation begins about the twelfth or fourteenth year' (translation of 1790), Maygrier (1819), writing on menstruation in the authoritative *Dictionnaire des sciences médicales* a generation later, said that in Paris most girls had their menarche between ages 13 and 14. F. C. Osiander (1795) of Göttingen, the first person it seems actually to record ages of menarche as recollected by adult women, lists values which give a mean of 16.6 years (see Chapter 11). He gives (pp. 380–8) the causes which hasten menarche: a southern climate, Marseille girls maturing at 12 as contrasted with Strassburg girls at 14; a sanguine temperament and therefore a quick growth of the body; inborn disposition; a soft upbringing with little work; early readings of romances and early caresses; stimulating drinks such as coffee and wine; various medicines; and untoward accidents. The list is almost word-for-word Guarinoni.

Age at voice-breaking in the Thomasschule, Leipzig

There is one study and one only which throws a small and perhaps uncertain light on the age of puberty in boys of the early eighteenth century. This concerns breaking of the voice in J. S. Bach's Leipzig choristers (Daw, 1970). Breaking of the voice is by no means as good an indicator as menarche for locating the age of puberty, because it is a gradual process taking a year or more and its stages are not entirely easy to define. Modern studies eschew it, preferring in boys to record the appearance of the pubic hair and genitalia.

Bach was employed in Leipzig from 1723 until his death in 1750, and was in charge of three choirs. All the singers were male, and nearly all were musical boarding scholars at the Thomasschule. Their dates and places of birth, periods of attendance and fathers' professions were recorded, and the records have been analysed for the years Bach was there; possibly they still exist for other years also. Musicians agree that in Bach's day soprano singers were those with unbroken voices, tenors and basses were those whose voices had completely broken and altos were the boys whose voices were in a state of transition, with high and low notes both inaccessible. Only in one year, 1744, do we have records of which voice was sung by each boy: out of ten altos, the youngest was 15 and the oldest 19.

Daw examined the situation in other years by applying the proportions of sopranos to altos to tenors-plus-basses of the 1744 choir (28, 17 and 55 per cent respectively) to the age distributions of other years, assuming that the youngest 28 per cent were all sopranos and the oldest 55 per cent were all tenors-plus-basses. Because in reality overlap occurs between the ages of each group the method is only approximate, but it does yield an average age for those with transitional voices of around 16½ to 17 years. A comparable modern figure for the UK is around 13 to 13½ years, a difference of 3½ years. On the same basis Leipzig girls of the same social and geographical provenance would be expected to have had menarche at around 16.5 years. Between 1740 and 1748, exactly the years of the War of the Austrian Succession, the average age of altos rose by six months to a year. This seems likely to have been due to the poverty and undernutrition caused in Saxony by the war, with its high imposition of taxes, and the occupation of Leipzig in 1745 by a Prussian army living off the land. Deaths at the Thomasschule increased markedly in these years.

J. G. Roederer

It was not long after the publication of Buffon's *Natural History* that the measurement of newborn infants was introduced by Roederer, with consequences continuing to the present day. J. G. Roederer (1726–1763) was a student of Fried's at Strassburg, where he was born. He studied briefly in Paris, was in London with Smellie (in 1748) and then visited Leiden, where he caught the attention of Haller, always in touch with his beloved school (Siebold, 1891, p. 387). In 1751 Haller was able to call him to Göttingen as head of the obstetrical clinic, the first in Germany to be established in a University Hospital. Between 1751 and 1762 he conducted 232 labours, an immense total for that time, and trained a whole generation of obstetricians. He died aged 37, while on the way to visit a patient in Paris.

Roederer made very precise measurements of the maternal pelvis; and he was the first to report the weights and lengths of a substantial number of newborn infants with clinical details for each individual. His paper was delivered in 1753, and published in the following year. It concerned nineteen male and eight female infants born at term plus five born early and a pair of twins. The weights, lengths, (except in two), sex, ages of mothers and placental weights are given. 'The pounds were those used by the people of Göttingen, and the feet were Rheinish, divided into twelve parts' (p. 411). The method of measurement is not given, but lines as well as inches were recorded in assessing length. Roederer was concerned to scotch the stories of birthweight so far put about in obstetrical texts. In France Mauriceau (d. 1709), author of the standard obstetrical text, thought it was '12 pounds or more' and Phillipson, a distinguished English man-midwife, thought it was 16 or 17 pounds (see note 4.12). Roederer's weights on the other hand are quite in line with those of today; they ranged from 5 lb 8 oz to just over 8 lb with a mean of $6\frac{1}{2}$ lb. Lengths (see Table 11.2) varied from 1 foot 6 inches (probably 43 cm: see note 4.9) to 1 foot 11 inches (probably 55 cm). Roederer's object in weighing and measuring was to detect immaturity in the baby (see Roederer, 1763); the same is generally true today. Roederer's approach is altogether modern, as is his insistence that others should also make measurements to confirm his and to add to knowledge. One of Roederer's pupils, J. F. G. Dietz, gave in his thesis (1757) similar values for the weights of seventy-five newborn infants, fifty-five of them mature and twenty immature, though without distinguishing the sexes. The mature also averaged about 6½ lb. A smaller number of lengths were given. However it seems the thesis was totally forgotten until discovered by Scammon (1927*a*). It took thirty years for Roederer's newborn weight to be publicly confirmed – by Clarke (1786) in the Dublin Lying-In Hospital – and it took a further twenty years for the true value to find its way into the textbooks (see the excellent account of Cone, 1961). Roederer, incidentally (an observer who can be trusted), says that menstruation in most girls began 'about the year 14–15 although this age is not common to all girls and there are some exceptions'.

5

The first growth studies: Montbeillard, the Carlschule and the recruiting officers

It was in relation to neither art nor medicine, however, that the practice of measuring grew up; it was because of military requirements. Tall soldiers were regarded as preferable to short ones. Not only were they generally stronger; they could cover more ground on the march because of a greater length of stride, and in combat they could reach further with the bayonet and load more easily the long-barrelled muskets of the time, in which the charge had to be rammed down the muzzle. The custom of having troops march in step, introduced about 1700, in Prussia, set a limit to permissible differences in height; and the requirements of ceremonial, like those of the present-day classical ballet, demanded a specialized corps whose heights had a still smaller range.

In the eighteenth century the possession of a battalion of tall household guards was a matter of prestige to numerous European rulers, and to one it was a passionate obsession. Frederick-William I, King of Prussia from 1713 to 1740 and father of Frederick the Great, had talent-spotting agents who scoured Europe for tall young men and brought them to join his Grenadier Guards, with transfer fees that modern football clubs could scarcely rival. Kidnapping was not ruled out either, and the purloining of specially tall Saxons, Hanoverians or Dutch caused considerable friction between Frederick-William and his royal neighbours, not to mention their long-suffering subjects, who several times resorted to assassinating the recruiting officers. Giant guardsmen made very acceptable presents too, and served as emollients at times of political crisis. Peter the Great made regular contributions of tall Russians, and on one occasion the Danish King augmented Frederick-William's collection with a dozen stalwarts in exchange for the refugee assassin of one of his ministers. The largest of this dozen, a Norwegian blacksmith named Jonas Heinricksohn, was said to have been 194 cm in height (Udjus, 1964). Frederick-William's guardsmen were given special privileges, and though the popular legend that they were married off to specially selected tall girls, to yield a supply of still taller children, seems to be without foundation, certainly they were encouraged to marry and their children often had the King as godfather (see note 5.1).

Indeed the Prussians in the eighteenth century seem to have been obsessed with physical size. Part of this stems from the idea, due to Julius Caesar and Tacitus, that in Ancient times the Germanic tribes had been much taller and stronger. Tacitus, a severe moralist, was idealizing these super-barbarians in order that they should stand in sharp and salutary contrast to what he felt were the degenerate Romans of his day (around A.D. 90). But in the seventeenth and eighteenth centuries Germans took his message at face value, and wondered why they had apparently become so puny and weak (see note 4.5). Hence Weise's thesis *Of Tallness of the Body*, Stöller's book with its insistence that 'even the tallest' must seek to live according to God's word, Frederick-William's more autocratic seizing of the bulls by the horns. The belief that contemporary man was inferior in physique to men of the past continued well into the nineteenth century: Virey, writing of giants in the *Dictionnaire des sciences médicales* in 1816 (see p. 119), still quoted Tacitus and extolled the virtues of the old Teutons.

Goethe, much to his disgust, was involved in recruiting as part of his duties for the Duke of Sachsen-Weimar, and he drew the sardonic little pen, pencil and wash drawing reproduced in Fig 5.1. He complained that when recruiting in the countryside 'the cripples all wanted to be called up and the stalwarts to get exemption certificates' (Bruford, 1962, p. 111). To the present-day auxologist the sketch is technically revealing, for it shows an altogether modern approach to measurement. The recruit is being made to stand up straight, aided by upward pressure after the manner of R. H. Whitehouse; his head is held correctly, and a correctly aligned block is being slid down the wall in contact with the fixed rule. There is even the requisite recorder; and the shoes have been removed, as shown by the figure in the foreground. To the literary historian the sketch is revealing too. In the decoration over the door of the room that the successful recruit is entering, with a soldierly pat on the back from the recruiting sergeant, Goethe has replaced what was presumably the regimental badge by a gallows, which neatly picks up the line of the stadiometer. The sketch was done in 1779, a year in which Goethe was in charge of recruiting, and the year too in which a levy was being sought by Prussia (and successfully resisted by Goethe) for fighting in a war against Austria that was called by soldiers the *Kartoffelkrieg* from the newly introduced potato which they uprooted to allay their hunger. The gallows seems to reflect Goethe's feelings about military life. The motto under the badge represents Goethe's comment, too, rather than the reality. It says 'Thor des Ruhms'. At first glance this means 'Gateway to Glory'. But Goethe evidently intended a pun, for *Thor* has two meanings, dependent on its gender; and the second meaning is 'fool'. Furthermore Goethe wrote *Thor des Ruhms* and not *Thor zum Ruhm*; fool/gateway *of* glory, not *to* glory. (The real motto of the Duke of Sachsen-Weimar was naturally a little different and said 'Pro fide, rege et lege'. One wonders where Goethe kept his subversive material.)

Tallness was not the exclusive prerogative of the chocolate soldiers of Archdukes' ceremonials. The conscripted recruits who formed the basis of national armies were subject to similar, though not so stringent, height requirements. It was this (together with the need to identify deserters) that spread anthropometry throughout Europe in the eighteenth and early nineteenth centuries. The earliest military data which have been retrieved and analysed come from Norway, and go back to 1741 (Kiil, 1939; Udjus, 1964).

Fig. 5.1. Goethe's drawing of the measurement of recruits for the Duke of Sachsen-Weimar's army in 1779. Reproduced by courtesy of the Nationale Forschungs- und Gedenkstätten der Klassischen Deutschen Literarur in Weimar, Goethe-Nationale-Museum.

In the United Kingdom, records (of Royal Marines) began in 1755; in other European countries somewhat later (see note 5.2). Measurements, if not records, were being made at an earlier date, however, as witness the height of the gift Norwegian blacksmith of the 1730s.

Indeed the Reverend Joseph Wasse (1672–1738), who was the rector of the parish of Aynho in Northamptonshire, wrote, in a letter which Dr Mead, the eminent iatromathematician (see p. 68) communicated to the Royal Society in 1724, that he had observed 'several soldiers discharged from being a little under the Standard and helped by telling the officer of the difference between the morning and later' (Wasse, 1724). 'Since that time', Wasse continued, 'I have measured Sir H. A., Mr C., and a great many sedentary people and day-labourers of all ages and shapes and find the difference to be near an inch'. He then quotes experiments on himself, apparently in relation to sitting height rather than stature. On 21 August 1723, at 11a.m., he sat down and fixed an iron pin above his head 'so as to touch it, and that but barely'. After half an hour rolling his rectory lawn he was $\frac{5}{10}$ inch below the mark. At 6.00 a.m. the following morning he touched it fully, but after riding 4 miles he was $\frac{6}{10}$ inch shorter. 'If I study closely, though I never stir from my writing desk, yet in 5 or 6 hours, I lose one inch. All the difference I find between labourers and sedentary people is that the former are longer in losing their Morning Height and sink rather less in the whole than the latter . . . I cannot perceive that when height is lost it can be regained by any rest that day or by the use of the cold bath . . . The alternation in the human stature, I imagine, proceeds from the yielding of the cartilage between the vertebrae'. A lively discussion followed this first demonstration of the diminution of stature during the day, with many words expended on arguing that since it *could* indeed be possible according to current theory, then it probably *had* indeed occurred. Isaac Newton, who was in the chair, happily came out strongly on Wasse's side.

Such refinement of method and resource in helping aspiring recruits can hardly have been general. The 'Standard' Wasse refers to is probably that specified in the Recruiting Act of 1708, which laid down a minimum of 5 feet 5 inches (Scouller, 1966, pp. 109, 114). Earlier, it seems no minimum was specified by law, but one of 5 feet 6 inches seems to have been applied in practice. The fact that out of a requirement of 18,657 men for Queen Anne's army only 868 were recruited, 37 as volunteers and 831 as pressed men (Scouller, 1966, appendix F), argues that the height limit must have been placed sufficiently low as to allow practically any stray volunteer to surmount it, and makes one wonder about the identity of Wasse's disappointed enthusiasts. Indeed, Wasse's discovery was put to quite another purpose. Before the days of photographs and fingerprints, height was used as a means of identification (as it was for the American slaves – see p. 165 – and in present-day passports). Knox, the translator of Quetelet's *Physique sociale* (see p. 126), remarks in a footnote: 'in bed, for example, the elastic fibro-cartilages connecting the spinal bones together seem to recover their full depth, and the stature may gain an inch or more thereby. Recruits for the army and deserters avail themselves of a knowledge of this fact, and occasionally succeed in making their identity difficult to establish' (Quetelet, 1842, p. 61). We shall return to consider some of the recruiting data towards the end of this chapter.

Montbeillard's son

The outstanding study of growth in the eighteenth century, made in the spirit of pure knowledge and not from any desire for military or ceremonial aggrandisement, was done at the behest of Buffon. It was the first longitudinal growth study of which we have a record, and it was made by Count Philibert Guéneau de Montbeillard (1720–1785) upon his son, in the years 1759–77. Between 1774 and 1789 a set of seven volumes by Buffon called *Supplements to the Natural History* were published, and it was in the fourth of these, issued in 1777, that the measurements of Montbeillard's son appeared.

Montbeillard was Buffon's closest family friend. He was a land-owner, lawyer and philosopher of science, and a devoted admirer of Buffon, whom he helped for many years in an unacknowledged capacity. He drafted much of the *Natural History of Birds* but his collaboration was kept secret until the appearance of the third volume, as Buffon, and it seems Montbeillard too, feared that no hand other than Buffon's could write, and sell, so well. Montbeillard was quiet but courageous; he was the first person in his area of France to carry out a vaccination (or more accurately a variolation, for this was much before the time of Jenner, and the inoculum used was pus from subjects with the disease). In the 1760s inoculation, though accepted and practised in England and Switzerland was frowned on in France and had actually been forbidden in city areas by Parliamentary decree. On 7 May 1766 Montbeillard inoculated his son, with, says his *mémoire* to the Dijon Academy 'the trembling hand of a father' (Flourens, 1860; Brunet, 1925). The inoculation was successful (or at least, not disastrous) and advanced greatly the cause of smallpox prevention in France. The success did not prevent the boy later falling victim to the guillotine under Robespierre, however (a fate shared also by Buffon's son). Montbeillard and Buffon were evidently very different sorts of men: Buffon indefatigable at work all day but given to familiarity and even boorishness in the evenings; Montbeillard 'of a delicate constitution' and starting each day with family madrigals.

Montbeillard's growth study was well known to scientists in the nineteenth century, being quoted by Quetelet in 1835 (see Chapter 6) and later by Roberts, Bowditch and Pagliani (see Chapters 8 and 9) in the 1870s. But then it became neglected, and passed into modern books only through the agency of Richard Scammon (1883–1952), professor of anatomy in the University of Minnesota (see p. 265). Scammon was the author of a celebrated book on fetal measurements (Scammon and Calkins, 1929) and of numerous pregnant and tantalizing abstracts referring to work never written up in full. He also began to write a history of growth studies and got as far as assembling a vast bibliography, partly analysed (1927*b*), before running out of impetus. In 1927 he reprinted Montbeillard's measurements from the Cuvier edition of Buffon's *Oeuvres complètes* (vol. 4, 1836: see Fig. 5.2), converted them to the metric system and plotted them as a graph of height against age (Scammon,

Fig. 5.2. Table of Montbeillard's measurements of his son (first page). Reproduced from Buffon (1777), by courtesy of the Trustees of the British Museum.

376 *SUPPLÉMENT*

enfans, & plus ſévèrement encore l'uſage des corps pour leurs filles, ſur-tout avant qu'elles aient atteint leur accroiſſement en entier.

III.

Sur l'accroiſſement ſucceſſif des Enfans, page 473.

Voici la Table de l'accroiſſement ſucceſſif d'un jeune homme de la plus belle venue, né le 11 avril 1759, & qui avoit,

	pieds	pouces	lignes
Au moment de ſa naiſſance.	1.	7.	"
A ſix mois, c'eſt-à-dire, le 11 octobre ſuivant, il avoit.	2.	"	"
Ainſi ſon accroiſſement depuis la naiſſance dans les premiers ſix mois a été de cinq pouces.			
A un an, c'eſt-à-dire, le 11 avril 1760, il avoit.	2.	3.	"
Ainſi ſon accroiſſement pendant ce ſecond ſemeſtre a été de trois pouces.			
A dix-huit mois, c'eſt-à-dire, le 11 octobre 1760, il avoit.	2.	6.	"
Ainſi il avoit augmenté dans le troiſième ſemeſtre de trois pouces.			
A deux ans, c'eſt-à-dire, le 11 avril 1761, il avoit.	2.	9.	3.
Et par conſéquent il a augmenté dans le quatrième ſemeſtre de trois pouces trois lignes.			
A deux ans & demi, c'eſt-à-dire, le 11 octobre 1761, il avoit.	2.	10.	3½.
Ainſi il n'a augmenté dans ce cinquième ſemeſtre que d'un pouce & une demi-ligne.			
A trois ans, c'eſt-à-dire, le 11 avril 1762, il avoit.	3.	"	6.
Il avoit par conſéquent augmenté dans ce ſixième ſemeſtre de deux pouces deux lignes & demie			

A trois

1927*a*). Later D'Arcy Thompson calculated the increments and plotted the height velocity curves in the revised edition of his *On Growth and Form* (1942). J. M. Tanner then used the combination of height against age and height velocity as the opening figure in his *Growth at Adolescence* (1955, 1962). This figure has been recopied many times since, and the growth of Montbeillard's son is one of the best-known illustrations in all human auxology.

The graphs are reproduced in Fig. 5.3. In the velocity chart 'rolling' yearly velocities are plotted, that is to say, yearly velocities calculated, like modern inflation rates, successively: thus from 0.0 to 1.0 years, plotted at 0.5; 0.5 to 1.5 years plotted at 1.0, and so on. In this way seasonal effects are removed. Where the incremental periods are uneven the appropriate adjustment has been made to obtain true velocities and each velocity has been plotted at the midpoint of the interval concerned. There is a gap in the measurements from 10.0 to 11.5 years for reasons disclosed neither in the report nor in the voluminous correspondence between Buffon and Montbeillard. Unfortunately nothing is said as to how the measurements were taken, except that

Fig. 5.3. Height and height velocity of Montbeillard's son. Velocities are 'rolling' whole-year velocities (see text). Bracketed points are 'corrected' ones (see text).

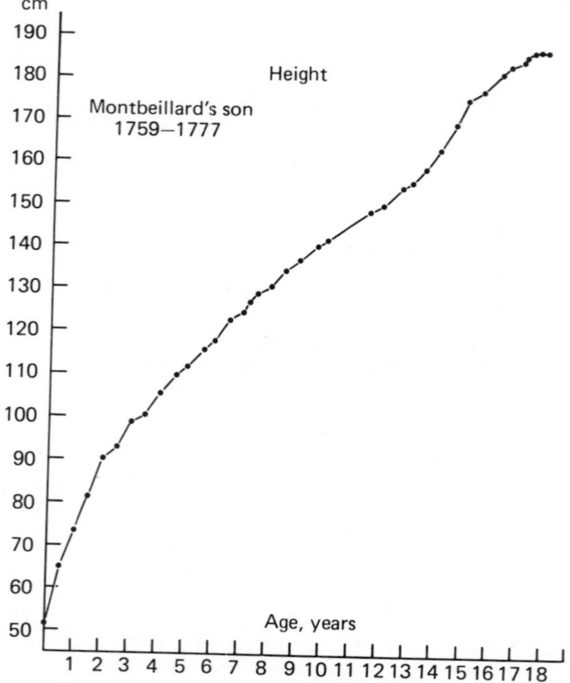

all were done by Montbeillard himself, with his son barefoot. At first the boy must have been lying down, but the point at which the change-over to standing took place is not indicated. Since standing height is usually about 1 cm less than supine length one would expect a particularly low increment at the point of change-over, and there is indeed one such, for the interval 3.0–3.5 years (1.6 cm in 0.5 years). If indeed the first standing height was taken at 3.5 years, then the two velocities 2.5–3.5 and 3.0–4.0 would be artificially reduced and should be augmented by about 1 cm/yr each. The tentatively 'corrected' points are shown in brackets in the velocity plot. The 'correction' does make the velocity from 3.5 to 4.5 look more consonant with the rest of the curve. Both distance and velocity plots show all the features of modern curves obtained in the same way, that is by a single measurer observing at six-monthly intervals, and indeed Montbeillard's record has never been surpassed, and seldom equalled, in elegance and presumed accuracy.

Montbeillard's son was very tall, ending above the 97th centile of modern standards. He also had a marked adolescent growth spurt (or pubertal growth

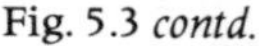

Fig. 5.3 *contd.*

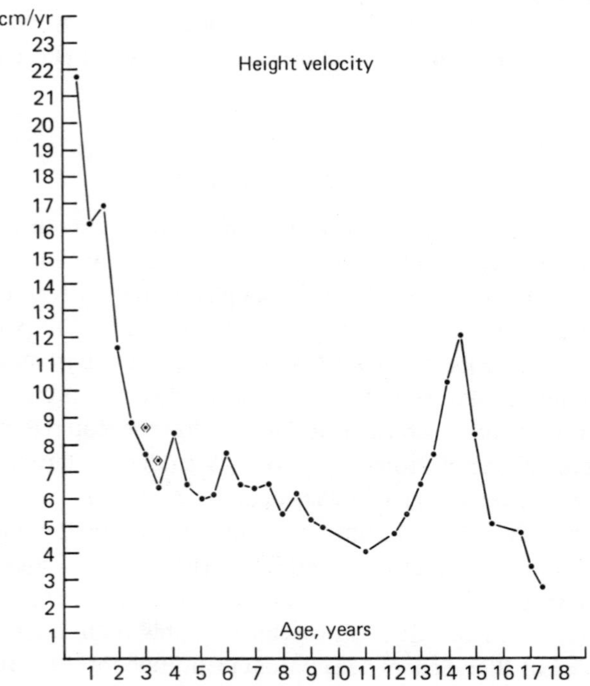

spurt: the terms are used here synonymously), with a peak velocity of 12.1 cm/yr and a total gain from beginning to end of about 31 cm, assuming 'take-off' at age 13.0. This gain is slightly above the present-day average (see Tanner and Whitehouse, 1976), as is the peak height velocity. The age of peak velocity is about 14.3 years (smoothing by eye), which is entirely in line with modern data, being about 0.3 years later than the modern mean. The whole curve shows a velocity falling rather strongly from birth to about age 4.5, then much more slowly (or not at all) till about 8.5. Between 8.5 and the beginning of the spurt at 13 the velocity declines noticeably.

Buffon did not remark on any of these features, however, in the *Supplement* where the record appeared, perhaps because he had foreseen the result already in 1749. He did report Montbeillard's observation that his son, when approaching maturity, suffered an apparent decrease of height when measured the morning after an all-night dance, a decrease that disappeared by the following morning (Buffon does not indicate whether he is aware of Wasse's earlier finding, though he certainly often read the *Philosophical Transactions* of the Royal Society, of which he was a member: see note 5.3). Buffon also remarked on the seasonal effect on growth rate, total growth in the age period 5 to 10 years being 7 *pouces* 1 *ligne* (19.2 cm) in the summer months (April to October) and only 4 *pouces* 1½ *lignes* (11.2 cm) in the winter ones. The seasonal difference agrees closely with modern values reported by W. A. Marshall (1971). Before 5 and after 10, Buffon says, the seasonal effect is much less obvious.

The Carlschule

There is a second source of individual longitudinal growth records in the eighteenth century, though it was only recently discovered. This is the collection of measurements made on the pupils of the Carlschule in Stuttgart from 1772 to 1794. The records, now in the Stuttgart City Archives, were discovered by Dr Robert Uhland (1953), the state archivist and historian of the Carlschule, and Professor Wilhelm Theopold (1967), professor of paediatrics in Stuttgart and a medical historian. Much of the material was published in the thesis of one of Professor Theopold's students (Hartmann, 1970); a short general account is given in Theopold *et al.* (1972).

The Carlschule was an institution unique in its time (Wagner, 1856–8). It was the beloved brain-child of Carl Eugen (1728–1794), Duke of Württemberg, a man who in his maturity attended in exemplary fashion to the welfare, and especially the education, of his subjects (in his youth it was quite otherwise). The great High School began as an institution for teaching the sons of soldiers how to be gardeners, such men being needed in the grounds of the Duke's newly built and favourite hunting lodge, *Solitude*. Soon the school was turning out plasterers as well, and from that it was short step to training musicians and dancers, also in demand at the lodge. At the same time Carl Eugen was making plans for a military orphanage to train a variety of artisans, and the two institutions were merged to form a Military Academy, which was opened in 1770. It became the Duke's favourite project and he himself acted as headmaster, dining with the pupils and taking the school assemblies. He called the pupils his 'dear sons', and placed himself firmly *in loco parentis*, which was perhaps just as well, since for many years neither pupils nor staff had any holidays whatsoever in which to visit their families; even later there were only two periods, of a week each, per year.

The school provided such an excellent education that officers, officials and even people of standing at court sent their sons there. Thus the original character changed. In 1775 the school moved into Stuttgart to be housed in a former barracks behind the castle. By this time it had three divisions: a basic stage into which boys were admitted at 7 years old; a *Gymnasium;* and an upper section which despite the opposition of the neighbouring University of Tübingen was soon declared of university status by the Emperor. The Carlschule included a medical school from 1775, and indeed boasted a wider selection of faculties than any German university of the time. Educational methods were advanced, and this attracted some of the best teachers in Germany. There were no more than twenty pupils to a class. Sons of nobility and of ordinary folk were taught together, and were subject to the same discipline, although each had their own dormitories and ate at separate tables in the dining hall (albeit the same food). Students wore uniforms, and used military ceremony and discipline. At the end of the yearly examinations there was a major assembly and prizegiving.

During its whole existence the school graduated 2,211 pupils. The majority were from Württemberg, but such was the reputation of the place that others were sent from all over Europe. The average yearly intake was about 350. From the start, the school was menaced by epidemics; the worst was typhus in 1784, when many parents took their sons away. The deficit was made up by day-boys from the town, and the school continued till Carl Eugen's death in 1794. But then it proved too much his own plaything and, with his death, it dissolved.

The graduates, not unlike the later *polytechniciens* of Napoleon's school, formed an elite that not only ruled Württemberg but spread far and wide beyond it. Amongst them were a number of famous people. Friedrich Schiller (1759–1805) was one; he graduated in jurisprudence and medicine, and at the prizegiving ceremony in 1779 saw Goethe for the first time, standing behind the young Duke of Sachsen-Weimar who was a guest that year. Schiller thoroughly disliked Carl Eugen and hated the school, military discipline not forming the most suitable environment for the new leader of the Romantic movement. He was overcome with humiliation at being seen there by the author of the *Sorrows of Young Werther* (albeit now quite a mature functionary). His humiliation was worse because while prizewinners amongst the nobles kissed the Duke's hand on receiving their prizes, the

Fig. 5.4. Height and height velocity of Friedrich Schiller while at the Carlschule. Drawn from data given in Hartmann (1970).

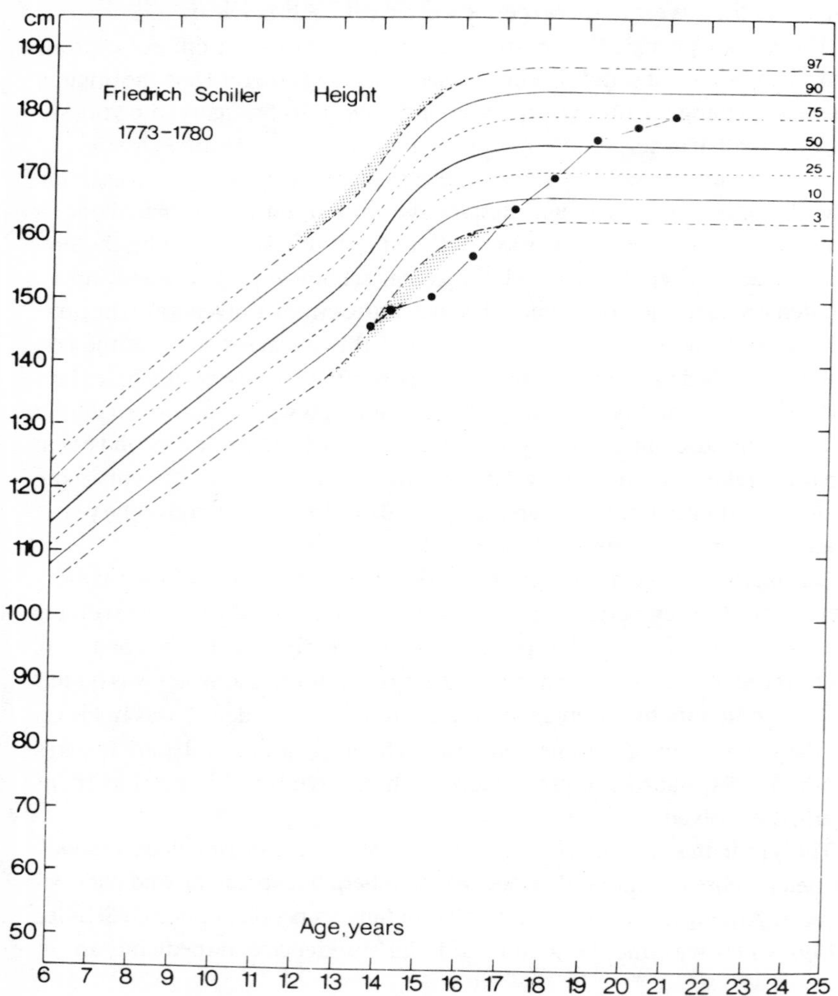

sons of bourgeois, amongst whom was Schiller, his father being an army officer, were only allowed a nibble at the hem of the Ducal robe (Theopold 1967, p. 31) Three times Schiller was called forth to receive prizes. Each time he hoped for a word or at least a glance from the greatest poet of the century, scarcely ten years his senior. No sign of recognition or interest came; and Schiller was not to know that the Duke had expressly forbidden his guests to talk with the bourgeois students (Burschell, 1968). (Perhaps this was because he feared the conversation might turn towards the subject of Schu-

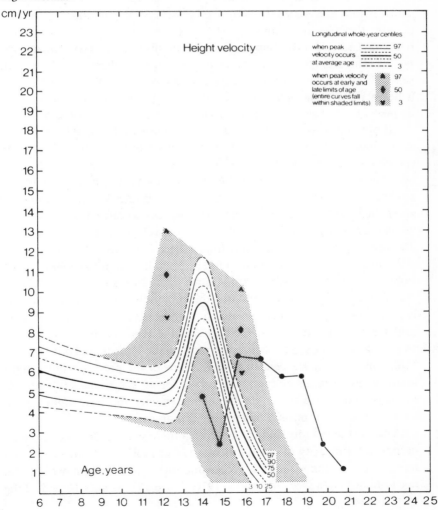

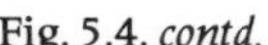

Fig. 5.4. *contd.*

bart, an unruly writer who openly referred to the school as a slave plantation and who had recently been put in prison by the Duke for a term of ten years; Schiller had known him since childhood and had taken from him the plot of his first and immediately successful drama, *Die Räuber*, which was already in first draft.)

Schiller's discontent and rebelliousness at the Carlschule may have been fed from another source, shared by so many thousand adolescents before and after him. He was a late maturer, with peak height velocity only at about 16.7 years, some nine months later than the average bourgeois and more than a year later than the average noble. He was a very tall man – indeed both before and after puberty the tallest-but-one amongst the bourgeois whose records Hartmann reproduces. At the frequent Carlschule parades the boys were lined up strictly according to height and Schiller must have started further up the line than all his bourgeois contemporaries. But he sank progressively down the order during his fifteenth and sixteenth years; when 15.0 he was only slightly above the bourgeois average and actually below the average of the nobles. With puberty, but only then, he regained his place. His height and height velocity curves are shown in Fig. 5.4.

All the pupils were measured at regular intervals, twice and sometimes three times a year (see note 5.4). This continued from 1772 to the time of the epidemics of 1782–4, when the frequency of measuring diminished. At each measuring session the previous as well as the present measurement was recorded, in Württemberg feet, inches (*zoll*) and lines (*strich*), and in later years the increment between the two occasions was calculated and written in also. Evidently Carl Eugen understood that velocity told him more about the pupil's health than did distance (that is, height attained).

An example of a record is given in Fig. 5.5, taken from Hartmann (1970). Nobility and bourgeois were distinguished in the lists, about one-third of the records being of the nobility and two-thirds of the bourgeois. Hartmann analysed the two sets of figures separately (her table C, p. 87). The number of students measured rises from 92 at age 8 to 442 at age 15, then diminishes to 155 at age 21. The series, then, is a typical mixed longitudinal one (Tanner, 1962), just like the Harpenden Growth Study of the present century. Hartmann gives distance and velocity curves for 60 noble and 60 bourgeois students; most have ten to twelve measurements and show quite typical adolescent growth spurts. Schiller's curve is an example.

Between the nobility and the bourgeois there was a difference in height, despite the similarity of environment and nutrition at the school. The difference relates almost entirely to advancement in growth, however, and not to final height (see Hartmann, 1970, figure 33). At ages 10 and 11 the nobles were on average 2.5 cm taller, and at age 15 nearly 7 cm taller; by 20 or 21 years, however, the difference was down to 1 cm, a quite insignificant amount. (The means at 21 are respectively 168.8 cm and 167.6 cm and represent very nearly mature height.) Peak height velocity was reached on average at a little under 16.0 in the bourgeois (Hartmann Table C, p. 87; note the error in age-plotting in the corresponding figure 34). In the nobility a curious chance in sampling makes somewhat uncertain what we ought to regard as the proper estimate of age at peak. In Table C, p. 87, Hartmann treats the total nobility data cross-sectionally and obtains a peak at 14.5 years, agreeing with the comparative values for height given above. But

Fig. 5.5. Original records of Carlschule measurements. Reproduced from Hartmann (1970).

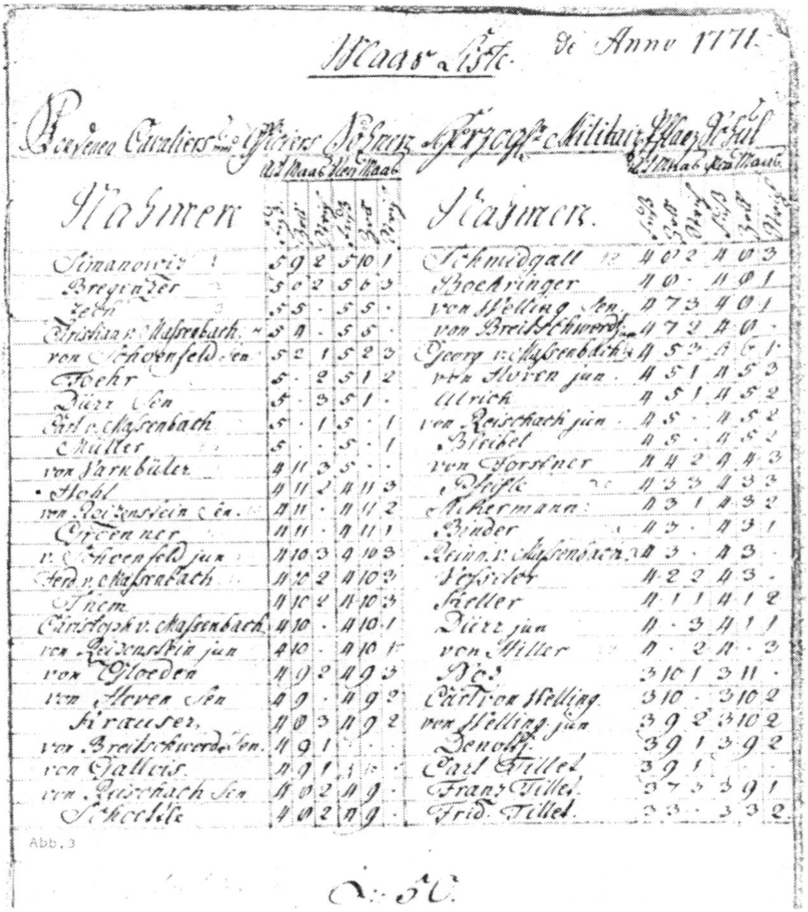

when one regards the 60 individual curves given (around half the total, at the relevant ages) their average age of peak is obviously later, and calculation from averaging these individual longitudinal data gives a value of 15.6 years. In the bourgeois, the selection of the 60 long-staying subjects has produced no such anomaly; their peak is at 15.9, exactly like that of the total series, which in this case is four times as large as the published sample.

Thus bourgeois, and perhaps also nobles, grew up eighteen months to two years later than their equivalents nowadays. (Peak height velocity is at approximately 14.0 in the British population as a whole, with the best-off section some three to six months earlier.) The mature height of the Carlschule students was lesss than nowadays, by about 6 to 7 cm.

The Marine Society

A unique set of data on boys' heights only recently discovered lies in the records of the Marine Society of the United Kingdom, now lodged in the National Maritime Museum at Greenwich. Though not longitudinal like the Carlschule records, the data are enormously greater in volume and stretch over a period of more than a century.

The Marine Society was founded about 1750 to provide young sailors for the Royal and Merchant navies. Boys were recruited from age 12 upwards and given suitable training. Records of Annual General Meetings of the Society go back to 1756, but measurement data begin only in 1786, when the Society acquired a training ship. On entry the boys' heights were measured and their places of birth and current occupations were recorded. The method for measuring height is not stated, but as army recruits of the time were measured barefoot (see Fig. 5.1) so, probably, were these naval cadets. Just why the measurements were taken is not clear. The navy had no height limits, nor had it the concern for ceremonial that characterized European land forces. Probably there was some question of identification involved (as with the American slaves: see below), for the Society would by no means admit runaway apprentices. Girls also were enrolled in the Society, to be trained as seamstresses, but their heights were not recorded.

The origin of the boys in the Marine Society was of course, diametrically opposite to that of the boys at the Carlshule. Nearly all were of the labouring class, and most probably out of work and on the edge of destitution (see note 5.5). Nearly all came from London. Thus the data mostly relate to the London poor of the last part of the eighteenth century. The material was discovered by Professor Roderick Floud of Birkbeck College, London. There are, of course, some problems in using these data. It is not clear whether the boys always knew their correct ages, though probably most had been baptized. Some boys might have deliberately given false ages, though only perhaps if their height was at the minimum acceptable, when it would have been politic to pretend to having more inches in store. The minimum height (on paper, never applied very vigorously) varied between 50 inches (127 cm) and 54 inches (137 cm) from 1757 to 1818; it then rose to 55 inches (140 cm) and further to 57 inches (145 cm) from 1824 to 1857. Sophisticated methods have been devised to allow for this in calculating means (putting back in the missing persons, so to speak, using the assumption of the population distribution being more or less Gaussian). However, plots of the actual uncorrected data seem to show the age and date trends quite as well as the corrected figures, and have been used here.

Age was recorded in years, and at this period is more likely to represent age-at-nearest-birthday than age-at-last-birthday (see discussion in note 11.8). For such a series we use the terminology age 13±, etc, meaning 12.5–13.5 (as opposed to 13+, meaning 13.0–14.0). Fig. 5.6 shows the means for boys aged 13±, 14±, 15± and 16± on recruitment between 1769 and 1860. The points plotted represented not single years of recruitment but cover half the years back towards the previous point and half the years onwards towards the subsequent one. Thus the second point of the 13-year-olds is the

Fig. 5.6. Heights of Marine Society recruits aged 13±, 14±, 15± and 16± from 1769 to 1860. Each point represents a 5–15 year cohort (see text). Data by kind permission of Professor Roderick Floud.

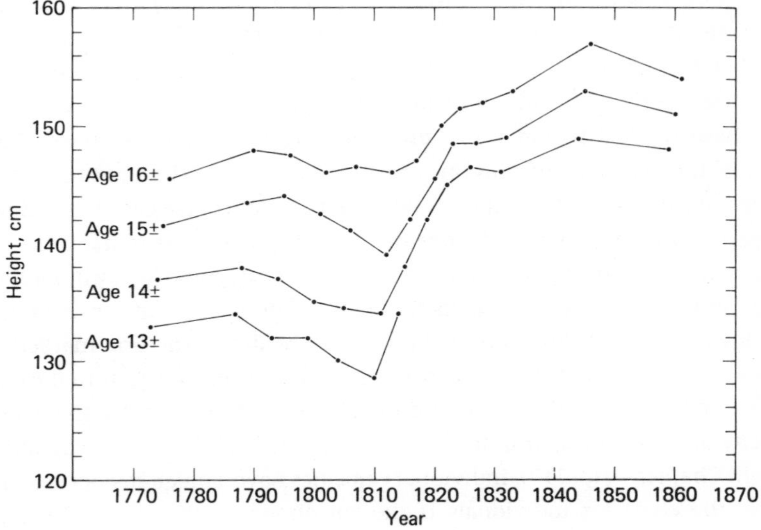

mean for the years 1782–91. The numbers for each point plotted are substantial; over 500 boys for nearly all ages from the beginning till about 1805, and subsequently over 100, except for some points of the 16-year-olds. Evidently marked changes in height took place during these hundred years in this destitute population. It seems that about 1800 something, presumably conditions of nutrition and infection primarily, began to worsen. By 1810, 13- and 14-year-olds were nearly 4 cm shorter than they had been twenty-five years before. But by 1815 things had improved and in 1820 boys at all these ages were actually taller than they had been during the period 1780–1800. It is in this way that the study of growth can throw light on aspects of economic and social history.

During this whole period, however, these boys were very much smaller than any normal boys of the same ages nowadays. In Fig. 5.7 the cohorts of about 1790, 1810, 1825 and 1860 are plotted on standard British population charts drawn to the data of Tanner, Whitehouse and Takaishi (1966). The upper section of the chart gives the cross-sectional centiles for British boys' height in 1960 (which were still appropriate in 1979). The means of the 1790 and 1810 cohorts are far below even the 3rd centile of modern standards (and a good deal below the mean of the bourgeois boys of the Carlschule, naturally, who average about the 10th modern centile, both before and after puberty). The 1825 and 1860 cohorts (the years are approximate) are above the 3rd modern centile at age 14, but thereafter drop below it. In the depth of the 1810 depression the 14-year-old recruit was the height of a modern boy scarcely past his ninth birthday. (The only boys recorded as equally small were those arraigned as criminals in London and transported to Australia in about 1840: see Chapter 7.) What we cannot tell from the data is whether this smallness represents growth delay or a stunting which would have been permanent. Malnutrition delays growth, so it is to be expected that these recruits would go on growing longer than modern boys. If the effect of the malnutrition was one of *pure* delay, then in the end they would reach the same height as modern boys; but this of course certainly did not happen. To the delay is added some degree of permanent stunting.

In principle some light should be thrown on the degree of delay by looking at the year-to-year measurements for signs of the adolescent growth spurt. The lower curve of the chart enables this to be done. The line marked 50 represents the differences between mean heights from one year to the next in the British population, cross-sectionally studied. This is not the height velocity of the average, or indeed of any, *individual*, for the reasons explained fully in Chapter 10 (p. 237). Individual velocities have a much steeper velocity curve; the chart for individuals is the one shown in the plot of Schiller's growth (Fig. 5.4). The population 'velocities', however, allow comparison between cross-sectionally studied populations, especially as regards the timing of the peak increment, or velocity. The age of peak gives an imprecise, but unbiased estimate of the age of the peak of the average individual in the population, whereas the height of the population peak is much below that of most individuals in the population. However, the height of the peak may be compared between populations; it reflects the variability of the timing of the adolescent spurt amongst individuals in the population, as well as the amplitude of individual spurts. We shall use these population comparison charts a good deal in the chapters which follow.

In Fig. 5.7 the increments for the two cohorts of 1790 and 1810 have been plotted. It looks perhaps as though the peak increment of the 1790 cohort is at 14.5, of the 1810 cohort at 15.5. These are only 0.5 and 1.5 years later than nowadays, and if such estimates could be believed, they would indicate

Fig. 5.7. Heights of Marine Society recruits in approximately 1790, 1810, 1825 and 1860, plotted on standard British population charts. Data as above.

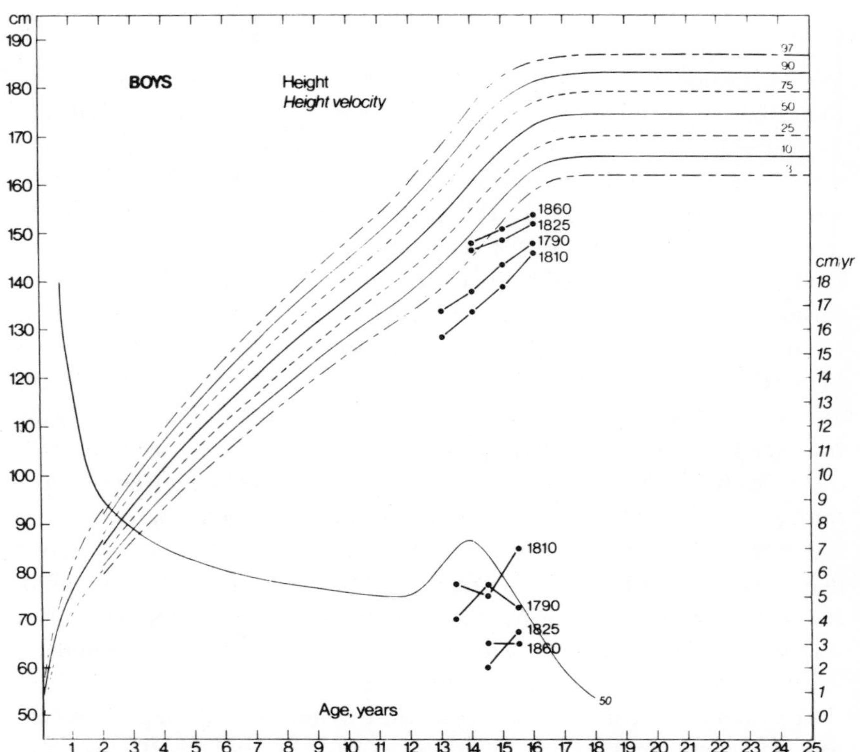

that these boys finished their growth around 18 or 19 and ended up very small indeed as adults, perhaps about 157 cm or 62 inches. The data are too fragile to support such a conclusion unequivocally; and the behaviour of the other two cohorts is merely confusing, perhaps because of distortion caused by the raising of the lower threshold for acceptance to 140 cm, and in 1824 to 145 cm (whereas the previous limits, even if applied, would scarcely have disqualified anyone aged 14 or over). But boys who worked in the industrial North in factories in 1833 had height curves, as we shall see in Chapter 7, that were not dissimilar; for them we have a fairly good end-point of growth and it was about 160cm, reached at about 19 (cf. p. 179). If so great a diminution of height compared with nowadays really did occur coupled with so relatively little growth delay then the situation must have differed from that seen nowadays in the undernourished children of the Third World shantytowns. It seems likely that birthweights especially were lower and that perhaps toxic substances as well as malnutrition were involved: the point is discussed further, in relation to the factory children, in Chapter 7.

The secular trend and recruits' statistics

We return now to consider the statistics of recruits. Recruits were smaller than soldiers nowadays, and they grew up with a slower tempo, so that even 21-year-olds were by no means all of mature height. The increase in size seen in children, and seen to a lesser degree in mature adults, has become known as the 'secular trend'. This rather curious phrase denotes both the tendency to get larger and the tendency to become more early-maturing, tendencies which are usually, though not invariably, linked. The tendency to earlier maturation has also been referred to as 'growth acceleration'. The word 'trend' is really inappropriate as it usually denotes a change always in one direction, whereas there is no reason why the growth trend should not go the other way too, as the Marine Society data clearly show. Following the recovery from the Napoleonic Wars, stature rose in France (see Chapter 7) and a lively debate took place as to whether the wars had lowered stature from a pre-existing higher level to which it was again returning, or whether the trend was a new phenomenon. Van Wieringen (1978), whose review is up-to-date and authoritative, stresses how in a historical period of diminishing comfort and increasing misery, delay in growth and shortening of stature may return. Oppers (1966) believes this may have occurred in Holland in 1820–60, associated with the famines and increases in the price of rye.

The military statistics seem to emphasize to how late an age growth continued in continental Europe 200 years ago, in all but the best-off households (such as Montbeillard's). A Norwegian archive enables us to estimate, with some precision, the age of peak velocity there in 1795. A national land and sea defence force was created in 1628 and all farms and estates had to send men to join it. From 1705 a royal decree required all men age 17 to 36 to parade at a muster each year. Their names, ages and heights were recorded, together with information as to whether they had done military service. From 1741 all were measured barefoot, to the last completed *ell*, an *ell* being 2.61 cm. In principle, therefore, these should be records of men individually followed from 17 to 36 years. Kiil (1939) in fact found in the company from Stryn in Nordfjord, north-west Norway, records with lower ages represented. He gives individual height curves covering the age span 16.0 to 21.0 years for twenty-two men, and from these the values in Table 5.1 were derived. Peak height velocity occurs in the interval 17.0 to 18.0. This is the age of peak velocity that characterized children living in the Highlands of New Guinea in 1950 (Malcolm, 1978); nowadays no group in the world has such a late maturation.

Kiil also reports a later pure longitudinal series of 238 men from Romsdal canton followed from age 17.3 to 22.3 in 1826–31 (his table 56, p. 118). The yearly increments were 4.0, 3.8, 2.7, 1.6 and 1.0 cm: the final height was 168.5 cm. It seems, therefore, that by the early nineteenth century age at peak height velocity had only fallen to about 17.0 years; furthermore mature height had only increased by 1 cm (Kiil, 1939). In the next fifty years the gain in adult stature was scarcely larger, being about 1.5 cm. Then things began to improve more rapidly, and in the seventy years from 1865 to 1935 mature height increased by about 4 cm and the age of peak height velocity dropped by nearly three years.

Udjus (1964) carried Kiil's work up to 1960. Norwegian conscripts mea-

Table 5.1. *Mean heights of twenty-two men from Stryn, Nordfjord, from age 16 to 21 (pure longitudinal, 1793–9 approximately)*

Age	Height (cm)	Height velocity (cm/yr)
16.0	148.2	
		4.5
17.0	152.7	
		5.0
18.0	157.7	
		4.5
19.0	162.2	
		2.0
20.0	164.2	
		1.4
21.0	165.6	

Calculated from curves of figure 30 in Kiil (1930, p. 93).

sured in that year were remeasured one year later. The average rate of growth was only 0.8 cm from 19.5 to 20.5 years and 0.5 cm from 20.5 to 21.5 years. Thus in 1960 most young men in Norway had completed their growth by age 20 or even 19, except for the very small increment in trunk length which continues even now in many men till about 25 or 30. (Modern researchers tend to use a gain of less than 1 cm in one year as indication of functional 'cessation' of growth.) Udjus' soldiers had measurements of sitting height as well as stature taken and he was able to show that more of the secular increase was due to increase in leg length than to increase in trunk length. (In general the better-off have relatively longer legs than the poor, a finding which agrees with this observation.)

Kiil makes the point that in the period 1818–23 the sons of cottagers were some 4 cm shorter on call-up than the sons of more wealthy farmers, and that the difference diminished sharply during the ensuing thirty or forty years. He attributes this to the large-scale introduction of the potato into the cottage diet and the resultant stabilization of energy input and output. By 1960 a city-dwelling class had arisen and Udjus found that only 0.6 cm of difference existed between young men born in towns and those born in the rural areas. But recruits from the higher socio-economic classes still had heights which averaged 2 cm more than those of the sons of unskilled labourers. This difference, as we shall see, has all but disappeared now in Scandinavia.

Oppers (1966) was similarly able to trace records of successive measurements on the same individuals, taken in Holland in the nineteenth century. He thus estimated the average amount of height gained from age 18.75 to full maturity. In the period 1820–80 this was about 7 cm; by 1900 it was down to 5 cm and by 1960 to 2 cm. Van Wieringen (1978) gives statistics showing that in Holland, as in Norway, the increase in mature height was slow in the nineteenth century and faster in the first half of the twentieth. During the fifty-two years from 1865 to 1917 only 3 cm of mature height was gained. The next 3 cm took just thirty-five years, from 1917 to 1952, and the 3 cm after that only eighteen years, from 1952 to 1970. The main effect in the nineteenth century was on the range, not the mean; the numbers of conscripts of very low stature diminished, but the upper half of the distribution remained static. In the twentieth century the whole distribution shifted upwards.

In the United Kingdom, statistics on recruits began about the same time as the activities of the Marine Society. Some volunteers, older than the Society boys but also mostly from London, were recruited to serve on Royal Navy ships in the period 1750–60, and their heights would seem to indicate cessation of growth at 19 or 20, apparently earlier than their Scandinavian counterparts. They averaged about 164 cm in height. Persons enlisting in the Royal Marines at about the same period seem to have been drawn from a population with an average height of around 166 cm, when allowance is made for the application of a lower threshold of height. No very detailed account of subsequent recruit statistics in the UK has been published, but Morant (1950) provides an overall review.

Ideas on growth in the early nineteenth century

Buffon's opinions on the causes of slow and fast growth and short and tall stature – which reflected Guavinoni, which reflected Arnald of Villanova, which reflected Hippocrates – remained unchanged for a further fifty years after his death. Dictionaries and encyclopaedias of medicine supplied much of the demand for medical textbooks in the nineteenth century, especially in France, and the most frequently quoted encyclopaedia article which concerns growth is that by Virey (1775–1846), published in 1816 in the *Dictionnaire des sciences médicales* (see note 5.6). Invited to contribute an article on giants, Virey managed to squeeze fifteen pages concerning factors which determine the heights of ordinary folk between a brief definition of gigantism and a coda about whether giants seven to eight feet tall existed as a race in Antiquity. (As to this latter, he thinks it indeed quite possible, especially in the light of the then-current version of the Abominable Snowman, whose gigantic and human footprints had been found in 1815 on the banks of the Swan river in Western Australia.)

Virey is still firmly in the humoural era: there is little sign (except, perhaps, in the elegant style) that he is writing after Buffon.

> Nourish a man or an animal parsimoniously, with dry and hard foods, smoked, salted, spiced or sharp and astringent; permit him to drink only a little and then a sharp and sour wine such as tartarous *vin rouge*, give him primarily acid and bitter things which harden and contract the fibres; it is very obvious that such a person will become thin, short, compact in all his organs. In contrast, stuff a child with soggy foods, get him used to taking milk and gruel and dough, to slimy drinks like beer, mead, whey and oily-chocolate, to warm and dilute liquids; cram him with all the foods apt to distend and enlarge him, in the way one fattens geese and pigs; then he is able to become colossal and gigantic in stature compared with a person nourished in the opposite way (p. 553).

This is Guarinoni again, more forcefully expressed. Jean-Jacques Rousseau and some philosophers, says Virey scornfully, may think savages are tall and strong and beautiful; but the truth, easy to see, is otherwise. The bad climate and lack of available nourishment in the humid and dark forests or on the barren windswept plains make the inhabitants unhealthy and small. It is Europeans who are tall. Or at least they used to be, in the times of the German tribes described by Tacitus. In a passage that exactly echoes Conring (see note 4.5) and would have done credit to Baden-Powell, Virey continues 'Nowadays we are soft and effeminate. The age of puberty is advanced because of a precocious awareness (*précocité du moral*), because of the pernicious solitary pleasures which bring on prematurely the sexual organs and exhaust youth. Thus turning the greater part of nutrition towards the excretion of sperm stops growth, and people stay short in stature. The promiscuity in towns and amongst the rich makes them feeble' (p. 560). The German tribes were strong and brave because of sexual abstinence, for Caesar said they thought it disgraceful for a man under the age of 20 to approach a woman. As for us, we live in the height of immoral and enervating luxury and no good can come of it, not even so far as stature is concerned.

Conring's and Virey's views continued to be quite widely held throughout the nineteenth century. In 1896, for example, an Italian doctor, Marina, wrote a book reviewing previous studies of growth and adding many thousand measurements that he had collected on young men, mostly conscripts, from all over Italy. Only after eighty pages does the real object of the book emerge: to warn about the effects on growth of masturbation, 'this saddest of vices, dreadfully prevalent amongst youths of all ages, including those who have not yet reached puberty . . . It consumes the precious force, proper to the most important years of growth . . . in any comparative study of physical and mental growth of children one cannot neglect this important cause of decline, of slowing up, and, as is well known, of inferiority' (1896, p. 78). Marina thought highly of the recuperative force of nature in the face of congenital or acquired defect: but it was this force itself which was drained away in the evacuated semen.

Virey also wrote the article on 'Enfance' in the *Dictionnaire* (1815). In it, he does say that children 'normally inherit the height of their parents; nevertheless good nourishment and a constantly hot and humid environment increase greatly the height of young people, in the same way as they *font végéter les plantes* (p. 234)'. A little later, Geoffrey de Saint-Hilaire (1805–1861) wrote a three-volume treatise (1832–7) which dealt with growth disorders, notably dwarfism, gigantism and precocious puberty, as well as congenital anomalies, such as conjoined twins. His tone is noticeably more scientific than Virey's, and his concern no less. Discussing the relation of height to puberty, he says previous authors are confused (a certainly correct statement except in so far as Buffon is concerned). When puberty is complete, growth ceases; and this is equally true of a pathologically precocious puberty as of a normal one. Furthermore, in boys a rapid increase in height invariably takes place at the same time as the genitalia develop, even if the development is premature. Sainte-Hilaire noticed that tooth eruption was seldom accelerated in cases of precocious puberty and developed in a different time sequence from the rest of the body. He has a long discussion of differences of heights between races, interestingly entitled 'Hereditary variations of height': but, even so, he has nothing to say about hereditary variation *within* a race, or in families.

About dwarfs Sainte-Hilaire writes a most telling paragraph (Vol. 1, p. 142):

> It was chiefly during the eighteenth century that a great number of researches and observations were made on dwarfs, certainly not because they were at that time less rare than in earlier centuries or nowadays, but because the attention of the public and in consequence of scholars was specially fixed upon them. The fashion of Court Madmen (*fous de cour*), to use the expression of the time, having gone out towards the end of the seventeenth century, it was necessary to dream up amusements of a special sort to occupy the leisure of princes and it was to dwarfs that fell the sad privilege of serving as the toys of the world's grandees. It was indeed natural, having exhausted the spectacle of the infirmities of our mental nature, to look in the vices of our physical being for the subjects of pleasure – of piquant novelty, as they said – and it was no less apt to arouse in the minds of the august spectators the sentiments of an egotistic and cruel joy ... Catherine de Medici [1519–1589] set the example of making marriages between dwarfs, which always remained without offspring ... There was also an Electress of Brandenburg, the wife of Joachim-Friedrich [Elector, 1598–1608], who many times gave her court the spectacle of these sad and dishonourable parodies without succeeding any more than Catherine de Medici in her ambitious aim of leaving a race of dwarfs for the pleasure of posterity.

Not only the tall, it seems, were collected by princes.

6

Adolphe Quetelet and the mathematics of growth

Quetelet and 'l'homme moyen'

Adolphe Quetelet (1796–1874) was one of the major figures in science and public health in the first half of the nineteenth century. By training a mathematician, he was the creator and director of the Brussels Observatory, where he lived from 1832 until his death. He contributed greatly to meteorology and earth science, was permanent secretary, from 1834 until his death, of the Brussels Royal Academy of Science and Letters, later (1845) the Belgian Royal Academy of Science, Letters and Fine Art, and held a quite extraordinary place in Brussels society in the 1840s and 1850s. He was in many ways the founder of modern statistics. It was he who introduced into practical work the Normal curve, discovered in another context by Laplace and Gauss. It was his energetic presence and the new type of data that he brought that caused members of the third annual meeting of the British Association for the Advancement of Science at Cambridge in 1833 to found a Statistical Section and, nine months later, the Statistical Society of London (see note 6.1). Babbage, Malthus, Richard Jones, Nassau Senior and Thomas Tooke were prime movers in both events, and at the Statistical Society's first meeting Quetelet was elected first Foreign Member in recognition of 'the part taken by him in the formation of the Statistical Section of the British Association ... to which the Society owes its establishment' (Royal Statistical Society, 1934, p. 13). Florence Nightingale was Quetelet's devoted admirer and Metternich was amongst the many expatriates of all countries and persuasions who sought solace and advice at the Brussels Observatory. John Stuart Mill and Friedrich Engels were contemporaries whose work would surely have made inescapable their reading his *Social Physics.* Quetelet made an extensive study of the growth of children, and is often, but wrongly, taken as the father of modern work on growth. That symbolic honour, for reasons that will become clear as we proceed, must be reserved for men of a later generation: for Bowditch, Roberts and, above all, Franz Boas.

But for all this, Quetelet was at heart an artist. Goethe and Gauss were his intellectual parents, with Goethe dominant. Quetelet was born at Ghent in

1796 and appointed teacher of mathematics at the Ghent Atheneum (high school) on his nineteenth birthday, just four months before the battle of Waterloo. He obtained his doctorate in mathematics in 1819 with a thesis on a new family of curves derived from conic sections, and was soon called to be professor of mathematics at the Atheneum at Brussels. But he had earlier thought of being a painter. 'I finished my schooling', he wrote, 'at the time of the events of 1814 which separated the Netherlands and France. I was dividing my time between art and science. My taste for art had already led to my being a student in a painter's studio, which I abandoned later when I accepted the chair of mathematics at the Royal Atheneum at Ghent. The taste for art continued to be associated with that for science in my moments of leisure' (1870, p. 6 n.). Quetelet also wrote poetry and once put on an opera of which he had written the libretto.

Paul Levy (1974), in a most perceptive essay that was a contribution to a memorial volume to mark the centenary of Quetelet's death, wrote 'It was the research of Beauty which obsessed him. It was that which drew him towards the mean (*moyenne*), which necessitated measurement, and from the mean towards the conception of the laws of social physics (*physique sociale).* Running through all his work one discerns a constant preoccupation with discovering, under common appearances, the intention of the Supreme Intelligence who has conceived it all. How to determine the ideal representation (*type*) of human beauty? Surely with a synthesis, in which all accidents would cancel out? It is at that point, of course, that number imposes itself . . . On the title-page of his little volume on the theory of probability, his pocket book one might call it, he wrote "*Mundum numeri regunt*" (Numbers rule the world)'. It sounds like a quotation, though Quetelet may have made it up. At all events it echoes Pythagoras' dictum *Arithmo de te pant' epeoiken* (perhaps best translated, in the modern idiom, 'numbers model everything') or Pico della Mirandola's Renaissance version 'By numbers a way is had to the searching out and understanding of everything able to be known' (see French, 1972, p. 105). So Quetelet, besides being an artist, was a Pythagorean. In the end, so far as growth was concerned, this tripped him up, for an excessive attachment to the symmetrical simplicity of number blinded him to the significance of the brute facts, and even, it seems, to the brute facts themselves.

Quetelet's meeting with Goethe, whom he greatly admired, must have strengthened, even if it did not actually initiate, his search for the ideal or representative 'type'. In 1823 Quetelet, then aged 27, was given the job of organizing the observatory to be set up in Brussels, and during the next few years he travelled extensively in Europe assessing astronomical equipment and methods. He spent the winter of 1823–4 in Paris learning how to make astronomical observations, and the brilliance of his doctoral thesis procured meetings with Laplace, Gauss and Fourier. Then or a little later Fourier introduced him to the great French epidemiologist Louis-René Villermé, with results we shall see later. In 1827 Quetelet toured Great Britain and met Babbage and Sir John Herschel, the Astronomer Royal, who became a lifelong friend. Two years later he travelled round Germany, and in late August was in Leipzig. Thence he went to Weimar to call on Goethe (see note 6.2; John, 1898; Collard, 1934). Goethe asked to see his instrument for measuring magnetic fields and Quetelet set it up in Goethe's garden. The old man was impressed and attracted by Quetelet, had him come every day for a week to his select evening receptions, gave him a copy of his book, *Zür Naturwissenschaft überhaupt, besonders zür Morphologie* (1820) and asked him to send details of the proceedings of the scientific congress in Heidelberg to which Quetelet was *en route*. Goethe, we now sometimes forget, rated his scientific works as highly as his poetry. He felt his views on optics to have been unjustly neglected and begged Quetelet to promote them, if possible, in Heidelberg.

Quetelet, for his part, shared Goethe's preoccupation with beauty. In his later years Goethe, turning towards classicism in art, developed an active interest in the study of proportions, writing in 1791 to a friend 'To work at a canon of masculine and femine proportions . . . and to seek out the beautiful forms that mean exterior perfection – to such difficult researches I wish you to contribute your share just as I, for my part, have made some preliminary investigations' (cited in Panofsky, 1955). Quetelet was deeply touched by Goethe's kindness and wrote later that his graciousness towards a young unknown scientist had influenced him throughout his career. In his whole life, he said, he had never passed such evenings, and when he left Weimar Goethe had bid him goodbye with the tenderness of a father (note 6.2). Goethe's idea of the 'type' clearly influenced Quetelet in his assessment of his own *homme moyen*. 'Goethe's Type was an ideal, to which individuals by their moral and intellectual development seek to attain . . . drawn by this monistic conception of the universe, Quetelet tried to found a new science of man, analogous to Goethe's Morphology of Organic Nature' (John, 1898, p. 328).

Quetelet's visit to Paris a few years earlier had also been of immense importance to him. From the mathematicians he obtained his central over-riding idea that is his lasting claim to statistical fame: that is, the applicability of the curve derived from the law of errors in describing the distributions of all kinds of things from children's heights to the ages of murderers. From Villermé (1782–1863), the chief founder of public health and public concern for health in France, he obtained a powerful impulse to apply his ideas to the benefit of society as well as to the perfection of man. Villermé was fourteen years older than Quetelet, and a very different man. Brought up in a small village outside Paris, his studies were interrupted in 1804 when he entered the army of the Napoleonic Wars as a surgical assistant (Astruc, 1933). He remained in the army for ten years, working for the most part under atrocious conditions. His thesis for the full doctorate had to wait till the fall of Napoleon; it was entitled 'Effects of famine on health in places which are theatres of war'. Experience of war and famine had affected him deeply; his thesis describes the Spanish campaign in Estremadura and the effects of starvation on people's behaviour in precisely the terms of Goya's *Horrors of the War*. In later life he avoided talking of his war years, but after only four years in medical practice he used his small family fortune as support while he devoted himself to examining the plight of the unfortunates in post-Napoleonic France – the weavers and cotton-spinners, the silk workers, the persons in prison. We shall follow this part of his career in more detail later.

Villermé was a humble man, very conscious that his education was far inferior to Quetelet's (although his experience of field-work was incomparably greater). He was an admirer and encourager of the younger man's work and 'if he disagreed he said so with a straight-forwardness and honesty for which he was known amongst his contemporaries' (Ackerknecht, 1952: the best paper on Villermé). Over the period 1823–62 Villermé, otherwise a poor, almost negligible correspondent, wrote eighty six letters to Quetelet, including the last one of his life. Quetelet, in his turn, was always prompt to acknowledge the debt he owed to Villermé, referring to him in the second edition of *Physique sociale* (1869*a*) no less than forty-eight times, twice as often as to any other author. One strong interest they had in common was the analysis of the causes of crime. Quetelet devoted much time and print to showing how regular were the statistics for the occurrence of crimes each year, and how each variety of crime had its specific age, sex and district prevalence. Indeed this led to one of his few disagreements with Villermé, who seems to have been a little shocked at the temporal regularity and uncertain whether its exposition might not imply on Quetelet's part an unwelcome materialism.

Quetelet and his wife were in Italy at a congress when the revolt of 1830 that separated Belgium from Holland broke out. The half-completed Observatory building was turned into a fortress and it was not until 1832 that Quetelet was able to move in and begin his astronomical and meteorological work. In the meantime he carried out the first cross-sectional population survey of children ever made. Altogether he did two separate surveys in 1831–2, the first of height only, the second of height and weight. Probably he was urged to do them by Villermé, who had only two years before published his classic analysis of the effects of poverty on the height of French recruits (see below). But he also had artists very much in mind, as the first twelve pages of his 1833 *mémoire* to the Brussels Academy show. Before presenting his tables of weight for age, he quotes Lavater, Gall and Rubens as well as Buffon, and explains at length the usefulness of the characterization of the *homme moyen* for art as well as for science. Quetelet never returned again to the study of children in the context of a population survey; after 1832 births, deaths, marriages, criminology and the search for ideal proportion occupied that portion of his mind that was concerned with statistics.

The results of the surveys were first reported to the Brussels Royal Academy (Quetelet, 1831, 1833) and subsequently reproduced in Quetelet's first and most important book, *Sur l'homme et le développement de ses facultés*, published in 1835 in Paris and seen through the press there by Villermé. The book won immediate acclaim; a German translation quickly followed in 1838 and an English one, made in Edinburgh under the direction of Dr R. Knox, lecturer in anatomy, in 1842. (The original English translation is now rare, but the whole book has recently been reprinted in Wall, 1973.) Quetelet always referred to the book as *Physique sociale*, or *Social Physics*, and the second, enlarged edition, issued in 1869, had this as its main title. Florence Nightingale thought it should be the basis of an honours course at Oxford for future administrators and Members of Parliament. A charming letter from her to Quetelet is appended to this chapter.

Before we turn to the somewhat complex task of unravelling the results of the surveys and placing them in a modern context, it is as well to continue the consideration of Quetelet's underlying, more philosophical, thoughts. As Hilts (1973) says, in a valuable essay contrasting especially the attitudes of Quetelet and Galton, 'Quetelet's lasting reputation is based upon his discovery that human stature is distributed in many populations according to the law of error. This was the first indication that the error distribution which had been used by Laplace [in 1812–14] and Gauss [contemporaneously with Laplace] in connection with errors of astronomical observations had a much more general application. Several historians have already realised, however, that Quetelet never broke with the traditional concept that the law of error really is a law of error and not a more general law of distribution'. This, as we shall see, was the great difference between Quetelet and Galton. Hilts attributes Quetelet's myopia in this respect to the hold that his concept of *l'homme moyen* obtained over him. Certainly it was powerful, almost mesmeric. 'Of the admirable laws which nature attaches to the preservation of the species, I think I may put in the first rank that of maintaining the type . . . The human type, for men of the same race and of the same age, is so well established that the differences between the results of observation and of calculation, notwithstanding the numerous accidental causes which might induce or exaggerate them, scarcely exceed those which unskilfulness may produce in a series of measurements taken on an individual' (Quetelet, 1849, p. 92).

This rather cryptic statement seems to imply that the errors of fit to the normal curve of an empirically found distribution are hardly greater than measuring error. What Quetelet does not grasp is that the two are related only when sampling error is taken into consideration. The errors of fit depend on the numbers entering into the empirical distribution; only if the number at each point is infinite are the errors of fit reduced to errors of measurement. Quetelet never distinguished sampling error from distributional variation. At that time what we call the standard deviation was called error, and the distinction, to us elementary, between the standard error of the mean and the standard deviation of the distribution did not occur to him. It was Galton who, realizing the difference, began to feel uncomfortable with the existing terminology and Karl Pearson who in 1894 actually introduced the term 'standard deviation' (see the excellent account of the history of statistics by Walker, 1929).

Galton was equally impressed by the law of error and described it in terms the poetical Quetelet would surely have admired. But instead of fixing his attention on the central tendency, Galton let his eye travel over the whole range of the curve. He wrote:

> I know of scarcely anything so apt to impress the imagination as the wonderful form of cosmic order expressed by the 'Law of Frequency of Error'. The law would have been personified by the Greeks and deified, if they had known of it. It reigns with serenity and in complete self-effacement amidst the wildest confusion. The huger the mob, and the greater the apparent anarchy, the more perfect is its sway. It is the supreme law of unreason. Whenever a large sample of chaotic elements are taken in hand and marshalled in the order of their magnitude, an unsuspected and most beautiful form of regularity proves to have been latent all along. The tops of the marshalled row form a flowing curve of invariable proportions, and each element, as it is sorted into place, finds, as it were, a pre-ordained niche accurately adapted to fit it (Gatton, 1889, p. 66).

Quetelet's discoveries about the applicability of the Normal curve were expressed in his second important book, *Lettres . . . sur la théorie des probabilités*, published in Brussels in 1846 and in translation in London in 1849. This small book, with another, *Du système social et des lois qui le régissent* (1848), grew out of the lessons he gave as private tutor to the two Princes of Saxe-Coburg in the period around 1837. He dedicated the first to the older brother Ernest, by that time Duke of Saxe-Coburg himself, and the second to Albert, who was to become Prince Consort to Victoria. It is uncertain to what extent the government of the Duchy was affected by Quetelet's book celebrating the Rule of Number; but one may well imagine that the idea of the *homme moyen* as *centrum mundi* struck an answering chord in Albert's breast. *Lettres sur la théorie des probabilités* had a considerable influence on Galton, who thought highly of Quetelet. It also attracted such a favourable and perceptive unsigned article in the *Edinburgh Review* in 1850 that Quetelet, discovering seven years later that its author was Sir John Herschel, published it in translation as an introduction to the second edition of *Physique sociale* and dedicated to Herschel his last work *Anthropométrie* (1870).

The *homme moyen* was, of course, that fictive individual who had the average value. But Quetelet distinguished clearly between averages proceeding from any old range of measurements (which Herschel proposed to call 'averages') and averages which proceeded from a distribution of measurements looking like the law of error. The latter he called '*moyens*' and Herschel 'means'. Hence the difficulty in translation: 'the mean man' has an awkward *double entendre* in English and, given Quetelet's high opinion of the *homme moyen,* the *entendre* is not made any clearer by Galton's perfectly reasonable use of the word 'mediocrity' to characterize aspects of the median. Thus the translation usually stands as 'the average man'. In *Sur l'homme* Quetelet wrote 'If the average man were completely determined, we might consider him as the type of perfection; and everything differing from his proportions would constitute deformity and disease; everything found dissimilar, not only as regards proportion and form but as exceeding the observed limits, would constitute monstrosity' (Quetelet, 1842, p. 99). 'An individual who combined in his own person all the qualities of the average man would ... represent all which is grand, beautiful and excellent' (p. 100). (Herschel, in a note added to his review, remarked simply that if this were the case the highest degrees of beauty would constitute the most common category of people, a conclusion absolutely contrary to experience).

Predictably, Quetelet gets lost trying to define a Great Man, and ends up saying he must be representative of his age in the sense of being comprehended by the masses. In this he slips sideways on the literary and statistical uses of the word 'representative': Goethe and Gauss have got inextricably intertwined. Elsewhere, in discussing limits, he writes 'One of the most interesting observations which I have had occasion to make is that they [the limits] narrow themselves through the influence of civilization, which affords, in my eyes the most convincing proof of human perfectibility' (Quetelet, 1842, preface).

Quetelet says there are three sorts of causes of variations around the mean: constant, variable and accidental. Variable causes are simply constant causes linked to changes in time, such as seasons, so have little theoretical importance. In astronomy the constant causes could be determined by finding the value at the centre of a range of observations subject to error, and Quetelet thought the same applied to man. It was the old error of belief in a universal constant; Darwin was, of course, a contemporary, but the *Origin of Species*, with its emphasis on variation in biological systems, only appeared in 1859, twenty-four years after Quetelet's first edition of *Sur l'homme* and when he was already ailing. 'The fixity [of the human type] is not one which cannot be influenced by constant causes', Quetelet wrote in *Anthropométrie* (p. 26). 'The Laplander is less tall than the ... Patagonian, but the law of accidental causes is no whit effaced. It simply happens that in each country the oscillations of height are made around a greater or lesser mean, and they are determined by the influence of climate, differences in nutrition and greater or lesser amounts of labour (*des fatigues plus ou moins grands*)'. He has absolutely nothing to say about children resembling their parents, any more than had Villermé (1829) in his 'Mémoire sur la taille de l'homme en France'. The omission may seem to us extraordinary, for family resemblance was surely a matter of common observation. Yet the history of academic attitudes to facts of common observation which fail to fit current theory is a continual one, and our own scotomata will doubtless appear equally wilful to our successors. Quetelet's silence does, however, show in its full light the originality of Galton, Boas, Pearson and their successors.

In 1855, when Quetelet was at the height of his powers, shining with a brilliance unequalled in Brussels for decades to come, he suffered a severe stroke. His biographer Mailly (1875) was present when it happened and describes the immediate and long-term effects with sympathetic honesty. Quetelet turned over the directorship of the Observatory to his son Ernest; he continued writing and was punctilious till his death twenty-one years later in his duties as secretary of the Academy. But his memory was damaged and his assistants had to help with his articles. The scientific and sociological world continued to heap honours upon him; and he recovered sufficiently to enjoy them, in particular the visit to St Petersburg referred to in Florence Nightingale's letter (see p. 141). Not long after that, in 1874, he died.

Quetelet's height and weight surveys

We now turn to Quetelet's measurements of children. The original series was collected in 1830–1. Quetelet had the lengths of fifty male and fifty female newborns at the Foundling Hospital in Brussels measured, using 'M. Chaussier's mécomètre' (see note 6.3). In *Sur l'homme* he gives frequency distributions and remarks, it seems to be for the first time, that boy babies are longer, as well as heavier, than girls 'by a trifle less than half an inch'. He continues: 'By adding these numbers to those which have been obtained in the junior schools of Brussels, the Orphan Hospital, boarding-houses and in public life, in respect of persons of different classes, I have been able to construct the following table, comprising the rate of growth from birth to 20 years; the height of the shoe is not included' (1842, p. 58). He adds that 'apparently but little interest is attached to the determination of the stature and weight of man or to his physical development at different ages . . . I do not think before Buffon any inquiries had been made to determine the rate of human growth successively from birth to maturity; and even this celebrated naturalist cites only a single particular example' (1842, pp. 57, 58).

The actual numbers of children that Quetelet had available in constructing his table (1842, p. 58: reproduced here as Table 6.1) are never stated. Judging by the fluctuations of the means and the absence of any measurements at all of boys aged 7, they were not large (see note 6.4). A footnote acknowledges the help of three people, perhaps the measurers. Quetelet had at first written: 'in order to succeed we must study the masses with a view to separating from our observations all that is fortuitous or individual. Everything being equal, the calculation of probabilities shows that we approach the nearer to the truth in direct ratio to the number of individuals' (1842, p. 7); but later, in *Anthropométrie*, he wrote that he had taken thirty individuals of the same age, regularly proportioned, and formed three groups of ten each. The differences between the means of the three groups were so small 'as to be less than the differences obtained by measuring a [single] model three times' (cf. discussion p. 127). In later studies he therefore took only ten persons of each sex and age: however there is no evidence that, as Roberts and Bowditch believed (see note 11.2), he was already doing this in the height and weight surveys.

Whatever the numbers involved, there is certainly something very odd, indeed almost unique, about this first Quetelet series. Due to girls having their adolescent growth spurt two years earlier than boys there is a period, in modern European and North American children from about 11.0 to 13.0 years, when girls, on average, are taller than boys. In modern malnourished populations the period of girls' ascendancy, as it is called, occurs later, around

13 to 15 years, since growth in both sexes is slower. But in Quetelet's data girls are at no time taller than boys. At ages 13 and 14 there is merely a diminution of the boys-larger difference to 1.4 cm, compared with 4.8 cm and 4.4 cm at 11 and 12 and 5.3 cm, 8.2 cm and 8.7 cm at ages 15, 16 and 17. The oddity can scarcely be ascribed to the poverty of the children, for Cowell and

Table 6.1. *Quetelet's data on height and weight*

	First series		Second series				Fitted curves	
	Heights		Heights		Weights		Heights	
Age (yr)[a]	Boys	Girls	Boys	Girls	Boys	Girls	Boys	Girls
0	50.0	49.0	49.6	48.3	3.20	2.91	50.0	49.0
1	69.8	—	69.6	69.0	10.00	9.30	69.8	69.0
2	79.6	78.0	79.7	78.0	12.00	11.40	79.1	78.1
3	86.7	85.3	86.0	85.0	13.21	12.45	86.4	85.2
4	93.0	91.3	93.2	91.0	15.07	14.18	92.8	91.5
5	98.6	97.8	99.0	97.4	16.70	15.50	98.8	97.4
6	104.5	103.5	104.6	103.2	18.04	16.74	104.7	103.1
7	—	109.1	111.2	109.6	20.16	18.45	110.5	108.6
8	116.0	115.4	117.0	113.9[b]	22.26	19.82	116.2	114.1
9	122.1	120.5	122.7	120.0	24.09	22.42	121.9	119.5
10	128.0	125.6	128.2	124.8	26.12	24.24	127.5	124.8
11	133.4	128.6	132.7	127.5	27.85	26.25	133.0	129.9
12	138.4	134.0	135.9	132.7	31.00	30.54	138.5	135.3
13	143.1	141.7	140.3	138.6	35.32	34.65	143.9	140.3
14	148.9	147.5	148.7	144.7	40.50	38.10	149.3	145.3
15	154.9	149.6	155.9	147.5	46.41	41.30	154.6	149.9
16	160.0	151.8	161.0	150.0	53.39	44.44	159.4	153.5
17	164.0	155.3	167.0	154.4	57.40	49.08	163.4	155.5
18	—	156.4	170.0	156.2	61.26	53.10	165.8	156.4
19	166.5	157.0	170.6	—	63.32	54.46	166.9	156.9
20	—	157.4	171.1	157.0	65.00	55.08	—	157.2
25 or 'terminated'	168.4	157.9	172.2	157.9	68.29	55.14	168.4	157.9

Heights (cm) of first (1831) series are from Quetelet (1842, p. 58). Heights (cm) and weights (clothed, [c] kg) of second (1832) series are from 1842, p. 64. Fitted curves are from 1842, p. 65.

[a] Centring of age not certain for empirical data but very probably 1.0, 2.0 (1 ±, 2 ± etc.); for fitted data certainly 1.0, 2.0 etc.

[b] Presumably misprint for 115.9.

[c] Quetelet says clothing accounts for about one-eighteenth of weight.

Stanway's contemporaneous sample of working children in England (see next chapter) shows the girls' ascendancy with the usual clarity, at ages 12, 13 and 14. Something has gone wrong, it seems, with Quetelet's sampling. He does mention in his first report on the series (1831) that the girls from 7 to 20 were all from the Hospice des Orphalines and also warns that as their numbers were less than those of the boys, less confidence should be placed in their results. It is true there had been a famine in Brussels in the years 1816 and 1817, and this could conceivably have affected the heights of adolescents in 1831; but in such circumstances boys are almost always worse affected than girls, which would produce the opposite result to that obtained (see note 6.5). As to the velocities, the girls have a fairly clear peak at 7.7 cm in the 12–13 interval. The boys' peak is scarcely to be seen, however; the increments are successively 5.4, 5.0, 4.7, 5.8, 6.0, 5.1 and 4.0 cm in the intervals 10–11, 11–12, 12–13, 13–14, 14–15, 15–16 and 16–17 respectively. The small and uncertain peak is located in the 13–14 and 14–15 intervals. Quetelet says absolutely nothing about these increases in rate at puberty, despite his acquaintance with Buffon's work. He merely observes that the average rate of growth of girls from 5 to 15 is 5.2 cm/yr, lower than that of boys which is 5.6 cm/yr.

Furthermore, in determining the age of peak height velocity for our own comparative purposes, a difficulty arises, often met with in the past and sometimes even now (see Eveleth and Tanner, 1976, p. 7). It is unclear whether Quetelet's ages refer to age at *last* birthday or age at *nearest* birthday; that is whether the age marked 13 includes children from 13.0 to 14.0 or children from 12.5 to 13.5. Modern surveys use the former system, with centres of age classes at 13.5 etc., but it seems much more probable that Quetelet used age at nearest birthday, centring at 13.0 etc. This follows from a comparison of his empirical and fitted curves (1842, pp. 61, 62 and see below). For the fitted curves 9 years should certainly represent exactly 9.0, and this is confirmed by the manner in which (appendix, p. 115) he compares the fitted Belgian curves with values derived from Horner's English data. Horner gave his values in half-yearly intervals and 'in order to get the height of a child of 9 years of age, we have taken the mean of the child's height in the age between 8.5 and 9 years and the height of the age between 9 and 9.5 years etc.' Quetelet writes, and when he compares empirical and fitted values he simply writes 9 years for each. (This use of age at nearest birthday was in fact fairly general at this period and for some time later. We shall meet it again in studying age at menarche in Chapter 11.) The modest peak shown by Quetelet's boys of the first series is most probably located, therefore, at 14.5. The girls' peak is at 12.5 years. Both are early in comparison with contemporary English data and indeed with data on schoolchildren later in the century. One wonders whether sampling problems have not intervened here too. About the age of the peaks, Quetelet himself is naturally silent, since he is not taking notice of the spurts, now or, as it turns out, ever. He remarks, however, that a girl at 16 is as relatively advanced in height as a boy at 18, something already noted by Bergmüller.

Quetelet goes on to discuss the possible differences in growth between urban and rural children. 'Already Dr Villermé has proved, contrary to the generally received notion [cf. Bacon, p. 26] that the inhabitants of towns are taller than those of the country. I have arrived at the same conclusion in respect of the inhabitants of Brabant' (1842, p. 59). This time he gives the numbers of his subjects, 19-year-old conscripts, and they are substantial. 'It could still happen', he very correctly remarks, 'that the inhabitants of the country might attain a greater eventual height than the inhabitants of the town if those in the town were closer to ending their growth at 19'. Villermé was of the same opinion about slowness of development in the country and Quetelet quotes him on poverty causing delay in, as well as stunting of, growth (the quotation is given below on p. 162). To determine when growth in stature in man ceased, Quetelet (1830) examined the Brussels conscript registers made about 1815 and found the mean height of 19-year-olds was 166.5 cm, of 25-year-olds 167.5 cm and of 30-year-olds 168.4 cm. He concluded that growth did not terminate in Brussels 'even at the age of 25, which is very much opposed to general opinion' (1842, p. 59).

After quoting the figures for Manchester and Stockport children that Cowell sent him (see below) – which show, incidentally, an absolutely clear spurt in boys at the 14–15 interval and a less clear one in girls in the 13–14 interval, – Quetelet fits a curve to his succession of means. This was the first time a mathematical expression had been used to express growth data.

> I have endeavoured [he says] to render the preceding results *sensible* [his italics] by the construction of a line which indicates the growth at different ages. We see that . . . the growth of a child diminishes as its age increases until towards the age of four or five years . . . Proceeding from the fourth or fifth years the increase of stature becomes almost exactly regular until about the sixteenth year, that is until the age of puberty . . . After the age of puberty the stature continues to increase but only inconsiderably . . . The curve representing the growth of females would be a little under that of males and would be nearly equidistant from it, until the age of 11 or 12 years, when it tends rapidly to become parallel to the [horizontal] axis . . . It remains for me to speak of the formula by which I have calculated the numbers shown in the table above (1842, p. 61).

Changing Quetelet's terminology to one consistent with our usage elsewhere, the formula reads:

$$h + \frac{h}{100\,(h_{max} - h)} = at + \frac{h_0 + t}{1 + 4t/3},$$

where h is height in metres, h_0 height at birth, h_{max} height at maturity, a a constant and t age in years. (For Brussels males Quetelet had $h_0 = 0.500$; $h_{max} = 1.684$ and $a = 0.545$; for females $h_0 = 0.490$; $h_{max} = 1.579$, and $a = 0.520$.) This is a curve which gives a continuously falling velocity of growth with deceleration fast at first, then slow, and faster again towards the end (see Fig. 6.1). Using the fitted curve, Quetelet gave a table of growth. He thus enshrined in a rule, as did Bergmüller, the monotonically descending velocity of growth, obliterating the adolescent spurt and confusing a number of later workers, some of great distinction in fields other than human auxology. D'Arcy Thompson used the table in the first edition of his famous *On Growth and Form* (1917), but in the revised edition (1942) substituted Quetelet's original empirical means, showing the spurt, probably because in the meantime Scammon had resurrected Montbeillard's curve. As late as 1945 Brody, whose book is a mine of information on the growth of farm animals, wrote 'The pre-pubertal acceleration, so conspicuous in the literature on growth of children, is usually found only in the curves of poorly nourished children' (legend to his figure 16.24, p. 511; also p. 539). Though Brody's figure shows the schoolchildren studied at the Horace Mann School by Baldwin (see p. 301) this opinion certainly does not derive from Baldwin, but is practically word-for-word Quetelet (see below) – who, however, is not actually quoted in Brody's book.

Those who themselves measured children, though, soon realized that something was wrong. Already in 1879 Dally, the author of the article on 'Croissance' in the authoritative *Dictionnaire encyclopédique des sciences médicales*, set forth in contrast the recent findings of Bowditch in Boston and Pagliani in Turin, both of whom found adolescent spurts. Though he found it impossible to say Quetelet was wrong ('it is all the same very vexing that nobody has confirmed the perfect regularity of his tables of measurements': p. 379), he does leave it to the reader, he says, to decide who is right. Bowditch (see p. 188) in the 1880s rejected Quetelet's smoothed growth tables and Stratz (1915) wrote: 'Like Weissenberg [1911] I have come to the conclusion that Quetelet's tables are so far removed from actual findings that one cannot consider them any longer as authoritative.'.

Quetelet and his associates made a second survey, in which both heights and weights were obtained (see Table 6.1). In estimating weight he 'generally used the balance of Sanctorius', though due to its inability to measure small weights, young children were weighed together with an adult. Again the sample sizes are unstated, but it seems that the source of subjects was the same as in the first survey up to age 12 (schools in Brussels and the orphan hospital). From 12 upwards most of the subjects were from the better-off sections of society, and it is to this that Quetelet attributes the fact that the

Fig. 6.1. Quetelet's 'growth table of height', that is, his fitted curve, plotted as velocity on modern British charts of Tanner and Whitehouse (1976). As a contrast, the velocity curve of Quetelet's son Ernest is also shown, illustrating his adolescent growth spurt.

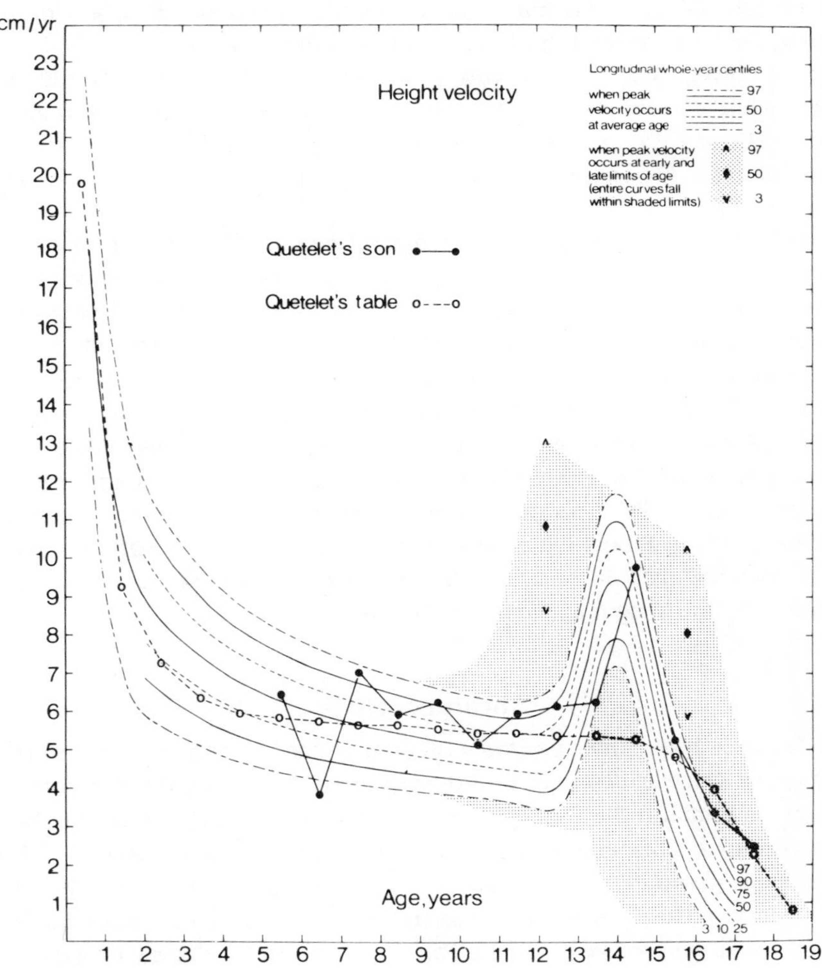

heights of boys from 15 upwards were greater than in the first survey. In giving a general curve for weight he wished to eliminate this bias, and thought to do so by observing that 'for individuals of the same age, the weight may be considered as having a pretty constant relation to the size of the body [i.e. height]. It will be sufficient then, to know the ratios . . . and to have a good general table of growth, to deduce the corresponding table of weight' (1842, p. 64). For the 'general table of growth' he takes the *fitted* height values previously obtained and applies to them, at each age, the values of the weight/height ratio obtained in this new survey, and thus obtains his weight table. Thus the weight tables as well as the height tables on his p. 65 are smoothed. Present-day workers wishing to compare Brussels weights with modern values should use the empirical means found in the p. 64 tables, bearing in mind the selection (and that he used clothed weight in these empirical tables, as did Cowell). The heights of this second series, incidently, show an adolescent spurt equally distributed in the 12–13 and 13–14 intervals in girls (thus 13.0 at peak) and a much clearer spurt in boys than in the first series, in the intervals 13–14 and 14–15 (thus 14.0 at peak). Once again Quetelet ignored this.

Quetelet also made studies of strength of muscles, and of heart rate and rate of breathing (1835, chapts. 3 and 4). Using Regnier's dynamometer he measured strength of hand grip, and of upward pull by straightening the back ('lumbar power'). For this investigation (the first tests of strength to be done in children) he used just ten boys and ten girls of each year of age from 6 upwards, thus already exemplifying the procedure he was at pains to justify in *Anthropométrie* later. His mean strengths show a considerable adolescent spurt in the 14–15 year interval in boys' grip; a less clear spurt in girls occurs a year earlier. This work stimulated considerable interest and Forbes (1836, 1837) in Edinburgh measured height, weight and strength of some 800 students aged 14 to 25, confirming Quetelet's height curve and the superiority in strength of Scots over Belgians.

Anthropométrie

In his later book *Anthropométrie* (1870), published only four years before his death, Quetelet describes a third series of measurements, this time made purely to establish the laws of development of human proportions. Thus he returned to this first and constant love. Again using only ten subjects in each yearly age group he took no less than eighty measurements, eighteen being of the face and head. He gives tables of means of the actual measurements and of each as a percentage of height. This therefore was the first series of measurements of children concerning more than height, weight and one or two other dimensions. Nothing is said about selection (except that he only took children of regular build), and as one looks at the means it becomes clear that the omission is important. The biacromial diameters of the boys have the following increments from 10–11 to 19–20: 1.1, 1.1, 1.1, 1.1, 1.0, 1.1, 0.9, 0.9, 0.9, 0.6 cm. No ordinary series of biacromial diameters could so almost totally fail to show the spurt. Indeed increments based on means from ten randomly selected subjects, different at each age, could scarcely be so regular. Either the sample is very special or else Quetelet has smoothed the means, employing something like his height growth function to do so. The height increments over the same ages are: 5.2, 5.0, 4.8, 4.6, 4.4, 4.1, 4.0, 3.6, 2.5, 1.4 cm. The girls' height increments show the same type of curve.

Quetelet senses there is something wrong, or something missing. 'At this time of puberty . . . there is a considerable change in the development of the human body', he writes (1870, p. 176). 'This change comes about by quick,

Table 6.2. *Quetelet's measurements (cm) of his son Ernest, his daughter Isaure, and the two daughters of friends. Montbeillard's son's measurements are also included, as given by Quetelet*

Age (yr)	Montbeillard's son	Ernest Q.	Isaure Q.	Antoinette	Amélie
0	51.4				
1	73.1				
2	90.0				
3	98.8				
4	105.3		90.3		
5	111.7	102.5	94.6		
6	117.9	109.0	99.6		
7	124.4	112.9	108.3		
8	129.9	120.0	116.7		
9	137.0	126.0	121.1		
10	141.9	132.3	128.4	126.0	
11	145.4	137.5	133.2	131.0	
12	148.8	143.5	141.8	137.5	129.8
13	155.3	149.7	146.0	143.5	132.0
14	162.9	156.0	153.5	152.0	—
15	175.0	165.8	159.0	157.4	—
16	180.0	171.1	159.5	159.4	152.4
17	184.5	174.5	—	161.0	155.6
18	188.0	177.0	—	—	—
25	—	179.5	—	—	—

Reproduced from Quetelet (1870), p. 185.

sudden and unforeseen transformations which do not take place at exactly the same age in all individuals, which is what makes difficult the appreciation of the size of the changes'. He sticks to his point, though, and in summarizing, says: 'Growth becomes less rapid the further one moves away from the time of birth' (p. 180). 'In considering a particular individual ... there are nearly always periods of arrest ... as also of more or less rapid growth. These anomalies are observed around the age of puberty and above all following illnesses ... When one works on a larger number of persons these little anomalies disappear in the general mean and what lacks in the development of the one is compensated by an excess of growth in another' (p. 183). In his first article Quetelet (1831) actually says that Montbeillard's son's slowing down at 10–12 followed by a reprise must be due to a disorder (*dérangement*) of growth, frequent at puberty and indeed more often seen in girls than boys (p. 112). At no time – except in a long quote from Buffon (p. 201) – does Quetelet mention anything about menarche either in individuals or *en masse*, nor about other signs of puberty (see, however, note 6.6).

There is, perhaps, a further or personal reason for Quetelet's unease. Partly because of Buffon, and partly perhaps because Dr Eduard Mallet (1805–1856), a distinguished analyser of the Geneva archives of births, deaths and marriages, had criticized him, when reviewing the 1832 edition of *Sur l'homme*, for not taking the measurements at successive ages in the same subjects, Quetelet made measurements of his own son and daughter from age 5 to maturity, as well as of the two daughters of a friend. The measurements are reproduced here (Table 6.2) just as he presented them, alongside those of Montbeillard's son (1870, p. 185). In Fig. 6.1 Quetelet's son's velocity curve is plotted, together with the velocity curve for boys from Quetelet's fitted 'table of growth', against the background of present-day British longitudinal standards. Ernest Quetelet shows a clear adolescent spurt at the expected age and of the expected amplitude. Quetelet's daughter Isaure's curve is less regular (though with an identifiable spurt at 13–14), but Antoinette, the friend, has an absolutely regular velocity curve, with the peak at 13–14, of amplitude 8.5 cm/yr. Yet even with these data in front of him, the oversimplified dream refused to fade; *l'homme moyen*, at least, should have a growth curve of perfect regularity and impeccable form.

J. H. W. Lehmann

Only one contemporary voice was raised in contradiction to Quetelet's views on adolescent growth. Rather surprisingly, for it belonged to a well-known astronomer and mathematician, it went almost totally unheard. The protester was J. H. W. Lehmann (1800–1863), who was born in Potsdam, graduated in both mathematics and theology, and took his PhD at Göttingen with Gauss in 1822. After teaching for a few years at the *Gymnasium* in Berlin he gave precedence to theology and spent most of his life as a pastor in the Brandenburg towns of Derwitz and Krilow. However, he continued his astronomical interests, and became one of Europe's leading experts on the paths of comets. He evidently first made his acquaintance with Quetelet's work through an article, contributed by Quetelet himself (1839), to a short-lived *Jahrbuch* series published from 1836 to 1844 by H. C. Schumacher, another, and very distinguished, astronomer. Quetelet had given his table of heights and weights, using the fitted mean curves. Lehmann at once realized what was wrong, and in two lengthy and detailed articles in the same *Jahrbuch* (Lehmann, 1841, 1843*b*) proposed a different curve consisting of two hyperbolae, the second riding on the back of the first, and representing the adolescent spurt. Lehmann fitted his curve to serial measurements of individuals, being the first person ever to do this.

Lehmann had access to measurements that Quetelet lacked. 'One finds', he wrote, 'in many families of the highest and lowest stations, in north as well as south Germany, measurements of individuals taken serially at 1, 2, 3 and up to 20 years, inscribed on a door-jamb. When measured later with an accurate rule, these permit us to discern a law, which only needs a little mathematical theory to bring it to exactitude' (Lehmann, 1844, p. 300). One such record, made from 1809 to 1823, on a boy aged, as he puts it, 9.934 to 23.822 years, was clearly of Lehmann himself, for his use of the decimal notation for age permits identification of the boy's birthday. Another record, the most complete of his collection, concerns measurements, this time not only of height but of hand length and arm width also, taken every Christmas Day from 1824 (the subject aged 1.238) to 1840 (aged 17.238): he was probably Lehmann's son. When Lehmann read Quetelet's work he increased his measuring activity, and in 1843, in a detailed account of his growth curve, he appended a table of the raw measurements of 122 different boys, most measured for two or three years, some longer, some only once (Lehmann, 1843*a*).

Lehmann was clear that a spurt of growth occurred in individuals at the beginning, as he put it, of puberty (*Mannbarkheit*), and he was clear, too, why it was that Quetelet had missed its occurrence in his cross-sectional data. He pointed out that K. F. Burdach (1776–1847), a leading physiologist, professor at Königsberg and author of an encyclopaedic textbook of physiology in the period 1826–38, had discussed the spurt in his textbook, where he wrote 'Growth during the first part of this period goes forwards very fast, and makes a new spurt (*Schuss*) of as much as four to six inches in one year'

(Burdach, 1838, p. 302). Curiously, neither Burdach nor Lehmann mentioned Buffon or Montbeillard.

More remarkable is Lehmann's criticism of Quetelet's methodology. 'No actual individual really grows according to this [Quetelet's] curve, which is throughout concave to the base and without a sharp angle-point. In the great majority [of boys] at the time of the beginning of puberty a sudden rapid spurt of growth occurs, which in the subsequent years fades away. Because the time of this spurt is very variable, taking place in some individuals at 10 years old and in others at 18 or 19, we cannot be astonished that in the mean curve the spurt completely vanishes. Really Quetelet . . . has not adduced a law for the development of human *individuals*; for he tried to discover in a twinkling that which takes nature 20 years and more' (1844, p. 300; his italics). Lehmann's two hyperbolae at first met in a sharp angle where the second takes off from the back of the first (in the pubertal spurt). He then modified this first diagram by smoothing out a little the angle, producing a curve indeed very similar to modern ones (Lehmann, 1843*a*). By fitting constants to these individual curves (he says by a least-squares method, but really the number of points available for most individuals was much too few for effective curve-fitting) he sought to predict final adult height as well as age at take-off for the adolescent spurt. Mostly, he thought, the spurt (called by now the *Hauptschuss*, or chief spurt) began between age 10 and age 16; but in some boys, especially those in the countryside, delay till 18 or even 18½ was 'by no means unusual'. Lehmann pointed out that Kepler had been one of the first to investigate the laws of the heart rate and Halley one of the first to construct Life Tables; he felt strongly that the astronomers' mathematics could be useful in biology and medicine and ended a talk to the Society of German Natural Scientists and Physicians by saying that the most important advances would be made by 'persons combining fundamental mathematical training with deep medical insight' (1844, p. 302).

Only one person subsequently ever referred to Lehmann. That was B. A. Gould (1824–1896), another astronomer and another of Gauss' pupils. An American from New England, Gould took his PhD at Göttingen twenty-six years later than Lehmann, in 1848. After the American Civil War he was persuaded to analyse the heights of soldiers in the army and in 1869 published *Investigations in the Military and Anthropological Statistics of American Soldiers*. He there referred to Lehmann's two articles in *Schumacher's Jahrbuch* and paraphrased Lehmann's clear explanation of the effect of the variation in time of spurt upon the presence of the spurt in cross-sectional data (see note 8.6). Gould's explanation was forgotten and had to be rediscovered by Galton and later by Boas (see Chapters 9 and 10).

Appendix. Letter from Florence Nightingale

Florence Nightingale wrote Quetelet a letter not long before his death, the substance of which is reproduced below. The original is in very correct French and beautifully written: it was found by Dr L. Martin (1951) of Brussels and brought to the writer's attention by Professor M. J. R. Healy. Its contents and tone bear out Karl Pearson's comment (1914–30, Vol. 2, p. 418): 'For her, Quetelet was the hero as scientist, and the presentation copy of his *Physique sociale* is annotated on every page'.

8 November 1872

Dear M. Quetelet,

I could not have been more touched and grateful for what you took the trouble to write to me, and for your good and kind letter of the 6th which arrived today.

It would be the greatest honour to receive from the illustrious author's own hand a copy of your *Physique sociale* as well as your *Anthropométrie*, as you have the extreme generosity to suggest . . .

However, I have a request which is much more weighty and important; it is for your *Physique sociale* not only for me but for all England. Would you make a second edition as promptly as possible, for the first is entirely exhausted. [F. N. clearly failed to understand that one of the books he was proposing to send her *was* the second edition] . . . I have no need to tell you, the author, inventor, Master of all the science of Statistics, and of all of us, of the extreme importance for the English legislature, the Statesmen, the members of Parliament, of becoming familiar with your research and discoveries early while still at University, for all the Science, I should say rather the Practice of Administration, Politics, Legislation, social Economics and Public Assistance should rest on the foundations which you have built . . . [She then says Oxford has enlarged its horizons and a friend of hers has become president of the Final Examination Board and that she has put in a word for social physics] . . . But there is an unforeseen difficulty. I have had all the bookshops in London asked; impossible to find a copy of *Physique sociale*. I made the same request in Paris with the same result; also in Brussels. The edition was exhausted. Finally they found me one copy, just one, in London. I had the glory of presenting the only copy of the famous work of M. Quetelet to the University of Oxford . . . it will be like the days when only having one book it was chained to the pulpit.

So my dear M. Quetelet, for reasons of social philanthropy, of which you are the foremost interpreter, you must not delay a single month, a single day, a single hour, if you can, in preparing a second edition .. I ask you in the name of humanity.

I asked our mutual friend Dr Farr, to tell you at St Petersburg of this need. I think he did so. But I see the need and not the result. Please write me a word, the kindest word ever written,.that you will not delay in setting the work in hand.

With much admiration.
Florence Nightingale

7

The rise of public health and the beginnings of auxological epidemiology

In the early part of the nineteenth century a new tradition of growth studies appeared, born of the reaction of humanitarians to the appalling conditions of the poor and their children. It developed amongst the conglomerate of Factory Legislation, Poor Law Commissions, and Sanitation and Housing Acts which embodied the new and powerful practice of public health. In Great Britain, its progenitors were such men as John Fielding, Jeremy Bentham, Robert Owen, John Howard and William Wilberforce; its pioneers were Senior, Chadwick and Farr. The studies of growth their activities initiated have continued in an unbroken though often tenuous chain through the measurements in schools and infant welfare clinics and the London County Council Surveys of 1905 to 1965, to the modern British growth standards and the National Survey of Health and Growth of the 1970s. We may call this activity auxological epidemiology, the use of growth data to search out, and later to define, sub-optimal conditions of health. In France, Villermé was one of the outstanding pioneers, and in Belgium Quetelet, a little younger than Senior and a little older than Chadwick, was a powerful early influence.

There was another, less humanitarian side to this development. As a result of the continual recurrence of European wars and the disruption, famine and epidemics that they entailed, statesmen in the eighteenth century became very much preoccupied with what we should call the manpower situation and the problem of human resources. Men were needed to fight wars, and at the beginning of the eighteenth century no country really knew how many men it had. Furthermore, men who were physically fit were needed, and as the censuses began to be done the authorities were appalled to discover how many men were undersized, consumptive and crippled, exactly as the American and British authorities were appalled by the same discovery in World War I (Anon., 1918; Ireland, Love and Davenport, 1919; Davenport and Love, 1920). Such men might give rise, it was thought, to inferior stock while the fit men were away engaging in more military pursuits. Indeed such men might give rise to no stock at all, and who would then fight the next war?

Sweden under Adolphus Frederick (1710–1771) established in 1748 the first regular census of a national population and the results were for some time regarded as an important state secret (Rosen, 1955: see note 7.1).

These worries gave considerable impulsion to the development of numerical methods of investigation and gradually turned men's minds towards the idea that 'political arithmetic', as it was named by William Petty (1623–1687), could be useful. All the same, the idea took nearly a century to become fully acceptable (see the valuable article by Rosen, 1955). In 1701 Arbuthnot (1667–1735), a younger member of the circle founded by William Petty and John Gaunt (1620–1674), wrote 'Those that would judge or reason truly about the state of any nation must go that way to work [i.e. the way of arithmetic], subjecting all the forementioned particulars to calculation,' (cited in Rosen, p. 28). The formidable Adam Smith, however, declared he had 'no great faith in political arithmetic' (Rosen, p. 28).

Gradually, the necessary mathematical bases were established. In the *Ars conjectandi*, published posthumously in 1713, the Swiss mathematician Jacques Bernoulli (1654–1705) laid the foundations of probability theory, with the theorem that an event occurring with a certain probability appears with a frequency approaching more closely that probability as the number of observations increases. In 1733 the Huguenot refugee De Moivre (1667–1754), working on games of chance in Soho, London (see note 7.2), published a technical paper that Walker (1929) calls 'The *fons et origo* of the Normal Curve'. Gottfried Achenwall (1719–1772), professor at Göttingen, replaced 'political arithmetic' by the term 'statistics' in 1749. In 1786 Laplace in Paris developed methods to solve the currently very important problem of whether variolation for smallpox (more or less begun in France, it will be remembered, by Montbeillard in 1766) actually was or was not effective in diminishing mortality; and he brilliantly used Bernoulli's theorem to deduce, from a study of the sex ratios of newborn infants in Paris and in the rest of France, that parents in the countryside sent a higher proportion of girls than boys to the Paris Hospital for Foundlings, where the majority perished (Hilts, 1973). Finally came the discovery by Laplace and Gauss, around 1810, of the curve of errors, and the extension of the curve to describe population distributions by Quetelet in the 1830s. Thus, just as a century before, the iatromathematicians had paved the way for the emergence of modern physiology, so now the statistical mathematicians paved the way for the emergence of public health.

In eighteenth-century Europe the condition of the poor, and especially of their children, was dreadful in the extreme. Descriptions of London, such as that given in such vivid and careful detail by George (1965), call to mind exactly the cities of the backward countries nowadays. Around the central urban core was a shanty-town of persons living in broken-down sheds, with few clothes, little food and less work. Abandoned and vagrant children were everywhere, and their mortality was such that, when from 1756 to 1760 the Foundling Hospital in Coram's Fields opened its doors to all homeless infants and children in London, out of 14,934 admitted just 4,400 survived to be apprenticed. In the Foundling Hospital in Paris the survival rate was about the same; in Dublin it was said to be about 10 per cent. In several London workhouses the survivors were even fewer (Pinchbeck and Hewitt, 1969–73, vol. 1, p. 181). In London, however, things began to improve about 1750 (George, 1965). Previously deaths there had far exceeded births; then gradually the burials began to decrease. In the countryside, where the improvement in conditions began perhaps a little earlier, the population started to rise about 1700, slowly at first and rapidly from the mid-century.

By the nineteenth century the surplus agricultural population was increasingly drawn into the new industrial towns of Lancashire and the North. To begin with, the spinning and weaving machinery was run from the shafts of millwheels, and the factories into which textile workers were concentrated necessarily remained in river valleys. Then after the introduction in 1813 (first in Stockport) of looms powered by steam, the textile factories crept out of the valleys and established themselves wherever coal was at hand and a supply of ordinary water was assured. The new towns thus created, built around the factories and lit by the new coal gas, were populated by immigrants from the countryside. By 1850 nearly half the English population lived in urban areas. Though hideous and cramped these towns at least provided wages for the workpeople, grown-up and otherwise. In the 1830s the Poor Law Board publicized widely the story of the peasant from Bledlow in Buckinghamshire with a family of nine children who by going to Lancashire raised his weekly income from 16s 3d to 41s 6d. Arthur Arnold, the contemporary historian of the cotton famine of the 1860s, remarked that it was uncertain whether the children benefited from the move as much as the parents; while in Buckinghamshire they might have received some education at the National School, in Lancashire they had to earn 70 per cent of the family income (quoted in Longmate, 1978, p. 34).

As for agricultural areas, where in the early part of the century the majority of the population still lived, there the economic situation was quite simply that of a present-day backward country, and it remained so well into the century, with much disguised under-employment, particularly in the southern counties. From 1795 to 1834 the wages of agricultural workers were so low that in the opinion of some economic historians they may have caused diminished productivity consequent on actual undernutrition (Blaug, 1964).

Burnet's (1966) history of nutrition in England paints a melancholy picture of much of the agricultural population as passing their lives in 'semi-starvation, existing on a scanty and monotonous diet of bread, potatoes, root vegetables and weak tea' with salt pork or bacon once a week and fresh meat practically never (p. 23). Furthermore this state of affairs continued in many parts till late in the century, the period 1830–50 being perhaps the worst of all.

When we see, therefore, that the growth of children of the manual labouring classes in England in the 1830s, and even the 1870s, was still more depressed than that of the poorer groups in some of the underdeveloped countries nowadays (Fig. 7.5 below) we should not be too surprised. Indeed children of slaves in the plantations of the southern states of America at this time (see p. 165) were taller (by some 2 to 5 cm) than contemporary children of manual labourers in England (although nowadays the heights of children of European and of African descent living under similar economic circumstances are almost identical: see Eveleth and Tanner, 1976). This also would have occasioned no surprise to the social critics of the 1830s. 'Writing about factory labour, Southey remarked in 1833 that "The slave trade is mercy compared to it" and Gibbins commented "The spectacle of England buying the freedom of the black slaves by riches drawn from the labour of her white ones affords an interesting study for the cynical philosopher"' (Pinchbeck and Hewitt, 1969–73, vol. 2, p. 407: see note 7.3). The growth curves fully bear out this comment (compare 7.1 and 7.2 with 7.7 and 7.8).

In the eighteenth century children worked, and worked far longer hours than any trade union nowadays would think permissible for adults. At the beginning of the century 'Public opinion', wrote Dr Marshall (1926) in her classic study of the English poor, 'strongly approved of the employment of children, and the ideas of a later age on the subject of education and the need to foster a child's self-development would have been met with blank incomprehension' (p. 24). Pinchbeck and Hewitt (1969–73) in their history of *Children in English Society* confirm this: 'From the days of Elizabeth onwards the labour of children was a social ideal explicitly encouraged both by the provisions for parish apprenticeships in the Poor Laws and also by the Statute of Artificers of 1563 —an ideal of which the extensive use of child labour in all types of industry was evidence' (p. 98).

By the eighteenth century, considerations of class had considerably diluted those of conscience. 'Few children', said Mandeville in 1723, 'make any progress at school but at the same time are capable of being employed in some business or other, so that every hour of those poor people spent at their books is so much time lost to Society. Going to school in comparison to working is Idleness and the longer boys continue in this easy sort of life, the more unfit they will be, when grown up, for downright labour, both as to Strength and Inclination. Men who are to remain and end their days in a laborious, tiresome and painful station in life, the sooner they are put upon it at first, the more patiently they will submit to it for ever after' (p. 328). Considerations of economics were also becoming paramount. 'The eighteenth century accounts of the children's workhouse movement bear eloquent testimony to the deterioration in social attitudes towards the children of the poor since the sixteenth century. Elizabethan reformers had set out in a burst of generosity and idealism to end poverty by education and training [in the Poor Law Acts of 1572, 1576 and 1597] ... When these early hopes failed, the mid-seventeenth century produced the children's workhouses, where the emphasis was to be on work, with only a modicum of education. Here too the original ideal of giving children some kind of skill was soon abandoned in favour of the principle of using the labour of children to reduce the Poor Rates. And when this could not be achieved because their earnings [from work done in the workhouse] were so small, children were left untrained and in idleness in the interests of 'economy'. From the end of the seventeenth century onwards, economy in relation to the children of the poor became the dominant idea in the minds of most administrators' (Pinchbeck and Hewitt, 1969–73, vol. 1, p. 175). Furthermore, confusion in the public mind of the problems presented by vagabond children on the one hand and by able-bodied poor adults on the other resulted in the low work output of children being ascribed to idleness and general incorrigibility. 'The opinion that the problem of training the young in the habits of industry must be solved, if the Poor Rates were to be reduced, was shared by all writers', said Marshall (1926, p. 25).

By the end of the eighteenth century this attitude had been greatly modified. It was the time of Rousseau's 'social contract', of Howard's advocation of prison reform, of Wilberforce's campaign to abolish the slave trade and, a little later, of the social and educational projects of Robert Owen and George Birkbeck. Children had their champion too, in the person of Jonas Hanway (1712–1786), the inventor of the umbrella, who from 1750 to 1770 continually complained to the House of Commons about the state of children in workhouses, which he called 'The greatest sinks of mortality in these kingdoms'. In the end his strictures led to complete discredit of the workhouse as an institution. He also got inserted into the Poor Law Act of 1767 a provision that 'the absurd tyrannical custom' of binding parish boys (i.e. orphans supported by the Poor Law Rates) to apprenticeships until the age of 24 be modified so they were free at 21.

Factory children: the first survey

Children, however, whether parish or not, sometimes began work in factories and even the mines at the age of 5 or less, and entry at age 8 was the usual thing. This is not as surprising as it might now seem. Even tiny children had for centuries worked in their homes at such things as cotton processing and, 'when, due to later inventions, the textile industry moved from the home to the factory, the children went with it. Small children of only three or four years of age were employed to pick up cotton waste, creeping under unguarded machines where bigger people could not go. The older children worked for fifteen hours a day, and on night work too, under conditions which were often enforced by fear and brutality' (Pinchbeck and Hewitt, 1969–73, vol. 2, p. 354). We should remember too, in passing, that conditions of agricultural labour for children were scarcely better; indeed the physical demands pass present-day belief, especially in the gangs system, in which a group of children and adults under an overseer would be sent to a farm with work to be done, the task often involving seven- or eight-mile journeys there and back on foot, usually before dawn and after sunset (see Pinchbeck and Hewitt, vol. 2, pp. 391ff.) Gradually a series of Acts was passed which mitigated these conditions and it was in connection with these Acts that the first large-scale studies of children's heights and weights were made.

The most important document in this development was the *Report of the Commissioners on the Employment of Children in Factories* (Parliamentary Papers, 1833, English Historical Documents, 1956–, vol. 12 (1962), pp. 934–949). The report resulted from an inquiry into 'the whole subject of the labour of children, as now enforced in the various Mills and Factories or places of work throughout the country' (Parliamentary Papers, 1833, p. 79). Edwin Chadwick (1800–1890), in his youth secretary to Bentham, and the chief architect of public health reform in England, was one of its signatories. With the *Instructions to Members of the Commission* we are quite suddenly in a new and wholly up-to-date world. Not only is the inquiry thorough-going, requesting data on morbidity, mortality, stillbirth and illegitimacy rates, accidents and so forth; it also lays great emphasis on investigating a suitable control population and on sampling by domiciliary visits to whole streets. 'A given amount of evil is experienced by a class placed under peculiar circumstances —a large portion of that evil is shared by other classes not under these peculiar circumstances; to attribute the whole of the evil experienced by the first class to those peculiar circumstances is obviously fallacious' it says (p. 84). In each area a Medical Commissioner was appointed and, amongst other things, he was specifically required to ascertain the stature of the children to see 'whether there be any difference at any age, and what age, and in either sex, between persons brought up from an early period in a Factory and persons of the same age and sex and station not brought up in a Factory'.

There is no indication of what were the antecedents of this request for height measurement; presumably Chadwick was in contact with the work of Villermé and his associates in France and knew of the 1829 *mémoire* on adult height and the conclusions Villermé had drawn from it. Very likely he knew also of Quetelet's survey of children's heights in Belgium, which was done two years earlier (see note 7.4). Quetelet must certainly have met Chadwick at the British Association for the Advancement of Science meeting in 1833, but that would have been after the inquiry instructions had been drawn up. At any rate, the instructions were too advanced for the Medical Commissioners; all that is, except a certain Samuel Stanway who worked in the area centred in Manchester, under the supervision of Assistant Commissioner J. W. Cowell, a veteran of Senior's Poor Law Commission. All other Commissioners ignored, it seems, the question about stature, but Stanway and Cowell visited the Bennett Street and St Augustine's Sunday schools in Manchester and two Sunday schools in Stockport and measured heights and

Table 7.1. *Mean heights and weights of children working in factories in the Manchester–Stockport area, 1833, and attending Sunday school*

	Boys			Girls		
Age	No.	Height	Weight	No.	Height	Weight
9 +	17	48.139	51.76	30	47.970	51.13
10 +	48	49.789	57.00	41	49.624	54.80
11 +	53	51.261	61.84	53	51.555	59.69
12 +	42	53.380	65.97	80	53.703	66.08
13 +	45	54.477	72.11	63	55.636	73.25
14 +	61	56.585	77.09	80	57.745	83.41
15 +	54	59.638	88.35	81	58.503	87.86
16 +	52	61.600	98.00	83	59.811	96.22
17 +	26	62.673	104.46	75	60.413	100.21
18 +	22	63.318	106.13	65	62.721[a]	106.35
	420			651		

[a] This value seems likely to be a misprint.

Measurements due to Cowell–Stanway, reported in Parliamentary Papers (1833, p. 697).

Heights, presumably with indoor footwear (see note 7.5), in inches; weights, lightly clothed with pockets emptied, in lb. Age presumably 9 +, 10 + etc.

weights of a total of 1,933 children aged 9 to 18. Some of these children, the majority, worked in factories; others did not. Their results on factory children are reproduced, in the original measurements, in Table 7.1, and plotted in Figs. 7.1, 7.2, 7.3 and 7.4. They do not state their technique, but Quetelet (1835) says that footwear was kept on and in comparing them with Belgian children subtracted what he thought was a suitable amount (see note 7.5).

The Cowell–Stanway figures, in contrast to Quetelet's, show just the sort of results one might expect. Amongst factory children, the numerically larger

Fig. 7.1. Mean heights of boys working in factories in the Manchester–Leeds area in 1833 (closed circles), measured by Cowell and Stanway (Parliamentary Papers, 1833), and in 1837 (open circles), measured by Horner (1837*a*). The 1833 means have been adjusted by subtraction of 1.0 cm for footwear: the 1837 children are presumed barefooted. Doubtful values are not joined up. The lower curve represents differences between successive yearly means. The data are plotted on cross-sectional British standards of 1965, in which the upper curves are centiles of height at successive ages and the lower line ('population height velocity') represents differences between successive annual means.

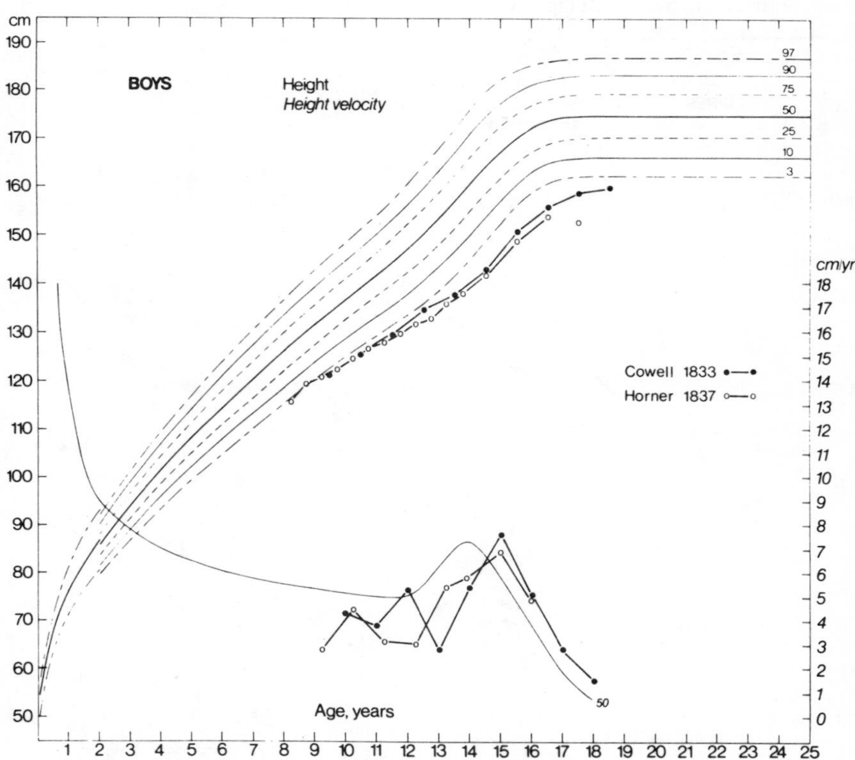

group, girls' heights exceed boys' at ages 12, 13 and 14. The boys have a clear-cut adolescent spurt with the year of maximum increment centred at 15.0 years, that is a year later than nowadays (Fig. 7.1, height velocity section). The velocity at the peak is 7.7 cm/yr, which is almost identical with the present value. The girls (Fig. 7.2) have no clearly identifiable spurt in these data (though see Horner's data below). Boy's weight (Fig. 7.3) has a maximum increment also about a year later than nowadays, and girls' weight (Fig. 7.4) a peak nearer two years later. The non-factory, or control, children were slightly taller, but in making the comparison Cowell's so-far very modern statistical technique breaks down. In calculating average heights of the two groups he pools all the ages together and computes the general means, ignoring the fact that the non-factory children were, on the average, younger. In consequence, the differences he gives (0.3 inches for boys and a negative

Fig. 7.2. Mean heights of girls working in factories in the Manchester–Leeds area in 1833 and 1837. Means for 1833 have been adjusted by subtraction of 0.5 cm for footwear. Sources and plots as in Fig. 7.1.

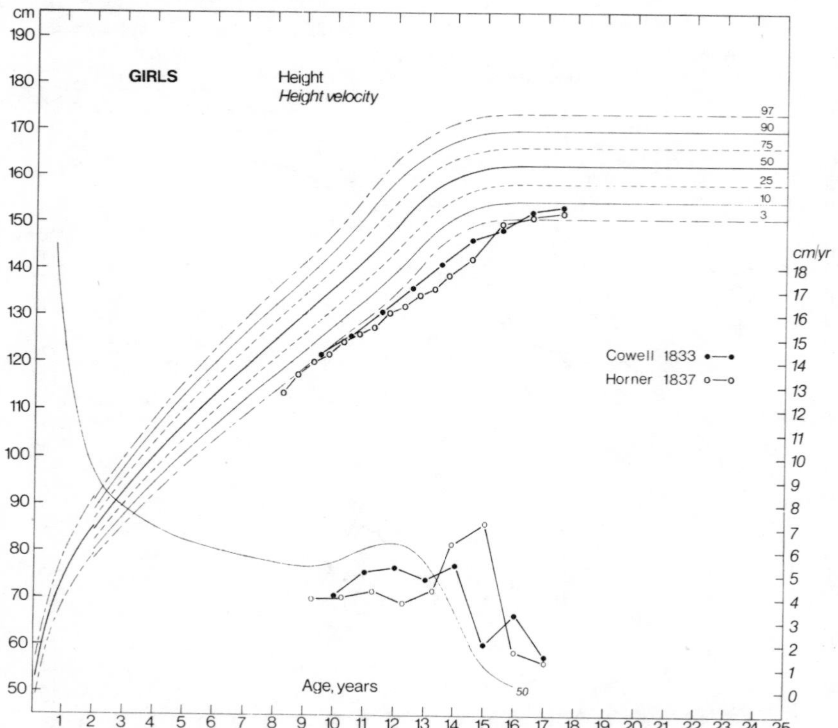

value for girls) are misleading. The average mean difference, calculated within years of age and weighted according to numbers, comes to approximately 0.25 inches (about 0.5 cm) both for girls (aged 9–15) and boys (aged 9–17), the non-factory children being taller.

Chadwick's 1842 *Report on the Sanitary Condition of the Labouring Population of Great Britain*, made for the Poor Law Commissioners, is justly a famous document, but his earlier 1833 *Report on the Employment of Children in Factories* (Parliamentary Papers, 1833) is just as detailed, thorough and well organized, a model in all respects. The 1833 report stressed that children worked ten to twelve hours a day in the factories and though an improvement in conditions had taken place during the previous few years, some ill-treatment still occurred, particularly in the smaller establishments. As a result of the report the Factories Regulation Act (1833) was passed which prohibited

Fig. 7.3. Mean weights of boys working in factories in the Manchester–Leeds area in 1833, measured by Cowell and Stanway (Parliamentary Papers, 1833). Means have been adjusted by subtraction of 6 lb for indoor clothing. Plots as in Fig. 7.1.

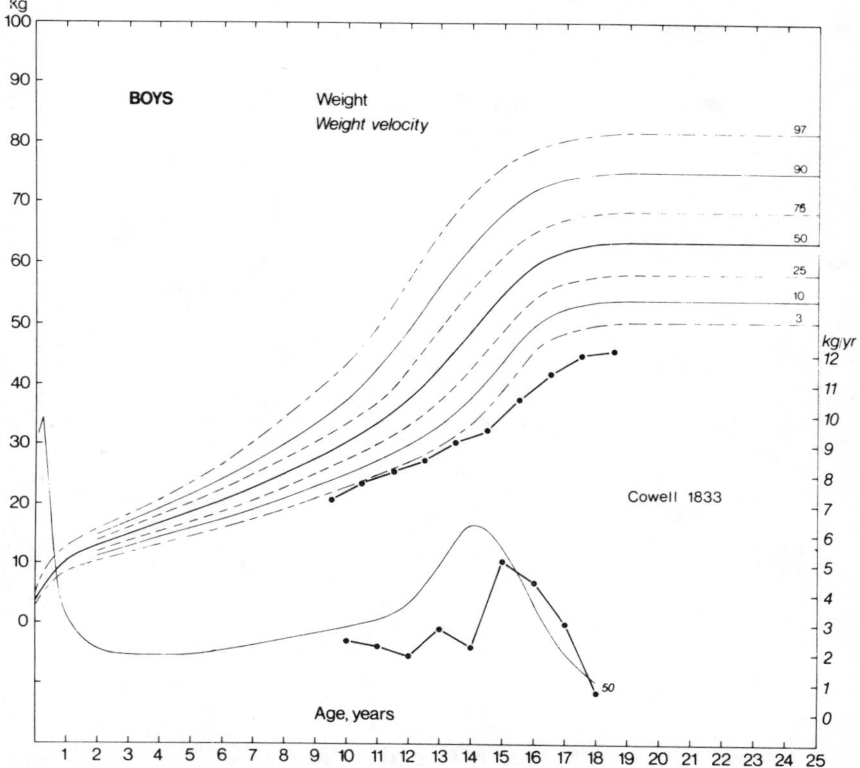

children under 9 years of age (i.e. before the ninth birthday) from working in various types of textile factories, and stipulated that children between the ninth and thirteenth birthdays should have one and a half hours each day for meals and rest. Such children had to have a certificate 'from a surgeon or medical man resident in the township . . . who shall certify on inspection of the child that he believes it to be of the full growth and usual condition of a child of the age prescribed . . . and fit for employment in a manufactory. This certificate should be given in the presence of a magistrate by whom it should be countersigned, provided he also were satisfied' (Young and Hoardcock, 1956, p. 944). Age is not enough however: 'The physical condition alone is the proper qualification for employment. Unless a discretion of this nature were given to the parties certifying, they might feel themselves bound to certify to the age of a child on the production of copies of the baptismal registers, which

Fig. 7.4. Mean weights of girls working in factories in the Manchester–Leeds area in 1833, measured by Cowell and Stanway (Parliamentary Papers, 1833). Means have been adjusted by subtraction of 4 lb for clothing. Plots as in Fig. 7.1.

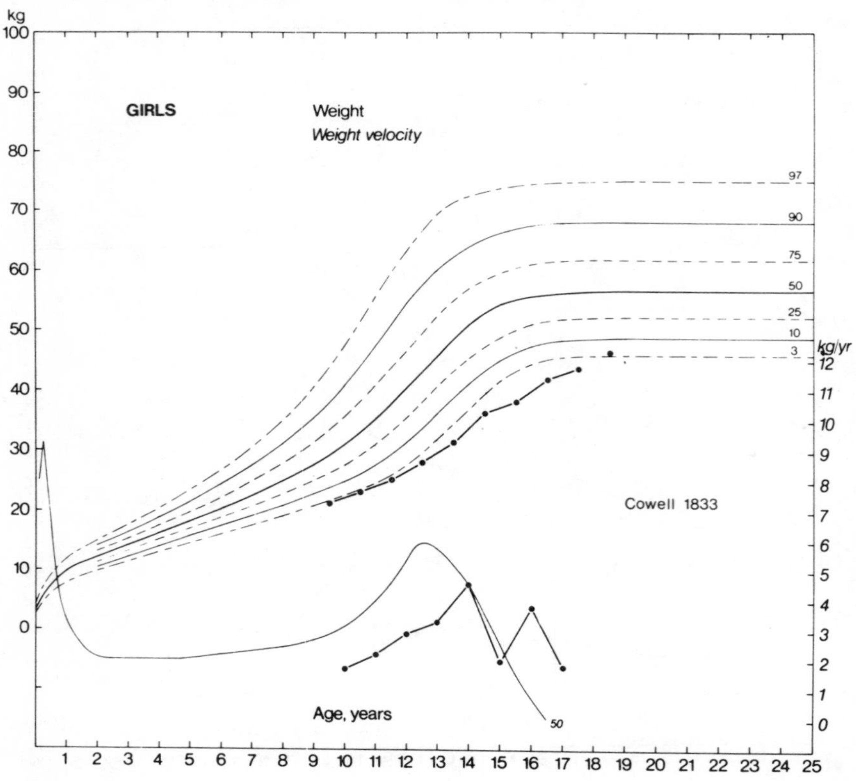

are easily forged, or on the evidence of parents who would be under temptation to perjure themselves' (p. 944). The flavour of the underdeveloped country is very strong. The report also adds that the local doctor's certificate would not be much use as 'his practice is liable to be dependent on the labouring classes' (thus pre-figuring the present-day certificates for sore throats and backache). Thus a more stringent certificate given by a factory inspector must be produced before entry to full-time work at 13.

Horner's survey

It was the belief that ages were being falsified that led directly to the second and much larger survey of children's heights and weights, organized by Cowell's colleague Leonard Horner. Horner (1785–1864) was for twenty-five years one of the four Government Inspectors of Factories. He was a man of wide interests and influence, especially amongst the scientists of the time. Son of a well-to-do wholesale linen merchant in Edinburgh, he took up geology at Edinburgh University, and though he went into the family business, geology remained a passion throughout his life. Fate was kind to him in this respect for his eldest daughter married Charles Lyell, who became the outstanding geologist of his generation, the author of the famous *Principles of Geology.* Horner was for fifty years a member of the Geological Society of London and twice its president, the second time with T. H. Huxley as secretary. He was a fellow of the Royal Society, a close friend of Francis Galton's father Tertius, of Babbage and of Hooker, and well known to Charles Darwin, who rather charmingly wrote to him in 1860: "I believe variations arise ... accidentally or spontaneously and these are naturally selected or preserved, from being beneficial to the successive individual animals in their struggle for life. I do not know whether I make myself clear' (Lyell, 1890, vol. 2, p. 300).

Horner's second passion was the education of working men and children. In 1821 he founded the first of the numerous Mechanics Institutions which sprang up in Great Britain in the 1820s and 1830s. This was the Edinburgh School of Arts, a 'College for the better education of the Mechanics of Edinburgh'. He also founded, jointly with Lord Cockburn, the Edinburgh Academy. Small wonder, then, that in 1826 he was invited to be the first Warden (equivalent of Principal) of the new University of London. He retired from business to take up the post, but the infant university was a turbulent place and he relinquished the Wardenship in 1830. He was appointed to the Factory Commission in 1833.

Horner took up office on 30 April, visited factories in May, and reported (with Cowell and two other colleagues) in June. The Act was passed on 29

August. Horner continued as factory inspector, at first in the Scottish region and from 1836 in the Lancashire area, until he was over 70. In 1837 he got twenty-seven surgeons appointed to factories in his district to measure 8,469 boys and 7,933 girls aged 8 to 14 inclusive in the towns of Manchester, Stockport, Bolton, Preston, Leeds, Halifax, Rochdale, Huddersfield and Skipton and the neighbouring rural districts. The surgeons were instructed to exclude children who did not appear to be in a good state of health, and to take the heights only of those with ages 'ascertained with tolerable certainty'. About half the children lived in rural areas, since many factories were located in quite small towns. The results of the survey were reported, anonymously, in the *Penny Magazine* of the Society for the Diffusion of Useful Knowledge ([Horner], 1837*a*) and reprinted (albeit with an important copying error) as appendix XI in the English (1842) translation of Quetelet's *Sur l'homme.* Horner's family letters (Lyell, 1890) unhappily make no reference to the circumstances of this large-scale survey, nor indeed do they mention Quetelet; Horner's international acquaintance was large, but for the most part geological. However, there is a letter extant from Horner to Nassau Senior, then chief of the Factory Inspectorate, in which Horner writes: 'From the great imperfections of the Act in all that relates to the determination of the age of the children, it is impossible for the Inspector to check the most palpable frauds, and to prevent the admission of children full-time long before they are 13 years of age. I have tried various checks – with but partial success – and I am persuaded that fully one-half of the children now working under surgeons' certificates of [being] 13 are in fact not more than 12, many not more than 11 years' (Horner, 1837*b*, p. 34). The reason for the alleged deception was the parents were paid according to the hours worked, and at age 13 adult hours were permissible.

The Factory Act of 1833 by no means satisfied Horner. He campaigned continually for better factory conditions, especially as regards accident prevention, and in a small book that still makes red-hot reading he urged reform in other areas besides textile manufacture. 'The inhumanity, injustice and impolicy of extorting labour from children unsuitable to their age and strength – of subjecting them, in truth, to the hardships of slavery (for they are not free agents) have been condemned by the public voice in other countries [also], he wrote (Horner, 1840). He goes on to cite the law in other countries, and appends a (rather complacent) letter from Horace Mann (1840), the Secretary of the Board of Education of the State of Massachusetts. In 1840 Horner and three other Commissioners reported on the situation of children working in the mines. The report (Young and Hoardcock, 1956, pp. 963–8) created such a scandal that it led to a total ban on women and children in mines. Some children of 8 or even younger were spending twelve hours a day standing in solitude and darkness (owing to the particular task they did). The Commissioners reported that 'The employment in these mines generally produces in the first instance an extraordinary degree of muscular development, accompanied by a corresponding degree of muscular strength; this preternatural development and strength being acquired at the expense of other organs, as is shown by the general stunted growth of the body' (p. 976). Unfortunately, there seems to be no evidence that the heights of children working in mines were actually measured. Horner's results are given in Table 7.2 and plotted with Cowell's in Figs. 7.1 and 7.2 above. The numbers in Horner's survey are substantial up till and including age 14. The values at 15, 16 and 17 are based on a smaller series, measured by one particular surgeon, Harrison, in Preston. The final boys' value is perhaps erroneous (whether through recording or printing error cannot be determined) and has not been joined to the others. Horner's means are given in half-yearly age groups and 'rolling' velocities have been plotted, that is 9.5 to 10.5 (plotted at 10.0),

Table 7.2. *Mean heights (probably barefoot) of children working in factories in the Manchester–Leeds area in 1837*

	Boys		Girls	
Age	No.	Height	No.	Height
8.0–	327	3/9 5/8	267	3/8 11/16
8.5–	339	3/11	272	3/10 3/8
9.0–	527	3/11 5/8	438	3/11 3/8
9.5–	418	4/0 1/8	375	4/0
10.0–	574	4/1	506	4/1
10.5–	550	4/1 7/8	421	4/1 5/8
11.0–	664	4/2 3/8	577	4/2 1/8
11.5–	559	4/3 1/16	478	4/3 1/4
12.0–	767	4/3 3/4	712	4/3 7/8
12.5–	660	4/4 1/4	618	4/4 3/4
13.0–	1,269	4/5 1/2	1,260	4/5 1/2
13.5–	864	4/6 3/8	980	4/6 1/2
14.0–	951	4/7 3/4	1,029	4/8
15.0–	82	4/10 1/2	106	4/10 3/4
16.0–	43	5/0 1/2	90	4/11 1/2
17.0–	47	5/0	112	5/0

Measurements due to Horner, as reported in *Penny Magazine* (1837), vol. 6, pp. 270–2. Heights in original measurements (ft/inches).

10.0 to 11.0 (plotted at 10.5), and so forth. Despite the larger numbers the trend of the means seems less regular than in the Cowell–Stanway series, perhaps because of the inaccuracies always associated with large numbers of measurers. The adolescent spurts are not very well marked and the girls' ascendancy, though present at 12.5 to 15, is very small in amount.

These factory children of the 1830s are quite extraordinarily small by present-day standards. Their mean heights are at about the modern 3rd centile from 8 till 10. During puberty they drop well below the modern 3rd centile. As maturity is reached the 3rd centile is nearly regained by the boys and slightly surpassed by the girls. The maximum height increment of Horner's boys is centred, like Cowell's, at age 15.0; Horner's girls have a seemingly clear-cut, indeed exaggerated maximum, probably really centred at about 14.0 or 14.5. Horner's series has slightly lower means than Cowell's. It is possible that some of his factory children were dissembling their ages despite what he wrote; if his 13-year-olds, and perhaps his 12-year-olds also, were plotted as though averaging six to nine months younger, they would approach, though not quite reach, the values obtained by Cowell. Cowell's children did not have the same reason to falsify their ages, because they were already working full-time in factories, before the Act was passed. A second possible influence is that Cowell's children were selected by being those who attended Sunday school, which was not compulsory. It is hard to know whether such attendance implies a disproportionate number of more intelligent and caring parents or of parents who wished to have their children out of the way. Horner's children (one assumes) were measured for the most part in factories and were not subject to this selection. Cowell's children wore Sunday shoes and indoor clothing (the plots have been corrected for this); Horner's we are uncertain about, but often children went without shoes in the factories, so it seems likely that all were measured barefoot.

This series represents the first properly documented cross-sectional survey of children's heights known to the writer. It is instructive and suprising – also perhaps encouraging – to compare the results with those for a typical underdeveloped country nowadays. In Ibadan, Nigeria, Dr Margaret Janes (1975) has for many years been engaged in a longitudinal study of two groups of children, one rich, the other very poor and living in a slum near the central market place. Fig. 7.5 shows that while the rich in Ibadan are exactly comparable with present-day British standards for height, the poor are considerably taller than Manchester factory children in the 1830s, their means being at the present British 10th centile.

The velocity curves of Cowell's and Horner's boys indicate that their adolescent growth spurt was about a year to eighteen months delayed compared with nowadays. The girls' curves are less definite. There is no indication of the puberty status of these particular children, but the later enquiries of James Whitehead, a Manchester surgeon, placed the average age of the menarche of factory girls about 1830 as late as 15.7 years, a delay of two and a half years compared with the present day (Whitehead, 1847: and see discussion in Chapter 11, p. 290). This is not necessarily at variance with the height data. Whitehead was inquiring of adults in the 1840s, and some of these would have grown up in the countryside, and in Ireland where conditions were still worse. Whitehead does not separate his 4,000 patients into those born and bred in the industrial north-west and those who came there as children or adolescents. Additionally, the interval between peak velocity and menarche might have been a little longer in these relatively slowly-growing girls than in the contemporary population. All in all, we might estimate the degree of delay as about one and a half year; perhaps six months less in boys and six months more in girls.

Fig. 7.5. Mean heights of rich and poor Nigerian boys living in Ibadan 1965–75 (longitudinal study of Janes, 1975; from Tanner, 1978). Plots as in Fig. 7.1.

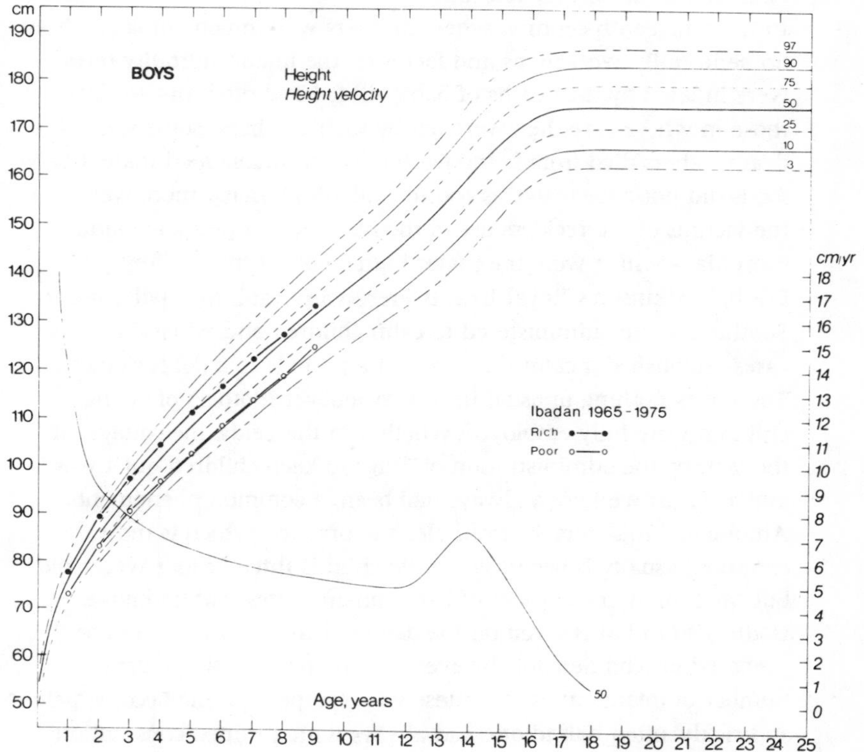

One thing is clear: most of the smallness of Cowell's children of the 1830s is persisting smallness, not the temporary smallness due to growth delay that characterizes the late-developing child seen often nowadays in the doctor's surgery or the hospital Growth Clinic (see Tanner, 1978*a*). The factory children's delay is indeed responsible for their sinking from about the 3rd centile in height and weight to below it in the pre-pubertal years, and this deficit is recovered as their own puberty develops. But that recovery only returns them to the 3rd centile again. The remarks of Horner and his colleagues about children in the mines, given above, seem very pertinent.

In the light of modern knowledge the causes of such short stature, persisting into adult life and not to a great extent accompanied by retardation of growth, have to be sought in early childhood and even the fetal period (see note 7.6 and Tanner, 1978*a*). Severe malnutrition of the pregnant mother followed by chronic and severe undernutrition of the infant could cause this result. More likely still is a low birthweight and/or a low weight gain in infancy caused by injurious substances breathed or eaten by the pregnant mother and the newborn child. The remarks of Pinchbeck and Hewitt (1969–73, vol. 2, p. 406) seem very relevant:

> In the nineteenth century, where mothers were much employed from home in mills, workshops and factories, the infant mortality rates were inflated by the deaths of babies ill-fed and often ill-used by those in whose care they were left by their mothers. Some starved to death; others died from being fed totally unsuitable food (patent baby foods did not appear in this country till 1867); many more were the victims of the reckless use of the narcotics – opium, laudanum, morphia – which were the major ingredients of the Godfrey's Cordial, Atkinson's Royal Infants Preservative and Mrs Wilkinson's Soothing Syrup, administered to calm children and which in many cases established a calm that was but a prelude to a deeper quiet. There was nothing unusual in this. Whenever mothers of young children were fully employed, whether in the fields, the cottage, or the factory, the administration of drugs to keep children quiet was, and as far as we know, always had been, a common phenomenon. Among embroiderers for example, 'the practice, which is most common, usually is begun when the child is three or four weeks old, but Mr Brown, the coroner of Nottingham, states that he knows Godfrey's Cordial is given on the day of birth, and that it is even prepared in readiness for the event . . . the result is that a great number of infants perish . . . those who escape with life become pale and sickly, often half idiotic and always with a ruined constitution'. Here as in so many other respects, the experience of factory industry,

far from being unique, was in fact the experience of cottage and workshop industry, writ large for all to see (see also Hewitt, 1958, pp. 141–52; Berridge, 1979).

This astonishing abuse of opium derivatives continued unchecked by the Poisons Act of 1868, which specifically excluded from control as unimportant the patent medicines used for children. Only in the 1890s was the practice finally made impossible by a series of court cases (Lomax, 1973).

Another contributory cause of the shortness might have been the undernutrition of the mothers much earlier, when they themselves were children, causing stunting of size and in consequence the production of small babies. Some of the mothers had grown up in the days before the Napoleonic Wars, when conditions in the countryside were still very bad. Some of the mothers too must have been immigrants from Ireland, an area Horner reports as greatly depressed even compared with north-west England.

There is another archive of boys' heights around 1840 which confirms the correctness of the Cowell–Stanway curves. This concerns boys transported to Australia as convicts. The first fleet of convicts arrived in Australia in 1788, and the history of children thereafter born or transported there has been outlined by Gandevia (1977*a*, *b*). The heights of the young offenders transported in two boys-only convict ships in 1837 and 1842 averaged about 2 cm less than in those in the Cowell–Stanway study at age 13 +, 6 cm less by 16 + and 17 +, and 5 cm less at 18 +. Boys convicted in London were separately identified, and averaged 3 cm lower still at ages 16 and 17; boys from other areas of England and from Scotland averaged 1.5 cm higher than the overall figures, thus some 4.5 cm less than the Cowell–Stanway values. The majority of the transported boys would have been convicted of stealing, and often of stealing food. They would have been drawn from the most deprived section of the population. Thus it is no surprise that they were smaller than the factory boys. But their growth was, still, more stunted than delayed; the difference between the means for 17- and 18-year-olds from outside London was only 2.8 cm, arguing cessation of growth around age 20. Sixty years later, about 1900, convict youths averaged 13 cm taller, which would place them then around the modern 10th centile, comparable with boys in manual workers' families at that time (Fig. 11.6, p. 275).

The smallness of children and adults was not confined to those who worked in the factory towns of the industrialized countries. From the archive of the Norwegian statistics on the heights of young men that stretches back to 1741 (discussed in Chapter 5, p. 117, Kiil was able to abstract longitudinal records of two groups of Romsdal peasants in north-west Norway measured during the years 1826 to 1837, just about the time of the Factory Commission surveys. One group consisted of 238 men measured each year from age 17.3 to 22.3, the other of 133 men measured each year from age 16.3 to 22.3 (Kiil, 1939, tables 56 and 57, p. 118). The mean heights of each pure longitudinal group, and the mean height velocities, are plotted in Fig. 7.6. (The modern average Norwegian, it must be remembered, is a little taller than the average Briton, with height at about the British 60th centile.) The mature height of these men is some 5 to 6 cm less than that of comparable Norwegians nowadays. Their maximum increment of height is not easy to estimate from these data but their velocity curve seems to be set about three years later than at present. The lateness is corroborated by the figure for age at menarche in contemporaneous Norwegian girls of similar provenance (see Chapter 11). This was about 16.0 years, also three years later than nowadays. Though these country dwellers could scarcely have been exposed to the adulterated atmosphere, food or even cordials of the urban factory workers, under-nutrition was clearly severe (until, Kiil thinks, the general introduction of

Fig. 7.6. Mean heights of peasants from Romsdal, north-west Norway, born 1809–15. There are two groups, each pure longitudinal. From Kiil (1939, p. 118). Plots as in Fig. 7.1.

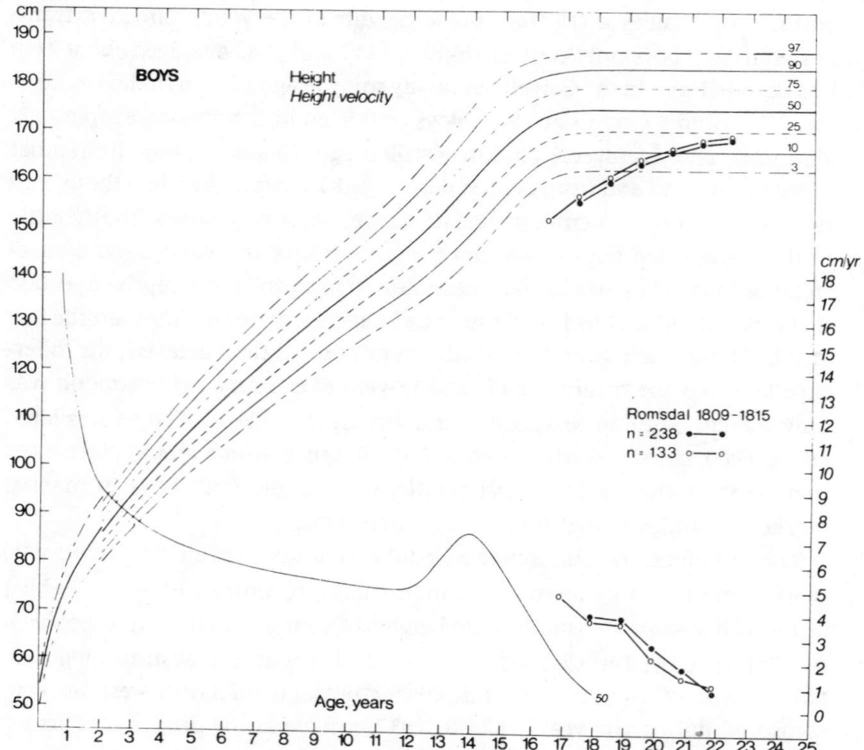

the potato in the 1830s). This undernutrition has produced, as we would expect, a greater delay than that seen in the factory children, but a somewhat smaller diminution of mature height. These data provide some corroboration then, for the interpretation given above.

To return to Great Britain: thirty years later, Danson (1862) gave the mean height of male prisoners in the Liverpool gaol as 163 cm for 18-year-olds (present-day 5th centile), rising to a plateau of about 168 cm at 25 or 26 (present-day 15th centile). On the other hand students at Edinburgh University in 1834–5 were about 175 cm (5 feet 9 inches) at maturity which was reached on average at about age 20 (Forbes, 1837). This is the present-day 50th centile for the general population, though about the 30th for students. University of Cambridge students had a very similar height (Whewell, cited by Quetelet, 1842). Aitken (1862), discussing the requirements of the army and deploring the high prevalence of consumption amongst recruits, says that in 1854 the army set a lower height limit for an infantryman recruit of 5 feet 4 inches (163 cm) and a minimum age of 18 years. But in 1804, under the stress of war, parents were given two guineas (£2.10) for each lad under 16 who was taller than 5 feet 2 inches (157 cm). The successful boys, or rather the boys with successful parents, were probably mostly early maturers; but even excluding these, the army could have expected to obtain soldiers considerably above average height for the time.

Horner's, then, was the first large-scale survey of the heights of children of which we have a record. It is perhaps a rather baleful reflection that just as the first surveys of adult height were made to see what steps could be taken to obtain better soldiers, so this survey of children's heights was made to see whether parents were cheating the law in the exploitation of their children. All the same, Horner had reason to be well pleased with his work, for when he retired (to make a translation of Villari's *Life and Times of Savonarola*) Britain led the world in factory legislation and the control of child labour.

France: Villermé

In France the same sort of development had been taking place, with Villermé in the position of Chadwick. Villermé's first large paper (1828) was on the mortality of rich and poor in France. It begins by saying the matter is disputed: some say the poor have a higher mortality rate because of their privation, but others that the rich have a highter rate because of their 'luxury, passions and excesses of all kinds' – surely an echo of the sentiments of Guarinoni two centuries before (p. 27). Villermé established that in Paris the rate of mortality for persons of all ages combined was nearly twice as great in the poorest as in the richest *arrondissement*; he concluded that 'the mortality is greater in the poor to a degree of which we had no inkling' (p. 75); 'Easy circumstances prolong life, poverty shortens it' (*l'aisance de fortune conserve notre vie: la misère l'abrège*) (p. 81). From this beginning Villermé went on to examine adult stature in relation to the same factors. His *mémoire* on this (1829) appeared in the first volume of the very influential journal *Annales d'hygiène publique*. He himself was one of the journal's founders, and very likely drafted its prospectus. 'Medicine', this said, 'does not only have as its object the study and treatment of sick people; it has close relationships with social organization; sometimes it helps the legislature in the drawing up of laws; often it clarifies their application, and always ... it watches over the public health'.

In his paper Villermé declared that his general aim was 'To demonstrate the height of men in France, the age at which full development of the body is achieved, and the general causes which advance or delay growth and which determine a large or small stature' (1829, p. 552). In 1812–13 there had been a large government enquiry about the heights of conscripts, and the proportion of conscripts rejected for either small size or illness or deformities. Villermé analysed the results. In Bouches-de-la-Meuse, later part of Holland, the mean height of conscripts of 1808–10 was 167.7 cm and 66 out of 1,000 were rejected (24 for stature less than 154.4 cm). In Chiavari, in the Apennine mountains, the mean height in some years only reached 156.0 cm, and 300 out of 1,000 were rejected. The prefect of Puy-de-Dôme, writing of the period 1804–10, remarked 'The young people develop late and many of them finish their growth at 24 years or even, 25 years. This slowness of development relates in part to race but more to poverty (*la misère*), to the smallness of the care that the poor give to the education of the children, to the precocious work which is necessary, to lack of healthy and abundant food and to the type of work done by the people' (1829, p. 363). Amongst the *arrondissements* of Paris there was an association between the percentage of persons owning houses and the mean stature of persons living in the district. (Villermé had used the same scaling of *arrondissements* in his *mémoire* on mortality, and it showed a close inverse association.) Poverty, said Villermé, was much more important than climate in influencing growth. 'Human height', he said, 'becomes greater and growth takes place more rapidly, other things being equal, in proportion as the country is richer, comfort more general, houses, clothes and nourishment better and labour, fatigue and privation during infancy and youth less; in other words, the circumstances which accompany poverty delay the age at which complete stature is reached and stunt adult height' (1829, p. 385).

In all France, Villermé found that conscripts taller than 165 cm (5 feet 1 inch) constituted 45 per cent of the total in 1816–17 and 50 per cent in 1826–7, with a regular increase in between. Growth from age 18.5 to 20.5 averaged about 2.5 cm in these data. Villermé commented on the discussion, then and later current in France, concerning the possible effect of the Napoleonic Wars on allegedly depressing the mean stature of Frenchmen in the early 1800s by virtue of the taller and fitter men being sent to fight far from home. He said he thought a diminution might be due to this selective factor; but no more than thought, for he had no proof of it. The increase in stature of 1816–27 could be due to other causes, such as a decrease, however slight, in misery.

Villermé has absolutely no word to say about the effects of heredity on stature, and his views of poverty as a cause of differences in height did not go for long unchallenged. In France, Dr M. Boudin (1806–1867), a distinguished army surgeon who in 1859 was a founder-member of the Société d'Anthropologie, published in the year of Villermé's death a massive account of recruitment procedures in various countries (Boudin, 1863). In this he showed that the increase in height of French recruits had continued over the period 1831–60 and attributed this to a diminution of the selective effects of foreign wars described above. Though not quoting Villermé's paper (they were colleagues on the editorial board of the *Annales*) he says the view that differences in heights between populations are due to differences in nutrition are quite mistaken: altitude, climate and race are the important factors (Perier, 1867). Indeed before Boudin, Dr Knox, the translator of Quetelet, had exploded in a footnote over Quetelet's quotation of Villermé's finding that the men in the Meuse region were taller than those in the Apennines (Quetelet, 1842, p. 60). 'The translator is persuaded', said Knox, 'that Dr Villermé and M. Quetelet have failed to detect the real cause of differences in stature in those two Departments: it is a question purely of race and not of feeding or locality. The taller conscripts were Saxons ... the shorter, found in the Apennines, ... were the descendants of the Ancient Celtic population of that country'.

Broca (1824–1880), the leading physical anthropologist in France, was admirably explicit on the point. 'I have recognized', he wrote, 'that the height of Frenchmen, considered generally, depends not on altitude, nor latitude, nor poverty or riches, neither on the nature of the soil nor on nutrition, nor on any of the other environmental conditions that can be invoked. After these have all been successively eliminated, I have been brought to consider only one general influence, that of ethnic heredity (*l'herédité ethnique*)' (Dally, 1879, p. 348). Broca and the anthropologists recognized that poverty and poor nutrition slowed down growth and postponed the age at which mature height was reached; but they denied that in France, at any rate, mature height itself, like the cranial index, was other than hereditarily determined (heredity being conceived of in a very imprecise way, for this was before Galton, let alone the rediscovery of Mendel). It was against this fixity of anthropological opinion that Boas had to fight in his studies, thirty years later, of physical changes of immigrants to the United States (see Chapter 10). It is probably true that mild undernutrition does indeed slow down growth without affecting final stature; but the line is a thin one, and it does not take much worsening of circumstances (so long as the bad circumstances are persistent rather than transient) to affect adult height. In the light of modern research (see Eveleth and Tanner, 1976), the truth lies between Broca's and Villermé's extremes of opinions; but Villermé's view is the nearer to it.

The condition of the factory workers and the poor was even worse in France after the Napoleonic Wars than it was in England and in the 1830s and 1840s Villermé in France played the part that Farr (1807–1883), a man twenty years younger, and Chadwick did in England. In 1832 the Académie des Sciences Morales et Politiques was reinstated in France after its suppression thirty years earlier by Napoleon and it set up the same sort of inquiry amongst the textile workers in France that the Factory Commission of 1833 had made in England. Villermé undertook the inquiry; ten years in the army had accustomed him to field-work and he did all the investigations himself (see Villermé, 1840, introduction), travelling from place to place and writing to Quetelet in June 1835: 'You ask me if I am satisfied with my tour: yes, in some respects, no in others. I was at Lyon, Avignon, Nismes, Montpellier, Marseilles, Toulon, Geneva, Lausanne, Berne, Zurich, Basle, Mulhouse and the neighbourhoods . . . If following my tour a maximum duration of work for children in factories should be adopted, I would certainly be well rewarded for all my trouble. This law, which would only be a copy of one passed in England not long ago, is absolutely demanded by conscience and humanity' (Ackerknecht, 1952). His work was issued in a report entitled *Tableau de l'état physique et morale des ouvriers* (Villermé, 1840; Deslandres and Michelin, 1938), which was successful in obtaining a law the following year forbidding children aged 8 to 12 from working more than eight hours a day or from doing any night work, and making schoolwork mandatory until the age of 12.

American negro slaves 1820–1860

It turns out that the greatest numbers of children measured in the first half of the nineteenth century were measured for purposes of identification and the prevention of smuggling rather than for the investigation of social ills and the conditions of factory life. In 1807 the Bill for the Abolition of the Slave Trade was passed in the United States. From that time on no new slaves could be imported into the country, though home-produced slaves continued to be a marketable commodity, and as such were shipped from place to place. Shipping often took place by coastal steamer, between places such as New Orleans (the greatest depot for both export and import) and Norfolk, Virginia. To prevent new slaves from Africa being added to the ship's cargo during the voyage, manifests had to be made out in which the individuals were identified by name, sex, age, colour and height. Those that got on had to get off; those and no others.

A large number of these manifests are lodged in the United States National Archives and a sample of 1,442 of them has been studied by Steckel (1979). There were 16,099 slaves named in the samples of the manifests – an average of 11.2 per manifest. This included 4,120 males and 3,583 females in the age range 8+ to 21+. Some, particularly the girls, were travelling as servants, but

Table 7.3. *Means and standard deviations of height (cm), sample sizes and population mean velocities, for male and female slaves*

	Males				Females			
Age	Mean	S.D.	*N*	Velocity	Mean	S.D.	*N*	Velocity
8+	116.0	11.22	123	7.3	117.2	12.40	127	5.5
9+	123.3	9.10	122	3.5	122.8	9.58	121	5.9
10+	126.8	11.51	200	6.0	128.7	10.24	158	4.9
11+	132.8	9.86	130	5.3	133.6	9.27	137	2.9
12+	138.1	10.41	237	3.8	136.5	9.96	210	6.6
13+	141.9	10.01	181	4.7	143.1	9.96	183	5.8
14+	146.5	10.41	253	5.6	148.9	9.14	263	3.9
15+	152.2	8.36	211	6.6	152.8	7.47	318	3.8
16+	158.8	7.75	266	4.1	156.6	7.21	461	1.5
17+	163.0	7.16	299	3.1	158.1	6.63	407	1.1
18+	166.0	6.38	496	2.4	159.2	7.01	565	0.2
19+	168.4	5.90	441	0.5	159.4	6.32	226	−1.3
20+	169.0	6.45	700	2.2	158.1	7.75	407	1.0
21+	171.1	6.96	461	−0.5	159.2	7.62	114	−0.8
22+	170.6	7.42	644	0.6	158.4	7.90	211	0.8
23+	171.2	7.09	383	−0.7	159.2	10.67	130	0.0
24+	170.5	6.48	400	−0.1	159.2	7.77	150	−0.7
25–29	170.3	7.16	1165		158.4	8.51	487	

From Steckel (1979, table 2).

most were being shipped for commercial purposes. The chief interest of this group for economic historians has been in determining the mean age at menarche and at first birth of a child, to confirm or refute the suggestion that slave owners deliberately manipulated the reproductive behaviour of the female slaves in order to increase their stock of slaves for use or sale (Trussell and Steckel, 1978). Additionally, the height curves tell us something about the conditions of life under which the slaves were kept, and especially about nutritional conditions.

Table 7.3 is taken from Steckel (1979). It shows the means and standard deviations for height at each age, together with the population mean velocities calculated cross-sectionally (hence centred at 9.0, 10.0 etc. years). It is not known whether height was measured barefoot, but analogy with military data suggests it probably was. The standard deviations are high and call into question the accuracy of the determinations; however, none of the statistics

Fig. 7.7 Mean heights of United States Negro slaves, 1820–60: boys. Plots as in Fig. 7.1. From data of Steckel (1979).

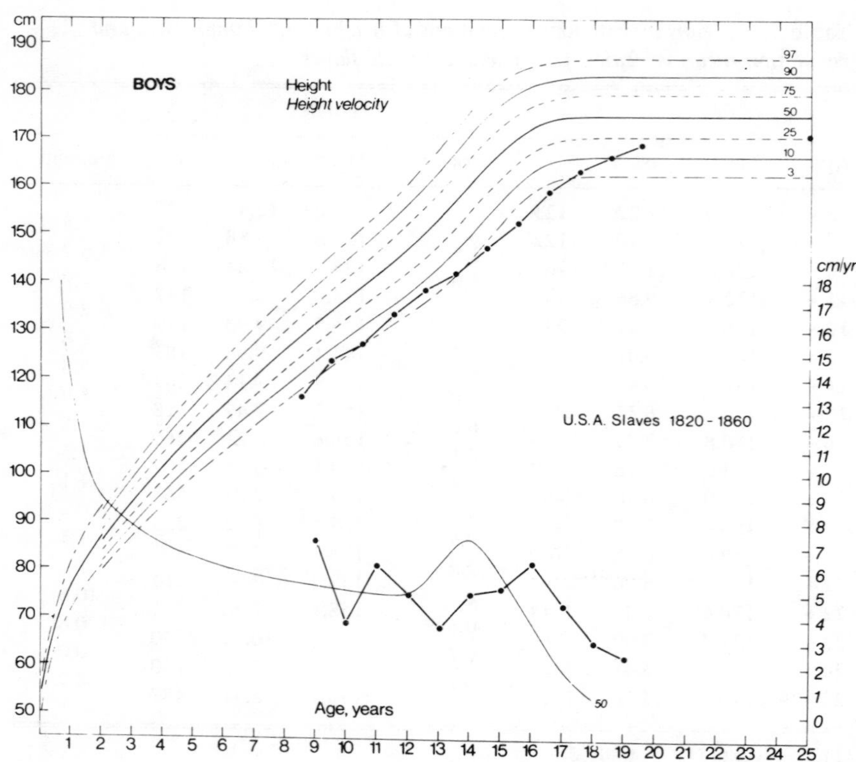

considered below is more than trivially altered by a 5 per cent trim (i.e. removing as suspected errors the largest and smallest $2\frac{1}{2}$ per cent of values at each age), a procedure which brings the standard deviations to their usual level. It seems, therefore, that the inaccuracy of the sea-captains was at least unbiased. The data are plotted in Figs. 7.7 and 7.8. Compared with the Cowell–Stanway factory children the slave children are taller, by about 1 cm at age 9, 2 cm at age 12 and at the end of growth by some 8 cm in boys and 5 cm in girls. They are at approximately the 3rd centile of the modern British standards from 8 to 14 in boys and 8 to 12 in girls. Much of this deficit is due to growth delay, for after dropping below the modern British 3rd centile during the pre-pubertal years they eventually reach the 25th to 30th, a very different picture from the factory children. Peak height velocity occurs a bit before 16.0 in boys and a bit after 13.0 in girls. The latter value would imply a mean figure for menarche of about 14.5 years.

Fig. 7.8. Mean heights of United States Negro slaves 1820–60: girls. Plots as in Fig. 7.1. From data of Steckel (1979).

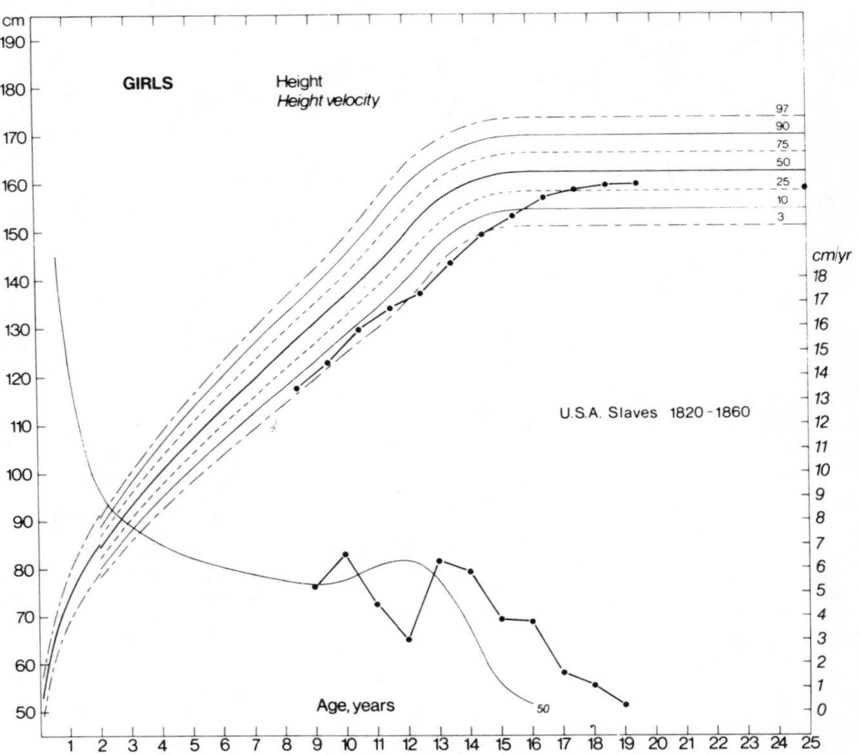

These manifests bear out entirely the comments of English observers of the 1830s and 1840s that the physical conditions of the English factory children were worse than the physical conditions of American slaves. Indeed the slaves were larger at all ages than the labouring-class children of the survey of 1872–3 reported by Roberts (see p. 173). Until the end of the century only Bowditch's sample of Boston children of 1875 were taller, and then only the American-born ones; the children of Boston Irish immigrant manual workers remained smaller than the children of negro slaves of a generation earlier (p.193). A similar study of the growth of slaves in Trinidad and other Caribbean islands has been made by Higman (1979). Trinidad slaves were not as tall as slaves in America and their age of peak height velocity was probably a little later.

8

Roberts, Galton and Bowditch: social class and family likeness

The Factory Acts continued to provide increasing protection for working children as the century progressed. Slowly conditions in the textile factories changed, partly as a result of the Acts but also because of the introduction of new machinery. Not all the changes were for the better: the machines were more complicated and ran faster, and the work was more monotonous. Children 8 years old still left home without brakfast at 5.30 on a winter's morning to begin work in the factory at 6, and pregnant women still stayed at their jobs till two weeks or even two days before confinement. Many of them returned within a month *post partum*, either leaving the infant in the care of some frequently ill-paid childminder, or bringing the baby to the factory where it could not be nursed sufficiently, for the mother could only break her work at five- or six-hour intervals.

The 1872–3 survey

In 1872 a Parliamentary Commission was set up, under a doctor, J. H. Bridges, (see note 8.1) and a factory inspector, T. Holmes, to inquire into all aspects of conditions of the work of women and children in textile factories (Parliamentary Papers, 1873). Evidence was taken from workpeople, employers and the factory surgeons. These last stressed particularly the plight of pregnant and *post partum* women. They remarked that growth of the fetus *in utero* was affected by the conditions of work, although they gave no evidence in the form of statistics of birthweight. In fact, as we have discussed above, a low birthweight, especially if combined with undernutrition in the first few months after birth, would explain the shape of the childhood growth curves of 1833 and 1873 (see on) a good deal more convincingly than continuing undernutrition during childhood. The curves show height greatly reduced at all ages, including maturity, but with only a relatively minor degree of retardation. Such curves are seen nowadays in babies with low birthweight due to disorder of fetus or placenta (see Tanner, Lejarraga and Cameron, 1975). If the factory mothers did indeed produce such babies the cause would most likely have been the combination of a poisonous atmosphere at work, the frequent adulteration of the food (see Burnett, 1966, Chapter. 5), and the relative undernutrition of the mother during pregnancy, in particular, perhaps, in respect of protein (see Tanner, 1978*a*). Undernutrition of the child itself during the first few months after birth (as opposed to later) may perhaps produce a not dissimilar growth curve or at least fix the child permanently in a curve of growth from which he might have escaped with better post-natal care. Such early undernutrition must have been general. In addition there was the use of laudanum by mothers and day-nurses, as described above. In the 1870s infant mortality was still appallingly high in factory towns, and deaths in the first years reached 250 per 1,000 live births in Manchester and Salford compared with 70 per 1,000 in the northern country areas (and 17 per 1,000 nowadays). Indeed despite the fall in the general death rate, infant mortality at this time 'was as high as it had ever been since records were made' (Rooff, 1957, p. 32). The surgeons were in nearly unanimous agreement that mothers should be prohibited from working in factories for some substantial portion of their pregnancy and for at least six weeks afterwards.

In addition to taking verbal evidence the Commission instituted:

> a careful and systematic examination of children upon an extensive scale, in a great variety of areas ... registering their height, weight and dimensions of the chest and recording all instances of malformation or disease ... Our object was to compare and contrast children employed in factories, first with children inhabiting factory districts but not employed in factories; secondly, with children from adjacent districts where no factories were situated. It was also necessary to distinguish between factories situated in large towns and those in suburban or semi-rural districts, to meet the obvious objections that whatever results were observed might be attributable to the child's locality rather than to its occupation. Lastly, the occupation ... of the child's parents seemed to us an element which it was essential to take into account (Parliamentary Papers, 1873, pp. 846–7).

There were five doctors employed to do the examinations and take the measurements. During the winter of 1872–3 they visited 'a large number of schools in Lancashire, Cheshire and the West Riding of Yorkshire' and examined nearly 10,000 'factory and other children of the working classes'. All children in a given school were taken, without selection. Height was measured in bare feet, weight in indoor clothes, and chest circumference with boys stripped to the waist. Bradford, Halifax, Preston, Rochdale, Stockport and Macclesfield were some of the urban factory districts; York, Kendal,

Sedbergh, Lancaster, Chester and Keswick some of the non-factory towns. 'The general impression made by the factory children was in many respects not unfavourable. As compared with the children of the East of London, or of the poorer parts of Liverpool, they were markedly superior. They did not appear to be more liable than ordinary country children to rickets or scrofula'; however 'very few were free from vermin' (Parliamentary Papers, 1873, p. 848) and most were flea-bitten. The 'personal habits of cleanliness of the factory children compared very unfavourably with those in the agricultural districts' (Parliamentary Papers, 1876, appendix D).

The tables in the report (Parliamentary Papers, 1873) show that urban children who worked in the factories, and had parents doing the same, were a little over an inch, that is about 3 cm, shorter at ages 8 to 11 than children who lived in non-factory rural or suburban districts (see Table 8.1). At age 12 the difference in height was 4 cm, and in weight about 2 kg. Boys and girls showed generally similar differences. The same factory children were smaller by almost as much when compared with children living in the factory towns but neither working in factories themselves nor having parents working there (though of the manual labourer class). The differences in this comparison averaged 2 to 3 cm. Hence simply living in the factory town was insuf-

Table 8.1. *Mean heights (inches) at successive years of age of factory children, non-factory children living in factory districts and non-factory children living in non-factory districts*

Age	F[a] children of F parents	Non-F children in F districts	Non-F children in non-F districts, urban or rural
Boys			
8+	45.75	46.72	46.66
9+	48.05	49.09	49.21
10+	49.77	51.02	51.00
11+	51.44	52.57	52.87
12+	52.82	53.56	54.05
Girls			
8+	46.48	47.40	46.73
9+	47.62	49.37	48.63
10+	49.52	49.76	50.07
11+	50.80	52.80	52.66
12+	53.13	53.39	54.41

From Parliamentary Papers (1873, pp. 851–2).
[a] F, factory.

ficient to cause the shortness; the social environment that went with factory work was the important thing. Factory children in the big towns were slightly smaller than factory children in the suburban areas, by about 1 cm. This report was soon followed by the Factory Act of 1874, which raised the minimum age of part-time employment – in the textile factories only – from 8 to 10 years and the minimum age of full employment from 13 to 14 years, unless a child of 13 could pass a standard of education. Meal- and rest-times were increased to two hours out of the working day of twelve, and schools, already by law attached to factories, had henceforth to be certified as efficient by examiners from the Department of Education.

Charles Roberts

One of the five doctors who did the inspections and measurements was a man whose interest in anthropometry after, if not before this experience, was profound. Charles Roberts (d. 1901) was a Yorkshireman by birth. He won numerous prizes in his career as a medical student at St George's Hospital Medical School in London, where he qualified in 1859. He then practised in York for a time, becoming surgeon to the North Riding prison (see note 8.2). He was elected Fellow of the Royal College of Surgeons in 1871 and in the same year moved back to London (see Power, 1930, vol. 2, p. 231). Probably the move followed Bridges' appointment as medical inspector to the Local Government Board the previous year. Bridges and Roberts had been students together at St George's Hospital and had worked together in the North Riding of Yorkshire where Bridges was factory inspector at the time Roberts was prison surgeon (see note 8.1). Roberts' interests were wide, and though he practised as a surgeon in London and had an appointment on the staff of the Victoria Hospital for Children he was also, in the phraseology of the time, possessed of independent means. He thus had ample opportunity to pursue his other interests, which centred on anthropometry and natural history. He edited the *Selborne Society's Magazine* and was secretary to the first Anthropometric Committee of the Anthropology Section of the British Association for the Advancement of Science (see note 8.3; he was not secretary to the Anthropology Section itself, as some of the obituaries allege). He was actively interested in physical training and himself had a 'spare, ascetic figure' (see Roberts, 1902). We cannot tell whether he read Quetelet's *Anthropométrie* (1870) before or after his involvement in the 1872–3 survey, but he certainly admired the work greatly, and set his heart on himself writing a book to be entitled *Physical Development and Proportions of the Human Body*. He never succeeded in this, but he did publish a shorter work, entitled *A Manual of*

Anthropometry (1878), which has much of Quetelet in it, including a section on body proportions addressed particularly to art students, and using Quetelet's own measurements of body segments. About Quetelet's survey of heights and weights, however, Roberts was critical.

Roberts' place in the history of growth studies rests on two papers, published between 1874 and 1876. He resembles, therefore, his contemporary and transatlantic collaborator Henry Bowditch rather than the later figure of Franz Boas, with his dozens of papers and decades of work. Roberts' report to the Statistical Society (1876) deals with the old question of how large and strong a child should be before being permitted to work in a factory. The provisions of the 1833 Act were still in force; and had been added to, so that in 1872 all children under the age of 16 applying for employment in a factory had to produce a certificate from a certifying surgeon saying they were not incapacitated by illness and had 'the ordinary strength and appearance' of children of at least 8 years, if they applied for part-time 5-hours-a-day work, or of young persons of at least 13 years, if they applied for full-time 10-hours-a-day work. (The provisions in the 1833 Act only applied to children under 13.) A Royal Commission appointed to inquire into the working of the Factory and Workshops Act recommended that all children employed under the provisions of the various Acts should produce certificates of birth and be given medical inspections of fitness for the occupation contemplated (Parliamentary Papers, 1876). However, a minority report, by the O'Conor Don, thought the surgeons could be dispensed with (they cost the manufacturers a good deal in fees), as in nearly all cases, he alleged, the age of a child could by this time be properly established by a birth certificate. As to certifying fitness, the 'previous factory legislation has rendered medical certificates useless and antiquated, as factories are more healthy, labour less fatiguing and bodily disease and infirmity uncommon, and excluded from employment by the simple commercial consideration that it would not pay for engaging [i.e. it would not pay to engage persons who were sick or infirm] (Roberts, 1876, p. 682; Parliamentary Papers, 1876).

Roberts therefore examined with exemplary care the large series of measurements of height, weight and chest circumference taken in the 1872–3 factory survey and established first, that 'there are no physical qualities sufficiently distinct and constant to indicate the age of a child within two and often three years of its actual birthday, and that the certificate of birth is the only evidence that can be relied on'; and second, 'that age is not an indication of any constant physical qualities, and a certificate of birth is not, therefore, sufficient evidence of fitness for factory work, and cannot supply the place of a medical examination' (1876, p. 682). As evidence of the first statement he gave frequency distributions of heights and weights of 771 boys 'of the exact age of 14 years', showing how large was the variation. He examined also whether or not the state of the eruption of the teeth could be used to validate age. Though he agreed that teeth were a better guide than body size, he wrote 'I am convinced that order of appearance of teeth varies so widely in different individuals and different classes of society that it cannot be trusted as a test of age' (1876, p. 684: see note 8.4). He then proposed using height, weight, chest circumference, and weight-for-height as criteria for fitness for employment, excluding the children at the lower ends of the frequency distributions of each of these variates. The actual values suggested as lower limits for height and weight are interesting: at age 8, 42 inches and 45 lb; at age 9, 44 inches and 57 lb; at age 12, 49 inches and 59 lb; and at age 13, 50 inches and 65 lb. Children as small as this would be quite impossible to find nowadays, except in cases of specific pathological disorders; the height limits are below the modern 1 in 1,000 line. The weights are a trifle higher, but still well below the present-day 1st centile.

Some at least of the certifying surgeons took their jobs seriously. Dr F. Ferguson of Bolton in the years 1872–3 rejected 237 boys aged 13 and over, and passed only 156. But with girls the proportions were reversed; he rejected 138 girls and passed 337 (Roberts, 1876, p. 692). It is not clear whether Dr Ferguson simply thought the place for a girl was in the factory, or whether perhaps he was suspicious that more boys than girls were overstating their ages in order to enter full-time work (and avoid the three-hours-a-day compulsory school attendance for children under the age of 13). In any case the effect on the statistics of the height and weight of the factory children in his area is spectacular. The weights of the 13-year-old girls he passed averaged 77.2 lb and of those he rejected 63.2 lb (p. 698). (We ought to view the 1872 results on factory children – column 1 of Table 8.1 – with this selection in mind. Even in the age range 8 to 12 the smallest children might have been rejected for part-time work, thus diminishing the differential between factory and non-factory children. Secondly the 12-year-old factory age group might have been shorn of its larger members, if they passed themselves off as 13 in order to get full-time work. The results themselves, however, show no evidence of a diminished 12 to 13 increment in the factory boys or girls.) Most certifying surgeons took a very different line from Dr Ferguson. A sub-inspector of factories in Glasgow in 1869 said certifying surgeons there 'with a few honourable exceptions, passed children regularly and systematically under age' (Parliamentary Papers, 1876, p. 64).

Though the survey of 1872–3 was a considerable affair, it concerned only children aged 8 to 12 inclusive. For this 1876 paper, therefore, Roberts in addition extracted measurements from the records of the Chelsea Royal

Military Asylum, a boarding school for sons of soldiers, and from the Royal Hospital School, Greenwich, a boarding school for sons of naval personnel, largely coastguards from rural areas. He thus extended the age range for boys to 13+. Roberts then divided the children into those living in rural and those in urban areas, with both groups subdivided into those working in factories and those not working in factories. Both means and modes are given for each year of age. (Roberts calls the modern mean the 'average' in his tables and the 'mathematical mean' at one place in his text; as in all his papers the word 'mean' in his tables indicates the modern 'mode'.)

The age range did not permit Roberts to say anything about adolescence. He does remark that 'girls from 8 to 12 are uniformly half an inch shorter, and from 2 to 3 lb lighter, than boys', but that '13-year-old girls are a little heavier and taller than boys in this country' (though actually he does not give any measurements of girls aged 13+ in his papers). He added in a footnote (1876, p. 691) that 'Professor Bowditch of Boston, USA, informs me that the girls are taller and heavier than boys in that city. In Belgium, according to Quetelet, the two sexes are the same weight at 12 but the boys are heavier at other ages'. There was a clear difference between the urban (factory plus non-factory) and rural boys, the rural being about 1.5 cm taller at 8 to 11, 2.0 cm taller at 12 and 3.0 cm taller at 13. This is a pattern which persisted in England for a further eighty years (see National Survey of Health and Development, p. 382 below), though the reverse is found in most modern continental European statistics (Eveleth and Tanner, 1976; Tanner and Eveleth, 1976).

The girls showed a similar but smaller difference, with rural girls taller by about 1.5 cm at 8 and 9, and by a little less at later ages.

As to the Minority Report's assertion that factories were so improved that medical examinations were unnecessary, Roberts declared 'It hardly deserves serious consideration . . . it would indeed be very deplorable if after forty years of protective legislation no decided improvements had taken place in the health and physical development of the factory operative' (1876, p. 691). Roberts compared his factory children's weights at ages 9, 10, 11 and 12 with those given by Cowell in 1833. The differences were 6.8, 4.5, 4.8 and 4.6 lb. 'From these figures', he wrote, 'it will be seen that a factory child of the present day of the age of 9 years, weighs as much as one of 10 did in 1833, one of 10 now as much as one of 11 then, and one of 11 now as much as one of 12 then; each age has gained one year in 40 years' (1876, p. 69). However, he said nothing about heights. (In this paper, that is; in the 1883 report of the British Association Anthropometry Committee — see note 8.3 — he admitted 'a slight but uniform increase in stature and a very large increase in weight'.) When we compare the heights of the 1872–3 factory boys with those of the 1833 factory boys of Cowell we find very little difference: at 9, 1.0 cm; at 10, 0.5 cm; at 11, 0.6 cm and at 12, nil. Even when all the 'labouring class' children, urban and rural, factory and non-factory, are pooled together, as in table 1 in Roberts' other paper, of 1874–6 (see below), the boys' heights exceed those of Cowell's boys only by 1.8, 1.8, 1.5, −0.7, 0.7 and 0.5 cm at ages 9 to 14. As to the girls, the 1872–3 factory girls were *less* tall at all ages than the 1833 ones, by a little less than 1 cm on average. Admittedly these comparisons ignore the fact that Cowell's children wore flat shoes and the 1872–3 children were barefoot. However this would only add 1 cm to the difference (see note 7.5). It seems that between 1833 and 1873 the heights of boys working in factories changed by only about 1.5 cm, or three months' growth, and that of girls changed perhaps not at all. Thus we have to question Roberts' interpretation of his results on weight. Some of the 4–6 lb weight difference, indeed, may be due to Roberts' children having worn more clothing. Roberts added 10 lb to the nude weights of the Chelsea and Greenwich boys to bring them equal to the rest, estimating that to be the weight of the older boys' winter indoor clothing. Cowell's boys were 'lightly clothed and had nothing in their pockets' and Quetelet suggested an allowance of only 5–6 lb for their clothing. Even this may be too much, for the British Association Anthropometric Committee (1879) actually weighed the clothes of Westminster schoolboys, after coats (i.e. jackets), waistcoats and boots had been removed, and found them to average 2½ lb. Thus in the case of Roberts' boys 3 to 4 of the 4–6 lb difference may be due to clothing, leaving a true difference of 1–2 lb which would agree well enough with the 1.5 cm difference in height.

It is just possible that the lack of difference between 1833 and 1873 is partly due to the 1833 children being selected by attending Sunday school. However, the difference between the unselected children of Horner's survey of 1837 and those of 1873 was just as small: a mere 0.6 cm in the case of boys and 1.2 cm in the case of girls. Furthermore, boys at the 'Friends' School, York' (presumably Bootham School), where records of height and weight were kept from at least 1853, showed no change in heights at ages 9 to 17 between 1853 and 1879. There was a very small increase in weight, at ages 15 and 16 only (British Association Anthropometric Committee, 1883, p. 47; see note 8.5). In the Friends' School heights were at the modern 25th centile throughout the age range except at 14 and 15, when they sank slightly below it. These boys were very much taller than the factory boys of 1833–73: the difference reached 4½ inches (11.5 cm) from ages 10 to 12.

Certainly the position of factory children was still deplorable. Roberts (1872–4) reported that the prevalence of severe flat feet reached 5 per cent in factory children aged 9 to 10, and 10 per cent and over in those aged 11 and 12. In children in country towns and agricultural districts the prevalence was only 1–2 per cent, as it was also in the factory children on entry at age 8. This, however, was the only difference between factory and non-factory children that Roberts thought was due to the work in the factory itself. 'Nearly the whole of the disadvantages the factory children labour under are to be attributed to social causes', he wrote, 'rather than factory work' (Parliamentary Papers, 1876, appendix D. p. 180).

Much of Roberts' other major paper (1874–6) is an exposition of Quetelet, though not an uncritical one. Quetelet had measured thirty individuals of the same age, divided them into three groups of ten each and observed that the means of each group were nearly the same; hence, he said, a group of ten persons was sufficient to characterize typical proportions. 'In this respect', says Roberts' (p. 15), 'my tables will show M. Quetelet was in error, and will account for the differences which occur between his results and mine'. In this paper Roberts gave a table of heights of boys which is something of a landmark in the literature of growth, for it represents the first occasion on which a frequency distribution for a body measurement appears since Quetelet's study of newborns in 1831.

Modes, means and frequency distributions are given for each year of age, effectively from 5 till 25 (boys only). Roberts pools all his 1872–3 values for the 'labouring class', which now includes the schoolboys of Chelsea and Greenwich, and boys in rural and urban areas, whether working in factories or not: all, however, belong to families of artisans or labourers, the manually employed. The numbers at age 5 to 15 are 175, 327, 581, 1,071, 1,252, 1,669, 1,246, 940, 735, 869 and 409; over age 15 the sample is biased, because it consists only of boys accepted for entry to the army and navy, which was governed by a lower limit for height. This bias is nicely shown in the relative progression of means and modes. Up to age 16 there is very little difference between them, but at age 16 and upwards the mean comes to exceed the mode considerably. In consequence, in Fig. 8.1 and in the interpretations below, the modes have been used for the age groups 16 and over. Viewed this way the adolescent spurt is clear, and the peak, represented by the largest difference between yearly means, is located in the 15.5–16.5 interval, that is at 16.0 years, two years later than today. This largest difference, or increment, amounts to 6.2 cm in the year, compared with 7.3 cm in modern data. Due to selection at the later ages it is not really possible to say what the final height of the manual worker group was, nor when it was reached, though the modes terminate at 168 cm (5 feet 6 inches), the same as that recorded for prisoners in 1862 by Danson (see p. 161). From 8 to 12 years this pooled group of sons of manual workers exceeds in height its subgroup of the 1872–3 factory-working boys, but only by about 1.5 cm.

Roberts' table also contains the heights of boys of the 'non-labouring' (i.e. non-manual, white collar) class, collected by Francis Galton and his associates at almost exactly the same time (see below). These were boys at Public (i.e. private, middle class) Schools, naval officer colleges, and universities. Fairly substantial numbers were again involved, the figures in successive age groups from 9 to 24 years being: 62, 235, 430, 745, 969, 990, 819, 462, 313, 300, 344, 262, 471, 600, 660 and 689. This time no differences between means and modes are seen in the older age groups; no cut-off selection occurred. The means have been plotted in Fig. 8.1. The maximum yearly height increment is 6.9 cm/yr, centred at 15.0 years, a year earlier than in the manual class.

Fig. 8.1. Heights of boys of manual workers and of the non-manual, mostly professional, class, in about 1870 in England. Values from Roberts (1874–6). Means are used for the manual group up to and including 15.0; modes thereafter (see text). Means are used for the non-manual group throughout.

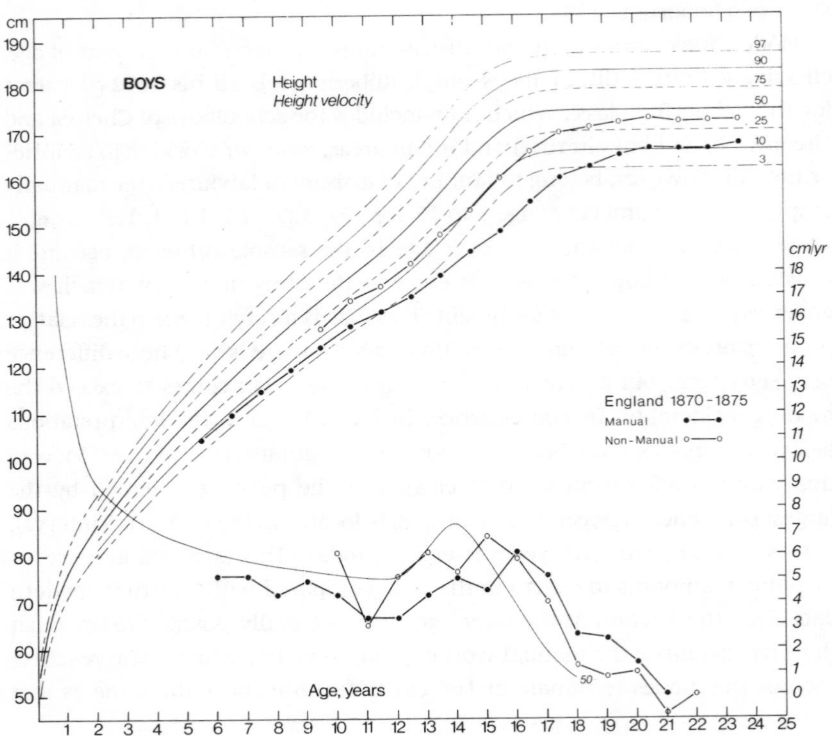

Final height is about 173 cm (5 feet 8 inches), which is some 5 cm shorter than the average for a similar social class nowadays. It was reached by the average boy at about age 20. The difference between manual and non-manual classes was about 6 cm at ages 9 to 11, 7 cm at 12, 9 cm at 13 and 14, and 11 cm at 15 and 16. It then gradually reduced to a final adult figure of between 5 and 6 cm. In 1965–75 the equivalent comparison gave a difference of about 3 cm at ages 6 and 7 rising to about 4 cm at age 11 (Goldstein, 1971; Rona, Swan and Altman, 1978), and dropping again to end in adulthood at 3 to 3.5 cm (Miller, Billewicz and Thomson, 1972).

Roberts noticed, and, in contrast to Quetelet, accepted, that 'with the accession of puberty there is, according to my observations, an increased rate of growth' (1874–6, p. 22). He saw this chiefly in the non-manual class, where it was more decided, and called it 'pre-pubertic growth'. The remainder of his paper returns to the exposition of Quetelet.

Roberts, again in contrast to Quetelet and indeed to all his predecessors, laid great emphasis on *variation* around the mean. 'A glance at the Table will show how impossible it is to study the progressive development of man, and the causes which check or promote it, by the uses of mere averages. The difference in height of the average Public School boys of the ages of 13.5 and 14.5 is about 2 inches but the difference between the tallest and shortest boys of either of these ages is 20 inches ... it is for this reason that I have given the whole range of heights which I have found to occur at each age' (1874–6, p. 20). Roberts gave also tables of weight against height, which, simultaneously with those of Bowditch, were the first bivariate tables of measurements ever to appear.

In the same year Steet (1874–6) reported the heights, weights and chest circumferences of 3,695 boys, aged 13 to 19, who applied for jobs as telegraph boys. Boots were removed. The maximum yearly increment was 7.0 cm/yr, reached at 15.0 years. Corresponding modern values are 7.3 cm/yr and 14.0 years. Final height, judging by the succession of increments, would have been reached by age 20 in these boys, and it would have been about 169 cm (5 feet $6\frac{1}{2}$ inches). At about the same time Colonel Lane Fox (1877), better known as an anthropologist under his later name of Pitt-Rivers, reported the heights of the territorial militia of mid-Surrey taken at their annual training camp. These were manual workers, and, Fox says, probably those in the lower part of the social scale, since the better-employed were not so attracted by the payment offered. They were nearly all country folk between the ages of 17 and 30. Mean height was 166 cm (5 feet $5\frac{1}{2}$ inches).

A few years later Roberts published his *Manual of Anthropometry* (1878). For the most part this merely repeats the material of his two big papers; but the tables he gives of growth of boys of the manual and non-manual classes have different numbers and somewhat different means from those given before. The non-manual group has been augmented by a large group of students at Oxford, Cambridge and St George's Hospital Medical School, some of the school-age subjects, however, have unaccountably disappeared. The non-manual boys' heights are all slightly greater, by about 1 cm till age 17 and by 2 cm at subsequent ages. Final height is now clearly 175 cm (5 feet 9 inches) instead of 173 cm (5 feet 8 inches). Maximum yearly increment remains centred at about 15.0 years. The manual boys' means are not much altered till 13, when the numbers increase and also the mean height, which is about 5 cm greater than before at 14, 15 and 16, returning to the earlier value at 23. Steet's (1874–6) telegraph boy's heights were included this time, but as they agree almost perfectly with the means of the earlier report they could not have caused the alteration. It seems best to rely on the first and most clearly presented tables and these are the ones plotted in Figs. 8.1 and 8.2. In his *Manual* Roberts also prints Bowditch's figures for Boston schoolchildren, saying that 'Professor Bowditch has, at my suggestion, tabulated his statistics in the manner I am now advocating, which I first adopted in my paper on Factory children . . . I have already insisted on the absolute necessity of making statistical tables complete in every detail' (1878, p. 87).

Roberts maintained his interest in growth, and served on the British Association Anthropometric Committee from 1879 to 1883. He wrote a book on *The Detection of Colour-Blindness and Imperfect Eyesight* (1881), a note for *Nature* commenting on sex differences in growth (1890) and a memorandum on physical education in schools for a Royal Commission on Secondary Education (1895). He retired at the turn of the century and died shortly after. 'He was a man', says his obituary, 'of that unassuming scholarly type which is now rarer in our ranks than was the case in former days'.

Francis Galton and family likeness

Roberts' insistence on the importance of variation certainly echoed, if it did not indeed stimulate, one of Francis Galton's many preoccupations. Galton (1822–1911: see Forrest, 1974), had a hand in everything, and growth was no exception. In 1873 he put a proposal to the recently formed Anthropological Institute asking them to sponsor a programme of body measurements in schools. Since schoolmasters were 'trustworthy and intelligent in no small degree' the statistics would be reliable, and the boys measured would grow up with favourable recollections of the procedure and hence would willingly submit to further measurement when grown up and working in universities, factories, etc. (Galton clearly had in mind his family record study, to be initiated some ten years later.) There is an echo of Villermé about the opening words of the proposal: 'We do not know whether the general physique of the nation remains year after year at the same level, or whether it is distinctly deteriorating or advancing in any respects. Still less are we able to ascertain how we stand in comparison with other nations, because the necessary statistical facts are, speaking generally, as deficient with them as with ourselves' (Galton, 1873–4). Galton made clear that he wanted to recruit all classes of schools 'from Public Schools to schools for pauper children' and then to obtain an idea of the overall situation in the whole country by weighting the different social classes according to their numbers in the census. This must have been one of the first proposals for the use of stratified sampling.

In the event only a number of Public Schools cooperated; they were Marlborough, Clifton, Haileybury, Wellington and Eton Colleges in the country; and City of London School, Christ's Hospital, King Edward's School, Birmingham and Liverpool College in the big towns. The survey in Marlborough College, a school then some thirty years old, attracting primarily the sons of parsons, doctors, officers and lawyers living in the south-west of England, was the first to get going, in 1874. This was because the headmaster, Dr Farrer, was 'honourably known to the scientific world by his early and outspoken advocacy of the introduction of science teaching into schools' (Galton, 1874). In consequence the College had already a full-time biology master, G. F. Rodwell, and in addition the first resident doctor to be appointed to an English school (in 1849), Walter Fergus (see Fergus, 1887).

Fergus and Rodwell (1874) submitted their own report to the Institute, where it was read and commented on by Galton (1874). They measured height, weight, and chest, upper arm and head circumferences in 550 boys aged 10 to 19. They constructed a height measurer much superior to any used before: or perhaps Galton did, for whereas they say they used a 'vertical board provided with a sliding square at right angles to it' Galton added blandly in a footnote: 'A bracket sliding between vertical guides, and balanced by a counter-weight, acting over two pulleys, as in the Figure [reproduced as Fig. 8.2 here] will be found easy, quick and sure in its action. The vertical board and foot-piece may be dispensed with, if the guides can be nailed to the wall' (1874, p. 126). He thus anticipated by seventy-five years R. H. Whitehouse's modern Harpenden stadiometer in everything except digital read-out. Height was taken without boots, and weight clothed and with boots. Frequency distributions were given for each year of age for height, weight and head circumference, and Galton added the 'probable error' values for each distribution, having remarked first that most of the distributions approxi-

Fig. 8.2. Galton's stadiometer, reproduced from Galton (1874). It is uncertain to what extent this was ever used.

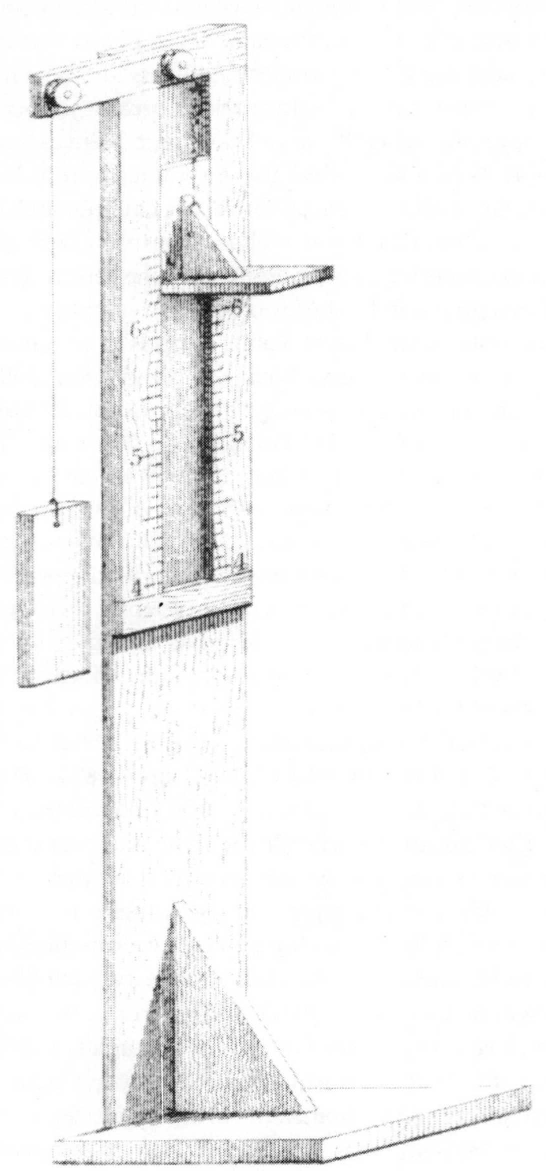

mated to the law of error described in Quetelet's Letters on the *Theory of Probabilities.* This is one of the first occasions on which the probable error appeared; it was later (1888) renamed 'probable deviation' and in 1894 replaced by Karl Pearson's 'standard deviation'. In the Marlborough figures there is a clear spurt in height, with a maximum yearly increment of 7.2 cm/yr centred at 15.0 years. Final height would be about 171 cm (5 feet 7 inches), reached at 20 years. Fergus and Rodwell remarked that head circumference was not related in any obvious way to ability in school.

Results from other schools came in more slowly and were never published in detail. However, Galton (1875–6) did contrast the heights of 14-year-old boys in the country schools with those in the city schools, in a paper chiefly remarkable for seeing the use of percentiles for the first time, a system invented by Galton himself in 1875. The country boys were some 3.5 cm taller; however, Galton does not mention that the country schools were also those with pupils from generally richer families. In none of these publications is Roberts mentioned, nor are the Commissioners or the Factory Acts, though Galton must have been aware from his youth of the earlier studies of the 1830s, for at that time he was a frequent visitor to the Horners' house. It seems that Roberts sought Galton out when he heard what was going on (through Dr Fergus, he implies) and got permission to use the entire statistical archive accumulated by the Anthropological Institute. The records of these Public Schools appear as the 'non-labouring' children in Roberts' 1874–6 paper and again in his 1878 book. Galton and Roberts actually worked together in the early 1880s on the Anthropometric Committee of the British Association for the Advancement of Science. This was set up at the Bristol meeting of the Association in 1875, and issued annual reports over the period 1878–83 (see note 8.3). Galton was a member from the beginning and chairman from 1881 to 1883. Roberts only joined the Committee in 1879, but thereafter played a very prominent part, and wrote much of the final report (Brabrook, 1904–5). Roberts' name, incidentally, fails to appear in the index of Karl Pearson's massive *Life, Letters and Labours of Francis Galton* (1914–30).

It was in the 1881 Anthropometric Committee report that Galton described, with his usual clarity, the distortion introduced into growth curves at puberty by averaging cross-sectional data – the so-called 'time-spreading' or 'out-of-phase' effect (see Tanner, 1962). Making characteristically light weather of findings that later confused Bowditch and Porter, and caused even Boas to labour mightily to comprehend, Galton wrote: 'Another very curious fact is a marked increased of range of height from about 14 years to 16 years of age, in Classes I and II ... which disappears afterwards. Probably the increase of rate takes place in different boys at slightly different ages, and therefore becomes smoothed down in the mean result. If so, it would be still more striking if the classes had been further subdivided'. (British Association Anthropometric Committee, 1881, p. 249). Nobody seems to have understood what he meant, and his statement has only once been quoted, by Backman in 1934 (see note 8.6). It took Boas forty years to close his hand fully on this particular nettle, starting with the dim realization in 1892 that a 'purely statistical phenomenon' was altering the curves, and passing to a full understanding and the construction of the famous 'time-spreading' diagram of Boas and Shuttleworth in the 1930s (see Tanner, 1959; and p. 236 below).

Galton also realized (like Pagliani before him: see Chapter 9) that tallness representing advancement in childhood was no guarantee of ultimate tallness. Perhaps it was quite the contrary, he reflected. In any case, to find out, 'the class of statistical researches in anthropometry that most deserves support at the present time is the preservation of records of the same individual throughout life ... A large collection of well-kept records of this kind would be of the highest value, not only from an anthropometric but from a sanitary point of view, using that term in its widest sense' (Galton in British Association Anthropometric Committee, 1881).

Galton later turned his attention to the totally neglected subject of the inheritance of height (Galton, 1885–6, 1886). 'I had to collect all my data myself', he says, 'as nothing existed, so far as I know, that would satisfy even my primary requirement. This was to obtain records of at least two successive generations of some population of considerable size' (Galton, 1889). He collected 150 family records of height simply by offering publicly a prize to the best ones sent in. Records arrived of 205 couples, with 930 children; but the data were poor, the heights taken some with and some without shoes; some, even, were only estimated. By personal correspondence he obtained rather better data on 783 brothers from 295 families, and he was able also to make use of the heights of nearly 10,000 unrelated persons measured in his Anthropometric Laboratory at the International Health Exhibition held in London in 1884. Percentiles for these latter were published in 1885 (Galton 1884–1885). Galton used a graphical method of estimation, and gave the 5th, 10th, 20th, 30th, 40th, 50th, 60th, 70th, 80th, 90th and 95th percentiles. The 5th, 50th and 95th for men's and women heights (in inches) were 63.2, 67.9 (172 cm), 72.4, and 58.8, 63.3 (161 cm), 67.3 respectively; for sitting heights 33.6, 36.0, 38.2, and 31.8, 33.9, 36.0; and for span 65.0, 69.9, 74.8, and 58.6, 63.0, 68.0.

In this paper, incidentally, Galton mentions, it seems for the first time, the difficulty occasioned by not knowing whether an observer has read the measuring instrument to the nearest unit or to the last completed unit (it making a difference of half a unit in the means). There are still many authors who fail to make this distinction, or make it wrongly.

It was in studying these family data that Galton introduced the regression coefficient and the coefficient of correlation (in 1885 and 1888 respectively, but the better known source of both is *Natural Inheritance*, published in 1889). Both with his emphasis on the inherited element in stature and his invention of the most important statistical tools since Quetelet's recognition of the general applicability of the law of Laplace and Gauss, Galton transformed the study of growth.

Henry Bowditch

This new burst of activity in England was matched, if not yet exceeded, in the United States. Galton in 1883 had written: 'The records of growth of numerous persons from childhood to age are required before it can be possible to rightly appraise the effect of external conditions upon development, and records of this kind are at present non-existent' (p. 41). He was wrong; besides the records of Montbeillard's son and Quetelet's children, and, had anyone known it, of the Carlschule, there were the slightly more numerous records kept by an extended New England family on the backs of its doors.

At a meeting of the Boston Society of Medical Sciences in 1872, H. P. Bowditch, then lately returned from postgraduate training in Germany to take over the teaching of physiology at Harvard Medical School, presented a paper that was the beginning of all the North American work on human growth. The paper was never published except in summary: "Dr Bowditch exhibited a diagram showing the rate of growth in height of the two sexes . . . The curves represented the average measurements on 13 individuals of the female and 12 individuals of the male sex. The measurements were all taken annually during the last 25 years, and the individuals were all nearly related to each other' (Bowditch, 1872, and 1877, p. 275). Sadly, we do not know the name of the Galtonian spirit in Boston who was responsible for the collecting. It seems quite likely it was someone in Bowditch's own family, which was certainly sufficiently scientific and perhaps sufficiently large, for he himself was one of six siblings. Bowditch's paternal grandfather translated Laplace's *Mécanique celeste* and wrote a classic seafarer's manual called the *Practical Navigator*. His father, a merchant, maintained a strong scientific interest and edited new editions of the *Practical Navigator* as they were needed. One of his uncles was professor of medicine at Harvard and founder of the Massachusetts State Board of Health. Perhaps these two men and their brothers and sisters produced between them the twenty-five children measured. At least we

Table 8.2. *Bowditch's records of the heights of twenty-five children in a Boston family, measured in 1850–70 and communicated privately to Roberts*

	Age-at-last-birthday										
	Birth	1	2	3	4	5	6	7	8	9	10
Females											
Lillie	—	—	—	—	—	—	—	—	—	—	—
Mary	—	—	—	—	—	—	—	:–	4/0.0	4/2.3	4/4.0
Alice	—	2/5.0	2/7.8	2/11.0	3/1.1	3/3.7	3/6.5	3.8.7	3/10.8	—	4/3.0
Charlotte	—	2/4.0	2/9.0	3/0.8	3/3.9	3/6.8	3/8.9	3/11.6	4/2.1	4/3.4	4/6.0
Lucy	—	2/4.5	2/9.3	3/0.7	3/3.7	3/6.6	3/8.8	3/10.6	4/1.1	4/3.7	4/6.0
Lily	1/11.0	2/7.1	2/10.0	3/1.2	3/4.1	3/6.4	3/9.2	3/11.6	4/2.0	4/4.3	4/6.2
Livy	—	—	—	—	3/1.8	3/4.2	3/6.8	3/8.5	3/11.0	4/2.0	4/4.1
Fanny	—	—	—	—	—	—	3/9.4	4/0.3	4/2.5	4/4.1	4/6.9
Esther	—	—	—	3/0.4	3/3.1	3/5.6	3/7.3	3/9.5	3/11.5	4/1.8	4/4.3
Susan	—	—	2/5.6	2/9.8	3/0.8	3/3.2	3/6.3	3/8.7	3/11.1	4/1.7	4/3.6
Arria	—	2/1.0	2/6.3	2/11.4	3/2.3	3/4.8	3/6.7	3/9.6	—	4/1.9	4/3.8
Mary	—	2/2.6	2/6.5	—	3/1.4	3/4.2	3/6.4	3/8.7	—	4/2.3	4/3.7
Annie	—	2/2.8	2/6.3	2/10.6	3/1.4	3/3.3	3/6.0	3/8.1	3/10.3	4/0.6	4/2.8
Average height	—	2/3.8	2/7.6	2/11.7	3/2.4	3/4.9	3/7.5	3/9.8	4/0.5	4/2.6	4/4.7
Annual increase (inches)	—	—	3.8	4.1	2.7	2.5	2.6	2.3	2.7	2.1	2.1
Males											
Frank	—	—	—	—	—	3/7.8	3/10.7	4/1.4	4/4.4	4/7.0	4/8.8
Henry	—	—	—	—	—	—	—	3/8.4	3/10.9	4/1.3	4/3.2
Charles	—	—	—	—	3/6.2	3/9.0	4/0.0	4/2.3	4/4.9	4/7.5	4/9.4
Alfred	—	2/8.3	3/0.6	3/3.8	3/6.7	3/9.0	3/11.8	4/2.2	4/4.7	4/7.3	4/9.3
Nat	—	—	—	—	—	—	—	3/11.0	4/1.7	4/3.2	4/5.0
Ned	—	2/4.3	2/9.0	3/0.0	3/2.3	3/5.6	3/7.7	3/9.8	4/0.0	4/2.3	4/4.2
Vin	—	2/8.2	3/0.0	3/3.2	3/5.6	3/7.3	3/10.0	4/0.3	4/2.7	4/5.4	4/7.0
James	—	2/2.2	2/4.7	2/10.0	3/2.4	3/5.5	3/7.8	3/10.7	4/0.8	4/4.8	4/6.6
Ernest	—	2/4.0	2/9.0	2/11.9	3/3.8	3/5.9	3/8.2	3/11.0	4/1.7	4/3.4	4/4.8
John	—	2/4.0	2/8.0	3/0.3	3/2.5	3/5.3	3/8.4	3/10.3	4/0.6	4/2.0	4/4.3
Arthur	—	2/5.0	2/7.5	2/10.0	2/11.4	3/1.6	3/2.7	3/4.0	3/7.6	3/10.0	4/0.0
Basil	—	2/5.0	2/8.0	2/11.8	3/2.5	3/4.0	3/6.7	3/8.6	3/11.4	4/1.4	4/3.1
Average height		2/5.1	2/9.5	3/0.3	3/3.5	3/6.1	3/8.6	3/10.6	4/1.3	4/3.6	4/5.5
Annual increase (inches)	—	—	4.4	2.8	3.2	2.6	2.5	2.0	2.7	2.3	4.9

From Report of the Anthropometric committee of the British Association (1880, pp. 142–3). Measurements are in feet, inches and tenths.

11	12	13	14	15	16	17	18	19	20	21	22
4/4.6	4/6.1	4/11.0	4/11.9	5/2.3	5/3.6	5/4.2	5/4.7	5/5.1	5/5.3	—	—
4/4.7	4/10.8	5/1.4	5/2.6	5/3.0	5/3.0	5/3.3	5/3.6	—	—	—	—
4/5.3	4/8.4	4/11.5	—	5/3.4	5/4.5	5/4.6	—	5/5.3	5/6.0	—	—
4/8.5	4/10.8	5/2.1	5/4.6	—	—	5/6.3	—	5/6.9	5/7.8	—	—
4/8.3	4/10.5	5/0.9	5/4.0	5/4.7	5/5.6	5/7.0	5/7.3	—	—	—	—
4/8.2	4/10.0	5/0.4	5/3.7	5/5.1	5/5.6	5/6.4	5/6.6	—	—	—	—
4/6.7	4/10.3	5/0.8	5/2.3	5/3.2	5/3.9	—	5/4.2	5/4.9	—	—	—
4/9.3	5/0.2	5/2.7	5/4.0	5/4.7	—	5/5.0	5/5.0	5/5.7	5/5.9	—	—
4/6.5	4/9.5	4/11.8	5/0.9	—	5/1.1	5/2.3	5/2.4	—	—	—	—
4/6.2	4/9.5	4/11.8	5/4.3	5/1.8	5/1.2	5/3.0	—	5/3.3	—	—	—
4/6.3	4/8.5	—	5/2.1	5/4.2	—	—	—	—	—	—	—
4/5.6	4/7.0	4/11.0	5/0.0	5/3.5	5/4.4	5/4.6	—	—	—	—	—
4/5.0	4/6.9	4/10.0	5/1.2	5/2.8	5/4.2	5/5.2	5/5.2	—	—	—	—
4/6.8	4/9.0	5/0.3	5/2.2	5/3.5	5/3.8	5/4.7	5/4.9	5/5.2	5/6.2	—	—
2.1	2.4	3.3	1.9	1.3	0.3	0.9	0.2	0.3	1.0	—	—

11	12	13	14	15	16	17	18	19	20	21	22
4/10.8	5/0.0	5/2.9	5/7.6	5/9.3	5/9.8	5/10.4	5/10.5	5/11.3	5/11.4	5/11.6	—
4/5.0	4/7.2	4/9.0	4/11.0	5/1.4	5/4.7	5/7.2	5/8.8	5/9.5	5/9.8	5/10.0	—
4/11.2	5/1.1	5/3.0	5/4.4	5/6.3	5/9.2	5/11.3	6/0.8	6/0.9	6/1.0	—	—
4/11.4	5/1.2	5/2.8	5/4.5	—	5/8.4	—	—	—	—	—	—
4/7.0	4/9.5	4/11.3	5/1.0	5/3.2	5/5.6	5/7.7	5/10.0	5/10.8	5/10.9	5/11.0	5/11.3
4/5.8	4/7.9	4/9.9	5/1.0	5/4.9	5/8.7	5/9.2	5/9.4	5/10.0	5/10.0	—	5/10.1
4/9.2	4/11.3	5/3.2	5/6.8	5/8.8	—	5/10.5	—	—	—	—	—
4/8.5	4/10.0	4/10.9	5/2.9	5/4.7	—	5/8.0	5/8.7	5/9.0	5/9.2	—	5/9.9
4/7.0	4/9.2	4/11.5	5/1.7	5/4.2	5/7.5	—	5/9.6	5/10.6	—	—	—
—	4/7.4	4/9.2	4/10.3	5/1.4	5/3.7	—	5/8.0	5/9.8	5/10.8	—	—
4/2.1	4/4.0	4/5.4	4/7.6	4/10.1	5/0.6	5/2.6	5/3.4	—	—	—	—
4/5.0	4/6.8	—	—	—	—	—	—	—	—	—	—
4/7.5	4/9.3	4/11.5	5/2.0	5/4.2	5/6.4	5/8.3	5/9.0	5/10.5	5/10.7	5/10.9	5/10.8
2.0	1.8	2.2	2.5	2.2	2.2	1.9	0.7	1.5	0.2	0.2	—

know the childrens' first names and, more importantly, their actual measurements. Both appeared in the 1880 report of the British Association Anthropometric Committee, placed there by Roberts, to whom Bowditch had presumably sent them during the course of their correspondence. They are reproduced in Table 8.2. The most interesting aspect of the inquiry perhaps was what struck Bowditch about the curves. He noted that boys were taller than girls from early childhood till 12, but then 'at about twelve and a half years of age girls begin to grow faster than boys, and, during the fourteenth year, are about 1 inch taller than boys of the same age. At fourteen and a half years of age boys again become the taller, girls having, at this period, very nearly completed their growth, while boys continue to grow rapidly till 19 years of age. The tables and curves of growth given by Quetelet show that, in Belgium, girls are, at no period of their lives, taller than boys of the same age' (1877, p. 275). Bowditch was intrigued, and disquieted. He noted that Cowell's figures for working children in Manchester and Stockport showed, like his own, girls taller at 13 and 14, but ended 'It would be interesting to determine . . . in what races and under what climatic conditions the growth of girls, at about the period of puberty, is the most rapid. It is possible that in this way, facts may be discovered bearing upon the alleged inferiority in physique of American women' (1877, p. 276). Quetelet's error had born dialectical fruit.

Henry Pickering Bowditch (1840–1911) was a prominent Bostonian who went to school with Oliver Wendell Holmes, studied with Claude Bernard and Carl Ludwig and was the teacher of William James, Stanley Hall, Charles Minot and Walter B. Cannon. He was the creator, practically speaking, of experimental medicine in North America, co-founder and second president (from 1888 to 1895) of the American Physiological Society and dean of Harvard Medical School from 1883 to 1893. He was the driving force behind the building of the new Harvard Medical School, which was opened in 1906, but by that time he had fallen victim to Parkinsonism; he retired after the dedication ceremony and died in 1911. He left a devoted succession of students; just as Bowditch had been attached to his teacher Ludwig, who in Leipzig had the most advanced as well, perhaps, as the most friendly physiological laboratory in the world, so also Bowditch's pupils spoke of their mentor with deep respect and affection. Prominent amongst them was his successor and memoirist W. B. Cannon (1924; see also the memoir of his youngest son: Bowditch, 1958).

We may suppose that the initial source of Bowditch's interest in growth was his family's height records. He was greatly beholden to Charles Roberts, as he was careful to say, for introducing him to frequency distributions, and also to Francis Galton for demonstrating how to calculate percentiles. But his first work was done quite independently of these two and in the same year (1872) as the Bridger–Holmes survey described above. Bowditch was in Germany at that time. His later contact with Charles Roberts was perhaps made through Galton, or perhaps through Michael Foster, the founder (in 1887) of the *Journal of Physiology*, of which Bowditch was American editor. Roberts and Bowditch exchanged unpublished data across the Atlantic and joined hands in rebutting Quetelet's ideas about the growth curve. Both were active in growth work during exactly the same span of dates, 1872–91. But whereas Roberts had no successor, Bowditch was the instigator of a considerable school, which included W. T. Porter, and, fractionally later, Franz Boas.

The second and more powerful source of Bowditch's interest makes interesting reading today. It is indicated by that curious phrase in his 1872 paper quoted above: 'It is possible that in this way facts may be discovered bearing upon the alleged inferiority in physique of American women'. In the 1870s in the Eastern United States a considerable movement arose against the way women's education was practised. It had in it a large element of what nowadays we would call backlash against women's liberation; but it also had a strain of genuine concern over the effects of too much school pressure on both boys and girls, which began in the 1830s and reached a climax about 1880, both in America and Europe (Duffy, 1968, and see next Chapter). It was articulated by Edwin Clarke, professor of materia medica at Harvard University, in a book entitled *Sex in Education, or A Fair Chance for Girls*, published in 1873. Amongst other things, Clarke defends Harvard College against the allegedly great regiment of women intent on occupying the best places in it. He directs his anger at Iowa College, Antioch, and Michigan University for their coeducational policies and raises high the standard of Harvard 'whose banner ... is the red flag that the bulls of female reform are now pitching into' (p. 149: this and subsequent quotations are from the 5th edn).

Clarke's language recalls Guarinoni and Virey at their best; even the anti-masturbators scarcely scaled his rhetorical heights. 'On the luxurious couches of Beacon Street;', he said, 'in the palaces of Fifth Avenue; amongst the classes of our private, common and normal schools; amongst the female graduates of our colleges; behind the counters of Washington Street and Broadway; in our factories, workshops and homes – may be found numberless pale, weak, neuralgic, dysmenorrheic girls and women that are living illustrations of the truth of this brief monograph ... If all this goes on much longer the sons of the New World will have to re-enact, on a magnificent scale, the old story of unwived Rome and the Sabines' (pp. 62–3). Clarke, and many other doctors of the time, showed the same sort of magical concentration on the ovary that earlier – and some contemporary – doctors had given to the testis. Clarke espoused the idea of a critical period in the growth of the female reproductive tract; the period around menarche was one of enormous importance and girls had to be guarded against overwork at this time or they would fail to develop healthy ovaries and their accessory organs. Too much schoolwork at this junction produced 'monstrous brains and puny bodies; abnormally active cerebration and abnormally weak digestion, flowing thought and constipated bowels' (p. 41). Girls and boys ruined by the enervating fever of repeated examinations, said Crichton-Browne (in Hertel, 1885) later 'sniff the Spring with nothing but the apprehension of catarrh' (p. xxvi). 'Dr Charles West' (the founder of British paediatrics and of The Hospital for Sick Children, Great Ormond Street), Clarke says, 'stresses that girls ought to take precautions to ensure that the second and subsequent periods occur on time . . . until at length the regular *habit* of menstruation is established' (his italics). There is an attribution of responsibility, even a note of moral reprobation.

It was true that the prevalence of consumption in girls aged 15 to 20 in England in 1860–80 was more than double the prevalence in boys of the same age, though after 20 and before age 5 boys had the higher risk (Hertel, 1885, p. xxxl); so it was perhaps not surprising that Clarke, in this book and a subsequent one (1874), roused considerable discussion and anxiety. The great Weir Mitchell supported him and inquiries were made into school hygiene in the state of Massachusetts. T. A. Emmett (1879) in his very influential *Principles and Practice of Gynaecology* wrote that at female puberty 'the whole nerve force is taxed for the full development of the organs of generation . . . to enable her to reach the highest physical development, the girl in the better classes of society should pass the year before puberty and some two years afterwards free from all exciting influences . . . such as music and light literature, which in a sensitive nervous system are capable of arresting the development of the uterus and ovaries . . . for the ovaries will always be arrested if the brain is forced'. Even as late as 1891 G. J. Engelmann (1847–1903: see Engelmann, 1904) gave a presidential address to the Southern Surgical and Gynecological Association full of frightening stories, strongly reminiscent of Guarinoni's on alcohol, as to the dire and ineradicable effects on a woman's life if one of the first periods should be stopped by a sudden shock.

Henry Bowditch found himself in the middle of this slightly hysterical controversy. He was a member of the Boston School Committee (and later chairman, from 1877 to 1881) and in 1874 wrote to Edward Jarvis, the founder of the American Statistical Society, saying he proposed 'to investigate the asserted physical inferiority of American women by comparing the rates of growth in height and weight of the two sexes in this country and in Europe. Quetelet gives certain data . . . ' (Bowditch, 1874–5). Several letters later he asks whom he should consult about the best way of working up the data, and on 14 October 1875 remarks that the head of the Statistical Bureau in Leipzig has 'gone into the matter with great energy' (see note 8.7).

In 1875 Bowditch persuaded the Boston School Committee to do the growth survey he had discussed, of pupils in the public (that is, state) schools; and it was extended into some of the private schools of the area as well. In all, 24,500 children were measured, more than twice the number in the British 1872–3 factory survey. Heights were taken without shoes, weight in indoor clothes. The birthplaces of the parents were recorded and their occupations obtained from school records. Bowditch's training as a meticulous and quantitative physiologist appears at once. He was the first person to go on record as editing growth data. 'In the progress of the work', he wrote (which was placed, incidentally 'in the hands of professional accountants'), 'many cases were met with of heights and weights differing so widely from the average measurement to which they belonged as to excite a suspicion of error in the observation' (p. 281). Forty errors were discovered and corrected by remeasuring the children concerned.

Bowditch's main contributions to growth were published in the 8th, 10th, 21st and 22nd annual reports of the Board of Health of the State of Massachusetts (which he chaired in the late 1870s) (Bowditch, 1877, 1879, 1890, 1891). All are of the highest quality and that in the 8th annual report (1877) is a classic of the international growth literature.

Besides presenting charts of the mean height at successive ages, Bowditch gave a table of the successive differences between the yearly means. He then discussed whether 'the rate of growth of children may be ascertained by computing at any one time the average height and weight of children of different ages, as well as by determining the average height and weight of a given set of children in successive years. This assumption is doubtless perfectly justified', he concludes, 'though certain theoretical objections may be urged against it' (1877, p. 280). It is a little surprising that Bowditch failed to realise that the objections were more than theoretical: or at least it would be so if he had read B. A. Gould's very clear account, in his *Investigations in the Military and Anthropological Statistics of American Soldiers*, published in 1869.

Something made Bowditch write to Gould about this issue in 1877 and Gould's reply (given in note 8.6) was certainly less enlightening than his earlier published account.

Bowditch's tables of mean increments show the adolescent growth spurts in both sexes very clearly. The children were divided into those of American-born parents and those of Irish-born parents (see below). The boys of the American-born parents had their greatest annual mean increment centred at age 14.0, the girls at age 12.0. These values are barely six months later than present-day North American ones. The mean increment over the peak year was 7.4 cm/yr in the boys, exactly the same as in modern data. In the groups of children with Irish-born parents the boys' peak is not well determined because of a change in social class in the samples at the critical ages (see below); the girls' peak was at 12.5 years. Bowditch spelled out the relation of the spurt and puberty more clearly than did Roberts (indeed the spurt was often referred to as 'Bowditch's Law of Growth' by American writers in the period 1880–1910). 'The age at which the rate of growth attains its maximum in the two sexes', Bowditch wrote, 'suggests a connection with the period of puberty which presents a similar difference in the time of its occurrence'. Perhaps because his observations only began at age 5 Bowditch did not stress the decrease of velocity before adolescence or its reversal in a 'spurt'. He talked of an achievement of maximal velocity just before puberty. He quotes recall data from 575 American-born women attending Boston City Hospital which put the average age of menarche at about 14.5 years; two years, as he says, after the maximum rate of growth (the modern figure is about 1.5 years). He remarked that whether an analogous relationship was true in boys remained to be seen. Bowditch's interpretation of the spurt was in a sense erroneous. He thought that growth and pubertal reproductive development were mutually antagonistic, and, quoting general physiologists of the time, wrote 'We may reasonably suppose that the age at which the organism becomes potent for reproduction will not be a period of excessive growth'. Nowadays we see the growth spurt as actually a part of the finalization of reproductive function, an adaptive secondary sex character (as did Bowditch's contemporary, Pagliani: see Chapter 9).

By the time the survey data were accumulated Bowditch was quite sure of his ground against Quetelet and levelled at him much the same criticisms as we have done in Chapter 6. He sums up the answer to his original query – as to the existence of the girls-taller phase – by saying that undoubtedly Quetelet would have obtained the same results in Belgium as Cowell and Roberts in England and he himself in Boston, if only he had used proper methods.

Bowditch analysed further the differences in growth between the children whose parents were born in America and those whose parents were born in Ireland. The former tended at all ages to be larger. In his second report (1879) Bowditch incorporated a test of whether this was only a reflection of the differences Roberts had demonstrated in England between children whose parents were manual or non-manual workers. The boys of the non-manual American-born parents were indeed taller than those of the manual American group, by about 1.5 cm at 6 to 11 years, 3 cm at 12 years and 2 cm at 18 years. The girls showed a smaller difference, with the non-manual only fractionally larger. The sons of Irish-born parents showed a similar occupational differential, which reached 2 cm at 12 and 3 cm at 13, above which age there were no data for non-manual parents. The daughters of Irish-born parents showed much the same social class differential as the sons: the non-manual girls averaged between 1 and 2 cm taller than the manual and ended 2 cm taller at age 16.

When the American and Irish were pooled, the non-manual boys were taller than the manual by about 1 cm at 8–10, 2 cm at 11–12, 3 cm at 13–14 and 1 cm at 18. The non-manual girls were taller than the manual by about 1 cm at 13 but ended practically the same. These were somewhat smaller occupational class differences than those which Roberts had obtained amongst children themselves not working; they were much smaller differences than those between Roberts' non-manual class and his factory-working children (see Fig. 8.1 and Table 8.1). When occupations were pooled in Bowditch's data the comparison between children of American and Irish parentage gave differences larger than the occupational ones, but Bowditch failed to realize the niceties of non-orthogonal analysis of variance (which was still fifty years away) and his analysis is biased by the larger numbers of manual Irish. In any case, citing both Villermé and Boudin, and siding with the former, he interpreted the American–Irish difference as being largely due to the higher standard of living of the American group. He pointed out that the boys of the American-born parents were about the same height as the English Public School boys of Galton's survey; but that boys from an American private school included in his material were a little taller still. Both Roberts and Bowditch plotted bivariate distributions of weight on height, though neither could analyse them further until Galton invented regression. Both realized, however, that the manual groups had a higher weight-for-height, a finding consistently found in all industrialized countries since (see Eveleth and Tanner, 1976). 'Deprivation of the comforts of life', wrote Bowditch; 'has a greater tendency to diminish the stature than the weight of a growing child' (1877).

Bowditch summed up the sex difference very clearly: 'Until the age of 11 or 12 years boys are both taller and heavier than girls of the same age. At this period of life girls begin to grow very rapidly and for the next two or three years surpass boys of the same age in both height and weight. Boys then acquire and retain a size superior to that of girls, who have now nearly completed their full growth. This statement is based on observations on several different races in various conditions of life' (1877). What neither Roberts nor Bowditch realized was the part played by early or late maturing in creating differences between occupational or national groups. Roberts partly saw the importance of this in relation to individuals of the same sex, and Bowditch went so far as to query the wisdom of coeducation in schools, given the different timing of the growth spurt in boys and girls (1877). But the real recognition of difference in tempo of growth had to wait for Boas.

In 1881 Bowditch was invited to address the section on children's diseases of the American Medical Association and emphasized to them his belief that the study of growth threw light not only on the comparative health of populations, but on that of individual children. 'It seems probable', he said, 'that the accurate determination of the normal rate of growth in children will not only throw light upon the nature of the diseases to which childhood is subject, but will also guide us in the application of therapeutic measures' (Bowditch, 1881, p. 373). 'It is interesting to enquire', he went on, 'whether [the] accelerated growth *after* the disease [i.e. 'certain fevers'] is anything more than a compensation for a retardation *during* the disease' (p. 373: his italics). He referred with approval to Percy Boulton, physician to the Samaritan Hospital for women and children, London, who in 1876 had examined Galton's statistics on Public Schools with a view to discovering a way to judge an individual's growth progress. 'If parents would once a year have the children weighed and measured by a competent person', Boulton wrote, 'they would frequently gain information which would be of the utmost prospective value. They would detect at an early age irregularities of development which would act as danger signals to give warning of approaching mischief, for arrest of growth, whether in latitude or longitude, is one of the earliest appreciable signs of disease' (1876, p. 280). (Boulton later (1880) promulgated rules for judging whether velocity of growth was within normal limits, being, it seems, the first person to do so. They were supposedly founded on a ten-year longitudinal study of an unspecified number of children of well-to-do parents. Growth rate was said to be constant in a given individual and between 2 inches and 3 inches a year. 'Every child should have its own regular rate of growth of 2, $2\frac{1}{2}$ or 3 inches a year, from which it has not the right to vary more than a quarter of an inch a year' (p. 611). Boulton ignored the adolescent spurt, saying that puberty simply 'at first slows and ultimately stops growth'. Tall children were those with the 3-inch growth rate; short children those with the 2-inch rate. Boulton's lack of grasp of the fundamentals of the growth curve rendered his system useless.)

Bowditch continued to be interested in using growth as a guide to health, and in his 1891 Report he applied to his data the method of percentile grades invented by Galton in 1875. In so doing he provided the first practical standards of growth by which the status of a given child could be judged *vis-à-vis* the population. The centiles given were 5th, 10th, 20th, 30th, 40th, 50th, 60th, 70th, 80th, 90th, 95th.

Bowditch (1891) pointed out an interesting feature of his centile curves, which had an influence on Boas' first analyses of growth data. The curves for each year of age showed symmetrical distributions only up to age 10 in girls and 12 in boys. Then the upper centiles moved further away, making the distributions asymmetric. At 16 in boys and 14 in girls exactly the reverse asymmetry occurred. Bowditch concluded that 'The pre-pubertal period of growth acceleration . . . which is such a distinct phenomenon in the growth of children, occurs at an earlier age in large than in small children' (p. 502). In thus interpreting cross-sectional distributions he was confounding them with longitudinally derived data (a point Boas made later). Bowditch did not realize that earlier maturers on average were naturally found at higher centiles in the early pubertal years and that it was their growth spurts which pulled the centiles askew.

In 1880 Bowditch got the Massachusetts State Board of Health to send out a circular suggesting to various institutions that they should systematically collect statistics on the inhabitants of the state (Bowditch, 1890). The problems on which this was supposed to throw light make very up-to-date-reading. They were:

1. Influence of geographical and climatic conditions on the growth of children and the physique of adults.
2. The number of generations necessary for the complete development of the influence of changed climatic conditions on the physique of a given race.
3. The comparative effect of city and country life on the growth and development of the human race.
4. The relation between disease and growth rate.
5. The effect of local hygienic conditions on the physique of children and adults.

Progress was slow and in 1890 Bowditch decided to publish what was available, since by then a more extensive programme of measurements of adults had been carried out by D. A. Sargent at the Hemenway Gymnasium of Harvard University (see Sargent, 1887, 1889). Sargent, a doctor totally devoted to the cause of physical education, measured over a thousand men and many hundreds of women in the years from 1870 onwards, constructing profiles, couched in terms of centiles, for no fewer than forty-nine measurements, including several of the strength of trunk and limbs. His aim was chiefly to assist in the assessment of the results of physical training, but he also believed his profiles would be of use to 'doctors, artists and sculptors, and superintendents of factories and prisons'.

The most useful of Bowditch's data on adults concerned women aged 17 to 24 in various colleges. Their average height was 158.8 cm, and the average percentage that sitting height contributed to this was 53.3 per cent. Their span averaged 100.4 per cent of height. Using Galton's and Sargent's data as well as his own Bowditch remarked on the important sex difference in relative length of limbs and trunk. Sargent's females averaged 159.1 cm in height with 52.7 per cent sitting height and 100.8 per cent span; Sargent's males averaged 52.2 per cent sitting height and 102.8 per cent span. Bowditch, harking back to the artistic tradition, remarked that it was popularly supposed that span equalled height; but this was more true of women than of men (1890, p. 303). Quetelet, in fact, had earlier pointed out the same thing; in *Anthropométrie* (1870) he gave span as 105 per cent of stature in mature men and 102 per cent in women. What neither Quetelet nor Bowditch pointed out was the relatively greater length of the forearm in proportion to the whole arm in the male (see Tanner, 1962, p. 41). Sargent (1889) described this in his student data.

9

Educational auxology: school surveys and school surveillance

Roberts and Bowditch were not alone in their investigations. There was one continental European whose work showed the same, if not greater, insight and exerted an over-riding influence on the educational auxologists of the 1880s and 1890s. This was Luigi Pagliani, professor of hygiene at the University of Turin. Pagliani (1847–1932) was an assistant in the Physiological Institute in Turin (working on the neural ganglia of the heart) when he began his investigations on growth. Like Roberts and Bowditch his reputation as an auxologist rests on just two papers; they were published in 1875–6 and 1879. The first anticipated Bowditch's great 1877 Report; the second, initially read at the Paris Anthropological Congress in August 1878, appeared just as the results of Bowditch's survey were becoming known in Europe. Pagliani was obviously delighted with Bowditch's '*bellissima lavoro*', made, he carefully points out, '*senza però alcun comune concerto*' (quite independently). Bowditch and he got the same results, and for anyone whose results differed from Quetelet's, such a confirmation was precious, indeed psychologically essential. Pagliani's work was a much-quoted model for the school surveys of the 1880s and 1890s but then his name unaccountably fell out of the literature and he has been virtually forgotten during the twentieth century. Like Boas, he was far ahead of his contemporaries in intuitive statistical understanding, and much of his 1879 paper makes perfectly up-to-date reading even now.

In 1881, when aged 34, he was made professor of hygiene and charged by the Italian government with the organization of the Italian Public Health Service. He became chief of the State Directorate of Public Health and founded its Central Laboratory. He retired from his chair in 1924, aged 77, having been president of the Turin faculty of medicine for fifteen years. Returning, like Boas, to his work of forty years before, he published in 1925 a little monograph on growth 'with its applications for educational and social hygiene'. He died in 1932.

What Pagliani refers to as his anthropological researches began in June

1872, just three months before Bowditch read his first paper on growth to the Boston Society of Medical Sciences, and only four to five months before Roberts began his factory survey. In 1872 an agricultural boarding school for orphaned and abandoned boys called the Institut Bonafous was opened outside Turin. Pagliani followed for three years the height, weight, chest circumference, vital capacity and muscular strength of all the boys admitted in the initial year. This, therefore, was the earliest short-term mixed longitudinal study, except for the unpublished one made in the Carlschule. The numbers were not large: six boys were followed from 10 to 13, ten from 11 to 14, eight from 12 to 15, five from 13 to 16 and more over shorter periods. In all 60 boys aged 10 to 15 were measured at the Institut and these were matched with 120 boys aged 8 to 18 from wealthy families, who were attending one of the best-appointed institutions in Turin (Pagliani, 1876; 1879, pp. 363–4).

As a physiologist Pagliani was principally interested in what a change of circumstances would do to growth. This was the reason he chose the longitudinal design for his study, whereas Quetelet and Villermé, with the same concerns, were content to compare the end-results of good and bad circumstances in cross-sectional samples. Pagliani was one of the first auxologists to understand precisely the relative advantages of what he called the '*individualizing*' and '*generalizing*' methods for studying growth (our 'longitudinal' and 'cross-sectional': see note 11.5). The individualizing method is the more natural way to study the growth process, he said, and reveals the oscil-

Table 9.1. *Pagliani's demonstration that in a mixed longitudinal study the means of individual yearly increments are more precise than differences between successive yearly means*

Age (centre)	Differences between successive measurements (cm)	Means of individual yearly increments (cm)
11.0	1.8	2.8
12.0	4.0	3.7
13.0	5.4	4.4
14.0	2.5	5.9
15.0	8.6	6.2
16.0	2.6	7.4
17.0	0.2	6.9
18.0	2.9	4.8

Data on boys of the Institut Bonafous (Pagliani, 1875–6).

lations of growth; but it takes a very long time. The generalizing method has not only the obvious strength that large numbers can rapidly be studied, but also the perhaps less obvious advantage that at each age the sample is independent of the sample at ages before and after (Pagliani, 1879, p. 360).

In working up the mixed longitudinal data of the Institut Bonafous Pagliani does not merely give cross-sectionally calculated means at each year of age and the mean increments derived from them. He starts this way, but then he calculates the actual increments shown by each individual, takes their means for each year, and constructs a growth curve of height by adding these mean increments successively onto the cross-sectional mean value of the 10-year-olds. He was thus the first to use the method for getting rid of sampling bias that was introduced, or reintroduced, by Tanner (1965; Tanner and Gupta, 1968) eighty years later. Pagliani notes the very considerable difference this makes to the increment curves. His figures for height increments are reproduced in Table 9.1. The middle column gives the differences between yearly means cross-sectionally calculated, and the right-hand column the mean increments calculated longitudinally. Peak height velocity is reached at about 16.0 years.

Pagliani also measured a similar number of girls in the Villa della Regina, a school for officers' daughters. This group was from the opposite end of the social spectrum; their homes were well-off and their school lavish. Cross-sectionally derived mean increments give peak velocity at about 14.0 years or a trifle later and peak weight at 14.5 years. In comparing the two sexes a curiosity arose, the girls being taller than the boys at all ages until 18, when they were equal. Pagliani attributed this, without doubt correctly, to the very different circumstances in which the boys and girls had spent their early youth; in a later study girls exceeded boys only at the usual ages. In weight also the Bonafous boys were lower than the girls, but nevertheless in vital capacity (the amount of air that could be inspired in a maximal inspiration) they exceeded the girls after puberty started, and they exerted twice the muscular strength at age 16 and later.

It was clearly Quetelet's work which served as Pagliani's inspiration; indeed in his 1875–6 paper Quetelet and Villermé are the only two authors quoted. But Pagliani recognized at once Quetelet's errors concerning puberty: 'Our results leave no doubt that at puberty the growth of the human is greater than before it', he wrote (p. 750). The acceleration occurs very regularly and is not, as Quetelet says, inconstant and variable. Pagliani called attention to the fact that 'the year before that in which growth is maximal is distinguished by a relatively very small growth', especially in boys. He emphasized, too, individual differences in growth status, if not explicitly in rates of maturation. Differences of height, he said, occurred not only between the sexes 'but also because of various other influences, which make growth rapid or slow, and lasting for a long time or a short time. We find individuals who at a certain date are taller than their contemporaries ... but I would not like to say that at the end of their growth the relationship is always the same; rather it may be reversed' (p. 751).

In his 1879 paper Pagliani pursued these thoughts much further. The disagreement of his early results with Quetelet's led him to seek confirmation in large numbers and in constructing a whole growth curve, as Quetelet had done. He persuaded the city authorities to help and in 1873–6 he measured 1,048 boys and 968 girls aged 3 to 19 in nursery schools, elementary schools and private institutions in Turin. This was the first cross-sectional study in continental Europe since Quetelet's (and probably the first ever with a reasonable number of subjects, though nobody knows the size of Quetelet's sample). Pagliani took height, weight, head length, breadth and circumference, chest circumference, vital capacity and the strength of the arms in pulling and pushing. He recorded father's occupation and, as a better index of ease of the family's life, the location and type of school attended. He was explicit about his age intervals, and used age-at-nearest-birthday as was most usual at this time. For height and weight he gave means (but no measure of dispersion) from 3 to 19, together with the average yearly increments derived from them. The greatest increment for boys (6.5 cm/yr) was centred at 14.5 and for girls (7.0 cm/yr) at 13.5. The girls' ascendancy in height lasted from 10.0 to 15.0 inclusive. Vital capacity became sharply greater in boys at 16.0; muscular strength, always somewhat greater, became much more so at 16.0 and 17.0.

Pagliani was one of the few auxologists who have successfully combined work in both longitudinal and cross-sectional studies. At the same time as the Turin school survey, he followed for five successive years the growth of thirty-nine girls. The exact ages of their first periods were recorded. Pagliani then produced the table shown here as Table 9.2. Four girls had their menarche at 12.0 (nearest birthday); the top row shows their mean annual height increments, with the increment for the year of menarche in bold type (as in the original). Twelve girls had menarche at age 13.0 and the second row shows their mean annual increments. Eleven girls had menarche at 14 and twelve at 15. Pagliani concluded – and he was the first to do so – that the adolescent growth spurt in girls is linked with menarche, having its peak some one or two years prior to menarche, whether menarche is itself early or late. He even recognized that the absolute value of the peak velocity was greater when menarche was earlier. Rejecting the Spencerian theory, still credited by Bowditch, that growth and reproduction were antithetical processes, he proposed, on the contrary, that both puberty and the growth spurt were manifestations of the same underlying influence (1879, p. 469).

Lastly, he added, 'However, evidently we do not obtain the same result from a generalizing mean as from individualizing means, because the measurements of a large number of children do not coincide in the same year, which leads to an adding together at a given year of age, increments some of which are at the moment of minimal, and some at the moment of maximal growth' (1879, p. 469). This clear recognition of the tempo effect came too late for Bowditch. We shall describe it in detail in the next chapter, for Boas is always given the credit for its discovery. Boas recognized its importance independently, for there is no evidence in his writings that the *Giornale della Società Italiana d'Igiene* was amongst his eclectic, but seldom medical reading.

Pagliani, of course, was not wholly ahead of his time. Though he maintained clearly that good conditions and adequate nutrition accelerated growth, the final limits were imposed, he thought, by sex and ethnic origin (Pagliani, 1877, p. 97). In the battle between Villermé and Quetelet on one side and Boudin and Broca (and Lombroso the criminologist) on the other, Pagliani sought an intermediate position. He realized that the comparison of the means of conscripts' heights, already much in vogue, was fallacious if members of one population grew up faster than members of another, and hence were nearer their final height when conscripted. But even though he himself had demonstrated the efficacy of an improved environment, he could

Table 9.2. *Pagliani's table of mean height velocities of four sets of girls followed longitudinally, grouped according to age of menarche*

	Mean growth in height (mm) at each year of age						
Age at first menstruation[a]	10	11	12	13	14	15	16
12	66	98	**31**	30	—	—	—
13	—	63	60	**36**	29	14	—
14	—	48	71	66	**35**	34	3
15	—	—	52	58	142	**29**	7

[a] Years are 12±; velocities are also centred.

Reproduced from Pagliani (1879, p. 468).

Numbers are 4, 12, 11 and 12 respectively in the four sets.

not escape the currents of the period; the most one could do, he said, for a small race (such as was found in Sardinia, for example) was to speed up their growth so that they would reach *più presto* their completed stature (1877, p. 120).

The school surveys

Bowditch and Pagliani were the first among many. Soon school committees got to work in various parts of Europe, and by the end of the century growth curves, of better or worse provenance, were available for many countries. Pagliani's initial purpose was to show that an improvement in the conditions of life of the orphans caused improved growth, and indeed, he interpreted his Bonafous data as showing some degree of catch-up in the first year after admission (though he did not use that phrase). The big school surveys were made with the same general objective: to see if the conditions of the schools were such as to encourage healthy growth.

At the centre of the early school work was the journal *Zeitschrift für Schulgesundheitspflege*, first published in 1888 in Hamburg. In the foreword to the first number, the editor, L. Kotelmann, declares its purpose. To date, he says, there is no journal devoted to studies of bodily development, or training (*körperliche Ausbildung*). The journal will supply this need. It will also publish studies concerning all matters pertaining to school buildings and classroom and furniture design and concerning the 'hygiene of teaching' (*Hygiene des Unterrichts*), under which heading come problems such as the time of the first lesson in the day, intervals between lessons, age at starting homework, and so forth. Health and disease are also the journal's concern. 'By thus uniting engineers, doctors and teachers in joint work *made from the standpoint of exact science and careful measurement* [his italics] we may hope to contribute to the attainment of the educational ideal . . . and by applying the principles not only of pedagogy but of physiology and hygiene to bring [the pupil] to a harmonious organization (*Gestaltung*)'.

Such an attitude became widespread in the last years of the nineteenth century. Maria Montessori is remembered still in educational circles, but scarcely in medical ones. Yet she was a medical doctor (Kramer, 1978) and one of her main works was called *Pedagogical Anthropology* (1913). The tradition of doctors helping over questions of schooling was perhaps strongest in Central Europe, and elements of this still remain, for example, in the emphasis laid on a medical examination for school readiness of the nursery child. Zeller (see Chapter 11) was a late flowering of this tradition, with his *Gestaltwandel*.

The *Zeitschrift* continued for just fifty years, changing its name to *Gesundheit und Erziehung* in 1932. Kotelmann, the first editor, had himself published, in 1879, a paper that was a pioneer work in the analysis of changes in body composition with age. He measured 515 Hamburg *Gymnasium* boys aged 9 to, essentially, 20. From 10 to 17 there were approximately 50 boys at each year of age. Kotelmann departed from custom in that he measured supine length rather than standing height; he does not give his reasons for this. The maximum yearly height increment was centred at 16.0 (perhaps surprisingly late in this group of high social class), as also were peak weight and chest circumference increments and the peak increment of vital capacity measured in the Hutchinson spirometer. Mean height is around the 10th centile of modern standards.

Kotelmann measured the circumference of the left upper arm over the biceps, with the muscle both relaxed and contracted: peak velocity was at 17.0 although there was a subsidiary high increment at 15.0. The circumference of the right calf was measured also, again both relaxed and contracted. Hand grip was measured with the Collin dynamometer and leg strength with a modification of it: they had peaks at 15.5 and 16.0. Kotelmann was the first to try to 'dissect' the body into fat, muscle and bone components and it was he who first pinched up a fold of subcutaneous fat over the biceps and measured its thickness, with the regular curved caliper used by anthropologists to measure head breadth. Halving his values, he obtained means, from age 9+ to 20+, of 1.15, 1.24, 1.39, 1.37, 1.32, 1.50, 1.44, 1.45, 1.68, 1.84, 2.51 and 2.48 mm. Biceps skinfold, as it is now called, is a difficult measurement and not a very good indicator of general subcutaneous fat; if Kotelmann had measured the fold at the back of the arm, over the triceps, he would have obtained clearer results. Nevertheless his figures do show the general type of curve that emerged in the studies of Franzen and McCloy fifty years later (see Chapter 12). Kotelmann also gave means each year of a number of ratios such as arm circumference per body weight. Though he seems to realize something further is needed to show how one measurement relates to another, he did not hit on the bivariate distribution table just being published at this time by Roberts and Bowditch.

In the first volume of the *Zeitschrift* there is a report on the heights of boys aged 6 to 16 at the Breslau Evangelical High School, made by the *Rektor* himself (Carstädt, 1888). Means and ranges are given for half-yearly intervals on 600 boys; the data are mixed longitudinal but the results are reported only cross-sectionally. Peak height velocity is located at 15.0 years (see Table 9.3 below). The increments are very regular and the peak well established. By 16.5 years height was 164.5 cm but growth was still continuing at about 2 cm/yr.

Also in volume 1 of the *Zeitschrift* was an account of the first really large-scale school survey made in continental Europe. In Denmark a School Commission was set up in 1882 to investigate throughout the whole country the condition of children in all types of school, including elementary and village schools. The main questions derived directly from the earlier studies of working children, and the widespread feeling that children in school were being subjected to a pressure of work all too reminiscent of the factory. Dr Hertel, the municipal medical officer in Copenhagen, in particular articulated this feeling and in 1880, or thereabouts, undertook a survey of the high schools of Copenhagen, to establish by questionnaire what percentage of the children were 'sickly'. Something of the conditions the Danish children were subjected to comes out in the introduction by Dr Crichton-Browne to the English translation of Hertel's book *Overpressure in High Schools in Denmark* (Hertel, 1885). Hertel is no rabid anti-educationalist, the introducer says: 'He who pleads for only half an hour for a mid-day meal out of a school session of six or seven hours duration, who asks no more than nine hours sleep in the twenty-four for children ten years old, and who would be content if boys of 16 were not called on to apply themselves to brain work for more than ten hours a day six days a week can scarcely be said to be ultra-sanitarian in his notions or exorbitant in his requirements' (1885, p. xii).

Hertel found a shocking amount of illness and in 1882 a Danish Schools Commission was set up to inquire into the prevalence of sickness in a sample of all the schools in the country. Some 17,600 boys and 11,600 girls were investigated, the results being given by Hertel himself (1888). The prevalence of ill-health was astonishingly high: 30 per cent of boys and 40 per cent of girls were noted as having some illness, often scrofula, anaemia or nervousness. A controversy arose as to the age at which the prevalence was greatest: Hertel and also Axel Key (1889, 1891: see below) thought the period of slow growth just prior to puberty was the worst for disease; Combe and some later authors (see Schmidt–Monnard, 1897) thought the prevalence rose at puberty. Burk (1898) quotes Key (1889) as saying 'The curve of disease, in boys, reaches its first summit directly before, or more correctly, at the beginning of, pubertal development. But as soon as the development sets in forcibly, the curve sinks year by year, so long as that accelerated growth continues ... Directly at the conclusion of the pubertal period, when the yearly increments in height and weight rapidly decrease, the curve of disease ... jumps to its second summit, which it reaches in the 18th and 19th year'. Key, it seems, would have been a supporter of Stöller's theory of cure by growth-acceleration, had he known of it. As to growth in the

Danish children, the maximum yearly height increment in boys was centred at 15.0 years, and in girls at 13.0 years. Boys aged 11 and 14 in *Gymnasia* averaged about 3 cm taller than boys in the *Realschulen*, and 5 to 6 cm taller than boys in the *Bürgerschulen* and the *Volksschulen*.

In the 1880s the question of 'over-pressure' at school was very much in the minds of all school doctors and educationists. As Duffy (1968) has shown, the concern arose in the United States in the 1830s, and from 1840 onwards the *Boston Medical and Surgical Journal* carried numerous editorials and contributions declaring that children in schools were pressed beyond their strength. The *Medical and Surgical Reporter* of Philadelphia joined in in 1859, and the New York Medical Society in 1873. The *Lancet* and *British Medical Journal* took up the cry in the 1870s, although Fergus (1884) of Marlborough College (see Chapter 8) and Farquaharson (1876) of Rugby School both argued strongly against the existence of any widespread ill-effects. In 1887 over-pressure was debated at a meeting of the French Academy of Medicine in 1887 (quoted in Carlier, 1892) and G. Carlier (dates unknown), a French army doctor, set up a study of the growth of boys in an army school at Montreuil-sur-Mer precisely because in this school over-pressure did not occur. In 1886 'our masters drew to the attention of the public authorities the over-pressure (*la surcharge*) of programmes in all educational establishments as something dangerous to the health of pupils. They criticized the too great part given to brain work and the more or less complete abandonment of physical training, particularly at a time when as a consequence of their living-in in the large educational establishments pupils are exposed to checks of physical and intellectual development' (Carlier, 1892, p. 265).

Carlier and one other physician took measurements of height, weight and chest circumference on 526 boys each six months from 1886 to 1890, over the age period 13 to 17. The study was longitudinal, 100 boys being measured four times, 108 six times and 93 eight times. Carlier, much influenced by Pagliani's Institut Bonafous study, compared the heights of 17-year-olds who had been four years at Montreuil with the heights of 17-year-olds who had just arrived. He found a difference of 1.6 cm in favour of the former, which he attributed to the physical exercise ($2\frac{3}{4}$ hours per day) at Montreuil, compared with the lack of it in the usual *lycée*. In only two respects, however, does he use the longitudinality of his admirable data. He discusses the seasonal effect, with reference to Buffon and Montbeillard, and shows that the growth in height of his pupils from April to October (the two measuring points) was 1.2 cm greater than their growth from October to April (averaged over seven measuring periods, i.e. $3\frac{1}{2}$ years). In chest circumference the summer excess was 4.2 cm, whereas in weight the gain was greater in the winter six months, by 1.8 kg. Carlier's was the first demonstration of a seasonal effect during puberty; indeed he worked at exactly the same time as Malling-Hansen (see below), who is usually credited with establishing the reality of seasonal differences in growth rate.

Carlier also investigated the then widely accepted notion that disease, especially childhood fevers, caused an increase in growth rate. Several of his pupils had such diseases and he compared their rates of growth with those of their companions. He was unimpressed: 'in general, the influence exercised by these pathological states does not seem to have been considerable', he wrote. He found some individual febrile cases where a spurt had occurred but lack of awareness of the variability of the timing of the normal adolescent spurt makes his analysis uncertain. The greatest increase of all in six months, he says, occurred in a perfectly healthy boy: and it was 11.5 cm (an only just credible figure). In respect of *la grippe*, however, he has enough cases to give a clear-cut table: flu is without influence on growth in height.

Carlier must have also been unimpressed by another view, which had a long history with a root in folklore, and a contemporary champion in Dally, a paediatrician. This was that excessive exercise stunted growth. 'One of the most powerful factors stopping growth is excessive premature exercise', wrote Dally (1879) in the *Dictionnaire encyclopédique des sciences médicales*. 'When growth is too rapid, that is to say when it exceeds 6 or 7 cm a year, it is best to have the children do exercises which have the effect of directing growth into the muscles, and favouring growth in thickness. If, on the contrary, growth is less than 5 cm, it is best to put the child at rest, to have him undertake only moderate walks, and never to let him become fatigued; for, as we have seen, rest is not opposed to growth; rather it favours its occurrence' (p. 396).

There are two separate ideas in what Dally writes: first that excessive exercise stunts growth and should be avoided in the childhood years; and second that a swift growth in height (though not in thickness) is, potentially at least, a source of illness. We are in the period of the delicate child, of the universal fear of the phthisical constitution, with its leptosomic body build and small chest circumference, of children out-growing their strength tending to fall victim to tuberculosis of the lungs. Stöller's idea that a rapid growth in height cures disease is losing ground to just the opposite view.

Dally calls for measurements to be done in all the state schools, especially in Paris 'where 100 medical inspectors have just been appointed'. Each six months height, weight and chest circumference should be taken, together with details of family size and conditions of life; the teacher should help to do this. Dally is a direct forerunner of Godin (see p. 223) and articulates well the concerns that led to the founding of the *Zeitschrift für Schulgesundheitspflege.*

The year after the big Danish school survey, that is, in 1883, a Swedish School Commission survey was carried out. This was directed by Axel Key, professor of pathological anatomy at the Karolinska Hospital and a friend of the famous Retzius. Key's report (1885, 1889), like Hertel's, is now a rare work (see note 9.1). Many more boys (14,590) than girls (3,209) were examined, this reflecting the preponderance of boys in the secondary school system. The maximal increment in boys' height (7 cm/yr) was centred at 15.0 years and in boys' weight at 16.0 years (Key, pp. 522, 526 : see also the treatment of Key's statistics in Ljung, Bergsten-Brucefors and Lindgren, 1974). The girls' height showed no clear year of maximum increment but the maximum weight increment was centred at 14.0 years (p. 527). All Sweden was sampled, but since only secondary schools were taken the better-off class would have been over-represented, as it tended to be in most early school statistics. Key's Swedish boys averaged a little less in height than the boys in the Danish *Gymnasia* and *Realschulen* and were only about 1 cm taller than the general mean of all Danish schoolboys. Key's later (1891) paper, in German, is an excellent and extensive review of the whole literature on growth at puberty.

In Norway school measurements were made in 1891 but in secondary schools only and chiefly in Oslo (Faye and Hald, 1896). The results are illustrated in Kiil (1939, p. 27); boys had their peak height velocity at 15.5, girls at 14.0 (see Table 9.3). The schoolchild-measuring tradition was signally revived in Norway later, in particular by Carl Schiötz when he became Medical Officer for Schools in Oslo in 1919 (see Schiötz, 1923). The Oslo archive of school measurements has continued into the present (Brundtland, Liestøl and Walløe, 1980; and see Chapter 14).

In 1888 the Society for the Furtherance of Industry in Moscow celebrated its twenty-fifth anniversary by commissioning a survey of the hygiene of schools in the Moscow region. Erisman (1888), professor of hygiene at Moscow University, reported the results in the same year. The usual statistics are given on the dimensions, lighting and heating of classrooms, and on sickness prevalence in children. Heights, weights and chest circumferences were taken, in the same way as in the survey of children and adults working in the Russian textile industry (see below). About 3,000 boys and 1,500 girls were measured in town schools and 4,300 boys and 700 girls in village schools. The town schools comprised both elementary and secondary, with an effective age range of 7 to 16 in boys and 7 to 12 in girls. Town-school boys had their peak height velocity at 15.0. They averaged 6 cm taller than the factory boys (Erismann, 1889) at ages 10, 11 and 12, 10 cm taller at 15, and again 6 cm taller at 17; the adult difference cannot be determined.

Town boys were taller than village boys by about the same amount; village and factory differed little. The girls' measurements were peculiar, as Erismann, always a careful worker, himself remarked. Town schoolgirls were actually smaller than factory girls, and never approached, let alone surpassed, the values for town schoolboys. It seems something went wrong with the arithmetic, for the girls' values are not really credible. A few years later, in 1889 and 1890, Sack (1893) measured some 6,600 boys in Moscow *Gymnasia* and *Realschulen.* Peak height velocity was reached at 14.0, a year earlier than in Erismann's boys. These high-school boys were 5 cm taller than the factory boys at 8, 8 cm taller at 11, 14 cm at 14 and 6 cm at 19. Schoolboys of Jewish parentage had their peak height velocity very slightly earlier than those with Russian parents.

Sack remarked that boys who made good school progress were on average taller than those whose progress was poor; he thus prefigured Porter's work (see p. 215 below). He was also concerned, as were many writers of his time, with the relation of chest circumference to height. The smaller this was, the more 'delicate', and likely to be overcome by tuberculosis and other diseases, was the child or young recruit. Sack sadly remarked that the percentage of Moscow high-school children with chest circumferences less than 50 per cent of height was higher than in Swiss conscripts; higher, indeed than in Russians in general. He was seeing the greater ectomorphy in the higher social classes, and did not think it at all a good sign.

In Lausanne, children at the *Volksschule* were measured every November from 1886 onwards, and Combe (1896) gave mean heights from age 8 to 14. Combe realized his 2,000 children were an example of the 'individualizing' method, but he failed to see that individual increments had to be calculated; the best he could do was to divide the children according to ages at the first November measuring, thus creating twelve mixed longitudinal cohorts, one for each month of age. He then treated each cohort cross-sectionally. He had evidence, he thought, that boys born in September to January were smaller than others; and girls born in December to May likewise. No tests for the significance of his differences could be applied, of course, and Combe seems not to have realized that something of this sort was needed. Lausanne girls had peak height velocity at 13.0, boys at some time beyond 14. Girls were taller at 12, 13 and 14.

In 1886 the school-director of Freiberg (modern Freyburg, near Leipzig) made a survey of some 21,000 children. The data were analysed by Geissler and Uhlitzsch of the Royal Saxon Statistical Bureau (see note 8.7), with a statistical thoroughness unusual at that time, though they elected to give height and weight distributions in terms of average (i.e. unsquared) deviations about the mean rather than Galton's 'probable deviations'. At about the same time Hasse (1891) made a study of schoolchildren in Gohlis-Leipzig (see also Geissler, 1892) and Schmidt (1892–3) reported on heights and weights in the region of Saalfeld, where at ages 6 to 13 inclusive rural children were still taller than children in the towns, by about 1.5 cm on average. Schmidt summarizes the data of previous authors, some of it difficult to find in the originals now. Rietz (1904) measured 5,134 Berlin schoolchildren aged 6 to 16; those in the boys' *Gymnasia* and the girls' high schools were on average 5 cm taller than boys and girls of the same ages (6 to 13) in the general or lower schools. Peak height velocity was located in the upper-school pupils at about 15.0 in boys and 12.0 in girls.

Kosmowski (1895) described the growth of Warsaw children going to summer holiday camps; they were taken from the 'poor and poorest' families 'living in basements and shanties', but their health was relatively good, as they were selected for this before being sent. These were very small children whose averages were well below the modern 3rd centile for height.

All these papers concern children of school age only. Though infants from birth to 1 year old were being measured by clinical paediatricians in the 1880s and 1890s there were practically no data on growth between the ages of 1 and 6. Even today, of course, pre-school children are much the most difficult to study *en masse*. Pagliani (1879) had been able to measure pre-school children from 3 upwards and Daffner (1884*a*), an army surgeon (whom we shall meet as a textbook writer in Chapter 11), had indeed covered the whole range from 1 to 6 but in an article so terse as to be not easily comprehended, which is perhaps why it was almost entirely ignored. Daffner measured heights and head circumferences of children in Augsburg and Munich, having some fifty of each sex at age 3 and over a hundred of each sex at 4, 5 and 6; at 1 and 2 the numbers were smaller. Daffner's height means (allowing for a probable misprint of 73 for 78 in the 1-year-old males) followed the 10th modern centile in both sexes. Daffner emphasized the tremendously fast growth in the first year, especially of head circumference, representing brain growth. The brain, he pointed out, nearly doubles in weight between birth and 1 year and this must be of great importance for mental abilities (actually the brain's weight more than doubles in this period). Despite these two papers, most textbooks were still using values for pre-school children from Quetelet and Zeising when Schmidt-Monnard, a paediatrician in Halle, published a paper in which the complete age range was at last covered (Schmidt-Monnard, 1901). Schmidt-Monnard simply gave average yearly increments calculated from cross-sectional means, and displayed them in a way still used by many authors today. Vertical bars for each year of age are drawn to represent the

Fig. 9.1. Schmidt-Monnard's representation of mean yearly growth increments by heights of vertical bars. Reproduced from Schmidt-Monnard (1901, p. 54).

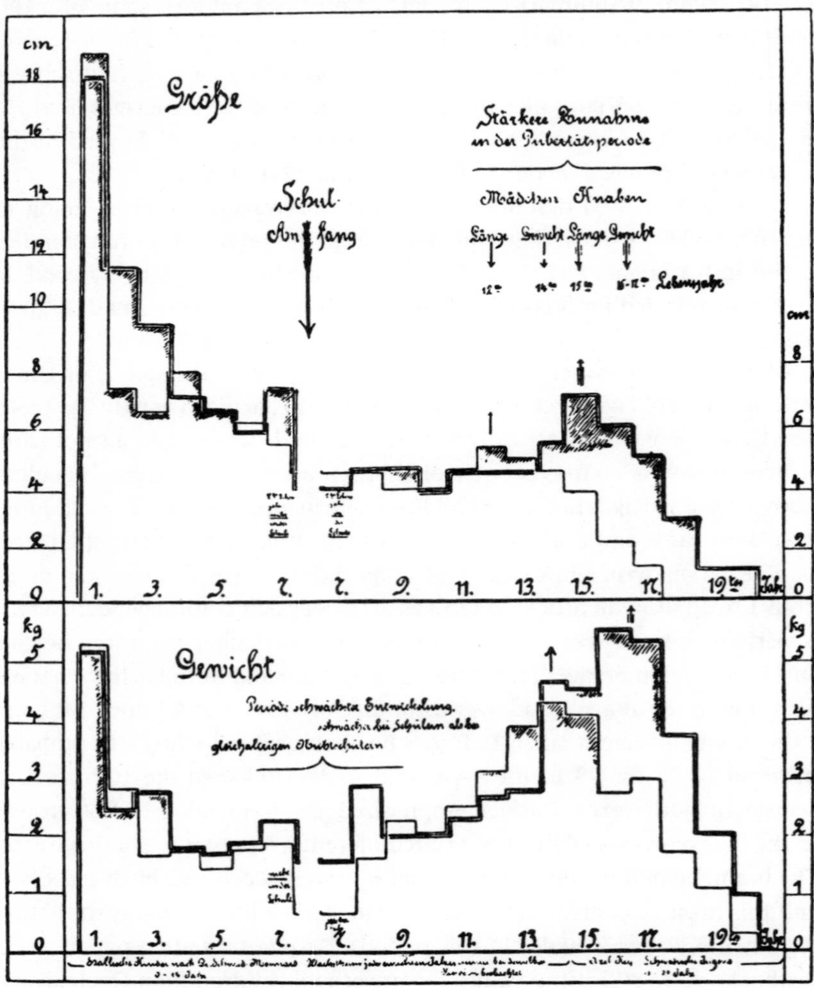

increments, thus giving the appearance of a sort of histogram in time. His figure is reproduced as Fig. 9.1.

In addition to these reports, there were investigations before World War I on children in Jena (see Rössle and Böning, 1924), Bonn (Schmidt and Lessenich, 1903), Dresden (Graupner, 1904), Kiel and Lübeck (Ranke, 1905), Zurich (Ernst and Meumann, 1906), Paris (Variot and Chaumet, 1906), Königsberg and Hamm (Ascher, 1912), Pomerania (Peiper, 1912), and Rostov on Don (Spielrein, 1916). Seggel (1904) followed some pupils longitudinally from 10 to 21 in Munich and gives means of actual annual increments for height and inter-pupillary distance. In Japan a German doctor, E. von Baelz (1849–1913), measured the heights and weights of children from age 3 upwards as well as of groups of soldiers, students and officials. He gives no numbers and his means, even with probable misprints corrected, give rather irregular curves. However, he did find that the Japanese ended their growth earlier than Europeans, a fact confirmed by twentieth-century studies. Mature heights of the Japanese were about 160 cm for men and 146 cm for women (Baelz, 1880–8).

Along with the school surveys of the 1880s and 1890s were naturally some survivors of an earlier time. One is a chatty paper by Landsberger (1888), a doctor of Posen (Poznan), who was very much a disciple of Quetelet. Landsberger has very pertinent remarks to make on the accuracy, or inaccuracy, of doctors' measurements of children, quoting the then-famous pathologist Beneke (1878) as saying 'Nothing is measured with greater error than the human body'. He goes on to point out that he, Landsberger, measured schoolboys serially over six years, every year at the same time of day, in the same room, with the same instrument and within seven days of the exact calendar date. He took most of the body measurements that Quetelet recommended, a total of twenty-five direct plus two derived. To begin with he had 104 boys; six years later 37 were left (thus migration and drop-out were worse than in the London studies of the 1950s). But he fails to use longitudinal methods or to calculate a single individual increment. Indeed it appears that to obtain his final table of each measurement at each year of age he followed Quetelet's post-1835 method of taking ten boys of 'normal' growth and simply giving their mean.

Factory children were still being measured too, up till 1880. The investigation of conditions in textile factories had spread throughout Europe and in 1879 the Russian government began a major inquiry. As part of this Erismann (1889) supervised the measuring of height (without shoes) and chest circumference in approximately 100,000 workers in factories in Moscow and other central Russian towns. About 2,000 of these subjects were aged 8 to 12;

9,000 12 to 14; and 15,000 15 to 17. Erismann's tables give the means of more than 2,000 of each sex for each year of age from 12 onwards. The boys have their peak height velocity at 16.0; the girls at 15.0. Boys finish growing around age 25 at an average height of 165 cm; girls finish around 19 at an average of 153 cm. Girls are taller than boys at 11, 12, 13, 14 and 15 with maximum difference at 13 and 14 years. Erismann deals also with the decrement of height in workers over 50 years old. By this time however, children in Europe had mostly come out of the factories and were settling into schools which all too often recalled their earlier employment.

In the United Kingdom work on growth slowed down after the early beginnings. In 1900 Rowntree (1913), in his classic study of poverty in the city of York, measured the heights and weights of 1,014 boys and 905 girls in manual working class families, whose living circumstances had been examined with exemplary care. Most were aged 5 to 13. Rowntree divided his families into three groups: the utterly poor (384 boys and 287 girls), the middling poor (393 boys and 397 girls), and the artisans, who were not very poor (237 boys and 221 girls). The mean heights of the poorest group were at the modern 3rd centile from 5 till 12, when they dropped below it; the mean heights of the artisan children were at the 10th centile, dropping to the 3rd at 12–13. Taking the years 8 to 12, the means for the poorest group average almost exactly the same, in both sexes, as those of the factory-working children of factory parents measured by Roberts in 1873 (Table 8.1). Children of the artisans averaged about 2 cm taller than Roberts' group of children living in non-factory towns. The improvement since 1873, then, was marginal, at least in the manual working class.

A few years later Thorne (1904) and Berry (1904) measured the heights and weights (both in boots) of boys and girls aged 11 to 15 admitted to the London County Council schools by competitive scholarships from poor homes. At the same time a survey was made of a number of schools in Glasgow (Kay, 1904–5); this was chiefly remarkable as the forerunner of the first survey in the UK covering a whole school population, which was done in Glasgow in 1905–6. The first analysis of the Glasgow survey contained errors of computation, corrected in the later report of Elderton (1914). In this survey there were some 66,000 children aged 6 to 14, divided according to the area of the city in which the school was located. The boys from schools in good areas averaged about 5 cm taller than those from schools in bad areas; in the girls the differences were only a trifle less.

Schoolchildren in Edinburgh were surveyed in the following year (City of Edinburgh Charity Organization Society, 1906) and those in Birmingham in 1910 (Auden, 1910). Differences between children attending a school in a poor part of Edinburgh and those attending a school for relatively well-off non-manual workers averaged about 8 cm at ages 10–14 (City of Edinburgh, 1906). In Birmingham, children in the best-off of the five city wards, as judged by infant mortality, averaged 5 cm taller from age 5 to 8 than those in the worst-off ward. The mean heights of the 7-year-olds in the best ward were at the present-day 10th centile, those in the worst at the present-day 3rd. The situation was exactly comparable, then, with that in York. School statistics for 1909–10 covering the whole of England were collected from school medical officers by Tuxford and Glegg (1911); only mean heights are given and they are at about the modern 5th centile. The rural children were still taller than the urban children, by about 1 cm, and children from the South were about 1 cm taller than those from the North. A particularly interesting paper is that by McGregor (1908), since it is one of the rare ones that deals largely with pre-school children. All children aged 2 to 10 admitted to a Glasgow fever hospital with scarlet fever, measles and whooping cough in 1907–8 were measured. A marked difference in height and weight was found between children whose parents had three rooms for their family to live in and children whose families lived in only one or two rooms. This study showed that differences in height of the order of 5 cm were already established by age 3 between children living in different conditions (all of them poor by comparison with modern housing).

Meanwhile the examinations of Boer War recruits, which revealed a disastrous level of disease and malfunction, were having a delayed effect; an Inter-Departmental Committee on Physical Deterioration was set up in 1904 and recommended regular and permanent anthropometric surveys of the population, especially of young people in schools and factories. Little was done, however, and in 1978, two wars and three-quarters of a century later, a government committee known appropriately as COMA (Committee on the Medical Applications of Nutritional Policy) found itself recommending exactly the same thing. However, the London County Council did initiate a survey in 1904–5, and measured 3,500 schoolchildren. In 1906 this was extended to 18,000. The results were circulated in a report (London County Council, 1908, cited in Cameron, 1977; Weir, 1952).

In 1907 an Education (Administrative Provisions) Act made the examination of elementary-school children mandatory over the whole of England and Wales. The results of the examinations, forwarded to the Chief Medical Officer of the Board of Education, caused him to say that 'Of the six million children registered on the books of the public elementary schools of England, about 10 per cent suffer from a serious defect of vision, 3 to 5 per cent from defective hearing, 1 to 3 per cent have suppurating ears, 6 to 8 per cent adenoids or enlarged tonsils of sufficient degree to obstruct the nose or throat and to require surgical treatment, about 40 per cent suffer from extensive and injurious decay of the teeth, about 30 to 40 per cent have unclean heads or bodies, about 1 per cent suffer from ringworm, 1 per cent from tuberculosis in readily recognisable form, from 1 to 2 per cent are afflicted with heart disease and a considerable percentage of children are suffering from a greater or less degree of malnutrition'. 'Such a statement', says R. H. Tawney in his introduction to Arthur Greenwood's *Health and Physique of Schoolchildren* (1915),' is likely to be read in the future with the sensation aroused today by a study of the reports of the early Commission on Child Labour in Factories and Mines' (p. xii). Greenwood, later a well-known Labour government minister, collected the country-wide statistics for 1908, 1909 and 1910. The total number of children exceeded 800,000 but the curves of the means are

Table 9.3. *Estimated age, to nearest six months, of peak height velocity in various sets of data from 1870 to 1910*

Place	Year	Boys	Girls	Author
Europe				
Turin	1873	14.5	13.5	Pagliani (1879)
Hamburg (*Gymnasium*)	1879	16.0	—	Kotelmann (1879)
Breslau (high school)	1888	15.0	—	Carstädt (1888)
Halle	1890	15.5	12.5	Schmidt-Monnard (1901)
Berlin	1903	15.0	12.0	Rietz (1904)
Denmark	1882	15.0	13.0	Hertel (1888)
Sweden	1883	15.0	?	Key (1885)
Oslo (high school)	1891	15.5	14.0	Faye and Hald (1896)
Moscow (schools)	1888	15.0	?	Erismann (1888)
Moscow (high school)	1889	14.0	—	Sack (1893)
Russia (factories)	1889	16.0	15.0	Erismann (1889)
Lausanne	1886	?	13.0	Combe (1896)
USA				
Milwaukee	1881	16.0	13.5	Peckham (1882)
St Louis	1892	14.5	12.5	Porter (1894)

not altogether easy to interpret, presumably because of the varying methods and care of the collectors. Girls were taller than boys at ages 11, 12, 13 and 14 but their point of peak height velocity is indistinct. The means are slightly below those of the Anthropometric Committee of the British Association of 1883, by about 2.5 cm at all ages in boys, and 2 cm in girls at ages 13–15; in girls of 3–12 there is no difference. Greenwood explains that the Anthropometric Committee was able to include rich children in private schools, excluded from his data, and, in contrast failed to obtain adequate representation of social class V (the lowest) which he thinks better represented in his survey. In modern terms Greenwood's mean heights are at about the 3rd centile for boys and girls at ages 5 to 10 and the 1st centile for both sexes at age 14.

Greenwood's values are very close to those of Tuxford and Glegg, as one might anticipate from the similar, sometimes identical sources of data. Rural children are clearly taller and heavier than children living in large towns, by 1.5 to 2 cm and 1.0 to 1.5 kg respectively. From 1907 onwards local authorities adopted various customs so far as measurements of schoolchildren were concerned (though a uniform one of burying any results in official archives). The London County Counicl took till 1938 actually to measure each child each year: it abandoned this custom with the war of 1939 and never revived it.

Before turning to examine the situation in the United States we may summarize, in Table 9.3, one aspect of the results of the European studies of 1870–1910. The mean age at peak height velocity in the bourgeois boys attending the Carlschule in the eighteenth century had been about 16.0 years. In Germany and Scandinavia in the 1880s and 1890s the mean ages for different series varied between 15.0 and 16.0 for boys, and 12.5 and 14.0 for girls. On average children were experiencing their adolescent growth spurt 1.0 to 1.5 years later than nowadays. This agrees well with the data on age at menarche, reviewed in Chapter 11. For English manual workers' children there was still little change from the values of the 1830s and 1870s. As for average heights, they were very low compared with nowadays; relatively well-off pre-pubertal children had heights around the modern 10th centile, and poor children were at or below the modern 3rd centile.

William Porter and the relation between size and ability

When Bowditch persuaded the Boston School Committee in 1875 to undertake a survey of the growth of Boston schoolchildren, he saw it as part of a larger undertaking spread over the whole of the United States. At the meeting of the Social Science Association in Detroit that year he read an account of his own plans and urged others to make similar studies. The only immediate response came from the town of Milwaukee, in Wisconsin, in the person of G. W. Peckham (1882), the teacher of biology in the Milwaukee High School.

In 1881 Peckham organized a survey of the heights, weights and sitting heights of 10,000 children in the city, aged between 4 and 18, copying Bowditch's protocol exactly. German immigrants, however, substitute for Irish ones in the comparison between children born to native Americans and those born to immigrant parents. Remarking that Quetelet 'had Platonic ideas' about the ideal type, Peckham reviews the field much as Bowditch had done ten years earlier. He gives excellent graphs of the growth of Milwaukee children. Girls were equal to boys in height at age 11 and taller at 12, 13, and 14; they appeared to reach peak height velocity at 13.5 and had ended growth by 17, at a final height of 158 cm (present 25th centile). Boys appeared to reach peak height velocity at 16.0. As in Boston, children of immigrants were a little smaller than children of native-born Americans. Milwaukee children as a whole were a little taller and heavier than Boston children, but the difference between the native-American groups in the two cities was quite small, averaging about 1.5 cm. German immigrant children in Milwaukee, however, were 3 cm or more taller than German immigrant children in Boston. (Germans and German Jews were not distinguished and their proportions in the two cities might have been different.) Peckham compared sitting height and leg length curves, realized that the spurt occurred earlier in the leg, and recognized that it was the longer growing period of the male, in conjunction with the relatively high rate of leg growth just before puberty, that accounted for the relatively longer legs of the male. Peckham quotes one of the earliest North American studies on age at menarche – a table by a doctor in Cleveland – which gives a mean of around 14.5 years, presumably from data obtained from the recollections of adult women.

The most important of the American studies of this time however (see note 9.2), was made by W. T. Porter in the schools of St Louis. William Townsend Porter (1862–1949: see Carlson, 1949; Landis, 1949) became, like Bowditch, one of the founding fathers of American physiology. Born in Ohio, he graduated in 1885 from what was then the St Louis Medical College, now Washington University, and was appointed professor of the new science, physiology, only three years later. One of these years, 1887, he spent in Breslau, Kiel and Berlin, where, to judge from his introduction of head measurements into his school survey and from his using an article by Stieda in the *Archiv für Anthropologie* of 1882 as his source of all statistical formulae, he evidently had contact with the physical anthropologists. In Breslau (modern Wroclaw) he probably became familiar with Carstädt's (1888) work on growth in the high school, which was in progress while he was there. His appointment in St Louis to the physiological department must have at once brought him into contact with Bowditch in Boston. He was friendly too, probably through his interest in physical anthropology, with Franz Boas at Clark University (see Chapter 10), and for one of his papers (1894) Boas supplied him with a sketch of a head with measurement landmarks. In 1891 Porter sought permission to make a growth study in the public schools of St Louis and in the following year this was carried out. Height, weight, sitting height, span, head and face measurements, vital capacity and grip strength were measured in 33,500 children aged 6 to 15.

Porter had scarcely had time to complete this work when Bowditch called him to Harvard to be his assistant professor of physiology. With Bowditch he founded the *American Journal of Physiology*, and guided it as managing editor as well as financial guarantor from its beginning in 1898 until 1914. Like Bowditch, he was a master craftsman and loved to build the sort of equipment that permitted clear-cut experimentation. In 1901 he founded the Harvard Apparatus Company, an organization which supplied equipment for physiological teaching throughout the world and used its profits to support research fellowships in physiology in American universities. Later he returned again to the study of growth and in 1910 initiated the first long-term longitudinal programme of school measurements undertaken in the United States. He retired in 1928 and died in 1949. In his long obituaries no one thought of mentioning his work on human growth.

What distinguishes Porter's St Louis survey from others is the relating of body size to apparent ability at school. In St Louis the schools were organized in grades, and a pupil moved up a grade when he had successfully completed the work of the previous grade, irrespective of his age. (The same system was in use in England at that time and the schools were known as Grade Schools; furthermore in England the teachers were paid according to the number of pupils who passed each grade examination each year.) Porter found that the pupils in the higher grades were heavier than the pupils of the same age in the lower grades. Not only was this true of the averages; Porter followed Galton's work closely and showed that the 20th and 80th percentiles of the higher and lower grades were equally distinct; the whole distribution of weight for age was shifted upward in the more scholastically advanced pupils. Fig. 9.2 is redrawn from the first of Porter's papers (1893*a*, plate 2). It shows the median weights of boys and girls divided into those in above-average grades for their age and those in below-average grades (Porter, incidently, like Roberts, calls the mode the 'mean', as did a number of authors up to about 1910. Others called the median the mean. The present-day mean was referred to as the 'arithmetical mean' or simply the 'average'.) In Figure 9.2 the difference in weight between the two sets of pupils is clear to see, and has been confirmed in scores of investigations since Porter's day (see Matiegka, 1898; Hastings, 1900; Rietz, 1906; Mumford, 1927, p. 272; Dearborn and Rothney, 1941, p. 270; review in Tanner, 1966). It was in fact first described by Sack (1893) in Russia, in work at the time unknown to Porter, who nevertheless very correctly acknowledged its priority in a footnote. Porter gives the equivalent statistics for height at only one year of age in each sex: 10 in boys and 12 in girls. The same distinctions hold; 10-year-olds in grade I averaged 5 cm shorter than those in grade IV; 12-year-olds in grade II averaged 5 cm shorter than those in grade VI.

Porter's findings raised incredulous but ill-documented opposition then,

Fig. 9.2. Median weights-for-age of boys (solid lines) and girls (dashed lines) divided into those in above-average school grades for their age and those in below-average grades. Schools in St Louis in 1892. Data from Porter (1893); figure reproduced from Tanner (1966).

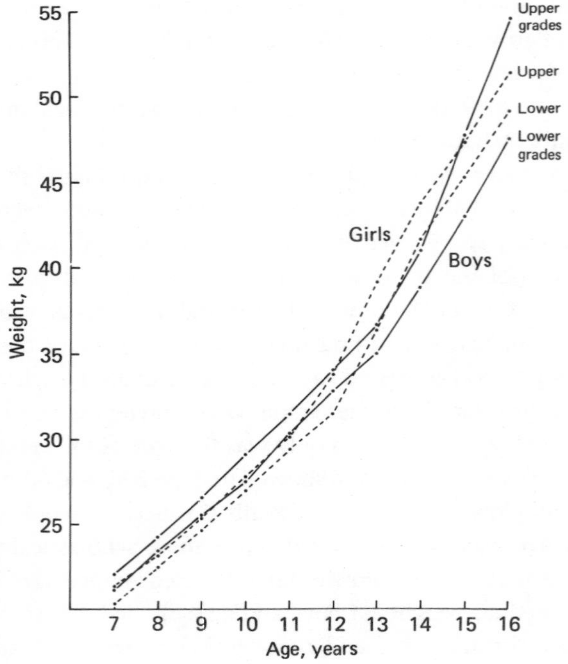

as sometimes now. Porter thought that physical strength, which he equated with height and weight, conditioned the amount of mental effort a child could make, and was much concerned, like all his contemporaries, with the question of school 'over-pressure'. Hertel's *Overpressure in High Schools in Denmark* had been published in English translation in 1885; and in 1887 in Breslau, Berlin and Kiel Porter must have felt the influence of the group connected with the nascent *Zeitschrift für Schulgesundheitspflege*. At the end of his paper Porter wrote 'No child whose weight is below the average of its age should be permitted to enter a school grade beyond the average of its age, except after such a physical examination as shall make it probable that the child's strength shall be equal to the strain'. This is pure Factory Commission concern, almost word for word.

Porter, however, went further. He pointed out that the growth curves of the two groups of children (he referred to one as 'precocious' – but meant 'bright' – and to the other as 'dull') followed parallel courses, with the adolescent increases in weight occurring at about the same time in each. He thus indirectly inferred (though never unequivocally claimed) that these differences of size and mental ability would persist into adult life. To this conclusion Franz Boas (1895*a*) took strong exception. 'I should prefer to call the less favourably developed grade of children *retarded* [his italics] not dull', he wrote, 'and these terms are by no means equivalent, as a retarded child may develop and become quite bright . . . furthermore I do not believe that the facts found by Dr Porter establish a basis for precocity and dullness, but only that precocious children are at the same time better developed physically . . . Dr Porter has shown that mental and physical growth are correlated, or depend upon common causes; not that mental development depends on physical growth'. Boas, as we shall see, was the first man really to appreciate that children grew at very different rates; thus naturally he saw the relation between ability and size as probably caused by differences in rate of development, some children being advanced both physically and mentally, others delayed. If so, then by adulthood the delayed child would have caught up the advanced and there would no longer be any difference between them either in size or in ability. After nearly a century the controversy continues and the data needed to resolve it are only beginning to be accumulated. Suffice it to say that in most countries, though perhaps not all, a very small positive relation between height and ability does indeed continue to exist in adulthood, probably generated by differential social mobility linked to size as well as intelligence, and perhaps generated also by early environmental influences.

Porter's next report on his growth study (1893*b*) was only the third paper to appear in which the 'probable deviations' of the distributions of heights and weights at each age were given. (The standard deviation did not appear till the following year: see p. 127.) Porter quotes extensively the preceding paper that used 'probable deviation', that of Geissler and Uhlitzsch (1888). He starts by invoking Quetelet, ascribing a considerably clearer and more modern meaning to his 'law' than is historically justified. As Porter put it, this 'law' holds that 'the median value of an anthropometric series expresses the physiological type of the series and each deviation from this value expresses the physiological difference between an individual and his type' (1893*b*, p. 233). Thus, 'If Quetelet's theory is true, the Probable Deviation is a measure of the degree of deviation of individuals from the Physical Type'. Things had moved on since Quetelet's time, however, and whereas Quetelet was confused about the difference between error and distribution, Porter was perfectly clear and at once splits up the variance, saying 'The Probable Deviation contains the Error of Observation, as well as the Physiological Difference of the Individual from the Type'. He then shows that the error of observation is relatively small; one boy aged 17 was measured seventy-eight times (whether by one or more measurers is unclear) and the probable deviation of the seventy-eight measurements was ±0.24 cm (giving an S.D. of 0.24 ÷ 0.6745 = 0.35 cm, a value in line with modern work if we assume that several different measurers were involved). The probable deviation of seventy-eight boys each measured once, however, was 5.15 cm.

Porter then showed, for the first time, that probable deviation, as an absolute value, increased with age. However, when considered as a percentage of the mean value at the same age (i.e. as the equivalent of the modern coefficient of variation), it changed little from 6 until puberty, when it rose; at maturity it fell to a lower value than during growth. The rise occurred at 11, 12, 13 and 14 in girls and at 13, 14, 15 and 16 in boys. This coincides, as Porter indicates in a graph, with the increases of height velocities, the ages of peak being two years earlier in girls than in boys. (Porter, like so many earlier writers, uses age-at-nearest-birthday not age-at-last-birthday, a potential source of confusion.) Porter concludes: 'The Physiological Difference between the individual children in an anthropometric series and the Physical Type of the series is directly related to the Quickness of Growth' (1893*b*, p. 247). In this generalization he missed the point, for it seems he was really thinking about *individuals'* rates of growth, and failed to realize that the increase in the probable deviation was much more due to differences in tempo *between* individuals. This is where we must proceed to the work of Boas, for this is one of the two points from which Boas started.

Boas' other starting point is also present in Porter's work, and before him in the work of Bowditch. Porter's definitive report on the St Louis survey was issued in 1894. In it he gives probable errors of the means (it seems for the first time ever in growth work), as well as probable deviations. He shows the boys' peak height velocity to occur at 14.5 and the girls' at 12.5. The girls' ascendancy in height lasted from 11.4 to 15.3 years; in head measurements girls never exceeded boys. For each sex Porter gave modes and means (modern terminology) for each year of age for his various measurements. Like Bowditch, he then presented the difference between these forms of average each year, this difference measuring the degree to which the distributions were skewed. Skewness increased sharply exactly when the pre-pubertal acceleration of growth (his phrase) occurred. This is the point at which we left Bowditch in 1891. It is the point at which Boas enters, as it was he who gave the correct interpretation of this apparently small technical nuance, which actually opened the door to an understanding of one of the most important features of human growth. Before considering the work of Boas, however, we will pause for a short intermezzo.

Porter took the whole question of school health very seriously and the third of his 1893 papers (1893*c*) is an interesting early attempt at a theory of growth standards. 'If we had sufficient data obtained by the individualizing method', Porter says, 'the deviation of children from the laws of normal growth could be quickly recognised and by timely treatment largely overcome; the evil effects of over-study could be watched and intelligently combated and systems of education, no longer exacting from all that which should only be exacted from the mean, could be rationally adapted to the special needs of the exceptionally weak and the exceptionally strong' (p. 578). Data (or standards) obtained by the generalizing method could not do this because (to translate Porter into modern language) they could not give standards for *velocity* of height, and that was what was needed. However, it would take a generation to produce such velocity standards and in their absence Porter suggested using standards of weight-for-height (irrespective of age) as the best measure of health. 'Overwork', he wrote, 'may cause a temporary or a permanent deviation in these [growth] curves . . . a prolonged strain in a growing child harms for life and leaves a mark which can never be effaced. The danger is greatest in the periods of quickest development, and particularly great in the prepubertal period. The child should be guarded against the possibility of harm. The anthropometrical system proposed in this article offers a means of doing this. It infallibly discovers those whose physical development is below the standard for their age' (p. 158).

Twenty years later Porter was less sure of this, and more certain than ever that velocity standards were necessary. He remained acutely conscious that his St Louis study was made by the generalizing method. 'Far different', he said, 'is the individualising method . . . The measurements must begin at the earliest age with many thousand children lest death and desertion so thin the ranks that the survivors will be too few for statistical treatment. The individualising method demands, therefore, a formidable expenditure in time and effort through many years. Toilsome as this task may be, it cannot be forgone. For the generalising method conceals a grave flaw; it does not give the growth of the individual child . . . [standards of weight] are not standards of growth, but merely standards of relative size' (Porter, 1920, pp. 121, 124). In 1909 Porter requested the Boston School Committee to carry out an individualizing study, and from 1910 to 1919 the school nurses measured the heights and weights of some 2,400 boys and 2,400 girls every month — a frequency of measurement that has been repeated on this scale only in Palmer's Hagerstown study of the 1920s (see Chapter 12).

Porter (1920) first analysed the seasonal differences in weight increment. His results closely paralleled Schmidt-Monnard's, though he does not say so, for by 1920 he seems to have lost contanct with the growth literature, quoting virtually nothing of earlier work. In the period September to January inclusive the boys (aged 6 to 13) gained an average of 3.4 lb, the girls 2.8 lb. In the period February to June inclusive the boys gained only 0.8 lb, the girls 0.7 lb. Admittedly the weighing was done with indoor clothes on, but Porter has no difficulty in showing that changes in clothing could not account for this magnitude of difference.

Two years later the results had been fully worked up, and Porter (1922) went to the heart of his argument. Having calculated the 1st, 5th, 10th, 20th, . . . , 90th, 95th and 99th centiles at each age and called each a 'grade', he showed that from one year to the next about 30 per cent of children changed their centile positions in weight by less than one grade. About 50 per cent changed by less than three grades, and 75 per cent by less than five grades (except at 12 and 13 where changes were more frequent). Thus about a quarter of the children changed by one or two grades, and a further quarter by three or four grades. The situation for height was only slightly more stable; about a quarter changed by one grade, and further quarters by two, three or four grades. Porter insists on the importance of looking at changes —at what we would call velocity of growth. Height, he thinks, should be watched more closely; and weight is correct when the child has the same centile position in the weight as in the height chart.

The centile charts themselves (Porter, 1923) were published in a format different from the one auxologists use nowadays : they were clearly modelled on Bowditch's presentations of 1890 and 1891. The horizontal axis was not age but centiles, 0 to 100. The vertical axis was height (taken, incidentally, with shoes on; heels averaged 1 cm except for girls aged 12–13, when 2 cm). The cumulative curves for each year of age were plotted, the line for 7 being above that for 6, and so on. The physician looks up the nearest birthday (not last birthday; Porter makes this clear) and sees the child's centile position; next year he does the same. Thus Porter revived Bowditch's centile charts of thirty years before, at the same time adding the 'outside', 1st and 99th, centiles. Porter was at pains to stress that the centiles he gave represented what *was*, not what should be. He quoted the early studies of Gray in private schools which showed both higher height at given age than in the Boston schools and lower weight for given height (Gray and Jacomb, 1921; Gray, 1921). He felt strongly that with adequate supervision of nutrition, exercise and conditions of life, the schoolchildren of Boston could be brought up to the standards of private schools such as Groton and Middlesex.

Paul Godin and growth surveillance

The large-scale surveys of heights and weights in schools were usually made by public health doctors, remote from the pupils themselves, in order to establish large-scale general facts – for example that children from overcrowded homes grew less well as a class than children from better homes. The action that followed, if the persons concerned were fortunate, was action directed towards ameliorating the environment, towards improving the homes, the conditions of work, or the low incomes. Surveys of this sort continue to be made (see Chapter 14) and continue, alas, to be necessary.

But growth measurements can also be used to follow the health of individual children, for, as Porter saw, one of the surest signs of ill-health is a gradual falling-off in growth rate. This use of measurements for individual surveillance is the province of family doctors, school doctors, or the teachers themselves. It was adopted with enthusiasm at the beginning of the century and then fell into almost total disuse, largely because later generations of school and family doctors had little training in how to interpret the measurements, failed to plot them or to calculate the essential increments, and usually thought their stethoscopes more practical. Recently, with the emphasis turning to prevention and family medicine and developmental assessment, growth surveillance is coming back, and child clinics, general practitioners and schools are beginning again to keep serial records of individual children.

One of the pioneers of such surveillance was a French physician, Paul Godin (1860–1935). Godin was an army doctor and his publications include works such as *L'earth système* (1901) which deals, after the manner of army doctors, with experimental researches on the earth latrine. But Godin was an educationist from the start and the thesis for his doctorate was on 'Physical education in the family and the school'. Like Maria Montessori, he was also greatly interested in physical anthropology. From 1891 to 1900 he worked in schools for children of army personnel and in 1893 he studied physical anthropology in Paris, with the celebrated Professor Manouvrier (Godin, 1914). In 1895 he was appointed doctor to the Military Preparatory School for sons of non-commissioned officers at St Hippolyte-du-Fort on the Mediterranean coast, where he had 400 boys aged 13 upwards in his care. An admirer of Carlier, his forerunner at Montreuil, he was already a convinced longitudinalist, declaring that only a study of '*la marche de la croissance individuelle*' would reveal the laws of growth, and deploring the fact that, Montbeillard and Carlier apart, nobody had yet concerned himself with this.

Accordingly, Godin took two consecutive entry classes, each comprised of 115 boys aged 13 or 14, and followed the boys until they left the school, most of them after four years, some, regrettably, sooner. Each boy was measured every six months. Godin was only able to measure three boys each day, because, trained by the Paris school of physical anthropology of the 1890s, he took 129 measurements (32 heights from the ground, 9 transverse diameters, 27 circumferences, 23 head and face measurements and 38 measurements on tracings of the hands and feet). In addition he made 46 ratings, for example of pubic and facial hair, skin colour, eye form, etc. He did this for five successive years, something which seems to have astonished even him slightly.

Paul Godin is a man after the heart of anyone who has personally undertaken longitudinal studies on children. In the book which reports the study (Godin, 1903; reprinted with an addition, 1935) he writes that although such activity demands a certain effort, 'I never yielded to physical lassitude, never allowed anyone to substitute for me; there is not a rating that I have not dictated to my corporal assistant; not a height above the ground, a diameter or a circumference which has been measured by anyone but myself' (1935, p. 6). He is a real operator too; he understands exactly. The span is always the worst measurement, but in the tyro's hands the circumferences are the next worst, despite their apparent simplicity. Godin knows why: 'The uninitiated, even if a doctor, always has a feeling of respect, whether he knows it or not, for the circular and sliding compasses. Already the stadiometer (*toise*) seems less imposing; despite its special shape and the disposition of the scale it is an instrument he has seen before ... But the uninitiated does not even pause to glance at the tape, the familiar object which is in everybody's hand and is used by all'. As to his devotion to the measurements, it has only been equalled by that of R. H. Whitehouse, also, and perhaps significantly, a man with a military training as background. Godin took 129 measurements on what must have been about 1,000 boy-occasions; thus a total of about 1.3×10^5 measurements. At the Harpenden Growth Study over the years 1949–70 Whitehouse took 15 measurements on approximately 9,000 child-occasions. The totals are about the same, but at least Whitehouse had time to put down his instruments.

Sadly, when it came to working up his data, Godin was less inspired. Despite his absolutely correct insistence on obtaining longitudinal data, his text, and above all his results, make it fairly clear that he worked everything out cross-sectionally. Instead of taking increments for individuals, he took 100 subjects at each six months of age from 13.5 to 17.5 years, calculated the means at each age and subtracted one from the next to get the mean increments. Had the series been pure longitudinal, with the same 100 boys at every age, this would have mattered less, but Godin says many boys left the school in the middle of the study; and the entry classes were not comprised of boys all of the same age either. Thus his increment curves have irregularities which it seems must be due to sampling bias.

For stature the two highest six-monthly increments are at 14.0–14.5 and 14.5–15.0 years, giving a peak centred at 14.5. For 15.0–15.5 the increment is small, but it increases again at 15.5–16.0 and 16.0–16.5. A similar small increment occurs at 15.0–15.5 in biacromial diameter and in leg length, though not in sitting height, whose peak is clear at 16.0–16.5, a year later than the rather indistinct peak for leg length. Godin was the first to demonstrate this well-known difference in timing in the two length components of stature at the adolescent growth spurt.

Godin produced the following values for six-monthly increments of thigh length and of calf length: thigh 9, 4, 8, 12, 2, 3, 3, 0 cm; calf 2, 9, 0, 6, 4, 10, 3, 1 cm. Inspection of the figures seems to reveal an inverse relation: 9 with 2, followed by 4 with 9, 8 with 0, 12 with 6, etc. The two measurements were made independently (that is, one was not obtained, as in some techniques, by subtraction of the other from a larger dimension). Thus Godin felt sure this alternation was real. His figures on upper arm and forearm lengths (3, 7, 8, 1, 10, 0, 4, 4 and 5, 4, 4, 10, 2, 7, 2, 1, respectively) seemed at least not to negate the generalization. Furthermore, figures for calf length and calf circumference (the latter 6, 2, 6, 2, 5, 2, 3, 0) alternated also. He thus enunciated a Law of Growth: The growth of the long bones of the limbs proceeds by alternating periods of activity and rest, which follow each other regularly. These periods are in opposition in the two body segments of the same limb' (1935, p. 98; and see Godin, 1914). There was also an opposition between length and breadth growth in the same limb segment.

Clearly such a generalization goes far beyond the evidence. Godin only had six-monthly periods to deal with and when he later (1935, p. 178) accepts that the *Loi des Alternances* has a time base of just six months, this proceeds from ignorance of any other, alternative base. (The seasons of the measurements are not given, nor the precise degree of longitudinality over each six months; it may be that Godin was partly seeing the seasonal effect already described by Buffon and Carlier, though this would not, of course, explain alternation between limb segments.) Godin, like most other doctors of his time, was evidently quite ignorant of statistics, which might have offered some barrier against over-enthusiasm. His alternations between limb segments seem exceedingly unlikely to be real; but it has to be said that as no modern test of his idea has ever been made, judgement should yet be suspended.

Godin was the first to use ratings for the development of the secondary sex characters in the way that is now standard. He had five grades for pubic hair and five for axillary hair. Though he does not adequately define what he means by the word puberty, he says that in the 'immense majority' of his boys it appeared in the 15.0 to 15.5 interval. Height reached its peak in the period immediately preceding puberty and weight its peak later, at the moment of puberty (1935, p. 158). Puberty, it seems, must be something later than the first appearance of pubic hair; perhaps it was seminal emission.

It was Godin who introduced the term 'auxology' into human growth. It is not used in the 1903 edition of his *Recherches* and seems to appear for the first time in 1919 in an article entitled 'La methode auxologique' (1919*b*). He seems to have taken it over from a physiologist, G. Bonnier (1900), who wrote the article 'Croissance' in Richet's *Dictionnaire de physiologie*. Bonnier wrote of plant growth being followed by means of an *auxanomètre enregistreur* in which a thread attached to the top of the plant ran over a pulley linked to a pen writing on a drum. The Greek root is *auxein*, to increase. (The word 'auxin', of similar origin, was introduced in the late 1920s to refer to plant hormones.) Godin defined auxology as 'The study of growth by the method of following the same subjects during numerous successive six-monthly periods (*semestres*) with a great number of measurements.' This seems a little egocentric and makes him the only true auxologist. Some French, and several subsequent Italian writers kept the use of the word alive in the human growth literature and in the 1970s Tanner and his associates adopted it and began its systematic deployment in the English-language journals. An

International Association of Human Auxologists was formed in 1977, as an offshoot of the multinational Society for the Study of Human Biology. In this later usage, however, convenient because it yields an elegant adjective and adverb, which the word 'growth' cannot do, 'auxology' covers the whole subject of physical and physiological growth and development whatever method may be used to study it.

Godin gradually elaborated a whole system for following the growth of individual children. In 1914 he retired from the French army and though he continued to live at Saint-Raphael and then Nice, he took up an appointment in Geneva at the already famous Institut Jean-Jacques Rousseau des Sciences de L'Education, founded by Claparède and Bovet and later to be graced by the life-time's work of Piaget (see Godin, 1921). In 1913 he published a book on the educational applications of the study of growth (translated into English in 1920) and in 1919 a *Manual of Pedagogical Anthropology* in which his full system is described. He begins the *Manual* by saying that he has evolved 'The individual formula for the child ... something wished for by all the great educators from Montaigne to Montessori and Baden-Powell' (1919*a*). His system consists essentially in obtaining a physiological age (something already popularized by Boas and others) by relating one body measurement or block of body measurements to others. (He thus, without knowing it, ran into all the difficulties of constructing a measure of shape age' described by Tanner (1962).)

The system owes a good deal to the mental-age concept of his countryman Binet, and a good deal to the once famous but now almost entirely forgotten Italian school of constitutional medicine (see De Giovanni, 1904–5, and Viola, 1932). The two most important indices reflect the Italian school closely. Multiply together the length, width and depth measurements of the trunk to give trunk volume (v). Add lengths of arm and of leg to give limb length (o). Multiply length, breadth and depth of head to give head volume (c). Then the ratio v/c is the index of growth and of energy and development *(conditionnement)* of the psychological power *(force psychique)*. The ratio v/o is the index of development of motor power *(force motrice)*. Multiply each ratio by chronological age and look up the result in tables supplied, which give normative values at (naturally) six-month intervals. If a given child's index is over seven semesters away from the standard for his age, then he is considered abnormal. Godin advises plotting these two indices on a chart (together with similar ones for mental development and health) but stresses that it is changes in the position of an individual that really matter, and not his absolute position. However, his use of this imperfect instrument is somewhat hair-raising. He gives several examples of his advice to parents; thus when a child was below par in the motor development index, the parents were advised to take him out of boarding school and send him back to the country air.

However this may be, Godin envisaged a whole new and useful area of preventive pedagogical medicine. In discussing a boy with retardation of his indices he writes 'The child is not sick, and that removes the usefulness of a clinical doctor; but it does not remove the usefulness of an educational doctor (*médecin éducateur*)'. Godin's work and even his name are quite forgotten nowadays. He was an enthusiast, perhaps an over-enthusiast, some might say an uncritical quack. But he saw further than most into the future of medicine in childhood, and though his methods are long since dead his approach is very much alive. He deserves a more enduring memory.

A. A. Mumford and G. E. Friend

There were other, less obsessional *médicins éducateurs*. In England there was Alfred Mumford (*c.*1860–*c.*1944), medical officer to (and historian of) Manchester Grammar School, a day school for fee-paying boys of high ability. Mumford's book *Healthy Growth* (1927; see also 1910) exemplifies all the concerns that Kotelmann had listed. In 1881 a gymnasium was established in the school and two years later regular measurements of height, weight and chest circumference begun. From 1910 to 1921 the boys' growth was monitored by reference to 'the strangely neglected but extremely helpful work of Mr Cecil Hawkins', master for mathematics and games at Haileybury College. Hawkins collected data of 40,000 measurements made in the Public Schools (see Chapter 8) and applied Galton's percentile method (Hawkins, 1899). Like Sargent, his counterpart in the United States, he provided tables and charts for following these measurements in terms of their successive percentile positions. At this stage Manchester Grammar School had perhaps the most efficient monitoring system anywhere in use. Then in 1921, after a brief period of actually using standard deviation scores of growth increments (it seems, from Mumford's not very clear text), they changed to a system of 'time increment charts'. These were not increments at all, but what later paediatricians called 'height age' tables. If a boy aged 14.0 was the height of the average boy of 15.0 then he was two standards above average (for Mumford and Caradog Jones, his mathematical mentor, used six months as a unit, just like Godin). But Jones and Mumford seem not to have understood Bowditch and Boas; Jones fitted a simple parabola to the mean heights to smooth them to provide nice-looking averages, ignoring totally the adolescent spurt and giving steadily decreasing mean increments from 11 to 18 (1927, p. 323). One can only suppose that common-sense prevailed over the bizarre recommendations that such charts might lead to at certain ages.

Mumford reported interesting data on the secular trend in these predominantly middle class well-off boys. Boys aged 13 + were 2.7 cm taller in 1907 than in 1885, and a further 1.6 cm taller in 1922. Boys aged 14 + were 3.1 cm taller in 1907 than in 1885, and a further 2.3 cm taller in 1922. Much of the increase was due to earlier puberty, for at age 17 the gain over the whole period from 1885 to 1922 was 3.0 cm.

G. E. Friend (1875–1956), was another school medical officer, to Christ's Hospital, an orphanage boarding school founded in Tudor times (Pearce, 1901). His book, *The Schoolboy* (1935), has a more modern, but less embracing outlook than Mumford's. He was chiefly interested in nutrition. He also used the Caradog Jones' 'time increment grading' but without fitting a parabola to obtain the means. Instead he used the actual means provided by Christ's Hospital measurements in 1926–9, which show a clear peak height increment centred at 14.0. Christ's Hospital boys also showed a secular trend in height. Boys aged 13 + were 1.5 cm taller in 1920 than in 1910, and a further 3.5 cm taller in 1930. Boys aged 14 + were 2.8 cm taller in 1920 than in 1910 and a further 3.5 cm taller in 1930. Again most of this represented earlier maturing; boys aged 18 were only 0.7 cm taller in 1920 than in 1910 and a further 1.0 cm taller in 1930. All the gains from 1920 to 1930 were greater than those in the earlier decade. In girls attending St Paul's School, London (representing well-off middle class families of high scholastic aspiration) the gain between 1907 and 1937 was considerably less. Over the thirty years, 11-year-olds increased in height by 2.5 cm, 13-year-olds by 1.8 cm and 16-year-olds by 1.5 cm (Jacob, 1938).

Winfield Hall and alternation in length and width

Godin was not the only doctor to describe supposed alternations in growth between different parts of the body. This particular hare was started, about the time Godin was making his myriad measurements, by Winfield Hall, a doctor who had worked in Sargent's laboratory at Harvard and who from 1889 to 1893 was medical examiner at Haverford College and in three Friends' Schools in the Philadelphia region. Evidently a trained physical anthropologist of the same ilk as Godin, Hall took twenty-five anthropometric measurements and measured vital capacity and the strength of the back, legs and arms. The age range covered was 9 to 23 years and some 2,400 boys were measured, once each. Hall took the measurements to Leipzig and worked them up there in 1894, presumably under the eye of

Geissler (see note 8.7). When Hall regarded the tables of medians at each year of age he was struck by the fact that the greatest increment of height was from 12 to 13. The increments for 11–12 and 13–14 were less, but that for 14–15 was again large. Body circumferences, however, showed no such thing and indeed had rather larger increases in the 13–14 interval. The actual yearly increments were, from 10 on: for height 3.0, 3.9, 10.6, 3.8, 7.5, 3.0, 1.7, 0.5 cm and for the sum of knee, ankle, elbow and wrist circumferences 2.6, 3.0, 3.5, 4.0, 2.7, 1.9, 0.9, 2.1 cm.

The girths show a perfectly ordinary curve with their peak velocity occurring at 14.0 years as expected. The heights, on the other hand, are quite peculiar, the third (12–13) increment being enormously high and the subsequent increment conspicuously low. The fifth value, 7.5 cm/yr, may represent the true peak height velocity, centred at 15.0. It seems likely that the 13-year-old median height has fallen victim to arithmetic error; but if so, so have those of knee height and pubes height, which show the same pattern. Perhaps some maverick or misread value consistent over all three measurements got in at 13. However this may be, to the modern eye the supposed alternation of velocities of length and width is most easily explained as artefact, statistical or otherwise. But Hall, like Godin, was an enthusiast. He enunciated at once and in italics a Law of Growth (thus beating Godin to it). 'When the vertical dimension of the human body is undergoing an acceleration of its rate of growth, the horizontal dimensions undergo a retardation; and conversely' (Hall, 1896). This was true of the bones rather than the muscles, he added, since it was best shown by alternation between lengths of bone and circumferences of joints.

Malling-Hansen and seasonal variation

Hall was probably influenced in this by the work of the pastor Malling-Hansen in the school for deaf and dumb children in Copenhagen. Malling-Hansen (1883) started by measuring the weights of some of his children eight times a day, but later settled for twice, at 06.00 and 21.00 hours. He espoused the notion that, considered over five-day periods, gain in weight paralleled the amount of radiant energy from the sun. He even alleged that children grew faster at certain times of the day than at others, views which when presented at an international medical congress in Copenhagen in 1884 drew some caustic comment (see Camerer, 1893). Subsequently, however, Malling-Hansen (1886) took daily weights and heights (at 09.00 h) over a period of two years on about seventy boys aged 9 to 15. The boys' rate of growth averaged 2.3 mm/month during September, October and November,

4.4 mm/month during December, January, February and March, and 5.6 mm/month during April, May and June (July and August were interrupted by holidays). Thus he remarked that in the maximum period, in spring-time, the rate was two-and-a-half times that in the minimum period, in autumn. Weight underwent still greater changes of rate, but in the opposite sense, with maximum rate in the autumn and minimum in spring. These observations have been amply confirmed since; first by Schmidt-Monnard (1895) in Halle and most recently, and in a detail which bears out Malling-Hansen's quantitative as well as qualitative findings, by W. A. Marshall (1971; see also Fitt, 1941, and Bransby and Gelling, 1946). Malling-Hansen wrote that 'In the maximal period of lengthening, the thickening of the body is at its minimum; and vice versa, the thickening has its maximum in the time of minimal lengthening' (1886, p. 41). However, the seasonal periodicities to which he referred were, of course, quite different from the supposed periodicities, extending over a whole year or more, postulated by Hall.

Malling-Hansen also measured the apparent diminution in height between measurements made at eight o'clock in the morning, shortly after getting up, and in the evening: it averaged 1.2 cm. Nowhere does he quote either Wasse or Montbeillard, and it seems his observations were entirely original. He warns measurers such as Pagliani about the effects on their values of time-of-day and day-to-day variation and as a practical step suggests re-scheduling holidays so they coincide with periods of maximal growth (it seems in weight). Towards the end of his long, and excessively rare, pamphlet (his own signed copy being the one in the United States National Medical Library), he reverts to his theme of radiant energy. Truly there used to be giants, he infers, and giant plants too — when the sun was younger and more energetic.

Christian Wiener and family longitudinal records

Amongst all this activity in north-west Europe, a number of people took the trouble, like Montbeillard and Quetelet, actually to measure their own children throughout their growth. The most distinguished of these measurers was Christian Wiener (1826–1896), professor of descriptive geometry and graphical statics in the Technical High School of Karlsruhe. Wiener taught in Karlsruhe from 1852 until his death and was a leading intellectual of the town, being interested not only in mathematics and science but in philosophy as well. He was the first person to give the true explanation of Brownian motion (see Wiener, 1927) and he wrote a book entitled *Grundzüge der Weltordnung* (1863) (Fundamentals of a World System) which D'Arcy Thompson (see Chapter 10) referred to as 'highly original'; in it he gave an explanation for the logarithimic spiral shape of certain sea-shells (which, together with his article *'On the beauty of lines'*, which directly followed the paper on his son's measurements, must have endeared his memory to D'Arcy: see Thompson, 1942, p. 757).

Wiener must have shared a common interest in conic sections with Quetelet, and perhaps he knew him personally. At all events it seems to have been Quetelet's work which stimulated his interest in growth. He began measuring the first of his four sons in 1856, taking his height, and head length, breadth, circumference and superior arc (root of nose to occiput, passing over the vertex) every year till he was 33.91 years. (Wiener used decimal age, being the second, after Lehmann, to do so, fifty years before Raymond Pearl and a hundred before Tanner. He clearly did not feel any explanation was necessary; he was, after all, going to fit curves to his observations.) The second son was born in 1857, the third in 1862. By the time the third was aged 12 Wiener had realized that the first two boys' growth by no means followed Quetelet's conic-section-derived curve (the only one he seems to have known of, having evidently missed Buffon). Both seemed to have clear spurts *(Schüsse)* at puberty. To make sure of it Wiener started measuring the third boy every six months. The graph he eventually gave is shown in Fig. 9.3. By the time his last son was born, in 1869, of a second wife, Wiener had

Fig. 9.3. Growth in height of Wiener's third son, born 1862. Reproduced from Wiener (1890).

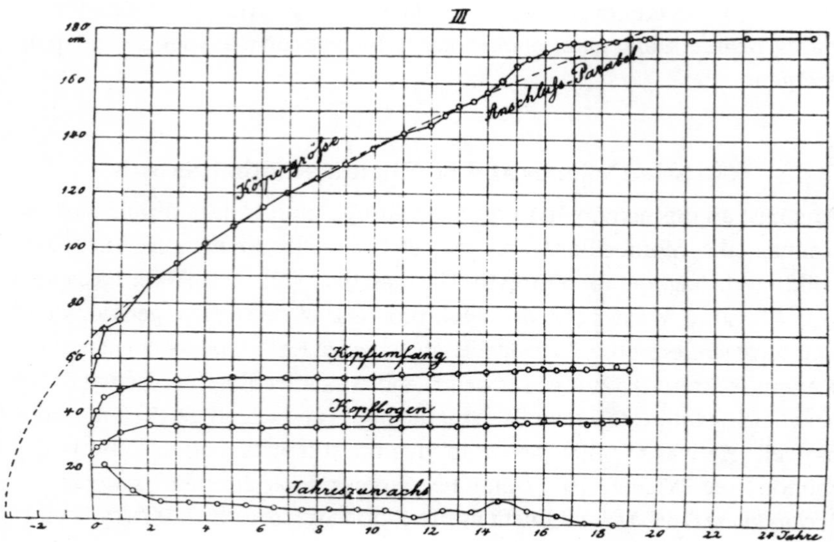

really got into his stride. This time he measured the boy's height on no fewer than eighty-nine separate occasions, not counting the considerable number on which both morning and evening heights were taken. This is the most complete record of an individual's height ever made. Wiener was careful too; he assessed his error (3 mm for height) and studied the decrement during the day (4 to 7 mm). He showed that a parabola ($h = a + bt + ct^2$) fitted the growth curves well from 2 to the beginning of puberty, but that after that the measurements accelerated away from the model (see Fig. 9.3). Wiener himself was a tall man, of 180 cm, and his sons were also tall. The first three had peak velocities at about 13.2, 13.5 and 14.5 years. The fourth matured strikingly early with a velocity of 9.9 cm/yr from 12.0 to 13.0.

By the time Wiener published his data and graphs, in 1890, Quetelet had recorded his own sons' measurements in *Anthropométrie* (1870). But Wiener's measurements were much more complete, almost certainly better done, and very much better interpreted. He is Buffon's and Montbeillard's real successor: like them his measurements proceeded from the spirit of pure enquiry. His paper, in the *Journal of the Karlsruhe Scientific Society*, was somewhat inaccessible, but Burk reprinted his measurements in 1898 (see note 9.2) and Lange knew of them in 1902. Wiener died in 1896: his considerable entry in the *Allgemeine Deutscher Biographie* does not mention the measurements he took: it is signed by a relative, most likely one of the measured (see also Wiener, 1927).

Wiener started his measurements in 1856. Perhaps there was some connection between him and the Boston family measurers whose results were available to Bowditch in 1872. Or more likely, we are seeing only the survivors (Ford, 1958, reports one) of an army of Victorian child-measurers, whose records, like those of the eighteenth-century military recruits, were pulped into Sunday newspapers or lie mouldering in the attics of half-demolished houses.

10

Franz Boas and the contribution of physical anthropology

Boas and the tempo of growth

Tradition has it that a railway train bore Franz Boas (1858–1942) into the field of growth. True or apocryphal, the story is that Boas, *en route* to the Cleveland meeting of the American Association for the Advancement of Science in 1888, became engaged in conversation with the man sitting next to him. As the train drew into Cleveland the man was revealed as G. Stanley Hall (1844–1924), a redoubtable figure in contemporary American academic life, who had just become the first president of, and professor of psychology and pedagogy at, a brand-new university named Clark in Worcester, Massachusetts (see Hall, 1925). Stanley Hall, 'the father of child study in America' (Grinder, 1969; and see Wallin, 1938), was the first American to study psychology at Wundt's laboratory in Leipzig, the first psychology department ever established, and from 1881–8 himself set up and directed the first American psychology department, at Johns Hopkins University. He was founder of the American Psychological Association and of both the *American Journal of Psychology* and the *Pedagogical Seminary* (later the *Journal of Genetic Psychology*). Nowadays he is chiefly remembered for his remarkably polysyllabic and mid-Victorian book on adolescence, which, though published in 1904, is full of phrases like 'the secret vice', 'enfeebled heredity' and 'the evolution of the soul'. But he studied with Ludwig as well as Wundt, and took his PhD under Bowditch. He was a firm champion of the view that psychology must be based on fundamental biological concepts and that the first thing in education was to understand the manner of growth of the body and the brain of the child. He was a brilliant teacher and a far-sighted man; it was he who as early as 1909 invited Freud to deliver the first lectures on psychoanalysis given outside Central Europe. He had a talent, too, for spotting men of exceptional promise, and he ended that railway journey by asking Boas to join him as head of a division of anthropology in his own department of psychology.

This was in 1888 and Boas was 30. He was already becoming known as an anthropologist interested especially in language and folklore, and formidably armed with a knowledge of scientific method. He was born and brought up in Minden in Germany where his father was a merchant; his mother founded the first Froebel kindergarten in the town. He attended university at Heidelberg, Bonn and Kiel (Herskovits, 1953). Destined originally for medicine, he studied physics, chemistry and biology; but he had always been fascinated by strange lands and places (he owned and revered a copy of Humboldt's *Cosmos*, and at Bonn bought forty volumes of Herder), and though he finally took his doctorate in physics, he combined this with a minor option in geography. After graduation he spent two years (1881–3) in Berlin, frequenting the Berlin Anthropological Society and receiving instruction in physical anthropology from Virchow, a man whom he greatly admired, both as a person and a scientist. In 1883–4 he went on his first field expedition, to the Arctic, where he studied the ethnology of the Eskimo as well as mapping the region, which was his ostensible objective. In 1886 he visited the Bella Coola Indians in the Pacific North-West and this time he stayed in the United States, joining the staff of the journal *Science*.

Despite his already wide interests and growing reputation in quite other branches of anthropology, Boas accepted Hall's offer and embarked on the study of human growth. Doubtless his interest was quickened by Stanley Hall, and probably by Henry Bowditch at Harvard University nearby; but subsequent events showed it needed little stimulation. Boas' first paper on growth appeared in 1892, his last in 1941. He was largely responsible for the discovery that some individuals are throughout their childhood further along the road to maturity than others, and for the introduction of the concept of physiological or developmental age. It was his studies which established growth and development securely as an item in the practice and teaching of physical anthropology in North America. Indeed, the first PhD ever given in anthropology by a North American university was to his student and successor A. F. Chamberlain in 1892, for a study on the heights and weights of Worcester children.

Boas approached the problem of tempo of growth – his own, exact and expressive phrase – in a very characteristic way. He was a contemporary of Karl Pearson, and for many years was in the forefront of those who were developing the new techniques of biometry. Bowditch (1891: see p. 195) found that the distribution of heights became skewed as the pubertal acceleration in growth developed and explained this by saying that the acceleration occurred at an earlier age in large than in small children. Boas (1892*a*) at once saw another, and more likely explanation. Though he couched it in obscure prose and almost impenetrable algebra, the gist is clear enough. Suppose there is an underlying variable of physiological status which is Normally distributed around an average value at any age. Then when the average rate of growth is not changing, the individuals who are advanced in physiological status and occupy on average the upper part of the height distribution gain the same amount as those who are retarded and occupy on average the lower part of the height distribution. Thus the distribution curve of height remains symmetrical, like the distribution curve of physiological status itself. But when the pubertal acceleration occurs, it happens first to those who are physiologically advanced. They therefore gain more than those who are physiologically retarded and thus pull out the height distribution into a positively skewed shape. Later they grow less than the retarded, so the skew reverses. Thus (though Boas in 1892 was in no position to put it this way) in early puberty there is a positive correlation between height and height gain and in late puberty a corresponding negative correlation. By 1906 he was able to show that such changes in correlation did actually occur (see below).

Though Boas indeed gave the correct explanation for Bowditch's finding, it is difficult to believe that he really came to discover the factor of acceleration/retardation of growth in this way. It seems far more likely that the idea came to him intuitively and was consolidated by his experience of measuring and observing children during the year 1891–2, when he conducted a short-lived longitudinal study in Worcester. This was the first longitudinal study set up in the United States and Boas (1892*b*) gave two reasons why he thought such a study essential. First, there is differential mortality. If children who are small die more frequently than children who are large then this will cause a spurious increase of growth rate if rate is calculated by subtraction of cross-sectional means. Boas pointed out that mortality is greater amongst the poor and the poor were indeed smaller. He carried the argument too far at one point. 'If, for instance', he wrote, many individuals of retarded growth should die during the period of adolescence, this might give the real explanation of the curious overlapping of the curves of the growth of girls and boys, the girls being heavier and taller than boys between the twelfth and fourteenth years' (we have already seen how much this worried Bowditch). 'Furthermore', said Boas, invoking unconsciously the shade of Quetelet's *homme moyen,* 'it would appear very likely that individuals far remote from the average, showing either too small or too large measurements ... approach the limits between physiological and pathological variation and are therefore more likely to die'.

Secondly, Boas realized, apparently independently of Galton and Pagliani before him (not to mention Lehmann and Gould), that variation between individuals in tempo of growth, combined with the fact of an adolescent growth acceleration, must mean that most children did not stay in their percentile channels during this phase of growth. This was contrary to what Bowditch had supposed. 'Consider for a moment', said Boas (1892*b*), 'all those children separately who will, as adults, have a certain percentile rank, and investigate their position during the period of rapidly decreasing growth, during adolescence. It seems reasonable to assume that the average individual (not the average of all individuals) will retain its percentile grade throughout life'. At 17, say, some of these individuals (let us say of the mature 80th centile) are advanced, others retarded. 'As the amount of growth is decreasing rapidly at this period, the number of retarded individuals will have a greater effect on the average than individuals of accelerated growth . . . thus the average of all observed values will be lower than the value belonging to the average boy of seventeen years of age'.

The explanation above sounds a bit obscure, even to the professional auxologist, because in 1892 Boas had not yet fully realized the very large implications of what he was struggling to say. He was not yet thinking in terms of velocity of growth. Thus the explanation that follows carries us beyond his position in 1892 to the threshold of his second period of papers upon growth. Boas gave somewhat clearer explanations three years later (1895*a*), when he commented on Porter's (1893*a*) identical results on the skew, and in 1897 in explaining his own results on Worcester children. Not till 1930, however, had he fully absorbed the implications of his discovery, and it was he who then initiated the use of the 'maximum growth age' (Boas, 1930), his second great contribution to the methodology and theory of human growth.

Fig. 10.1 explains the effect of differences in tempo of growth on mass 'velocity' curves. It shows the velocity curves from 5 to 18 years of five individuals, each of whom starts his spurt at a different time. The mean of these curves, taking the average value at each age, is shown by the heavy broken line. It is obvious that this line characterizes the average velocity curve very poorly; in fact, it travesties it, smoothing out the adolescent spurt and spreading it along the time axis ('the average of all observed values will be lower than the value belonging to the average boy'). When the same curves are plotted against years before and after maximum growth age (Fig. 10.1*b*) then the average value *is* characteristic of individuals. Fig. 10.2 shows the same thing in terms of height 'distance', or height for age. There are two implications. The early-developing child at first rises through a system of standard percentiles, if these are calculated from cross-sectional data, and later drops back: the late developer does the reverse. Secondly even the *average* individual departs from the average centile if the centiles are calculated cross-

Fig. 10.1. (a) Relation between the individual velocity curves and mean velocity curve (heavy broken line) during the adolescent growth spurt. (b) The same individual curves plotted against years before and after age of maximum growth velocity. The mean is again shown by the heavy broken line. From Tanner (1962), after Shuttleworth (1937) and Boas (1930).

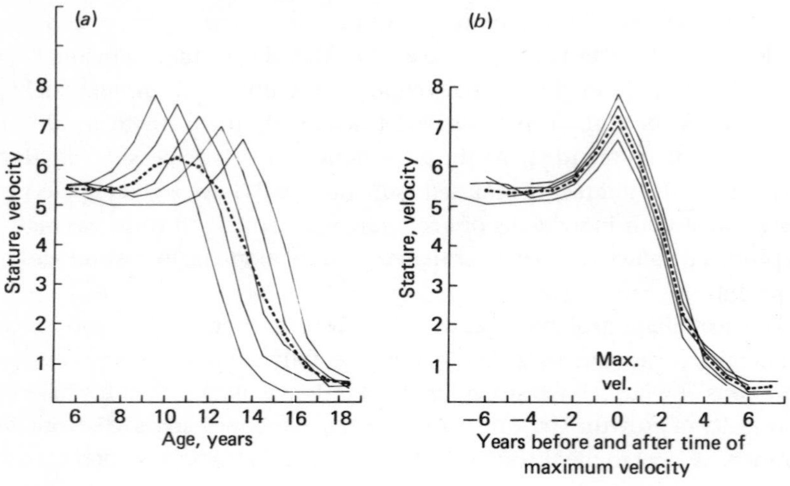

Fig. 10.2. The same individuals as in Fig. 10.1, plotted as 'distance' curves (height-for-age). The heavy broken line is again the mean of the five; note how the central thin line rises faster than the heavy line. Sources as for Fig. 10.1.

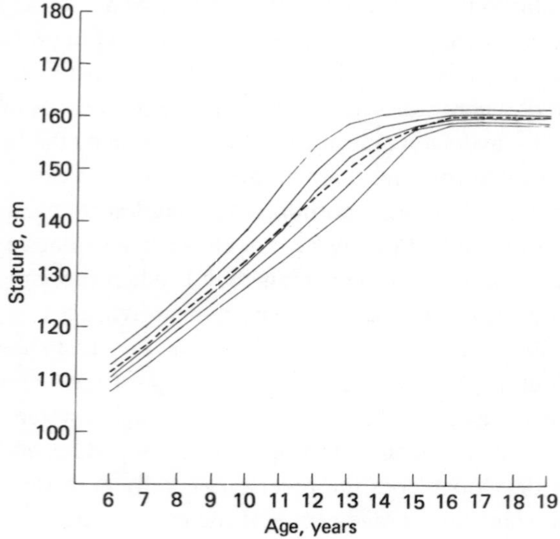

sectionally, for the slope of his curve is greater than the cross-sectionally derived average. Thus in Fig. 10.2 the central individual (third thin line) rises more sharply than the mean curve for all five individuals (heavy broken line). It was the realization of this effect which prompted Boas in 1892 to insist on the importance of following individuals. Nevertheless, it was ignored by nearly all students of growth until revived and clarified by Boas himself in 1930, and restated by Davenport in 1931 and Shuttleworth in 1937. Even then standards of growth for individuals continued to be based on cross-sectional data, despite efforts by Bayley (1956*b*) and Bayer and Bayley (1959) to provide separate curves for early and late maturers. It was not till 1966 that Tanner, Whitehouse and Takaishi introduced longitudinally based standards in which the boy of average height and average tempo actually followed the average curve (see also Tanner and Whitehouse, 1976).

These differences in tempo, combined with the pubertal acceleration, cause an increase in the variance of height and other measurements at puberty. Bowditch noticed this, as well as the skew, in his data, and Boas correctly explained it as another 'purely statistical phenomenon' created by adding together figures for individuals of different degrees of acceleration and retardation. It is an interesting comment on the biometrical atmosphere of the 1890s that the very minor skewing effect (actually forgotten and uncommented on from 1892 to 1959) was the one which primarily caught Bowditch's and Boas' attention rather than the major variance increase which is used nowadays by every student of auxology to locate peak velocity. The belief in the universality of the 'law of errors' was so strong that a small departure from it appeared more obvious than a large but systematic increase in the variance itself.

The first American longitudinal growth study

In May 1891 Boas began a long-term longitudinal study of Worcester schoolchildren. In addition to height and weight, sitting height, forearm length, hand breadth and head length and breadth were taken, and later reported by his assistant, F. M. West (1893, 1894), and by Boas himself (Boas and Wissler, 1906). The study was ill-fated however. Just over a year after it started Boas resigned from Clark University, partly because he was offered a job under F. W. Putnam at the Chicago Columbian Exposition, and partly because of a disagreement with Stanley Hall, a disagreement which did not, however, prevent Hall from writing of Boas a dozen years later as 'our best American authority on the treatment of these [growth] statistics'.

Only one year's growth had been completed by his subjects, but when he wrote up his results in 1897 Boas used his scanty material to excellent effect. The experience of actually handling his own longitudinal data seems to have clarified considerably his conceptions and in this paper one feels he has really got a grip on the acceleration/retardation problem. He gives means and standard deviations of the yearly increments, and comments that 'young children grow more uniformly than older children. The increase in variability [of growth rate] is very great during the years of adolescence . . . this increase must be considered due to the effects of retardation and acceleration'. He then divides his data into two halves, one containing all the children who are shorter than average for their age and the other all the children who are taller. He found the shorter children grew less than the taller in the years before adolescence. But during adolescence the shorter grew more; that is, they continued to grow at a time when the taller were stopping. He thus demonstrated for the first time the relative independence of final adult height and the speed with which it is reached. He overplayed his hand, perhaps, in concluding that 'small children are throughout their period of growth retarded in development, and smallness at any given period as compared to the average must in most cases be interpreted as due to slowness of development'. He had a special reason for doing this, in the shape of his controversy with Porter (see p. 219). In line with his position in this controversy he added also: 'the differences in development between social classes are to a great extent the results of acceleration and retardation of growth' (1897). Boas did not challenge Porter's figures; indeed he himself got similar results in his Worcester children (Boas and Wissler, 1906). But he felt himself to be championing the cause of the late developer, and he never lost interest in the subject, returning to it in his very last paper on growth, published in 1941.

In the 1897 paper he extended the application of tempo difference to explain differences in height between different social and ethnic groups. Returning to Bowditch's Boston data, he took the thirteen different classes constituted by five different nationalities of parents and four different occupational classes in Americans and Irish, and showed that differences between them reached their maximum at the fourteenth year in boys and the twelfth year in girls. 'The figures prove', he wrote, 'that the differences in development between various social classes are, to a great extent, results of acceleration and retardation of growth, which act in such a way that the social groups which show higher values of measurements do so on account of accelerated growth and that they cease to grow earlier than those whose growth is in the beginning less rapid, so that there is a tendency to decreasing differences between these groups during the last years of growth'.

In an address given in 1935 he discussed the tendency for children of the

Horace Mann School of Columbia University to become larger between 1909 and 1935 (the secular trend) in similar terms. 'It would seem that the changed conditions [of life] result in a change of the tempo of development', he wrote. 'We find here the proof that the tempo of the life cycle in youth may be modified by conditions of life.' This time he added, however: 'I do not venture to speculate on the causes that may underlie these changes, for it is not apparent that the social and economic conditions of the groups concerned have changed noticeably during the interval of twenty-five years (Boas, 1940, pp. 91, 122).

Standards of height and weight for American children

Besides starting what was to have been the first substantial longitudinal growth study, Boas produced also the first national standards for height and weight of North American children. 'In 1891, when active preparations for the World's Columbian Exposition were being made, Professor F. W. Putnam ... placed me in charge of the section of Physical Anthropology ... We agreed upon a plan to represent as fully as possible the growth and the development of American children' (Boas, 1898). School authorities were approached and those in Oakland, California, and in Toronto agreed to contribute figures to add to those already collected by Bowditch in Boston, Peckham (1882) in Milwaukee, Boas and West in Worcester, and Porter in St Louis. A total of nearly 90,000 children between 5 and 18 years old were involved. Boas pooled the results from these six cities to produce the composite North American standards, remarking that, though he would have liked to weight each series in accordance with the percentage of the total North American population it sampled, the lack of accurate census data prevented his doing so.

Though larger than most contemporary European children, these subjects were still small by modern standards. Both boys' and girls' average heights correspond to the modern British 20th centile at pre-pubertal ages, drop to the 5th centile at mid-puberty and end at the British 30th with growth complete or nearly so. In relation to modern North American standards they would be about 5 centile points lower. Peak height velocity is at 15.0 in boys and 13.0 in girls, around fifteen months later than in the United States nowadays. Oakland children were taller and heavier than children in any of the other towns (Oakland, 1893). Later, surveys of Washington, DC (MacDonald, 1899), Nebraska (Hastings, 1900) and Chicago (Christopher, 1900) became available. Boas (1895*b*) used the Oakland and Toronto data to show that there was a relation between birth rank and height and weight at all ages in boys from 6 to 16 (where the data terminated) and in girls from 6 to 18; the first-born was largest, and others successively smaller. This seems to have been the first demonstration of this sibling-number and birth-rank effect: Boas treated it only as a difference between first-born and all later-born, but his data show a graded regression.

Boas published the standards for height in 1898 in a *Report of the US Commissioner of Education.* As one might expect, the work-up is a model for all to follow; and as one might fear, it held at least one point which subsequent authors for sixty years ignored entirely. Boas gave the mean height for each sex at each half year from $5\frac{1}{2}$ to $18\frac{1}{2}$ and then corrected this figure by linear interpolation to the exact half-year point (since the mean age of his 5-year-olds, for example, was a little over 5.50). He then gave not only the standard deviations calculated from all the values for the whole year-group, but also the 'corrected' standard deviations appertaining to an exact age 5.0, 5.5 and so on.

This correction for the artefact due to having in cross-sectional studies data coarsely grouped in yearly intervals seems to have escaped all subsequent workers till 1958 (see Tanner, 1958*c*; Healy, 1962), despite repetition of its importance by Boas in a review of Woodbury's standards in 1922. Woodbury (1921) gave standards for height and weight based on 172,000 American children 'selected on the basis of reliability of the data from among two million cards received as a result of the Children's Year campaign' (see Chapter 12). The statistical work-up was poor and Boas had every reason to feel nettled that his thoughts of twenty and thirty years before had been so completely ignored. Yet no trace of annoyance shows through the fairly severe statistical criticism that the review makes. Boas does not even quote his earlier articles; he simply explains over again. 'Another error in these calculations is introduced by the neglect to consider the annual growth. This effect is considerable, particularly when growth is rapid. The variability at a given moment may be called σ^2. When we call the annual rate of growth d, we have a total variability of $\sigma^2 + \frac{1}{12} d^2$. Consequently, to obtain the "instantaneous variance" $\frac{1}{12} d^2$ must be subtracted from the grouped variance.'

Boas left the Chicago Exposition in 1894 and in 1896 settled in New York, at the American Museum of National History, till 1905, and simultaneously at Columbia University, where he was to spend the rest of his working life. We may regard this move as bringing to a close his first period of the study of growth. His interest had been increasingly engaged by the problems of the effect of heredity and environment on body form, and in 1908 he began a famous series of studies on immigrants. Until 1930, when he embarked on a second period of the study of human growth curves, his interest in children's measurements related rather to their family relationships than to the process of growth itself. During the first years of the century, however, a further *Report of the US Commissioner of Education* was issued (Boas and Wissler, 1906) and a paper (1912*a*) assimilating to his thoughts the new methods for measuring developmental age then being evolved.

The 1906 paper, with Wissler, is chiefly remarkable for presenting what seem to be the first correlation coefficients calculated upon human growth data. Boas, as we have seen already, was in the fore-front of biometrical advance; all his life he taught and encouraged anthropologists in the use of statistical methods. As early as 1894 he wrote a paper, expository in character and placed in the *American Anthropologist*, describing the correlations between two anthropometric measurements. Galton (whom Boas does not mention in this article, but whose work must certainly have inspired it) had only introduced the correlation coefficient in December 1888, and the book that brought Karl Pearson to him, *Natural Inheritance*, appeared in 1889. Pearson's method for calculating the correlation coefficient, which replaced Galton's, was published in 1896. So Boas came to the biometric fold as early as the great K. P. himself. In the 1894 paper Boas presented no coefficients of correlation, but gave diagrams of the regression of two anthropometric measurements, each on the other, in Sioux and Crow Indians. He remarked that the greater closeness of the two lines for stature and span than those for head length and head breadth indicated that the former pair of variables were the more closely associated. He also understood clearly the reduction in variance of one measurement which occurs when it is standardized for a fixed value of a correlated measurement. He remarked that the discovery of ultimate causes of correlations was a futile endeavour and wrote, nevertheless, that 'the results of anthropometric statistics are a means of describing in exact terms a certain variety [of man] and its variability ... it is clear that a *biometric* method would undoubtedly open new ways of attacking problems of variation' (his italics).

The Worcester longitudinal data are the ones used for the correlational analysis of 1906, supplemented by new data from the files of the Newark Academy, a private school near New York. Boas calculated the correlations of stature and other measurements at year t with their respective increments from t to $t + 1$. He demonstrated in these terms what he had really already shown by his division into short and tall groups in the 1897 analysis: that 'the expected results [occurred], that is a maximum of positive correlations during the periods of most rapid growth, and a sudden drop to negative correlations when growth is nearly completed'. For the Newark Academy data, which encompassed a longitudinal series of seventy boys from 12 to 17 years, he calculated the correlations between the increment t to $t + 1$ and the increment $t + 1$ to $t + 2$, and so on, showing them to be positive till 14–15 and negative thereafter.

Boas, with his unbounded regard for scientific integrity and the ethics of research, made a practice of publishing all his raw data whenever possible, so that others also could use them to further knowledge. The 1906 paper starts this habit; the end of the paper consists of the full records of the Worcester study, giving the individual measurements of the children, their school grade, birth order, sibling number and so forth.

It was just about this time that Pryor (1905) and Rotch (1910*a*, *b*) demonstrated that children of the same chronological age showed considerable differences in the degree of ossification of the bones of the hand, and that girls were at all ages ahead of boys in their ossification (see Chapter 12). It is not now clear whether these researches into developmental age were stimulated by contact with Boas and his ideas on tempo of growth, but it seems very likely. The other advance in developmental age measurement, that made by Crampton, was certainly stimulated, though not actually begun, by Boas himself. Crampton was a school physician in New York who felt the need of some more realistic criterion than chronological age for classifying his adolescent schoolboys for fitness to take part in athletics programmes. About 1901, under the influence of Boas and of Stanley Hall, he began studying the growth of pubic hair as providing such a criterion, and in 1908 published a paper (reprinted in 1944) in which he described a three-fold classification according to absence, initial appearance, and full development of the pubic hair. Godin, it will be remembered, had already been using a five-fold classification of the same since 1895 and published an account of it in 1903. It is not clear if Crampton (or, more likely, Stanley Hall) knew of this: Godin's work was known in Italy and Spain, but very little, it seems, in Germany and the English-speaking countries.

Boas welcomed these advances and additions to his basic conception and in his 1912 article wrote: 'A study of the eruption of the teeth which I made a number of years ago, and the more recent interesting investigations by Rotch and Pryor on the ossification of the carpus, show that the difference in physiological development between the two sexes begins at a very early time and that in the fifth year it has already reached a value of more than a year and a half . . . The condition of the bones . . . gives us a better insight into the physiological development of the individual than his actual chronological age and may therefore be advantageously used for the regulation of child labour and school entrance, as Rotch and Crampton advocate' (1912*a*).

Second period of research on growth, 1930–2

From 1912 to 1930 Boas published little on growth itself, but in 1930, at the age of 72, he produced the first of two papers embodying a second major contribution to the understanding of human growth. A few individual growth curves had accumulated in the literature since 1906. In particular there was the work of Baldwin, an educational psychologist who published a number of longitudinal curves in 1914 and 1921, mostly taken from the records of the Horace Mann School of Teachers College, the school of education of Boas' own university. Baldwin, whose work is discussed in Chapter 12, became in 1917 the first director of the Iowa Child Welfare Research Station. Baldwin had data on age at menarche in the girls and drew the first graph of height velocities in different age-at-menarche groups since the initial and totally forgotten demonstration by Pagliani (1879). But there he stopped.

Boas, we may presume, had pondered over these longitudinal data and come to realize that they held more possibilities for analysis than their authors had exploited. He also realized that his own correlational methods of 1906, though useful indeed for analysing a longitudinal series consisting of only two values a year apart, were insufficient to extract more than a fraction of the information contained in the growth curves of individuals followed over many years. Their inadequacy was felt especially in dealing with events of adolescence. Boas (1930) wrote that he 'demonstrated a number of years ago [in the 1906 paper] that ... the actual individual growth rate must show a much more decided decline during childhood and a much more decided increase during adolescence, and that at the same time the period of increased rapidity of growth during adolescence must be much shorter than would appear from the generalized curve'. He now wished to study the details of these individual curves more closely and he wished, in particular, to see if final adult height was related in any way to the time of occurrence of the spurt, or its rapidity or intensity. Initially he used the same longitudinal records as in 1906, from boys at Newark Academy, but this time supplemented where possible by records of their adult stature. For the later paper, chiefly published in 1932, but with continuations in 1933 and, less importantly, in 1935(*a*), he added similar longitudinal though retrospective material from the Horace Mann School, the Ethical Culture Schools, and the College of the City of New York.

Because 'the fundamental questions [at issue] are not well expressed by the coefficient of correlation' Boas used graphical methods, subdividing his subjects into groups on the basis of height at 14, age at maximum velocity, or age at menarche, and then plotting the growth of the means of each of the groups. The method, in Boas' hands, was a powerful and productive one. It inspired Shuttleworth's more detailed but essentially very similar studies on the more plentiful Harvard Growth Study data some five years later (Shuttleworth, 1937, 1938*a*, 1939: see Chapter 12.) Nowadays curve-fitting to the individuals' data would be the appropriate, and indeed the equivalent approach. But in 1930 computers were not available and efficient methods of curve-fitting were in their infancy; Boas liked to handle his data himself and not leave them to multitudes of assistants working desk machines.

The essence of his findings is shown in Figs. 10.3 to 10.5, redrawn from illustrations in the 1930 and 1932 papers. In Fig. 10.3 the distance curves for stature are given for five groups of boys, the grouping being by age at maximum velocity of growth. This is the first introduction of this method of classification. Two effects were clearly demonstrated. The first was that on average those who had an early adolescent spurt were already taller by age

Fig. 10.3. Mean curves of height for each of five sets of Newark Academy boys, grouped by age at maximum growth velocity. Set 1 contains all those with maximum velocities (thick portion of line) between ages 12.0 and 13.0; set 2 all those with maximum velocities between ages 13.0 and 14.0, and so on. From Tanner (1959); redrawn from Boas (1930, Fig. 4).

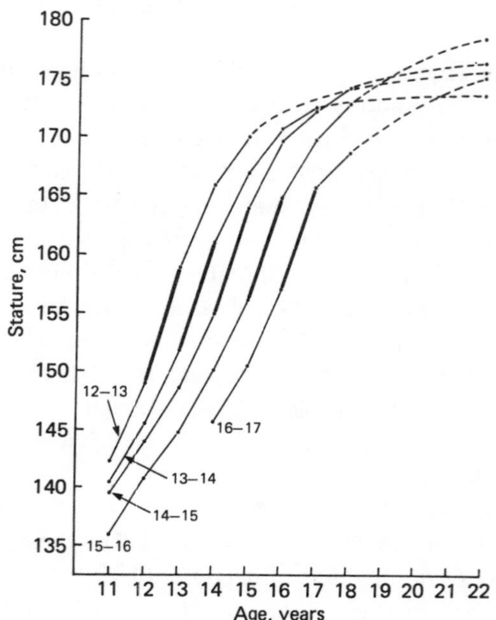

11, but not necessarily taller when adult stature was achieved. Having regard to the sampling errors involved in the relatively small numbers in the groups and the apparently random ordering of the adult values, Boas wrote 'we may well assume that there is no relation between the time when the greatest rapidity of growth sets in and the stature finally attained'. Secondly, Boas discovered from these curves that the earlier the spurt occurs, the higher was the peak reached.

The 1932 paper documents these conclusions with more exactitude and detail. This time the data are mostly for girls, who are classified both in relation to age at maximum velocity, as before, and age at menarche. Fig. 10.4 is the equivalent for girls, classified by age at menarche, of Fig. 10.3 for boys, classified by age at maximum velocity. It extends the conclusions further back into the growth period, for already at 7 years the early maturers are larger than the later maturers.

Fig. 10.5, also redrawn from the 1932 paper, shows the velocity curves for groups of girls classified this time by age at maximum velocity. Boas divides

Fig. 10.4. Mean curves of height of six sets of Horace Mann schoolgirls, grouped by age at menarche. From Tanner (1959); redrawn from Boas (1932, Fig. 2).

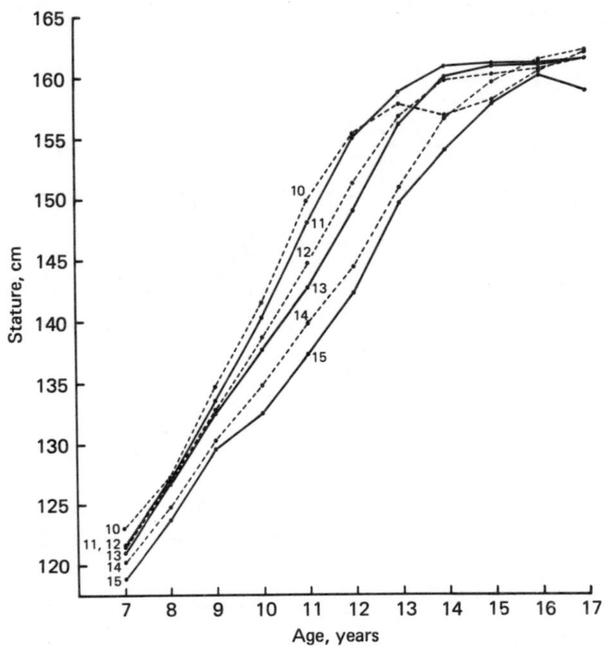

his subjects into 'Hebrew and non-Hebrew' to get an idea of genetic influences, but finds no difference. The data bring out well the way in which the intensity of the peak velocity declines in individuals whose peaks occur at the later ages.

The 1933 and 1935(*a*) additions to this paper contain little new information, except for a rather tentative investigation of the relation between dental maturity and advancement of growth in height. Boas showed that the time of eruption of the permanent second molars was to some extent related to the time of maximum growth, the regression being about four months' advance in molar eruption for one year's advance in peak velocity occurrence. For eruption of the permanent canines and premolars the relation was less marked, and the regression only one month for each year of maximum growth rate. This result was a refinement of earlier observations made by Boas in Puerto Rico and reported by Spier (1918), which demonstrated that at all ages from 7 to 13 boys who were advanced in the development of the permanent dentition were taller than those whose dentition was delayed.

Later, in 1935(*b*), Boas showed that the siblings he had studied in the

Fig. 10.5. Height velocity curves of Horace Mann schoolgirls, Hebrew and non-Hebrew, grouped by age at maximum velocity of growth. The curves are spaced out along the time axis for clarity. From Tanner (1959); redrawn from Boas (1932, Fig. 3).

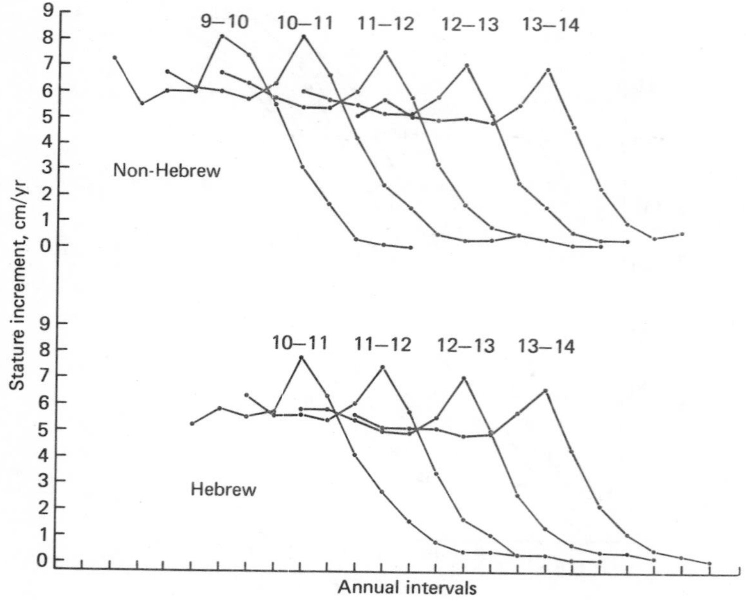

Horace Mann School and the Hebrew Orphange resembled each other in their times of maximum growth velocity. 'The observations', he wrote, 'may be summarized in the statement that each individual has by heredity a certain tempo of development that may be modified by outer conditions. The gross, generalized observations available at the present time suggest that in a socially uniform group the tempo of development may be considered as a hereditary characteristic of individuals' (1935*c*).

Environment and heredity

Between these two 'growth study' periods, Boas' interest was engaged principally in the question of the relative influence of environment and heredity upon body size and form. (That is, the portion of Boas' interest allotted to physical anthropology was so engaged. So adequate to one man's entire life study were Boas' contributions to this section of anthropology alone that it is almost impossible to bear in mind that he was at least equally active in two or three other fields, as witness the accounts in his memorial volume (Goldschmidt, 1959).) A full account of his contributions to the dispute between Pearson and the Mendelian School, and of his biometrical studies of 'family lines' in physical characteristics will be found in Tanner's (1959) commemorative article, from which much of this chapter is taken. A word should be said, however, on his study of immigrants to America and their children, a study vast in scope, unexpected in outcome, and staggering in professional and public reception. Begun in 1908, it was made for the United States Immigration Commission, whose object seems to have been to determine whether the influx of immigrants was resulting in deterioration of the physique of the American population. Boas, with some thirteen assistants, prominent among whom was Crampton, measured the stature, head length and breadth, bizygomatic diameter and in some cases the weight, of a total of nearly 18,000 immigrants and American-born children of immigrants in New York. About 5,500 were aged 25 and upwards, and the remainder children over the age of 4. The material was almost exclusively family groups, the main comparison being between children born before the parents immigrated ('foreign-born') and hence growing up in Europe, and children born in America after the parents had come to New York.

The main report was issued in book form in 1912 under the title *Changes in Bodily Form of Descendants of Immigrants* (1912*b*). In the same year (1912*c*) a summary of the main findings was published in the *American Anthropologist,* together with answers to some of the criticisms already encountered. The criticisms were to continue, off and on, for years, and in 1928 all of the raw data were issued, constituting the largest collection of family measurements ever published. Boas' findings, that stature was a little greater in children growing up in America and that head length and breadth were a few millimetres more, would scarcely be regarded as surprising nowadays. But anthropologists of the time had an astonishing belief in the fixity of what they called human types or races, and when Boas showed that even the central tabernacle of the doctrine – the cephalic index – was built on sand they chorused their displeasure and disbelief.

It is hard for us to realize the degree to which the anthropologists of the last century were obsessed with the constancy of the cephalic index. Head form appeared to them a method of tracing the origins and distributions of peoples; if it altered in response to environment, a major prop was kicked out from under the anthropological pile-dwelling. Long after the fuss had died down, Boas (1930) summarized the work as a whole in a statement with which nobody nowadays could quarrel:

> It has been known for a long time that the bulk of the body as expressed by stature and weight is easily modified by more or less favourable conditions of life. In Europe there has been a gradual increase in bulk between 1850 and 1914. Adult immigrants who came to America from south and east Europe have not taken part in the general increase ... presumably because they were always selected from a body the social condition of which has not materially changed. Their children however, born in America or who came here young, have participated in the general increase of stature of our native population ... With this go hand in hand appreciable differences in body form ... These changes do not obliterate the differences between genetic types but they show that the type as we see it contains elements that are not genetic but an expression of the influence of environment.

Boas was a remarkable innovator, who nevertheless made few mistakes. He was a man of enormous width of interest, yet his writings in the small sector of auxology show no trace of dispersion of talent. On the contrary, most of his work makes the other writers of his time look like dilettantes; he it is who penetrates below the surface. To a large extent this is due to his biometrical skill and biological understanding. In a cogent address given in 1904 he described anthropology as 'partly a branch of biology, partly a branch of the mental sciences'. A portion will split off as linguistics; a portion will remain as the study of cultural change; but anthropologists must always be familiar with the main lines of all the branches. On the physical side of anthropology he here, as elsewhere, declares an indebtedness and allegiance to Galton and Pearson, and before them to Quetelet. In the bridge between the older concepts of physical anthropology and newer disciplines of human biology, Boas' work constitutes much of the framework. (The bridge showed some signs of incompleteness when Boas' papers were collected as *Race, Language and Culture* in 1940. None of his contributions to biometrical genetics appeared, and other papers which were reprinted had the biometry cut out of them.)

When Crampton's 1908 paper on physiological age was reprinted in 1944, Crampton wrote a short postscript. It said 'The author [i.e. Crampton] is commonly given credit for originating the term "physiological age". This, I believe, properly belongs to the much respected and beloved Franz Boas.' Many others might have written the same of their contributions to human biology in the first half of this century.

D'Arcy Thompson

No account of the history of growth studies could ignore the author of *On Growth and Form*, and in placing D'Arcy Thompson next to Boas there is a deeper logic than of dates. D'Arcy Thompson (1860–1948) and Boas were both giants, intellectually and personally, but giants of totally different shapes. Boas was a meticulous research worker, cast recognizably in the modern mould, who picked his way carefully forward with letters to *Science* and a multitude of postgraduate theses. Thompson was a late flower of the Renaissance, a man of vast scholarship, whose footnotes alone are a delight and an education. *On Growth and Form* (1917, 1942) is above all a Renaissance book, crammed with all sorts of considerations and written in a prose of such clarity and elegance that one reviewer compared its writer, and justly, with the author of the *Anatomy of Melancholy*.

D'Arcy, as he was universally called, was brought up in the Classics, wrote *A Glossary of Greek Birds* and *A Glossary of Greek Fishes*, and translated Aristotle's *Historia animalium* with an understanding only possible to a professional zoologist who was also president of the Classical Associations of Scotland, England and Wales. The scientific establishment, already grown unaccustomed to such scholarship, did not know what to make of him, and he spent his whole career rather peripheral to the main stream of scientific activity, as professor of natural history, first in Dundee (1884–1917) and then St Andrews (1917–48). *On Growth and Form* is a massive book, and its running page-headings include 'Of modes of flight' and 'The comparative anatomy of bridges' as well as 'Of growth in infancy'. Two chapters only, out of seventeen, deal with matters of growth in the strict sense. One is entitled

'On the theory of transformations, or the comparisons of related forms'; in it appear the Dürer drawings and the application of transformed coordinates to depict the evolution of a species or the relations between different genera. D'Arcy's method was graphical, and it was fifty years before the use of computers made analytical applications of the technique generally possible; geographers and systematists (see Sneath, 1967) in particular are now amongst its users.

The second chapter that concerns us is entitled 'The rate of growth'. Form is produced by growth of varying rates in different directions:

> when in a two-dimensional diagram we represent a magnitude (for instance, length) in relation to time . . . we get that kind of vector diagram which is known as the *curve of growth.* We see that the phenomenon we are studying is a *velocity* . . . By measuring the slope or steepness of one curve of growth at successive epochs, we shall obtain a picture of the successive *velocities* or *growth rates.* Plotting these successive differences against time we obtain a curve each point on which represents a certain rate at a certain time; while the former curve (height against time) showed a continuous succession of varying *magnitudes,* this shows a succession of varying *velocities.* The mathematicians call it a *curve of first differences*; we may call it a curve of the rate (or rates) of growth, or still more simply a *velocity curve* (1942, p. 95: his italics).

Though others before D'Arcy Thompson had plotted annual increments (for example, Schmidt-Monnard), it was D'Arcy who emphasized the continuous nature of the *process* of velocity. For D'Arcy, 'to say that children of a given age vary in the rate at which they are growing would seem to be a more fundamental statement than that they vary in the size to which they have grown' (1942, p. 128). These quotations are the origin of the term velocity curve, and of the emphasis laid on growth velocity by Tanner (1951 *et seq.*) if not also by Boas and Shuttleworth (who do not quote D'Arcy).

D'Arcy's biography has been well written by his daughter (Thompson, 1958), but *On Growth and Form* will ever remain his epitaph. The generation who heard D'Arcy lecture (which includes the writer as a schoolboy) witnessed not only an actor-scholar of superlative skill, but a sort of beneficent magus, large and bearded, surrounded always by the objects about which he was talking – dodecahedrons and other examples of the Platonic Figures in the writer's case, shells, fishes and in at least one instance a live chicken (Thompson, 1958, p. 213). D'Arcy was one of the relatively few British scientists honoured with a Festschrift (Clark and Medawar, 1945) and when a symposium on aspects of growth was organized at the Royal

Society so late as 1950, it was inevitable that its title shculd be 'A discussion on the measurement of growth and form' (Zuckerman, 1950). Though he made no explicit contributions to the facts of human growth D'Arcy Thompson, like Franz Boas, elevated the whole intellectual level of the field and broadened the outlook of all who worked in it.

11

The contribution from clinical practice: fetal and infant growth; age at menarche

Naturally the school doctors and hygienists had little to say about growth before age 6, but at much the same time as they were amassing their first great archive, a number of obstetricians and paediatricians began a corresponding work in the course of their daily practice. The first object of their attention, partly because it was the most vulnerable and partly because the most subject to medico-legal dispute, was the newborn infant.

Newborn infants: length, weight and head circumference

The practice of measuring the weight of the newborn infant, now almost universal, took a long time to develop after its initiation by Roederer in the 1750s. Indeed at Göttingen, at Roederer's clinic, it only became a matter of routine some eighty years later, when Eduard von Siebold (1801–1861) was appointed to the chair of obstetrics in 1833 (Siebold, 1860). It was in Dublin that the most important step after Roederer's was taken. The Dublin Lying-In Hospital, a famous centre for obstetrics throughout the nineteenth century, was founded in 1757, three years after Roederer's publication. Its immense obstetric practice was supervised by a Master, appointed always for a term of seven years, assisted by one or two junior men. The Master from 1786 to 1793 was Joseph Clarke (1758–1834), a man who in addition to having a celebrated private midwifery practice became vice-president of the Royal Irish Academy. Clarke spent his life seeking ways to make childbirth safer, and he died attending the 1834 meeting of the British Association for the Advancement of Science in Edinburgh, where he had gone to contribute a paper on the lowering of infant mortality occasioned by proper ventilation of hospital wards (Collins, 1849). (This was the year after the 'Quetelet' meeting of the British Association at Cambridge.)

When Clarke was an assistant at the Dublin Lying-In Hospital 50 per cent of all children born there died before they reached the age of 3. Most died either at birth or soon after, and of these the great majority were boys. The fact of this differential mortality had been recognized for some time, and Clarke put forward two possible reasons to account for it. First, the larger size of the male made for a more difficult delivery, and second, the larger size implied a greater requirement of nourishment during fetal life, and this might not always be satisfied. Clarke determined to document properly the larger male size, and particularly the larger size of the boy's head, which he saw as of prime importance, naturally, in difficult deliveries. He thus measured the weights and head circumferences of sixty newborn infants of each sex (Clarke, 1786). The average weights (weight of clothing subtracted) were 7 lb 5 oz (3.35 kg) and 6 lb 11 oz (3.09 kg).

For measuring head circumference, Clarke made use of a linen tape, varnished to prevent its length being affected by changes in the humidity of the atmosphere (a very necessary precaution frequently omitted right down to the present day). The tape was divided into inches, halves and quarters. 'I took first', Clarke wrote, 'the greatest circumference of the head from the most prominent part of the occiput around the frontal sinuses; and, secondly, the transverse dimensions from the superior and anterior part of one ear across the fontanelle to a similar part of the other ear. The measurements appeared to me the most likely to afford data for determining the respective sizes of the brain in the different sexes' (1786, p. 358). His newborns averaged 13 inches $11\frac{4}{5}$ lines (35.5 cm) and 13 inches $7\frac{2}{5}$ lines (34.6 cm) in circumference, and 7 inches $5\frac{1}{7}$ lines (18.8 cm) and 7 inches $2\frac{3}{4}$ lines (18.3 cm) from ear to ear across the top of the calvaria, male and female respectively. To the writer's knowledge this was only the second time head circumference was measured in newborn infants (Jampert being the first, and forgotten, measurer). Gradually the measurement found its way into the routine of neonatal paediatrics, until today – when the causes of the greater male mortality are known to be much more complex – it is continued as a ritual of obscure provenance and uncertain efficacy.

Weighing was introduced as a routine at the Maternité de Port-Royal in Paris not later than 1802. This was probably at the behest of François Chaussier, the constructor of the *mécomètre* (see note 6.3). Chaussier was only put in charge of care of the newborn at the Maternité in 1804, but he had been professor of anatomy and physiology at the faculty of medicine sine 1796 and, as a forensic medicine specialist he had been particularly concerned with the weight at which newborns had a chance of been born alive, and thereafter of surviving. The Paris weights were reported by Michel Friedlander (1769–1824: see note 11.1) in one of the earliest books on growth written by a physician for teachers (*De l'éducation physique de l'homme,* 1815). Friedlander gives the frequency distribution of the birthweights of 7,077 infants, both sexes together, born in the Maternité between 1802 and 1806. The numbers are reproduced in Table 11.1. They have a mean of 6.48 lb (2.94 kg). (This value is not given by Friedlander, who as usual for the period quotes only the mode, 6 + lb.) The following year Murat, writing the article 'Foetus' in the *Dictionnaire des sciences médicales*, gave the average weight (*poids moyen*, by which he also probably means the mode, not arithmetic mean) of over 20,000 newborns weighed at the Maternité. The figure was $6\frac{1}{4}$ lb (2.84 kg); again the sexes were not differentiated. This is at about the 10th centile of the weight distribution of present-day full-term newborns, but one has to remember that, as Friedlander put it, 'the Maternité is the refuge of the distressed (*l'asile du malheur*), of mothers fraught with anxiety, malnourished, exhausted, mostly hiding their pregnancies, and exposed to continual hurt and suffering'. The patients of the Maternité were the poorest people in Paris. In addition, these early weight statistics, as can be seen from Friedlander's frequency distribution, include very light, presumably pre-term, infants.

Friedlander also reports what is perhaps the first longitudinal study of infancy. 'M. Schwartz', he says (1815, p. 26) 'one of the most highly regarded *instituteurs* in Germany, has measured and weighed his child at various ages after birth. At birth, 1, 3, 4, 5, 7, 11, 13 weeks, 5, 6, and 12 months the lengths were [in *pouces* and *lignes*]: 18p8L, 20p2L, 20p8L, 20p11L, 21p3L, 21p8L, 22p0L, 23p5L, 23p7L, 24p0L, 24p7L, 28p6L'. If the inches used were Rheinish ones then the child would have been at about the modern 10th centile; if French then at about the 75th. But M. Schwartz's measurements jump from one centile to another and back rather irregularly. Perhaps he did not have a *mécomètre*; at least his technique of measurement left something to be desired. Friedlander calls for a multiplication of this sort of observation so as to give averages for mothers to go by. Thus, logically enough, the first call for standards of growth appeared in a treatise on physical education or hygiene addressed to the laity, and not in a book on clinical medicine.

When Quetelet collected data on weights and lengths of newborns in a Brussels hospital in 1831 (see note 11.2) he reported boys and girls separately and a decade later the almost equally famous James Young Simpson (1811–

Table 11.1. *Weight distribution of newborn infants at the Maternité, Paris, 1802–6*

Weight (lb)	1+	2+	3+	4+	5+	6+	7+	8+	9+	10+
Number	34	69	164	396	1,317	2,799	1,750	463	82	3

From Friedlander (1815).

1870), from 1839 professor of midwifery at Edinburgh and the man who introduced chloroform anaesthesia (in 1847), did likewise (Simpson, 1844). Simpson had his assistant measure the weight and length of fifty newborn infants of each sex and confirmed the larger male size reported by Clarke and Quetelet (see Table 11.2).

Table 11.2. *Weights and lengths of newborn infants in the eighteenth and nineteenth centuries (all parities; public clinics and maternity hospitals)*

		Weight (kg)		Length (cm)	
Approx. date	Author and place	Male	Female	Male	Female
1753	Roederer Göttingen[a]	3.09	2.93	49.1	48.2
1786	Clarke, Dublin Lying-In	3.35	3.09	—	—
1804	Friedlander (1815), Maternité, Paris	2.94		—	
1830	Quetelet (1833), Brussels	3.20	2.91	49.6	48.3
1842	Simpson (1844), Edinburgh	3.47	3.12	51.9	50.4
1855	Veit (1855), Berlin	3.22	3.13	—	—
1860	Hecker (1865), Munich[b]	3.34	3.22	51.1	50.4
1860	Duncan (1864–5), Edinburgh	3.31	3.26	48.8	48.7
1875	Roberts (1878), Edinburgh	3.43	3.29	49.1	48.2
1885	Issmer (1887), Dresden, Munich	3.32	3.21	—	—
1895	Pearson (1900), London	3.32	3.22	52.1	51.1
1910	Benestad (1914), Oslo[b]	3.52	3.41	—	—
1915	Bruce-Murray (1924), London	3.30	3.18	51.4	50.6
1925	Low (1950) Aberdeen[b]	3.48	3.43	49.8	49.3

[a] Assuming the Göttingen lb was 3 per cent greater than the present English lb (Cone, 1961) and the Göttingen inch was the same as the contemporary Würtemburg inch and equal to 2.38 cm (Theopold *et al.*, 1972).
[b] Infants weighing less than 2.50 kg omitted.

Simpson's paper was a most comprehensive and cogently argued one. He returned to Clarke's ideas about the higher male mortality, analysing a set of data published by Clarke's own son-in-law, Robert Collins, himself Master of the Dublin Lying-In from 1826 to 1833. During Collins' Mastership 16,654 children were born, and Collins, presumably urged on by his father-in-law, presented extensive statistics on all these cases, separating girls and boys, something which had not generally been done before (Collins, 1835). Quetelet (1835) meanwhile had filled in an important detail by showing that the excess mortality in boys was greatest in the weeks immediately following birth and disappeared after a few months. Collins' data gave the same result. Furthermore the time when boys were at the greatest risk relative to girls was during birth itself. In that short period, according to the Dublin statistics, three boys died for every two girls. Amongst stillbirths, however, infants who died before being born, boys outnumbered girls by only six to five. Of the mothers who died in childbirth bearing singleton infants (that is, one only) 105 had had boys and 49 girls. Simpson concluded that the cause of the higher intrapartum and neonatal mortality of males was the slightly larger male head, which made birth more difficult and more damaging to both infant and mother.

With weighing now established as a routine in a number of clinics the pace of epidemiological discovery quickened. In 1855 Gustav Veit, professor at Rostock, analysed the data accumulated in the Berlin policlinic, where he had worked as assistant. He showed that even at the same body weights more boys died than girls, but as no head measurements had been made he could go no further than this. At Munich, however, head circumferences had been taken 'for a long time, with great care' by Carl von Hecker (1827–1882), professor of obstetrics and ultimately *Rektor* of the Ludwig-Maximilian University, and since 1859 chief of the Obstetrical Clinic. Munich at this time and for several decades after was a major centre for the acquisition and dissemination of data on growth, in the post-natal as well as the fetal period. Much of this was due to Hecker (the son of the famous historian of the epidemics of the Middle Ages, Justus Friedrich Carl). Hecker found mean head circumferences of 34.9 cm in 528 newborn boys and 34.3 cm in 465 newborn girls (Hecker and Buhl, 1861), a difference of 0.6 cm, compared with Clarke's 0.9 cm. Soon after, Breslau (1862) in Zurich and later Pfannkuch (1872) in Marburg and Fasbender (1878) in Berlin presented evidence that boys' heads were larger than girls' even at the same body weight.

In Breslau's 576 cases the average sex difference in head circumference at a given weight was about 0.6 cm; in both Pfannkuch's and Fasbender's cases it was 0.4 cm. These figures are slightly less, naturally, than the overall mean differences recorded by Clarke and by Hecker. Thus Simpson's hypothesis stood. Pfannkuch endeavoured to see if head shape, as well as size, differed between different weight classes, but obtained a negative result. He ended his paper by calling for more measurements – such as shoulder breadth and chest circumference – to be taken on newborns so that the ideal bodily proportions for successful birth could be established (quoting Quetelet to the effect that the measurement that varied least would be the best guide to this ideal, a lasting echo of the tyranny of the *homme moyen*). A number of obstetricians followed up this suggestion (bibliography in Scammon and Calkins, 1929) without arriving at any substantial ideal. Fasbender (1878) showed that the infant's head size was related to the head size of the mother, but this line was not pursued further.

Veit and Hecker also analysed their birth weights by sex and by whether the children were first-born or not. Veit found that second and subsequent children ('later-born') weighed less than first-born in both sexes. Indeed the difference between first- and later-born infants of the same sex (0.28 Prussian civil *Pfunden*) slightly exceeded the difference between boys and girls of the same birth rank (0.22 *Pfunden*). The absolute Berlin figures were equivalent to 3.18 kg and 3.31 kg in boys and 3.08 kg and 3.21 kg in girls. Hecker's Munich figures gave a similar difference of 0.14 kg between first- and later-born, the sexes considered together. A dialogue then developed between Hecker and Matthews Duncan (1826–1890), an Edinburgh obstetrician whose prime interest was in the problem of over-population. Duncan showed for the first time that the weight of the newborn infant increased with the age of the mother, at any rate up to the age of 30; and he ascribed a greater importance to this factor than to the mothers' parity (Duncan, 1864–5). His number of subjects was considerably greater than Hecker's (1,011 primiparous mothers and 1,042 multiparous mothers, that is, respectively those with first-born and later-born babies). Hecker (1865) rejoined at once, adducing this time a series of 4,449 observations, which indicated that both factors operated, seemingly in independence. Duncan (1866) then acknowledged this was correct. Birthlengths behaved in the same way as birthweights, though the effects were less marked. Hecker omitted infants whose weights were below 2.50 kg as premature or pathological; his mean weights and lengths, including children of all parities (see Table 11.2), are at about the 35th centile of modern standards. Hecker's first-born averaged 3.23 kg and 3.14 kg, with lengths 50.8 cm and 50.2 cm.

Meanwhile, Frankenhauser (1859) in Jena had showed that birthweight, and also head circumference, was related to mother's height, infants of tall mothers (greater than 160 cm) being 0.3 kg heavier than those of short mothers (less than 146 cm). Hecker's assistant, U.K. Gassner (1862), followed the increase in maternal weight throughout pregnancy, and the loss of weight following the birth of the child (Gassner, 1862). The weight of the products of conception (fetus, fluid and placenta) was proportional to the mother's weight. Gassner mentions, incidentally, where his vital weighing machine came from: it was Schoenemann's *Patentwage* from the workshop of J. Pintus in Brandenburg.

Soon reports on birthweight became almost commonplace (see Table 11.2). Ingerslev (1876) measured some 6,000 newborns in Copenhagen; F. Faye and Vogt (1866) gave values for Oslo covering the years 1846–63, and Benestad (1914) similar data for the years 1909–11. Issmer (1877), reporting from Munich and Dresden, gives a list of previous authors' findings, as does Benestad. Benestad's own means, 3.52 kg for boys and 3.41 kg for girls, but exclusively on mature (*nur Reife*) singletons, were the highest reported up to that time.

In the United Kingdom Charles Roberts seems to have been the first to have organized collection of data on the newborn since the times of Simpson and Duncan. Edinburgh was still the source. Roberts had 100 newborn boys and 100 newborn girls weighed and measured, obtaining the equivalent of 3.43 kg and 3.29 kg for weight and 49.1 cm and 48.2 cm for length (Roberts, 1878). He says the mothers were 'of the artisan class', confined in the Edinburgh Royal Maternity Hospital. The means that the Anthropometric Committee of the British Association collected – again through Roberts – on 451 boys and 466 girls at Queen Charlotte's Lying-In Hospital in London and the Royal Maternity at Edinburgh were considerably lower: 3.24 kg for boys and 3.16 kg for girls, with lengths of 49.6 cm and 49.1 cm (British Association, 1883). It seems that the London values for weight at this time were surprisingly low, as the report of Karl Pearson (1900) on 1,000 boys and 1,000 girls delivered at Lambeth Lying-In Hospital about 1895 also shows (Table 11.2). Even in 1915 that situation was unchanged (Bruce-Murray, 1924).

In general in the 1880s and 1890s birthweights averaged some 3.40 – 3.45 kg for later- born full-term males, which is at the 35th – 40th centile of modern European standards, whose mean for such infants is about 3.55 kg. However, even in most modern data birthweight is related to height of the mother, and if we allow for the difference between heights of women in the nineteenth century and now – using the allowance appropriate for modern women (about 0.2 kg for 5 cm of height: Tanner and Thomson, 1970) – then the old birthweights are closely equivalent to present-day values. The sex difference, and the difference between first- and later-born, are both similar to modern figures in the data from 1850 onwards. But it is noticeable in Table 11.2 that the sex difference in weight before 1850 is considerably greater (200 to 350 g) than that after 1850 (50 to 140 g). The same is true of the sex difference in birth length (1.3 to 1.5 cm before 1850; 0.5 to 1.0 cm after 1850). Part of the reason may be the generally post-1850 practice of excluding infants who weighed less than 2.5 kg. This will eliminate more girls than boys, hence raising the girls' means more than the boys' means. It seems unlikely, however, this could account for the whole, rather large, effect. Perhaps before 1850 only the larger boys survived.

As well as measuring birthweight, obstetricians and paediatricians followed the weight changes of the infant in the first few weeks with interest and concern. During the first few days after birth the infant loses weight, a phenonemon that Quetelet says was first pointed out by Chaussier. Quetelet (1833) followed seven infants for seven days and found minimum weight on the third day, with some deficit still persisting at the end of the week. Quetelet was ahead of his time, for nearly thirty years later Siebold (1860), Breslau (1860) and Brunniche (1866) were still among the first to report on this weight loss in the first three to seven days. Ingerslev (1876) followed carefully day by day a number of healthy children born to healthy mothers and found that on average the weight loss continued for three days only, and was mainly made good by the end of the first week. Benestad's infants, too, had recovered nearly all their lost weight by the end of the first week and he strongly criticized the celebrated paediatrician Camerer for promulgating the idea that recovery might take longer.

Fetal growth

The systematic weighing of newborn infants did not come only from a desire to assess the chances of survival. There was also the question of assessing the possibility that infanticide had occurred in a situation where compulsory notification of births had not yet been instituted. The determination of the age of an alleged stillbirth and thus of the infant's possible viability was a problem felt to be of much importance; it was surely no coincidence that Chaussier was by training and vocation a forensic medical scientist. The solution of this problem demanded growth curves for the fetal period and in the 1860s efforts to construct them began to be made by obstetricians, embryologists and anatomists.

Carl von Hecker was again in the forefront of this activity. In the second volume of his *Klinik der Gerburtskunde* (1864) and in a later paper (1866) he reported in some detail the weights and lengths of 486 stillborns, of whom 371 were fresh, that is, not subjected to post-mortem change. Then, as now, the length of gestation was calculated from the first day of the last menstrual period in terms of weeks completed, a 'month' being four weeks. Though this gives a gestation period on average two weeks longer than the actual period since fertilization, it is a practical method, if all too often fallacious (due especially to menstrual irregularities and periods half-missed).

There were nine fetuses supposedly of 3 months post-menstrual age, seventeen of 4 months, twenty-seven of 5 months, and so on. Hecker's weight values were much quoted, and indeed figured as the 'best series' in C. S. Minot's well-known textbook of *Human Embryology* (Minot, 1892). But actually, as Hecker himself says, 'As a basis for the diagnosis of the month of gestation I used the length of the child, and after many comparisons made the following suppositions: a child from the 4th month is 10–17 cm long, from the 5th 18–27 cm, 6th 28–34 cm, 7th 35–38 cm, 8th 39–41 cm, 9th 42–44 cm and 10th 45–47 cm' (Hecker, 1864). (The fourth month here means between 3.0 and 4.0 post-menstrual months or 12–16 post-menstrual weeks: in the later paper Hecker designated 3-monthly fetuses as 4–9 cm.) Thus there is a circularity in Hecker's argument; all his values really yield is a curve of weight against length (see note 11.3). Taken thus, the weight increments, month by month, showed a velocity gradually accelerating right up to term (whereas modern figures give a peak velocity at about 36 weeks). The lengths give a curve with a maximum rate of gain between the fourth and fifth month, a still high velocity in the fifth to sixth month and a rather steep fall from that point onwards.

Several other studies were made (see e.g. Calderini, 1875), but it was over forty years later before convincing curves for fetal length and weight were published. These appeared in two papers (1909*a*, 1910) by the gynaecologist and writer C. H. Stratz (see p. 282). Stratz summarized his own and his predecessors' work and suggested the curves reproduced in Fig. 11.1. The peak velocity of length is almost evenly placed between the two intervals 4.0 to 5.0 post-menstrual months and 5.0 to 6.0 post-menstrual months, hence approximately at 5.0 months or 20 post-menstrual weeks (see note 11.4). Stratz puts the peak for weight (for which, surprisingly, he says there are fewer data than for length) considerably later, at about 30 weeks post-menstrual age. The final or term length he takes as 50 cm, a reasonable value; but the term weight is placed at 3.00 kg, which is oddly low. All the same these curves have proved correct in their essentials; modern data themselves still leave a good deal to be desired, and even with the newer methods of measurement of the living fetus by ultrasound it is at present impossible to locate the peak velocity of length with precision. Stratz's age for it coincides well enough with our best modern guess (see Tanner, 1978*a*).

As for the weight, if Stratz had taken a term value of 3.30 kg as more adequately representative, then the high velocity of weight would have continued until a few weeks before birth, just as it does in modern curves.

Stratz was particularly interested in change of shape during development; he found that at the end of the second post-menstrual month the height of the head, from chin to vertex, was about one half of the whole length of the fetus; at five months it was one third, and at birth one quarter (Fig. 11.2). Two years after birth it constituted only one fifth of the total length, at age 6 one sixth and at maturity (in the male) the Vitruvian one eighth (Fig. 11.3).

Stratz's figure has been reproduced on several occasions. It is one of the few diagrams of growth to have found its way into a textbook of anatomy; C. M. Jackson (1875–1947), professor of anatomy at Minnesota, used it in the fifth edition of *Morris's Human Anatomy*, which he edited (Jackson, 1915). Medawar (1945) took it from Jackson and used it most penetratingly in D'Arcy Thomson's Festschrift, where it served as the basis for an equation characterizing changing proportions. In one respect, however, the diagram

Fig. 11.1. Stratz's (1909*a*) curves for length (Längenwachstum) and weight (Gewichtszunahme) growth in the fetal period. From Stratz (1909*a*, Fig. 2).

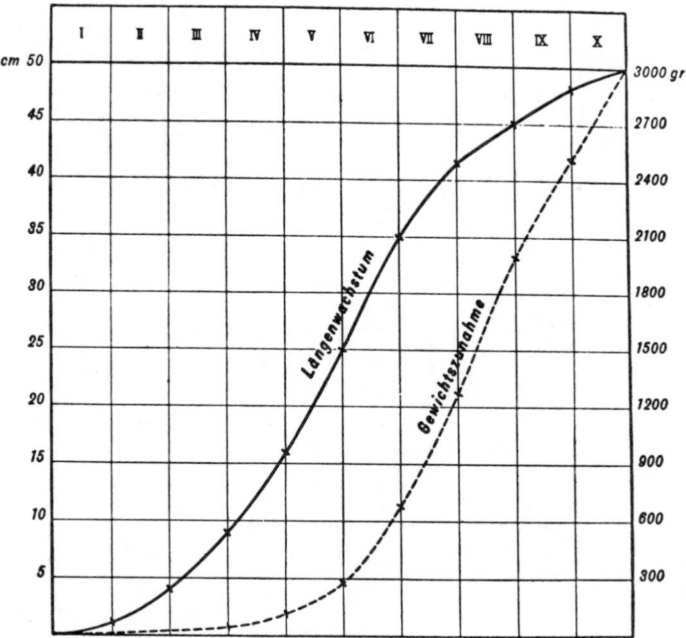

is misleading. Superficially it looks as if the ratio of trunk length to leg length continuously falls throughout the whole growing period, as does indeed the ratio of head height to leg length. Actually the decrease in the ratio of trunk length to leg length is stopped, then reversed, during the pubertal growth period when a different growth pattern, under different hormonal control, emerges. The trunk grows faster than the legs and the ratio rises (a particularly good illustration is given in Meredith, 1939).

A considerable literature concerning growth in shape of the fetus arose during the period 1900–30. The contributors were mostly anatomists, of whom one of the most famous was Retzius (1904) in Stockholm. By far the most extensive and carefully done study, however, was made in 1919–21 in the departments of anatomy and obstetrics of the University of Minnesota (Scammon and Calkins, 1929). Some 400 fetuses of 2 to 54 cm in length, all in good condition, were measured by L. A. Calkins (1894–1960), later professor of obstetrics at the University of Kansas. Calkins took seventy-one measurements; Scammon, who planned the exercise, probably acted as recorder. In all there were some 35,000 measurements taken; sufficient perhaps to qualify Calkins for entry to the 'Club Godin'.

Fig. 11.2. Stratz's diagram of change in body proportions during fetal growth. From Stratz (1909*a*, Fig. 3).

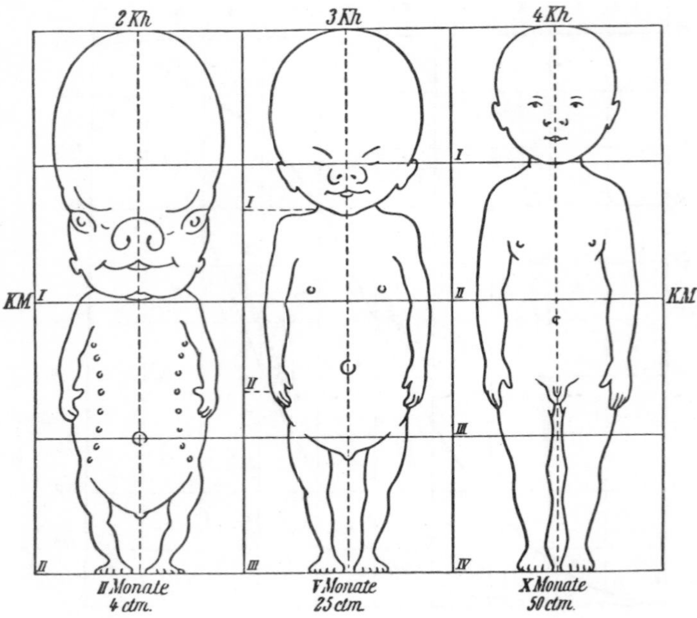

Richard Scammon (1883–1952) was born in Kansas City, and took his PhD at Harvard in 1909 under C. S. Minot (1852–1914). He never qualified as a doctor, but still, like Minot, pursued his career almost entirely in a school of medicine (see Wilmer, 1952; Boyd, 1953). From 1914 to 1930 he was professor of anatomy at the University of Minnesota Medical School where he established a laboratory of fetal anatomy under the aegis of Jackson, who was head of the department. Jackson himself was much interested in growth and the book written jointly by him, Scammon, Harris and Paterson, entitled *The Measurement of Man* (Harris *et al.*, 1930) is a minor classic of the growth literature. After a year (1930–1) spent as professor of anatomy and dean of biological sciences at the University of Chicago, Scammon became dean of the Minnesota Medical School. However the skill which earned him universal praise and respect as a teacher and scholar seems not to have transferred to his administrative post and he gave up the deanship in 1935 to become Distinguished Service Professor in the Graduate School. He retired in 1949.

Scammon found it hard to complete any work to his satisfaction and apart from the book with Calkins his publications are mostly snippets drawn from proceedings of various societies. They make a tantalizing, but

Fig. 11.3. Stratz' diagram of change in body proportions during post-natal growth. From Stratz (1909*a*, Fig. 6).

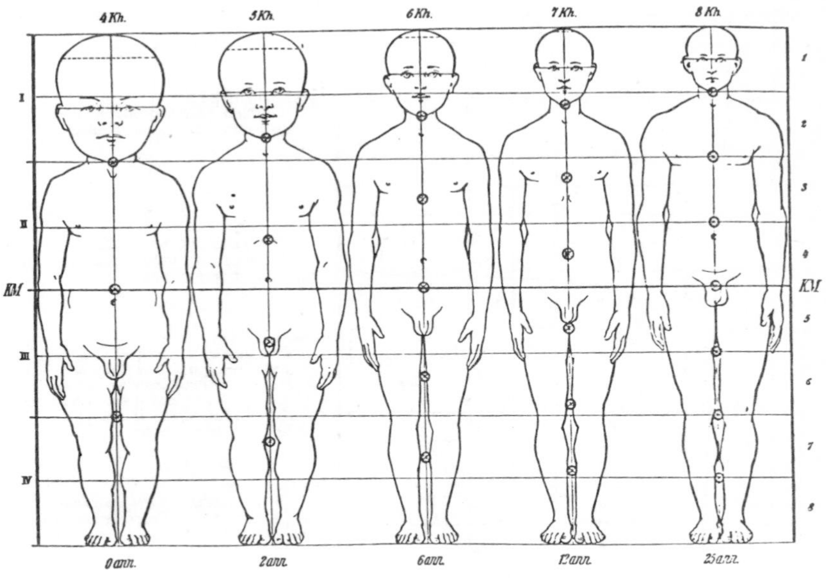

almost useless archive, a memorial to a creative but obviously retiring mind. Scammon's section in *The Measurement of Man* (1930) is one of his longest and best statements about growth. His plot of Montbeillard's son appears, as do plots of the growth of Martin Guttman's four sons (see below). This paper contains the famous diagram, often reproduced, of the four types of growth. It is shown as Fig. 11.4. Scammon, with Edith Boyd (see Boyd, 1936), had made a cross-sectional study of the size of thymus and lymph glands in children killed in accidents. The other curves came from data in the literature, such as the organ weights given in Vierordt's *Daten und Tabellen* (1906), in Rössle and Böning (1924) and in Hammar (1906, 1926). Scammon also gives curves of the growth of the endocrine organs, but the data at that time were poor, and the curves are erroneous.

Besides his interest in fetal growth Scammon had a passion for history and, above all, for bibliographic research. As we have seen (p. 103), it was he who raised Montbeillard's son from an ill-deserved oblivion. He gathered materials for an immense history of growth, but after 1930 he brought little scientific work to a conclusion, and the papers eventually passed to Edith Boyd, his closest colleague. Boyd seems to have absorbed Scammon's work-

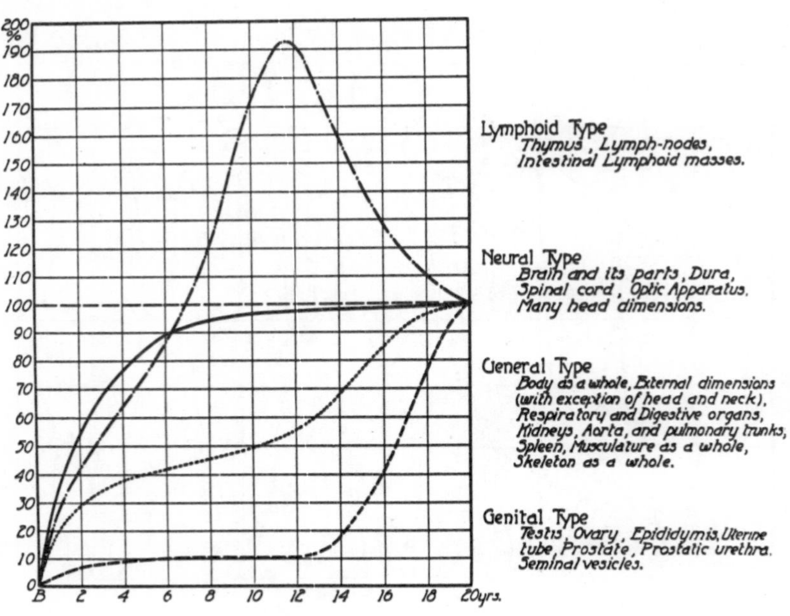

Fig. 11.4. Scammon's curves illustrating the four main types of growth. From Scammon (1930).

ing characteristics all too well, and though she did rewrite the material into publishable form, she died before publication could be arranged.

In view of this it is not surprising that Scammon and Calkins' *Growth in the Fetal Period* (as it is usually called) contains the most extensive list of references about the subject ever assembled; many of the items are not even in Baldwin's great bibliography (Baldwin, 1921 : see below, p. 300). Most of the papers are annotated. But the use made of the references, in the many tables of the book, is somewhat uncritical (Liharzik and Weissenberg get equal weight) and in a few instances misleading (Clarke's head circumference figures are only given sexes-pooled, for example (p. 95), thus ignoring the point of Clarke's work). It seems unlikely that such a massive study as this will ever be repeated; it belongs more in the tradition of the late-nineteenth-century physical anthropologists. Scammon and Calkins showed that throughout the fetal period the length of nearly every external dimension was linearly related to the crown–heel length of the fetus. Dimensions differed, of course, in their rate of growth, but such differences remained constant for the whole period. There was a gradient of growth, as Stratz and others had shown, and it was a continual one, with no periods of reversal. After Scammon and Calkins, interest in fetal growth declined for a time. The fundamental problems of obtaining fetal material that was reasonably healthy, and of knowing the true ages of the fetuses, remained. Interest revived in the 1960s when the introduction of abortions purely for social reasons made healthy fetuses up to 20 weeks or so readily available. Modern studies are summarized in Southgate (1978).

Growth of infants

The growth of children in the first year or two after birth was naturally the preserve of paediatricians. Monsieur Schwartz's example seems not to have been followed for many decades. In the 1850s Guillot (1852) advocated regular weighing of infants as a guide to the amount of breast milk being provided, but it was Cnopf (1872) in Nuremberg, Fleischmann (1877) in Vienna and Russow (1880–1) in St Petersburg who first systematically extended for a longer period the weighings by then frequently done during the first ten days after birth. Fleischmann emphasized the importance of systematic weighing during the whole first year, and in his paper gave data on several individuals weighed each week during this period, with a commentary on the implications of the various gains and losses for the infants' feeding and handling.

Russow (1880–1), like Fleischmann, was concerned with how to diminish the appalling infant mortality of the time. Since he thought artificial feeding injurious he compared the lengths and weights of infants who were breast-fed with those fed by bottle. His data were cross-sectional (or perhaps mixed longitudinal in reality) and between 1873 and 1878 he and one nurse carried out, he says, 5,000 weighings. The children he measured suffered from nothing worse than coughs and colds; he wished to establish standards – the first ones – for well, breast-fed infants. He gives mean lengths and weights each month from birth to 1 year and each year from age 1.0 to 8.0 for a total of 900 children. Russow's comparison of breast-fed and bottle-fed infants is largely vitiated by lack of control of social class, birthweight and other factors. All the same, Russow was very much a pioneer of growth standards and of the idea that growth is a guide to well-being in infancy.

In Germany C. Lorey, of Frankfurt-am-Main, weighed and measured a large number of children aged 1 to 30 months in his clinic practice. He himself only published tables of means for each month based on about a dozen children of each sex at each occasion (Lorey, 1888), but after his death Schmidt-Monnard (1892) gave a more extensive analysis of his data, which actually included 2,300 children; Schmidt-Monnard provided averages to serve as a standard for children of artisans and manual workers (reproduced in Fig. 11.5 below). Schmidt-Monnard divided the infants into those whose fathers were fit enough to be accepted for military service, and those whose fathers were rejected, thinking that some difference might appear (which it did not). A generally similar consideration had motivated an earlier study by Brunniche (1866), the professor of paediatrics in Copenhagen. Brunniche measured height, weight, and chest and head circumference in some 300 children aged 2 to 8 in the Children's Hospital. He, arbitrarily it seems, divided them into well-developed and ill-developed, and compared the scrofulous, rachitic and tuberculous with the remainder. (The differences, in fact, are far from striking.) Brunniche formulated the rule – still used occasionally today in judging the approximate age of children in non-literate societies – that chest circumference should exceed head circumference by age $3\frac{1}{2}$.

Gradually longitudinal records accumulated. Karl Vierordt (1818–1884), the professor of physiology at Tübingen, in the second edition of the section *Physiologie des Kindesalters* in Gerhardt's *Handbuch der Kinderkrankheiten* (1881, p. 241) listed the weekly weights of thirty-eight infants; of these thirteen were from Cnopf (1872), eight from Fleischmann (1877) and nine his own. Vierordt regarded these data as so precious that from his sick-bed he specificially asked his former pupil, Camerer, to publish them in more detail; Camerer (1882) did so, adding twelve cases he himself had followed.

Evitsky (1881) in New York reported the mean weekly gains of some fifty children from birth to 2 months, about twenty of whom were followed till 6 months.

Wilhelm Camerer (1842–1909) – or more accurately Johann Friedrich Wilhelm, to distinguish him from his son, also Wilhelm – was a dominant figure in German paediatrics and scientific medicine in the last years of the nineteenth century. He was born in Stuttgart, and studied first at Tübingen under Karl Vierordt and later in Vienna. In 1878 he began work on the physiology of childhood; or, to be more precise, on the chemical contents of 24-hour urine collections from his own five children (Camerer, 1880). But in 1883 he went to settle in the small Württembergische town of Urach as country doctor and public health official, and stayed there until he died. He set up a laboratory in his bedroom and from this retreat he continued to lead a considerable area of medical advance, simply through the force of his scholarship and the power of his personal warmth (see Camerer, 1910).

In 1893 Camerer published a further paper on infant growth, but even then, at a time when studies of thousands of schoolchildren were in full swing, he was able to muster the records of only 116 infants who had been followed throughout their first year; of these 66 were newly collected by himself. He added that 'to date very little is known as to the increase in weight in the second year' (p. 258); for this period he quotes just six cases.

This paper is something of a classic in its own right, as well as forming the foundation for his son's very influential chapter on growth of children in Pfaundler and Schlossman's textbook of paediatrics published in 1906, and in English translation in 1908 (see below, p. 280). Camerer senior discussed with clarity and understanding the differences between the 'generalizing' and 'individualizing' methods of studying growth, basing his account on Vierordt's in *Physiologie des Kindesalters* (1877 : see p. 277). The individualizing method, Camerer says, is to be preferred in the study of growth itself, and in general gives the best results; but in determining what differences exist between children of different groups such as rich and poor, the generalizing method is the one to use. Camerer gives the growth curve of his son (from 13 to 19 only) and his four daughters and remarks that he has confirmed by the individualizing method what he alleges (incorrectly) Malling-Hansen showed by the generalizing method: that growth in height is faster in the spring and slower in the autumn. (Like other German authors, he was apparently ignorant of Buffon and Montbeillard's work.)

By this time a number of authors were appearing who followed Quetelet in trying to fit mathematical curves to growth data. Camerer was quite unsympathetic, perhaps because he had been understandably put off by the fantasies of Liharzik, whose table of growth he castigated as embodying '*ganz unrichtige Resultate*'. He poked fun at the sober Raudnitz (1892) of Prague, who had modestly sought to fit the parabola $W = a + bt + ct^2$ (W, weight; t, age) to growth in weight. This, said Camerer, was mere mathematical *Spielerei*, not serious scientific progress.

Infant welfare clinics

The impetus to collect data on infant growth was greatly intensified when in 1892 infant welfare clinics began to be set up (see note 11.5). Pierre-Constant Budin (1846–1907), professor of obstetrics at the Charité Hospital in Paris, was the pioneer in this, initiating a *Consultation des Nourrissons*, the first well-baby clinic. Mothers delivered at the Charité were encouraged to bring their babies back for weekly weighing and supervision over a period of two years. When artificial feeding proved necessary, bottles of sterilized milk were supplied. Similar clinics began to spring up elsewhere and in all of them the infant's weight was used as the prime measure of healthy growth. Gaston Variot (1855–1930), a paediatrician, established a second *Consultation* in Paris in 1893, and in 1894 Leon Dufour, a general practitioner, actually without knowledge of what Budin and Variot had done, established a third, in Fécamp. Dufour called his centre a *Goutte de Lait* because he, also, supplied sterilized milk. The name caught on and *Gouttes de Lait* sprang up all over France.

In England, too, infant weighing was introduced through the medium of the infant welfare clinics, or milk depots. These were modelled exactly on French practice. The first was set up in St Helens, Lancashire, by F. Drew Harris, the medical officer of health. Other municipal authorities followed suit, amongst them (in 1902) Battersea Borough in London, where the medical officer of health was G. F. McCleary, the historian of the infant welfare movement (McCleary, 1933, 1945). A second London milk depot and infant consultation was started in Finsbury in 1904 by George Newman, who later became chief medical officer to the Ministry of Health. The Corporation of Glasgow was another pioneer body in this field, setting up a milk depot in 1903 and an infant consultation in 1906. The infant welfare movement reflected the concern caused by the appallingly high infant mortality rate, which, far from falling, in 1899 reached the highest figure ever recorded in England: 163 deaths in the first year of life per 1,000 live births (McCleary, 1945). In York in 1899, infant mortality amongst the poorest of the manual working class was 247 per 1,000, in the artisan class 173 per 1,000, and in the servant-keeping class 94 per 1,000 (Rowntree, 1913). There were no statistics on causes of death in early infancy until 1905 (Rooff, 1957, p. 33) and notification of births only became compulsory in 1915. Thus in England in the early part of the century health visitors and Medical Officers of Health often did not know of the arrival of a newborn infant till after it was dead. The welfare movement was given a further powerful impetus by the revelations of the Inter-Departmental Committee on Physical Deterioration set up following the shock of the discovery of the poor physical condition of recruits in the Boer War (Parliamentary Papers, 1904).

The infant consultation centres, then as now, needed standards of growth by which the adequacy of the infants' health could be judged. Newman published the first of such charts in the 1906 edition of his book *Infant Mortality.* This was the chart used in London's Finsbury milk depot and indeed, under the name of the Newman Standard, in most other infant welfare clinics in the period 1905–15 (see Robertson, 1916). The Standard, however, was actually derived not from Newman's own children at Finsbury but from the French Standard: 'Dufour's Standard has been used in the Finsbury Depot as in the French Depots', wrote Newman, 'and charts appearing in the present volume are drawn to that scale' (Newman, 1906, p. 303). Thus the weighings done in the original *Goutte de Lait* in Fécamp served as the first British Standards for babies' growth. In fact the Fécamp babies were very small, even compared with the London ones of the time. When the weights of London children were at last studied – by an Australian biochemist – they were considerably above the mean values given (sexes-mixed) by Dufour.

The biochemist was T. Brailsford Robertson (1884–1930), a man born in Edinburgh but educated in Australia from the age of 8. After graduating at Adelaide in 1905 he went to the University of California and eventually became professor of biochemistry and pharmacology there. In 1920 he returned to occupy the chair of biochemistry at Adelaide, where he died of pneumonia at the early age of 45 (see Robertson, 1932, for memoir and bibliography). He was the founder and first editor of the *Australian Journal of Experimental Biology and Medical Science.* In the period 1915–20 he attempted to stimulate mouse growth by means of ox pituitary extracts. All his attempts failed, but it was at the same medical school only a few years later that H.M. Evans announced the successful extraction of growth hormone. Robertson was always interested in human growth, and after his return to Australia he published the first height charts to give means with lines above and below representing one and two standard deviations (see Chapter 14).

Robertson made an important early attempt at describing the mammalian weight growth curve in mathematical terms (Robertson, 1908). He was struck by the generally sigmoid shape of growth curves, whether of rats or humans, and supposed this was because in some way growth resembled the type of biochemical reaction known as autocatalytic. He fitted the equation of this reaction to growth data, thus resuming the *Spielerei* that Camerer had castigated. Where W is weight and t time the equation is: $\log W(A-W) = K(t - t_1)$, A being the ultimate value of W reached at the end of the process, K a constant giving the average rate of the process and t_1 a constant giving the time at which the process is half completed. It was clear such a formula would not fit the whole human growth curve, so Robertson split the weight curve into three sections. The first 'cycle' as he called it, began soon after fertilization and ended about 12 months after birth; the second began as the first ended and had its maximal rate at 5.5 years in both sexes; and the third cycle had its maximum rate at 'about 16.5 years in boys and 14.5 years in girls' (Robertson, 1908). Robertson thought that the points of junction of the cycles might well be periods of special vulnerability to environmental upsets, as he believed had been shown in laboratory animals. (Whereas most contemporaries thought the period when each organ was developing fastest was its time of special susceptibility: see e.g. Bean, 1914.) We shall pick up the trail of mathematical curve-fitting again, in Chapters 12 and 13.

Having fitted his equation to the growth in weight of infants from birth to 1 year in Adelaide (Robertson, 1915*a*) Robertson wanted to do the same for English children, especially because the average birthweight of Australians was much above the value reported for English. He obtained the records of weighings made, from 1911 up to the outbreak of war in 1914, in the Pimlico Road and Golden Square branches of the Chelsea and Westminster Health Societies, in London, and in the Babies Welcome in Leeds. These were three of the infant consultation clinics described above. Mothers brought their babies to be examined at intervals, then as now, but there was of course no compulsion and only a small faction of the population availed themselves of these services. Thus the babies were certainly a selected population and probably in the main selected for having mothers who were particularly concerned to obtain good care for them. Robertson collected only the weights of babies who were stated to be 'well' at their consultations. Twins and malformed infants were excluded. Robertson does not say to what degree the data are longitudinal, and reports them simply as though cross-sectional. The values are plotted in Fig 11.5, together with the curves for infants seen at a similar Paris clinic in about 1910 (Variot and Fliniaux, 1914), and infants of working class families measured in Frankfurt-am-Main somewhat earlier, in 1884–6 (Schmidt-Monnard, 1892).

A month after birth the English infants weighed on average as much as present-day ones. But soon the curve falls, and by six months they are at the

Fig. 11.5. Mean weights of London and Leeds well-babies 1911–14, collected by Robertson. Data from Robertson (1916) plotted on present-day charts of Tanner and Whitehouse (1973). Other curves are of babies of manual workers in Frankfurt-am-Main about 1885 (Schmidt-Monnard, 1892), and breast-fed babies from a Paris clinic about 1910 (Variot and Fliniaux, 1914).

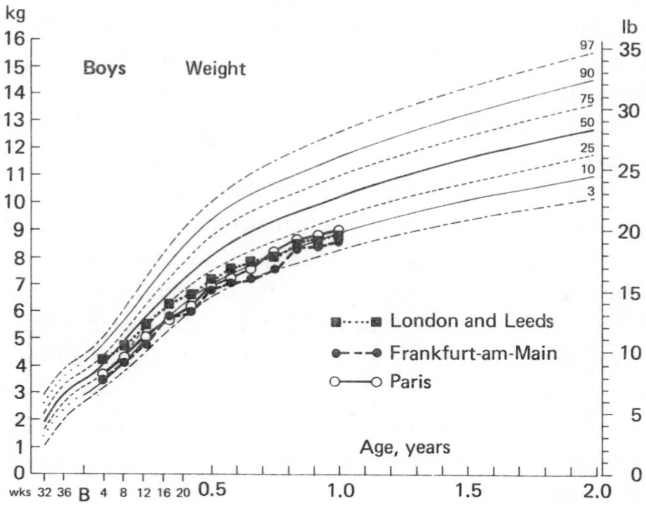

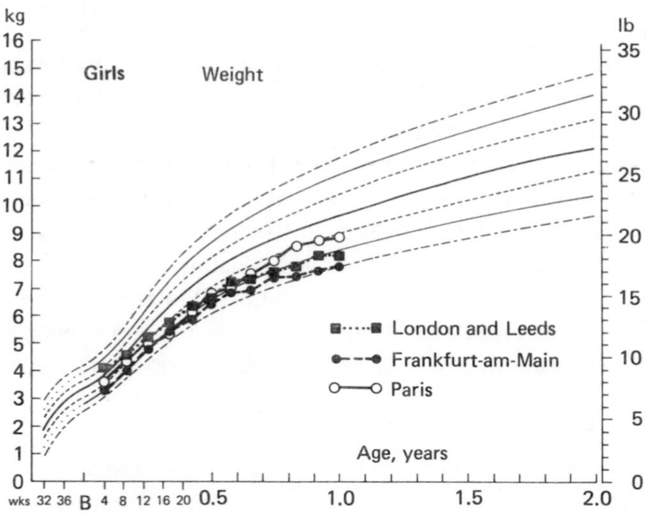

modern 10th centile, where they stay for the rest of the first year. The data are not pure longitudinal, so some of this fall may be due to selective loss of the heavier babies; mothers whose children were doing well might have thought it unnecessary to attend after the first few months, while those whose children caused anxiety perhaps continued to attend. But the picture is not inconsistent with that given by the surveys of schoolchildren done at about the same time (see Fig. 11.6). Conditions for children were only just beginning to change. Newman (1906) says that infant mortality (that is, in this context, deaths during the first three months after birth) was still high, and much higher in urban than in rural areas. One-third of it was comprised of deaths in the first week due to what he calls 'immaturity', by which he means low birthweight. He attributes this low weight to bad pre-natal conditions and the (unsuccessful) use of abortifacients containing lead, especially in Middlesex. Evidently some of the factors operative in the nineteenth century (see p. 170) remained unchanged. The Parisian children initially weighed less than the Leeds and London ones, but by 9 months the boys had caught up; the girls had caught up by 6 months and thereafter weighed more. The German infants of twenty years earlier weighed a little less than the English or French. The boys remained around the 5th centile of modern standards, and the girls dropped from the 10th at 4 months to the 3rd at 1 year.

In England no suitable growth standards for pre-school children were available till after World War I. But in France Variot and Chaumet (1906) carried out a considerable survey of Parisian children in 1905, measuring them by means of a specially constructed *paediomètre* (about which they unfortunately give no details). About 150 children of each sex at each year of age were measured from age 1 to 16 (4,400 children in all). Variot and Chaumet began by remarking that tables existed for the first year but that beyond that paediatricians were still relying on the table published by Quetelet over half a century before. In the new survey all measurements were made by the same person (presumably Chaumet). Shoes were removed, and the weight of the clothes worn was subtracted to give nude weight. The children were from the schools of the ninth *arrondissement* (Opéra, a well-off area) and the twentieth *arrondissement* (Bastille, a working class area): from the technical schools at ages 13–16, and from various nursery schools, crêches and some orphanages at the younger ages. The means are plotted in Fig. 11.6. They lie at about the 5th centile of modern British standards. The maximum increment of height means was at 12.5 for girls (7.1 cm/yr) and 14.0 for boys (8.7 cm/yr). The height curves agree with data on menarche (see below) in placing Parisian children as relatively early maturers compared with the children of north-west and Central Europe. Variot and Chaumet pointed out

Fig. 11.6. Mean heights of Paris children in 1905 (closed circles) and London children of the same year (open circles) plotted on British population charts. Data of Variot and Chaumet (1906), and London County Council (1905 : and see Cameron, 1979).

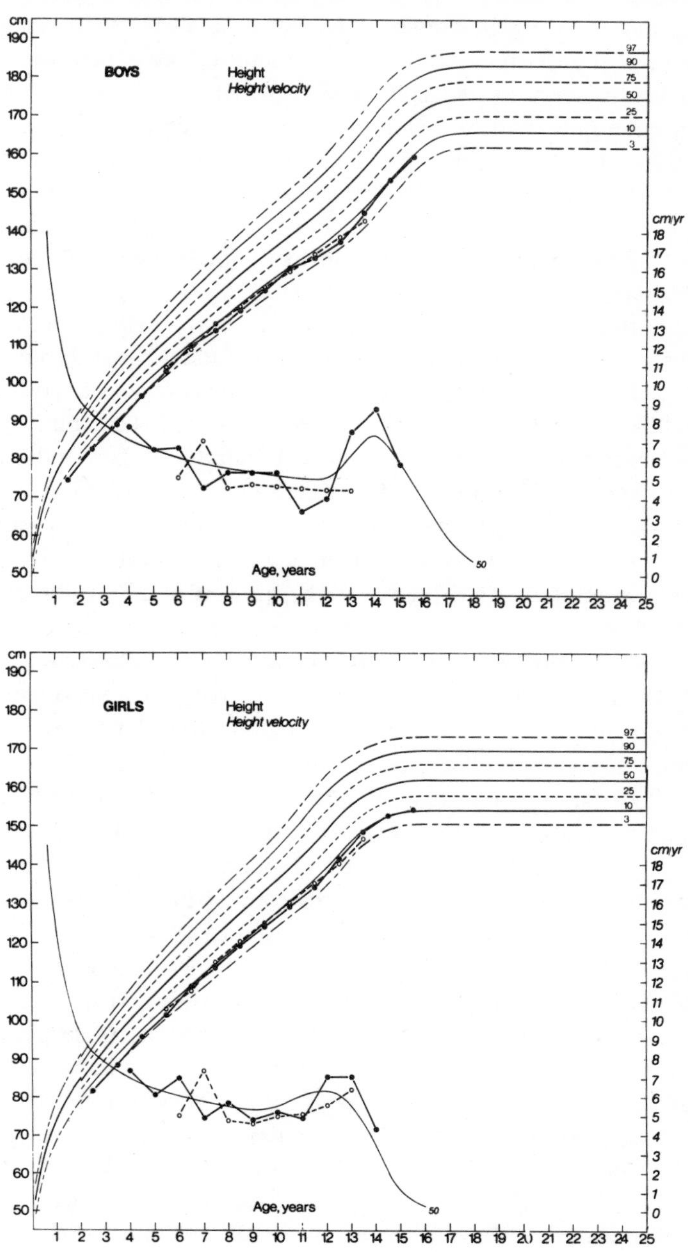

yet again how erroneous Quetelet's tables were and remarked how closely their own values approached those of Bowditch. Though they do not give percentiles at each age, as Galton and Roberts would have liked, they at least give minimum and maximum values as well as means. In this they were in advance of the German standards of Camerer (1906) which gave means only, but not as sophisticated as Boas who in 1898 published standards, for schoolchildren only, with standard deviations (though not a very practical way of using them). Also plotted in Fig. 11.6 are the means of the London County Council survey of schoolchildren made in 1905 (see Cameron, 1979). The values are almost identical to the Parisian ones except at the last, pubertal, age (13–14) when they become distinctly lower, owing to the earlier puberty in Paris.

Shortly after World War I a massive survey of the heights and weights of Scottish pre-school children was carried out under the direction of the professors of physiology and paediatrics in the University of Glasgow (Paton and Findlay, 1926). During the years 1919–23 some 10,000 children were examined in the Child Welfare Centres of Edinburgh, Glasgow and Dundee, and a further 11,000 by home visits in the same cities. Additionally some 6,000 children were measured in rural areas, where the fathers were engaged in farming or rural mining. Paton and Findlay's monograph is heavy-weight, but somehow uninformative. It gives the best review of the weights of clothing in childhood ever published, and it makes it clear that the measurements were taken (in Edinburgh all by one individual) with care and accuracy. Only manual workers' children were measured, however, and the correlations of growth with income and diet within this group were low. Maternal efficiency, together with number of persons in the family, appeared to exert the greatest effect. But the rural children showed the clearest possible advantage: at age 4 an average of 6 cm for boys and 5 cm for girls. The maximum heights at each age differed little between town and country; but the minimum in the town was far below the minimum in the countryside. The average of the urban children is at about the present 30th centile.

The urban slums are still little better than in Charles Roberts' day: and we shall meet nearly the same conditions in Newcastle twenty years later. Before that, Boyd-Orr (1936) made a comparison between the Christ's Hospital (middle class) schoolboys of 1926–9 (from Friend, 1935: see p. 228 and working class young employed men. The boys of lower social class were 6 cm shorter at 17. The values had scarcely changed since 1883 and Boyd-Orr, a nutritionist, wrote: 'it appears that in the last 50 years, though the average height for all classes has risen, there has been no marked change in the order of differences between the classes'. Even in 1937 a sample of Public School (upper middle class) boys were some 7.5 cm taller at age 14 than boys in ordinary state schools (Norman, 1939). Only after the war did this difference begin really to diminish (see Chapter 13). The Paton and Findlay and the Boyd-Orr studies constitute a rather precise link between the surveys we have looked at in Chapter 8 and the ones we shall meet in Chapter 14.

Textbooks and growth standards: 1850–1925

Textbooks on growth, and chapters on growth in textbooks on paediatrics, began to appear towards the end of the nineteenth century, chiefly in Germany. Their appearance was encouraged by the view of professors of anatomy and pathology – who then dominated medicine – that, in the words of Beneke (1878), one of the most famous: 'In all cases careful and regular surveillance of length growth has a very practical importance. It seems that the surest evidence of an improvement in health is a regular improvement of growth in length'. 'Weight', he added, with a perspicacity denied to generations of his successors, 'has scarcely the same value for judging the state of health.'

The first text to appear was by Karl Vierordt, who included a chapter on growth in his *Grundriss der Physiologie des Menschen* (1st edn, 1861; 2nd edn, 1871). The chapter is by no means sophisticated and relies entirely on the suspect figures of Quetelet and Zeising. Karl Vierordt also wrote the sections on 'Physiologie des Kindesalters' in the first two editions of Gerhardt's *Handbuch der Kinderkrankheiten* (Vierordt, 1877, 1881). He described quite clearly the differences between the 'generalizing method' ('giving averages for each age class of the population or population sample') and the 'individualizing method' ('which follows the development of individuals stage by stage') (1877, p. 50; and see note 11.6). He still gives Quetelet's smoothed 'table of growth' rather than Quetelet's original values (citing Quetelet as the pioneer of the generalizing method). He explains the absence of the adolescent growth spurt in Quetelet's table by saying that the spurt 'which is very important to doctors' can only be shown by the individualizing method. On growth in general he quotes only Quetelet, Zeising, Liharzik (though with unfavourable comment), Pfannkuch, Cnopf, Frobelius (another paediatrician) and Fleischmann. He is unfamiliar with the reports of Roberts (1874–6), Pagliani (1875–6) and Bowditch (1877). Much worse, he fails to quote Buffon and Montbeillard, who seem to have been totally forgotten in Germany for more than a century.

His son, Hermann Vierordt (1853–*c.* 1933), who was also professor at Tübingen, but of internal medicine and therapeutics, repaired his father's omission so far as Roberts, Pagliani and Bowditch were concerned. He quotes their data, together with those of the German school physicians, in a celebrated work called *Anatomische, physiologische and physikalische Daten und Tabellen,* whose three editions were issued in 1888, 1893 and 1906. Though still not mentioning Buffon, this is otherwise an excellent bibliographical source, not only of growth data but of the whole study of growth up to the end of the nineteenth century. Hermann Vierordt's own contribution concerned the weights of the internal organs in childhood, but he seems to have taken a delight in tracking down unusual and out-of-the-way data (he was later professor of the history of medicine as well as of therapeutics) and his book gives, for example, all extant information on the weights of clothes, an important detail but one authors more often than not forget to mention. *Daten und Tabellen* is not a textbook in any sense, but a prolegomenon to one.

F. Daffner (*b.* 1844), an army doctor who retired to Munich in 1891 because of gout, seems to have been the first person to write a book actually entitled *Human Growth.* When *Das Wachstum des Menschen* first appeared, in 1897, it was a slimmish volume of 129 pages. But five years later Daffner had got a real grip on his subject and the 1902 edition is nearly four times as large. Daffner subtitled his book 'An Anthropological Study', and he evidently regarded human auxology in this light, for he published his own studies, even one on pre-school children, in the *Archiv für Anthropologie.* This paper (1884*a*), discussed in Chapter 9, was very much a pioneering effort, and filled an almost total void. The fact that it was hardly ever quoted presumably tells us about the contemporary demarcation between paediatrics and physical anthropology. As well as making this study, Daffner, like his French army colleague Carlier, had himself done a longitudinal study on Army cadets aged 11 to 20, noting particularly the seasonal differences in height velocity (1902, p. 329; see also 1884*b*). Thus he was no stranger to practical work on growth, and his book is much more sophisticated than the chapters of the Vierordts. All the same he remains at times very uncritical. Discussing age at menarche, for example, he says that the girls in towns are younger than country girls at menarche; heredity is a factor; but climate predominates, so that in the tropics 'girls are already menstruating at eight years of age'.

Hermann Vierordt's figures in *Daten und Tabellen* were used as the basis for the first attempt to provide practical growth charts for clinical paediatric use. This was by E. von Lange (dates unknown), 'Klinik-Direktor' in Munich, but, it seems from his writings, a teacher rather than a doctor. Lange, in an eighty-page article, gives almost a textbook treatment of growth (Lange, 1903). He knew Daffner in Munich, and gives Daffner's son's height measurements from 6 months to 20 years (a tall man, with peak height velocity at 14.5

years), in company with the measurements of his own three daughters (peak height velocities at 13.5, 12.5 and 12.5 but no mention of age at menarche) and two of his sons (peak height velocities both 14.5). In all there are individual data on sixteen boys and nine girls followed from 4 years or earlier to age 20.

Lange's chief interest, however, was in providing standards of growth in height to be used by paediatricians, parents and teachers. In 1896 he had published tables for this purpose, together with a book on the normal height of man. But the sources of his tables then were Quetelet, Liharzik and Zeising, and at that time he still did not believe, he said, in the temporary female ascendancy in height at puberty 'apparently demonstrated by recent authors' (he meant Bowditch, 1877, in particular). By 1903 he had accepted it; hence his new standards. Lange has no data of his own, and relies almost entirely on Vierordt's *Daten und Tabellen.* He provides growth charts with channels for tall and short children, the lines being drawn backwards from starting points at

Fig. 11.7. Lange's chart of height growth. Reproduced from Lange (1903).

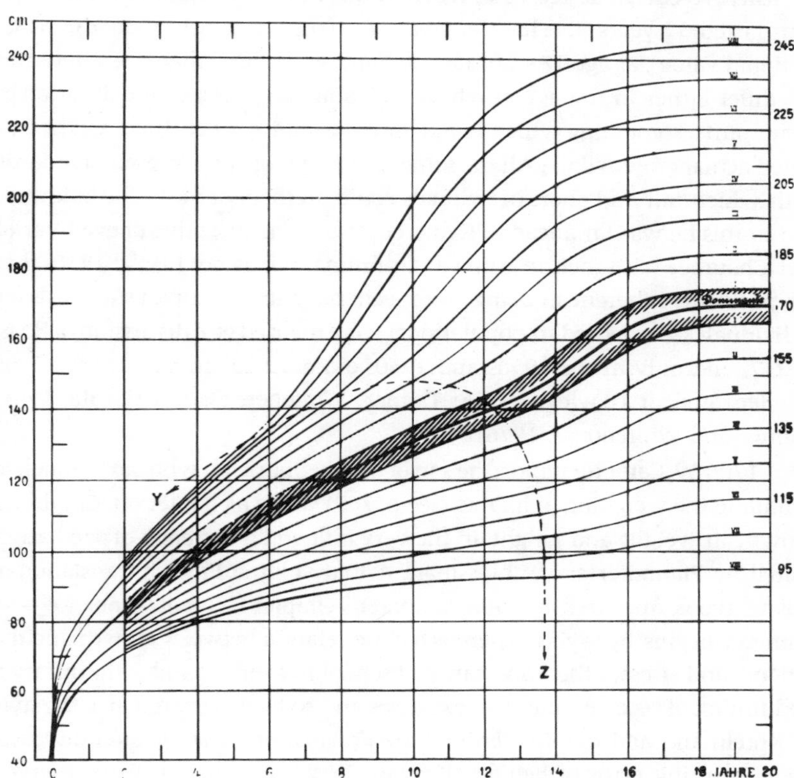

each 10 cm of adult stature (see Fig. 11.7). The mean curve, oddly called the 'dominant', ends in the case of men at 170 cm; the next curve above ends at 180 cm, and so on. The basis on which these outside lines are drawn is quite unclear. Lange evidently knew nothing of Galton's method of percentiles, and would scarcely have been able to calculate from the values given in Vierordt the correct positon of, for example, the 190 cm line at age 6. So he drew his lines by eye, it seems, to have the same general shape as his dominant line, and to end, near birth, at the known measurements of large and small babies.

Lange did not realize the significance of the increase in variability of height at adolescence that was demonstrated by Bowditch, if, indeed, he even knew of it. So he did not realize that tempo differences would affect the way individuals followed – or failed to follow – his curves. He therefore supposed, erroneously, that these cross-sectional curves really presented the growth of tall and short individuals, and looking at them, concluded rather bizarrely, that the adolescent spurt of the tall man lasted ten years and had its greatest intensity, 10 cm/yr, at age 12 +, whereas the adolescent spurt of the average man lasted $3\frac{1}{2}$ years and had its greatest intensity, 7 cm/yr (note the cross-sectional value), at age 15 +. In fact, tall and short men (that is as adults) do not differ either in the age at which their spurts occur or in the duration of their spurts, something which Lange surely would have realized had he collected actual growth data. All the same, Lange did construct growth standards with which tall and short individuals could, in theory at least, be followed; and in this he was far ahead of Camerer with his mean values or even Variot and Chaumet with their maxima and minima. It was not till the 1930s that the problems inherent in using cross-sectional data as growth standards for individuals (as opposed to populations) were realized (see discussion of Boas, p. 237, and only in the 1950s and 1960s did practical attempts to solve the problem appear (Bayley, 1956*b*; Tanner, Whitehouse and Takaishi, 1966; Tanner and Whitehouse, 1976).

In 1906 W. Camerer junior, the son of the paediatrician who did so much to stimulate the weighing of infants (see p. 269) wrote a chapter on 'Children's growth in weight and height' in the very influential textbook of paediatrics edited by Pfaundler and Schlossman. A four-volume English translation of this textbook appeared in 1908; Camerer's chapter is in vol. 1, pp. 409–24. Camerer begins by saying 'There is a close relation between growth and the school', and stresses the importance of school hygienists, and of the duration and timing of recesses, physical exercises and so forth, very much in the style of Kotelmann and the *Zeitschrift für Schulgesundheitspflege*. He later discusses the measuring error of height: 'The late Dr Wiener, an excellent observer, found differences amounting to 3 mm, measuring repeatedly within short intervals; observers of less skill will easily make errors of 5 to 7 mm even if their method is good' (p. 412). Children and adults measure between 1 and 3 cm longer immediately after a night's rest. There are large chunks of text taken directly from his father's 1893 paper, and finally tables of weight and length at each year from birth to 18. The gains are displayed in the manner of Schmidt-Monnard (see p. 210). The tables give peak height velocity quite late: 15.5 in boys and 13.0 in girls. A considerable degree of smoothing seems to have gone into the tables, for annual height gains are constant at 5 cm each year from 4.0 to 13.0 in boys, except that 6.0–7.0 is 6 cm; and constant at 5 cm a year from 4.0–11.0 in girls, except that 5.0–6.0 is 4 cm and 6.0–7.0 is 6 cm.

Camerer's chapter, a prime source of paediatricians' knowledge of growth for many years, is sound enough, but the divide between clinicians and auxologists and anthropologists is clear to see. Camerer does not mention the already classical work of Boas: paediatricians were scarcely to become acquainted with that till the 1950s. Indeed when Camerer died the chapter on growth was dropped entirely and nothing on the subject appears in Pfaundler and Schlossman's 1923 and 1931 editions. In contrast textbooks of paediatrics in the United States began to include larger chapters on normal growth. L. Emmett Holt (1855–1924) had provided sixteen pages on the subject in the first two editions of his textbook (1897 and 1902). Holt gave only a chart of mean weight from birth to 1 year but stressed that there was a range of healthy normality extending about 1 lb above and below his line. For the age range 2 to 5 he did little except deplore the absence of data, and from 5 onwards he gave Bowditch's means. The best chapter on growth for many years was the one by T. B. Robertson (1923) in I. A. Abt's great textbook. (This book included also an excellent chapter by Scammon (1923) on the anatomy of the child, considered chiefly by organ systems, a distant ancestor perhaps of Davis and Dobbing's *Scientific Foundations of Paediatrics (1974.)*

The first book on growth with something approaching a modern feel is that of Samuel Abramovitch Weissenberg (1867–1926), who worked as a physician in Odessa. Weissenberg was born in southern Russia, educated in Karlsruhe and Baden and qualified as an MD at Heidelberg in 1890. His papers, published mainly in anthropological journals, concern both the physical characteristics (Weissenberg, 1895) and the folklore and customs of the Russian Jewish community. Weissenberg's book on growth, *Das Wachstum des Menschen nach Alter, Geschlecht und Rasse,* was published in 1911 in Stuttgart and contains chapters on fetal growth, the newborn, growth in childhood, the adult, sex differences and growth, environmental influences, race and growth, and growth disorders. There is nothing, however, on heredity and no mention of Galton. Quetelet is quoted as 'epoch-making' but still Buffon and Montbeillard are omitted. Daffner is regarded as the best text since Quetelet; there is no mention of Bowditch, and Roberts is quoted only in the copying of a graph. Apart from the Boas and Wissler report of 1906 American work is represented only by the paper West (1894) wrote in German. The language barrier operated both ways: few English or American authors have ever quoted Weissenberg (though D'Arcy Thompson did so and also Baldwin and Gray).

In the same year as Lange's article, 1903, there appeared a book on growth that was to become for a while very influential. It was by the best-selling medical writer of the whole Edwardian era. Carl Heinrich Stratz (1858–1924), a doctor born in Odessa, combined gynaecology with the study of beauty from an anthropological – and photographic – viewpoint. He studied medicine at Heidelberg and Leipzig and trained as a gynaecologist in Berlin. He then departed to Java and served there five years as a medical officer in the Dutch army. From 1892 to 1898 he travelled in Japan, China and the United States, eventually returning to Europe in 1899, to set up in The Hague as a gynaecologist. He was given the Prussian state title of Professor in 1914. Between 1897 and 1914 he published twelve books, one of which, *Die Schönheit des weiblichen Körpers* (The Beauty of the Female Form), issued in 1899, ran through thirty-nine editions before its author's death and a further seven between 1924 and 1941 (see p. 64). *Die Körperpflege der Frau* (The Care of Woman's Body), a book of practical hygiene, diet, care of teeth, posture, breathing and so forth, in rather the style of an Edwardian Guarinoni, went through thirteen editions between 1901 and 1927. *Die Rassenschönheit des Weibes* (Female Beauty in All Races), also issued in 1901, had its twenty-second edition in 1941. *Die Körperformen in Kunst und Leben der Japaner* (1902), a more specialized horse from the same stable, had only four editions. It is true that most of these editions were really re-impressions with little or no textual change, but all the same Stratz's books certainly enjoyed a telling popularity.

Compared with these best-sellers Stratz's book on children, *Der Körper des Kindes* (The Child's Body), was something of a flop; still, it ran through eleven editions by 1928. The first and second editions were identical and appeared in 1903; but the third edition, entitled *Der Körper des Kindes und seine Pflege* (... and its Care, or Hygiene), issued in 1909, represented a considerable enlargement.

This book of Stratz's has rather the style one might anticipate from his sales figures. In chapter 2 he discusses the newborn child: 'After hours of pain and woe, of anxiety and care, there rings out the first cry with which the newborn child greets life. The first cry! How many thousands of times have I heard it, and seen the blissful smile that it conjures up on the face of the young mother.' But he can also be appropriately scathing, especially when writing about art, where he is thoroughly at home. 'In most religious representations the Christ-child is a pretty dwarf, but no child . . . the situation with angel-children is no better'. As for the cupids of Boucher, Fragonard and even Rubens and Titian, 'they are all lusty little thickset men-about-town, and certainly no infants'. (All the same he missed Botticelli's illustration of a classic Babinski reflex in the Christchild, about 1468: see Cone and Khoshbin, 1978.)

A feature of all Stratz's books are the photographs, and *Der Körper des Kindes* includes children posing in the nude, children dressed up as pierrots, and adolescents draped over trees. Some of them illustrate his points well: he talks about variation in the proportion of legs to trunk and shows three girls, aged 6, 8 and 10, sitting in the nude on a tree-trunk, with their legs hanging down; their heads are at the same level. In a second picture they stand, and now their heads are at very different levels. A second matching pair of photographs shows them aged 10, 12 and 14. Stratz's interest in art and body proportions led him very seriously to advocate the inclusion of regular nude photographs in any longitudinal study of child growth, something which links him with Ernst Kretschmer, William Sheldon and later longitudinal studies in the USA and England (see Chapters 12 and 13).

Stratz began a theory that plagued human auxology for fifty years. He gave the mean height increments 0–1, 1–2, etc., as 25, 10, 8, 5, 6, 8, 10, 4, 3, 2, 5, . . . cm. Thus from 5 to 7 it seemed there was a sharp increase in height velocity, from 7 to 10 a decrease. Out of this Stratz fashioned the system of growth illustrated in Fig. 11.8. From 1 to 4+ is the first *Fülle* or filling-out; from 5 to 7+ is the first *Streckung* or stretching-up; from 8 to 10+ the second *Fülle*, and from 11 to 15 (in boys) the second *Streckung*, followed by the *Reifung*, or puberal development. Where Stratz obtained his figures for height increment is a mystery; he says (3rd edn) that they come from Camerer (presumably 1893 or 1906) and Monti (1897–9), but no source for them there can be found. There are echoes of Zeising (1858: see Chapter 3) in the low 8 to 9 velocity; this may be an unacknowledged, even unconscious source. Stratz makes no mention of Hall and his periods of alternating length and breadth growth and indeed seems ignorant of or uninterested in previous studies of growth (though he read them up later for a chapter on growth in Kruse and Selter's textbook (Stratz, 1914*b*)). Though Stratz makes the growth periods depend on the relative increments of height and weight (see Schmidt–Monnard, 1900), in fact it seems the idea flowed from pictures rather than measurements, and probably reflects little more than the variations in subcutaneous

Fig. 11.8. Stratz's scheme of human growth. Reproduced from Stratz (1909*a*, Fig. 5).

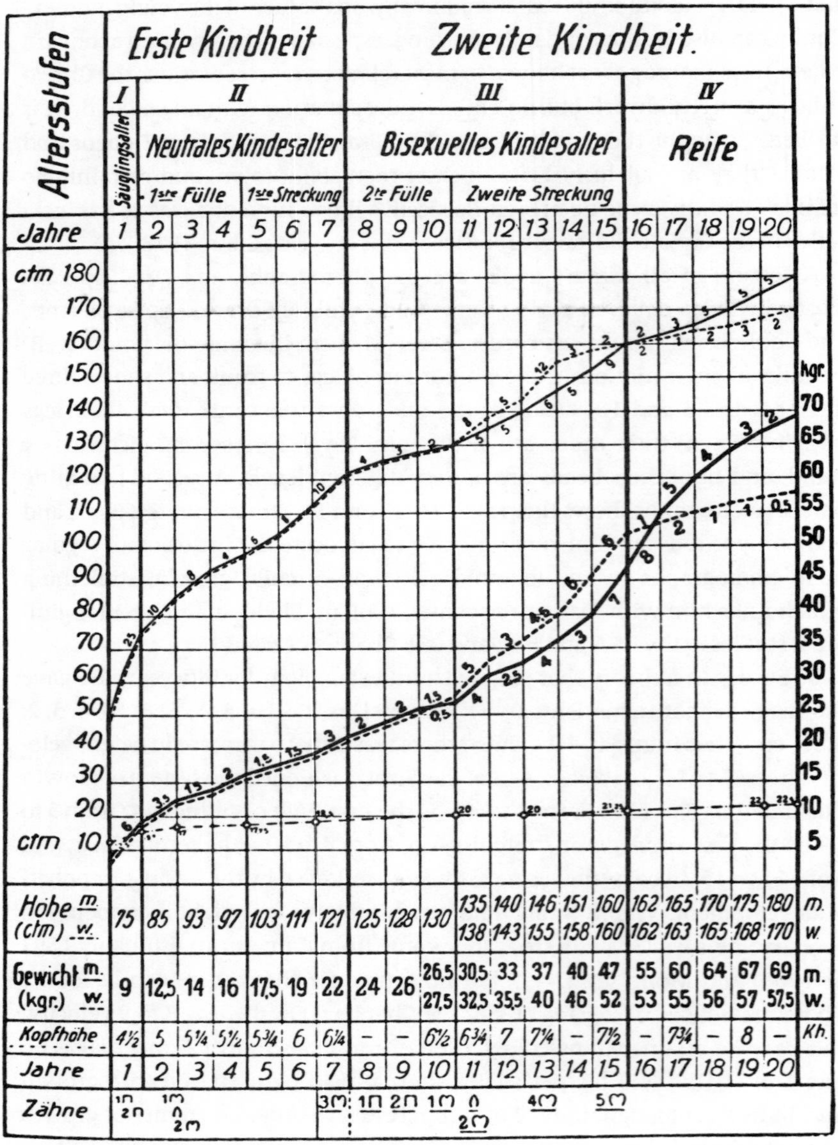

Jahre	1	2	3	4	5	6	7	8	9	10	11	12	13	14	15	16	17	18	19	20	
Höhe (ctm) m.	75	85	93	97	103	111	121	125	128	130	135	140	146	151	160	162	165	170	175	180	m.
Höhe (ctm) w.											138	143	155	158	160	162	163	165	168	170	w.
Gewicht (kgr.) m.	9	12,5	14	16	17,5	19	22	24	26	26,5	30,5	33	37	40	47	55	60	64	67	69	m.
Gewicht (kgr.) w.										27,5	32,5	35,5	40	46	52	53	55	56	57	57,5	w.
Kopfhöhe	4½	5	5¼	5½	5¾	6	6¼	–	–	6½	6¾	7	7¼	–	7½	–	7¾	–	8	–	Kh.
Jahre	1	2	3	4	5	6	7	8	9	10	11	12	13	14	15	16	17	18	19	20	
Zähne	1n 2n	1m n 2m					3m	1n	2n	1m	n 2m		4m		5m						

fat. In the light of modern data we can say Stratz's first *Fülle* reflects the increase of fat in the first year and his first *Streckung* its loss in the period 5 to 8. The second *Fülle* reflects the gradual increase of fat before puberty and the second *Streckung* the loss of fat and increase in height at puberty (see Tanner, 1978*a*).

Stratz set great store by his stretching-up and filling-out periods and reacted sharply (Stratz, 1915) to the very lukewarm reception accorded the idea by the more critical Weissenberg (1911). Despite its confused and subjective foundation, Stratz's idea caught on and appeared in textbooks, mostly of anatomy, for half a century. It was the basis of a system for judging school readiness promulgated by W. Zeller, a school physician in Berlin in the 1920s and 1930s. Zeller (1936, 1938, 1952) talked of the change of form (*Gestaltwandel*) which had to occur before the child was ready for school. This change, a result of the first stretching-up (really slimming) was used by Zeller as an indication of a maturational age supposed to be linked to psychological development (see Simon, 1959).

Stratz was a supporter of the idea that Roberton so carefully and explicitly opposed (see p. 289) that the 'white race' had a longer period of youth and a later maturation than other races. Stratz presented no data on Africans or Indians (which at that time would have certainly destroyed his thesis) but he does quote measurements of Japanese by Baelz, a German doctor who settled in Japan in the 1880s. These data indeed showed an earlier age for cessation of growth in Japanese, as do modern values, given equivalence of socio-economic status in the groups compared. Along with this Stratz in the hygiene section of the third and later editions of *Der Körper des Kindes* advises the longest possible postponement of sexual experience in youth, this being 'the first and most important finding of sex education' (p. 372). But he is more sensible than most of his contemporaries in that he advocates the frank answering of questions, and fails to say anything at all about the perils of masturbation, still a generally very popular subject.

Stratz was a man of some charisma, and even today still has his supporters (Schott, 1974*a, b*), who see in him a figure comparable to Ernst Kretschmer, the Tübingen constitutional psychiatrist. Indeed it is not hard to feel sympathy for a man who introduces (1903, p. 35) his use of children's pictures with Goethe's famous advice from Mephistopheles to the student in search of a faculty: *'Grau, teurer Freund, ist alle Theorie, und grün des Lebens goldner Baum'* (grey, my friend, is all theory, but green Life's golden tree).

Not till 1914 did the first book supposedly on the physiology of growth appear. Though a modest 161 pages it was rather grandiloquently entitled *Allgemeine and spezielle Physiologie des Menschenwachstums*. The author, Hans

Friedenthal (b. 1870) taught anthropology in Berlin; his own work, reported in the book, concerned the measurement of fetuses. He begins by saying that in the four-volume Handbook of Physiology edited by Nagel there is not a word about the physiology of growth. Nor indeed is the subject touched on in medical or physiological courses. Then, after a description of amino-acids, carbohydrates and other basic building blocks, Friedenthal describes weight growth in man. He uses only Hermann Vierordt's *Daten und Tabellen* of 1906; and when it comes to growth in height the table he gives is both approximate and erroneous. The velocity of growth falls, he says, *so gut wie unterbrochen* (more or less continuously) from birth on. As to the endocrinology of growth, which was soon to assume so great an importance, it is discussed in half a dozen lines. Not light, but darkness visible, is what Friedenthal sheds on nascent auxophysiology.

Age at menarche in the nineteeth century

It was, perhaps, natural that the first major inquiries into the age of menarche (see note 11.7) should have been done in Manchester, the centre of the Industrial Revolution, and done at just the same time as Cowell and Stanway's first measurements of heights and weights (see p. 147). In 1827 John Roberton (1797–1876) was appointed surgeon to the Lying-In Hospital of the Manchester Charity; between then and 1838, when he resigned – his health, he says, broken – he and a few colleagues had supervised the delivery of 43,500 women (Roberton, 1851, Preface). He asked a number of these women at what age they recalled their menstrual periods as having started, and tabled the results for 450 who thought they could remember with some degree of accuracy (Roberton, 1830, 1832).

Obviously such recollections are unreliable, as actual studies have shown (see note 11.8). This way of estimating the mean age of menarche of a population was almost always used, however, until 1950, when Wilson and Sutherland (1950) introduced the method of probit (later also, and equivalently, logit) analysis. In the probit, or status-quo, method no act of recollection is involved. A sample of girls covering all ages at which menarche may occur in the population under discussion (say 9 to 18) is selected, and each girl asked simply her date of birth and whether or not she has experienced her first menstrual period. From the resulting cumulative frequency curve of the percentage of girls who are post-menarcheal at each successive year of age the mean and standard deviation of age at menarche can be derived. This method also has the advantage that a good random sample of schoolgirls in a population can usually be readily obtained. In the older, recollection, studies pregnant women were the ones usually chosen as the sample since they were at least more representative of the general population than women attending a gynaecological clinic, say, or women seen in a general medical practice. However, unmarried and infertile women were excluded, and this may be a source of bias, albeit a relatively minor one. Data collected since 1950 are always of the status-quo type, and there are a few sets of data on school populations in the 1930s which can be recalculated by probit analysis (see Tanner, 1962, p. 155), as well as one surprising table published by a doctor at the British Consulate in Siam in 1862 (see below).

There seem to have been three main reasons for the initiation of studies of menarcheal age in the nineteenth century. The first was the same as that underlying the collection of heights and weights of children: to see whether the nature of employment, the diet, and social conditions generally were reflected in such data, and, if so, to use the data as an argument in the fight for social reform (including the abolition of slavery). This was the chief concern of Roberton, a man from the Scottish lowlands who qualified as a surgeon in Edinburgh and went to Liverpool with the intention of travelling the world as a ship's surgeon, but instead settled into practice first in Warrington and then in Manchester. 'For many years', says his obituary notice, 'he threw himself heart and soul into every philanthropic endeavour to ameliorate the condition of the working classes. The shortening of the then excessive hours of labour . . . occupied a large portion of his time . . . his private charities were as numerous as unostentatious' (Roberton, 1876). Thus Roberton must certainly have known Cowell and Stanway, the local men, and Leonard Horner too, a fellow-lowlander who from 1836 was factory inspector in the area which included Manchester (see p. 153).

The second impulse to study the age of menarche sprang from a concern about over-population, stimulated in particular by the then-recent works of Malthus. Matthews Duncan, of Edinburgh, whose studies on the length and weight of the newborn we have just discussed (p. 259), had this issue especially at heart. Though he made no studies of menarche himself, he drew extensively on others' recent work in his discussion of the ages at which procreation was possible and desirable. The third impulse seems to have been the simple clinical one of understanding the causes of sterility; clearly, primary amenorrhoea (lack of menstrual periods) was one such cause, and in order to diagnose it correctly one had to know the range of ages over which menarche occurred in normal, though perhaps undernourished and overworked, women.

Menarche in the United Kingdom

Roberton presented the results of his inquiry, made during the years 1828–9, in a table of frequency distribution of age of menarche; that is, as the numbers of women having their first period at successive years of age. Though he was interested in the average age, he emphasized more the very wide range of ages – 11 to 20 – at which menarche could occur. His mean is probably best calculated as 15.2 years, though this involves assuming that in his inquiry he asked his patients about age-at-nearest-birthday rather than age-at-last-birthday, as would normally be the custom nowadays. On balance this seems the more likely interpretation (see discussion in note 11.9) and it has been adopted here for all nineteenth-century tables unless there was reason to do otherwise; it may however lead to a mean age that is a little too low, perhaps by 0.1 or 0.2 years.

Roberton's long series of papers (1830, 1832, 1842, 1844, 1845*a*, *b*, 1846, 1848), culminating in a monograph published in 1851, marked a profound change in attitude toward received opinion, not only about age of menarche but concerning general causes of differences between populations. At the outset of his second, and most notable, paper Roberton (1832) summed up current opinion: 'The age of puberty, as indicated by the eruption of the menstrual secretion, or the age at which the aptitude for generation commences (for the two phenomena are regarded as occurring at the same period) is a point concerning which there exists little or no difference of opinion. Physiologists concur in teaching that the period is not uniform, but that it is influenced by many circumstances, as climate, constitution, and mode of living; that in warm climates it occurs often as early as the eighth or ninth year; in temperate climates is usually postponed to the thirteenth or fourteenth year; and in arctic regions to the nineteenth or twentieth (1832, p. 227)'.

He undertook to show this view was nonsense, and depended on travellers' tales and the authority of Haller. Haller, he suspects, took it from Montesquieu's *Spirit of Laws* (1757), which circulated widely in the years before the volume of Haller's *Elements* which dealt with human growth was published (1766). Actually both Montesquieu and Haller were only repeating and embellishing established doctrine; Demol's 1731 Montpelier thesis (see p. 79) presents the climatic view in considerable detail, as do other, earlier writings. But Roberton brings out, with some acerbity, one of the social bases on which such notions rested. The slower the growth, it was thought, the more the perfection. Growing up fast, having an early menarche, and an early start to procreation was the sign of an inferior race (presumably by analogy with monkeys who grow up more rapidly than humans). The slavers, says

Roberton (1851), use the supposed earliness of maturing of Africans as an argument to support their activities. In which case, he adds caustically, the Eskimos, with an age of menarche supposedly retarded by the cold, must by the same token be much superior to Europeans. Montesquieu, moreover, justified the inferior position of Eastern women on the basis of their supposed early maturing. Women in hot climates, he said, are marriageable at 8 or 9. They are physically old at 20. But their minds are not precocious, so reason does not accompany beauty; by the time reason is attained, beauty has fled. It was therefore natural that in such places a man should leave one wife to take another and that polygamy should be the rule. Roberton made short shrift of this, as also of the 'analogy said to exist between the influence of heat upon fruits and the female constitution' (1832, p. 234).

Having cleared the decks somewhat, Roberton, clearly an anthropologist *manqué* who made the voyage he never took from Liverpool through the medium of books and letters, set about gathering real statistics on populations in hot and cold places. This he did by enlisting the help of reliable physicians working in the Caribbean Islands, Tahiti, Madeira, Greece and India, and amongst the Labrador Eskimo and the North American Indians. In most of these places the mean ages of menarche were much the same as in Manchester; in Madeira around 15.4, in Guyana 15.6, in Jamaica and Antigua the fourteenth and fifteenth years (thus 14.5 say) with no difference between Whites and Blacks living there. Greek girls averaged the same and sixteen Labrador Eskimo girls gave values directly comparable with the means of three samples of sixteen women, each independently drawn from the Manchester archive. Apart from the Tahitians, whose menarche was said to be around 10 or 11, the Hindus alone were somewhat earlier, with means of 12.3 years in a sample in Calcutta and 12.6 years in a sample in Bombay.

But as to India (and *a fortiori* Tahiti) he remarks, quoting an Indian scholar at the Calcutta Medical School, writing in 1845 : 'Most of the females of this country begin to menstruate after the 12th year, or at the beginning of the 13th . . . it is the custom of our country, in our early marriages, to send the girl at perhaps nine years occasionally to the house of her husband; but if the husband be so distant that this cannot be done, menstruation is generally delayed until the 13th year. I have been informed by several women, that when the menstrual flux begins as early as the 11th or 12th year, it does not, in many cases, recur for a year after this first appearance, but after that period the secretion again takes its natural course. It may therefore be fairly questioned whether or not this which is supposed by them to be a first appearance, may not be rather caused by the first act of sexual intercourse and the result of a ruptured hymen. I believe the catamenia appear sooner or later according to the mode of living of the females and the sexual excitement to which they may be subjected.'

A decade later, James Campbell, a naval surgeon attached to the British Consulate in Siam, inquired of 104 Siamese girls of known and clearly specified age whether or not they had begun to menstruate. His report (Campbell, 1862) provides the earliest data to which a probit analysis can be applied. The mean age, estimated this way, comes to 15.1 ± 0.3 years with a standard deviation 1.8.

Roberton's investigation in Manchester was soon followed by others. E. W. Murphy, professor of obstetrics at University College London, added to Roberton's 450 cases 552 from his clinic and 1,160 from the clinic of Dr Robert Lee (1793–1877), an Edinburgh-trained physician in London who had the distinction of spending the years 1824–7 in Russia (Schuster, 1968). The whole 2,162 women together gave a mean of 14.9 years, assuming age-at-nearest-birthday (Murphy, 1845; and see Roberton, 1845*b*, p. 165).

In the same year W. A. Guy (1845) of King's College, London, who actually states that he used age-at-nearest-birthday (see note 11.9) gave a frequency distribution for 1,500 women. The mean is 15.1 years. The distribution has a longish tail to the right, with 3 per cent of the women first menstruating at age 20 or over. This tail characterizes most of the frequency distributions in the nineteenth century. Presumably it represents the effect of undernutrition in a section of the population, though one has to remember that these are not random surveys, and in at least some of the studies the samples were selected by the fact of attending a gynaecological clinic. The standard deviation of Guy's series is 2.1 years, which is about double the value derived from modern status-quo inquiries. The difference doubtless partly reflects the unreliability of recollection, but also the reality of greater variation and skew.

Another Manchester surgeon, James Whitehead (1812–1885), made a still more extensive inquiry, this time linked specifically with circumstances of life (Whitehead, 1847). He obtained data from 4,000 pregnant patients, of whom 2,127 were employed in mills and warehouses, 1,349 were domestic servants, 363 seamstresses and 161 'educated ladies'. The mean age of menarche for the whole group comes to 15.6 years (his table 1, p. 46; the standard deviation is 1.9 years; Whitehead actually gives the sum total of age in months so evidently inquired to the nearest month). But the mean for the 'educated ladies' was 14.3 years. That for servants, incidentally (15.5 years), was scarcely lower than the value for the factory hands.

Twenty years later Rigden (1869) reported a new series of inquiries at University College Hopital obstetric clinic, this time on 2,696 women. The

Table 11.3. *Average ages of menarche in United Kingdom, Scandinavia, Germany and Russia in the nineteenth century*

	Year of menarche (approx.)	Mean age at menarche	Place	Author
Working women				
UK	1815	15.2	Manchester	Roberton (1830)
	1835	15.6	Manchester	Whitehead (1847)
	1830	15.1	London	Guy (1845)
	1830	14.9	London mostly	Murphy (1844–5)
	1855	15.0	London	Rigden (1869)
	1910	15.0	Edinburgh	Kennedy (1933)
Scandinavia, Germany and Russia	1785	16.6	Göttingen	Osiander (1795)
	1835	16.4	Copenhagen	Ravn (1850)
	1850	16.8	Copenhagen	Hannover (1869)
	1850	16.4	Berlin	Krieger (1869)
	1850	16.8	Munich	Hecker (1864)
	1865	16.6	Bavaria	Schlichting (1880)
	1870	15.6	Oslo	Brudevoll *et al.* (1979)
	1875	15.7	Russia	Grüsdeff (1894)
	1875	16.5	Helsinki	Malmio (1919)
	1890	15.7	Stockholm	Essen-Möller; in Lenner (1944)
	1895	16.2	Berlin	Schaeffer (1908)
	1900	16.2	Schleswig	Heyn (1920)
	1900	16.0	Helsinki	Malmio (1919)
	1900	14.6	Oslo	Brudevoll *et al.* (1979)
Middle class				
UK	1835	14.3	Manchester	Whitehead (1847)
	1890	14.4	London	Giles (1901*a*)
Scandinavia, Germany and Russia	1820	15.0	Norway	Brundtland and Walløe (1976)
	1835	14.4	Copenhagen	Ravn (1850)
	1875	14.4	Russia	Grüsdeff (1894)
	1895	14.4	Berlin	Schaeffer (1908)

Note: The average date of year of menarche has been calculated from the probable mean age of the women studied: it has an error of up to five years.

mean value came to 15.0 years. These would mostly be working class women. In contrast Giles (1901*a*) gave a figure of 14.4 years for 1,000 women 'drawn mainly from the middle classes' in London. Giles gives the frequency distribution; even in these patients there is a long tail to the right indicating the presence of women who, at least according to their recollection, had their menarche delayed into the late teens or early twenties.

These and other, continental, studies are summarized in Table 11.3. The 'year of menarche' is calculated from the known or assumed average age of the women questioned. It is thus not accurate, even as an average, to nearer than five years. It also carries a built-in snag. When Whitehead, for instance, inquired of 4,000 women in Manchester at what age they first menstruated he did not inquire where they were living at that time, or in the years before. Thus when in the table the location is given as Manchester (or London, or any other town) only a proportion of the women will actually have grown up in that town; many will have passed their infancy, or childhood, or even adolescence in the countryside or elsewhere (for example, in Ireland). A few studies were made of the difference between those born and bred in the town – for example in Paris (Boismont, 1841) – and those who were immigrants to the town, and the differences were not great. However it is probable that city-born-and-bred women would have an average menarcheal age some half year lower than women immigrant to the city.

Clearly one should be cautious about generalizing from these data, dependent as they are on the women's memories and the presumed manner of recording age. However, as Table 11.3 shows, it seems that working class women in the nineteenth century had an average age of menarche of about 15.4 years in Manchester and about 15.0 in London. Middle class ladies in both cities had averages of about 14.4 years. Little if any change in these averages can be discerned from 1815 to 1855 or even to 1890. Even as late as 1910 it seems likely that these values were still more or less the same. Kennedy (1933) reported the average for 10,219 women who attended the gynaecological clinic of Edinburgh Royal Infirmary in the 1920s. Their mean age of menarche, experienced on average in about 1910, was 15.0 years. Though they are admittedly a selected group, their standard deviation, (1.7 years) is not out of line with that shown by other groups; it lies just half-way between Guy's 1830s figure and modern ones.

In the North of England about 1960, mean values were 13.4 for middle class and 13.6 for manual working class girls (Miller, Billewicz and Thomson, 1972; Miller *et al.*, 1974; Roberts, Rozner and Swan, 1971). In southern England, a more prosperous area, the equivalent values for 1960 were 12.9 and 13.1 years (since when there has been no change: see Tanner, 1973).

There was, therefore, a drop in age at menarche in working class girls of about twenty-two months in the fifty years from 1910 to 1960; and in middle class girls during the same period a drop of about fourteen months (rates of 4.4 months per decade and 2.8 months per decade respectively).

Age at menarche in continental Europe

Inquiries similar to Roberton's were made in other European countries. The results up to the 1860s were summarized by Raciborski (1868) and up to the end of the century by Giles (1901*b*). The earliest study seems to have been by F. C. Osiander (1759–1822), a celebrated obstetrician and director of the Poliklinik at Göttingen. It concerns only 137 women, and assuming *im 12ten Jahre* means 12 to 13 (see note 11.9), gives a mean of 16.6 years, appropriate to about 1785 (Osiander, 1795: there is a copying error in Raciborski, 1844). The next German data come from 3,114 patients attending Hecker's obstetrical clinic in Munich about 1860. The mean, calculated from the tables given by Schlichting (1880), is 16.8 years and the standard deviation 2.1 (assuming age-at-nearest-birthday). This very high figure would refer to menarche in about 1850 (see Table 11.3). Schlichting's own investigation on 8,881 women and concerning menarche about 1865, also in Bavaria, gives a mean of 16.6 years, with a standard deviation of 2.2 years (assuming age-at-last-birthday, which his phraseology makes more probable than age-at-nearest-birthday). In Berlin about 1850 the average for 6,550 clinic patients studied by Mayer and Krieger (Krieger, 1869) was 16.4 years, standard deviation 2.3 years.

In Berlin, as in London and Manchester, the change was slow. About 1895, clinic patients still had an average age of menarche of 16.2 years (S.D. 2.1) according to Schaeffer's (1908) study of 10,500 women. Well-off Berliners, however, judging by a questionnaire Schaeffer sent to fifty doctors regarding such of their patients as qualified (in all 1,801 women), had an average menarcheal age of 14.4 years (S.D. 1.7). Amongst these the 420 women of Jewish origin and culture averaged 14.0, the remainder 14.6. Similar social class, and perhaps also genetic, differences in Poland were communicated to Raciborski (1868, pp. xiii–xiv) by a Dr Dropsy in Zaclaw (now Zaslaw, USSR). Slavic peasants had an average age of 15.9 years, 100 Jewish women 14.2 years and 70 'nobly-born ladies living in very easy circumstances' 13.9 years. Heyn's (1920) values (see table) for 1900 relate to women working in a textile mill in Schleswig.

Table 11.3 also shows the values for Scandinavia and Russia. Ravn (1850) found a mean of 16.4 for the poor in Copenhagen about 1835, but a figure of 14.4 for the wealthy. In the countryside landowners' daughters had menarche at just the same age as the wealthy in the city; and labourers' daughters at just the same age as the poor in the city. Hannover's (1865, 1869) means of 16.8 years in 1850 for persons living in Copenhagen and 17.0 for those living in the countryside around it are the highest seen in Europe. In Helsinki values of 16.5 (Henricius, cited in Malmio, 1919) were obtained for clinic patients about 1875, which reduced to 16.0 by 1900 (Malmio, 1919). Patients attending obstetricians privately at the same period had an average of 15.0 (Malmio, p. 128). In Stockholm about 1890 the average for clinic patients was 15.7 (Essen-Möller cited in Lenner, 1944); by 1915 the value had fallen to 14.7 years. Patients coming to a private practice at this time averaged 14.5 and in 1930, 13.9 years (Lenner, 1944). The standard deviations in all these last three series were about 1.4 years, so by this time the long tail at the upper ages had disappeared, even in the clinic patients.

A number of studies were made in Russia. Weber (1883) recorded in his day-book during 1879–81 the recollected dates of menarche of 2,371 women seen in his practice in St Petersburg; he emphasizes half were born elsewhere. His average figure, relevant to about 1860, is surprisingly low, being 14.6 years on the assumption that he used age-at-nearest birthday (as he includes *Mädchen* as well as *Frauen* he may possibly have recorded young girls who had menstruated and ignored those of the same age who had not, a potent source of error for nearly a hundred years). Patients from the peasantry *(Bauernstand)* averaged 14.8, townspeople 14.6, traders 14.1, nobles and officer class 14.1, soldiery 14.8. Grüsdeff (1894), collecting some 10,000 records from all over Russia regarding menarche about 1875, obtained higher values, at least for the lower social class groups: peasants 15.7 years, townsfolk 14.8 and the well-to-do 14.4. In Elizabethgrad (modern Kirovgrad in the Ukraine) about 1885, 768 Ukrainian townsfolk in Weissenberg's practice had an average of 14.9 years and 1,273 Jews an average of 14.2 years (Weissenberg, 1909: where age is *stated* as at-nearest-birthday).

By far the most complete archive, however, comes from Norway. Brudevoll, Liestøl and Walløe (1979) discovered the medical records of pregnant women attending the Women's Clinic in Oslo back to 1837; they were preserved in the clinic cellars. These records all gave recollected age at menarche, and Brudevoll and his colleagues selected at random fifty cases for each year from 1861 onwards. The results are shown in Fig. 11.9, redrawn from Brudevoll's paper. Up to 1945 the graph relates to working class women almost exclusively, after that to most of the population. The squares represent means of modern status-quo inquiries on Oslo schoolchildren, covering all classes. In 1860 and 1870 the mean menarcheal age was about 15.6 years, but by 1900 it had decreased to about 14.6 (a decline of four months per decade); there it remained till 1920. From 1920 to 1960 there was a decrease amounting to about sixteen months (also about four months per decade); from 1960 to 1975 there was no further change. These investigators also noted the long tail extending into the higher ages in the distribution in the 1860s, compared with modern distributions. In an earlier paper, Brundtland and Walløe (1976) estimated that in the 1840s the average age of menarche was around 16.0 years. In Norway, therefore, in the mid-nineteenth century, the working class women had menarche about six months later than working women in Manchester. There is one record of middle class Norwegian women; F. C. Faye, head of the Oslo Women's Clinic and instigator of most of the early researches on menarche in Norway, had a private medical practice amongst middle class families in the town of Skien, outside Oslo. In 1839 he assembled data on their menarche; the figures were reported in Raciborski (1844) but Brundtland and Walløe (1976) give reasons for thinking Raciborski misinterpreted Faye's method for ages. They conclude that the average menarcheal age of middle class Norwegians in about 1820 was 15.0, about one year earlier than contemporary working class women.

Fig. 11.9. Mean menarcheal age for working class women (up to 1945; thereafter middle class also) in Oslo from 1860 to 1975. Recollection data; each point represents the average of about fifty maternity clinic patients. The squares represent status-quo probit-fitted data on Oslo schoolgirls. The curve is a 21-term moving average until 1960. Redrawn from Brudevoll, Liestøl and Walløe (1979).

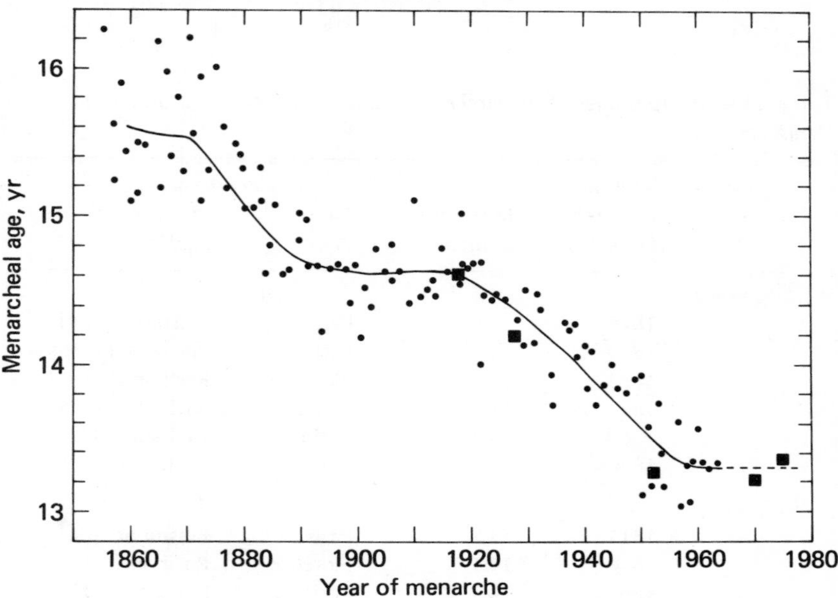

Thus the data for northern Europe are very consistent. In the Mediterranean countries however – if France may be reckoned as such – a difference does appear, with menarche distinctly earlier (see Table 11.4).

The first major study was by Brière de Boismont (1797–1881), who himself obtained details regarding various facets of menstruation from 830 women in Paris hospitals in the period 1830–40. To these were added the results of 370 inquiries by Dr Menière (of Menière's disease). Boismont, in a long and finely produced paper (Boismont, 1841) reports a mean age for the poor, hospital-attending class of 14.8 years, and for the well-off, private clientele class of 13.7 years. This would be referable to menarche occurring about 1815 (the year that the poor in Manchester averaged 15.2 with the rich probably about 14.3). Though Boismont writes that the notion that the first menstruation occurs earlier in town girls than in country girls 'has the force of an axiom' there are nevertheless many exceptions; and he attributes much importance to nutrition and social conditions. He endeavoured to find differences in menarcheal age between girls of different temperaments (lymphatic, lymphatic-sanguine, sanguine and lymphatic-nervous) and remarked that the periods begin earlier 'in little women and later in those who are tall' (an impression not borne out as correct by modern work).

Raciborski (1809–1871), also in Paris, organized a series of inquiries all over Europe in the 1840s. The physicians concerned met to compare results in Paris in 1843 and Raciborski incorporated the findings in his two books

Table 11.4. *Average ages of menarche in France, Spain and Italy in the nineteenth century*

	Year of menarche (approx.)	Mean age at menarche	Place	Author
Working women				
	1815	14.8	Paris	Boismont (1841)
	1830	14.6	Paris	Raciborski (1868)
	1830	14.5	Madrid	Raciborski (1868)
	1830	14.6	Florence	Raciborski (1868)
	1860	15.4	Turin	Pagliani (1879)
	1890	14.5	Rome	Rossi-Doria (1908)
Middle class				
	1815	13.7	Paris	Boismont (1841)
	1830	13.5	Florence	Raciborski (1868)
	1860	14.3	Turin	Pagliani (1879)

(1844, 1868). His own inquiries of 500 patients in Paris hospitals give a mean age of 14.6 (see Ducros and Pasquet, 1978, for correct figures) referable to about 1830. Patients in Lyon, Montpelier and Toulon, about 1830, gave values little different from those in Paris (Raciborski, 1868). The studies Raciborski organized in Madrid and Florence gave still lower means: 14.5 in hospital patients in Madrid, and in Florence 14.6 amongst the poor patients and 13.5 amongst the rich. Some twenty years later de Soyre (1863) reported the rather higher mean of 15.2 years for 1,000 women admitted to hospital in Paris for the birth of their first baby. Raciborski gives a huge table of results in his 1868 book, but even at that date he had not heard of Roberton's 1851 monograph.

Pagliani (1879) had inquiries made in the Turin obstetrical clinic and obtained a mean of 15.4 years (assuming age-at-nearest-birthday, which he himself used in his anthropometric survey) for menarche experienced around 1860. He also published the value for 146 girls in three institutions of high standing and good care in and about Turin, where the girls had been followed longitudinally and the exact date of menarche was known. In these girls the average age was 14.3 years. In Rome Rossi-Doria (1908) found a mean of 14.5 years for 31,659 patients attending the obstetrical and gynaecological clinics; this would be referable to about 1890.

Thus the Mediterranean means are about six months lower than the English ones of the same period and a year less than the values in north-west Europe. Modern values in fact still show a difference: the mean for London is 13.0, for Paris about 12.8 and for Italian cities, even in the north, about 12.6.

In general, however, the situation on the European continent was not dissimilar to that in England. The difference in age at menarche between the well-off and the poor was about a year in England and eighteen months in Germany. In Norway a considerable decline took place in the age at menarche of working class women during the last one-third of the nineteenth century. In other countries the evidence points to a smaller decline, if any. Certainly a very rapid decline took place from 1910 or 1920 to around 1970, and this occurred in most or all European countries, and in both rich and poor families though in somewhat different degree (see Brundtland and Walløe, 1976; Tanner, 1962, 1973). The relation to economic activity and standard of living seems clearly implicit in these figures.

USA

In the eastern United States, as expected on these grounds, the average age of menarche at the turn of the century was a little below that in England and the differences between social groups was less. Emmett (1879) reported a mean of 14.2 years for women in New York City about 1875 and Engelmann (1901*a, b*) a mean of 14.3 in about 1890 for 2,503 of Chadwick's Boston dispensary patients. His own St Louis dispensary patients had a mean of 14.2 years. Engelmann found no difference between his dispensary patients and patients in his private practice; girls who had attended women's colleges, however, had a mean of 13.5 years; the modern mean for similar women is 12.8 years.

Japan

At this same time in Japan, that is about 1890, the average for a large collection of clinic patients from all over the country was 14.8 years (Yamasaki, 1909). The modern value is 12.8.

To summarize: age at menarche provides a convenient measure of tempo of growth of a population at a given time. Modern data are all obtained by status-quo methods. The older material inevitably incorporates errors of recollection, or worse, unconscious conformity to a received type. A century ago there was a striking difference between social classes, amounting to a year or eighteen months, and also between north Europe and the Mediterranean countries. Both these differences progressively diminished during this century, the former to nothing in some countries and a matter of two to three months in others. The geographical difference has reduced to four to six months, presumably as girls in Mediterranean countries reach a biological limit beyond which earliness can go no further.

12

The influence of educational psychology and child development: the North American longitudinal growth studies

Franz Boas' work, as we have seen, was divided into two phases: the first ended in 1910 and the second began in 1930. During the intervening years the whole situation in North America regarding child growth and development was transformed.

Even as late as 1915 very few longitudinal records extending over any considerable portion of the whole growth period were available. Paediatricians restricted their speciality to children under the age of 14 or even, in some places, 12. The adolescents who were studied in the boarding schools by Carlier, Godin and their colleagues arrived without any prior history of growth. Von Pirquet, the celebrated professor of paediatrics in Vienna, went so far as to invite a contribution to his journal, the *Zeitschrift für Kinderheilkunde*, from a teacher at the Vienna *Gymnasium*, Martin Guttman, who for twenty-eight years had been making serial measurements on children and young adults. Guttman (1915) says his data, which included birth-to-maturity measurements on his four sons (illustrated in Scammon, 1930), are special in two respects: each subject was measured within ± 8 days of his birthday, and each subject was followed over a prolonged period. Camerer and Vierordt had advocated this, he says, but their data stopped at age 14, just before the most important part of development. In all relevant literature he found only 'about 127 serviceable cases'. (Unfortunately he gives no references so we do not know which cases were known to him.)

In Europe this situation remained unchanged until after World War II; but in North America a powerful child welfare movement arose in the 1920s and provided the soil for a crop of longitudinal studies, whose harvesting shaped the whole pattern of human auxology in the years 1935–55. Many people and interests combined to create the movement, but the man who perhaps played the largest single part throughout the whole period was Lawrence K. Frank (1890–1968), assistant director of the General Education Board of the Rockefeller Foundation and guiding spirit of the Laura Spelman Rockefeller Memorial Fund. Later, when the Rockefeller Foundation moved away from the child development field, he became assistant to the president of the Macey Foundation. He has himself given a useful account of the main events of this period (Frank, 1962; see also Frank, 1935, 1943).

The origins of the movement went far back of course – to Pestalozzi and Froebel, not to mention Thomas Arnold in England and Horace Mann in Massachusetts. But for the United States Stanley Hall was in many ways the immediate fountainhead, and much of the stream was channelled through Harvard University, where Hall took his PhD in 1878 and subsequently held his first job as lecturer. Hall went to Clark University in 1888 (see p. 234). With Boas he got off to a false start so far as longitudinal studies of growth were concerned (see p. 239); but one of his graduate students, Arnold Gesell, went on to establish the first American longitudinal study of mental development in young children, at the Yale Psycho-Clinic in 1911.

Two other very influential figures, both connected with Harvard, were W. F. Dearborn (1878–1955) and B. T. Baldwin (1875–1928). Dearborn, a New Englander, took his PhD in education at Columbia in 1905, taught at Madison and Chicago for a period, and then went to Germany where, at Munich, he added, most unusually, a degree in medicine to his PhD. He never practised medicine; but the degree was symptomatic of his orientation. In 1912 he went to the Harvard Graduate School of Education as assistant professor, and in 1917 he became professor, a post he held till his retirement in 1947. Dearborn had successive generations of graduate students take anthropometric measurements of schoolchildren as a routine part of their training and in 1922 he started the largest longitudinal study of the growth of schoolchildren ever made (see below). B. T. Baldwin took his PhD at Harvard, also in 1905. Both W. T. Porter and Henry Bowditch were teaching at Harvard Medical School when Baldwin was a graduate student and when Dearborn was establishing his Psycho-educational Clinic in the School of Education. It was Porter's work (see Chapter 9) which had so much impressed educationists with the importance of physical growth. Baldwin in particular was struck by Porter's findings, and replicated them in his first published work (see below); and Dearborn carried on in his growth study the work just completed by Porter in the Boston public schools.

From the beginning the child development movement was devoted, philosophically, to the study of the 'whole child' (see note 12.1). When the journal *Child Development* was founded in 1930, as the organ of the Society for Research in Child Development, only three of the six editors were psychologists: besides them there was T. Wingate Todd, an anatomist (see below); E. V. McCollum, a biochemist; and Edward A. Park, a paediatrician. (The last two were professors of their subjects at Johns Hopkins, as was the chief editor, B. Johnson, a psychologist.) As the years went by, the concept of the whole child failed to bear the expected operational fruit and gradually the different disciplines moved apart. By 1950 *Child Development* was largely devoted to psychological and sociological studies, and by 1960 almost exclusively so. Reports on physical growth had moved into *Human Biology* (see below).

Nevertheless, all the North American longitudinal studies (as well as all the International Children's Centre ones, described in the next chapter) had an interdisciplinary approach. The predominant discipline varied from study to study; so perhaps the title of this chapter, with its emphasis on educational psychology, should be qualified a little. Most of the North American studies had this orientation indeed; but there were exceptions. The Child Research Council at Denver and the Center for Research in Child Health and Development at Harvard were directed by paediatricians and their orientation was primarily towards medicine. The Fels Research Institute laid a fairly equal emphasis on the medical and the behavioural sciences; and so, in a smaller way, did the Brush Foundation at Western Reserve: both of these were also directed by medical men.

Baldwin and the University of Iowa Child Welfare Research Station

The first of the longitudinal studies, however, was set up by an educational psychologist with impeccable whole-child credentials. This was Bird T. Baldwin, a Pennsylvanian Quaker who graduated from Swarthmore College, was principal of a Friends' School in New Jersey for a couple of years, and then went to Harvard to take his PhD (1905). For most of the period 1906–16 he lectured in psychology and education at Swarthmore, with summer teaching in the School of Education at Johns Hopkins University. Baldwin's first substantial publication on growth, in 1914, was an account of the relation between school marks, height, weight and vital capacity in 861 boys and 1,063 girls in the Horace Mann School of Teachers College, Columbia University, and, in Chicago, the Francis Parker School and the University School. These children had been followed longitudinally for three to twelve years, with measurements at annual intervals in Horace Mann and six-monthly in the other schools. Baldwin's findings entirely confirmed what Porter had found twenty years before. Baldwin understood that 'the composite curve of the average heights based on simple means of different individuals at different ages does not represent the growth of any individual or any group of individuals' (1914, p. 34) and gave twenty-eight charts in which were plotted the individual growth curves of 170 children. In addition, tables of the raw measurements for 100 boys and 100 girls were given. Baldwin's charts, some of which were repeated in his 1921 monograph, provided, in 1914, the first opportunity for most educationists to see and appreciate, in a practical way, the great differences between individuals in the timing and shapes of their growth curves. Taking his cue from the work of Boas, Rotch and Crampton on physiological age (see Chapter 10) Baldwin talked of pedagogical age, the tempo of a child's passage through the educational process. He discussed physiological age also, giving a graph of the growth in height from 5 to 16 of eight girls whose ages of menarche were known and remarking on the earlier cessation of growth in those with an early menarche.

Baldwin contributed also to the growing sophistication of methods of growth surveillance. Standards of growth, especially those that were applicable to individuals, tended to be more the concern of schoolteachers than of medical men. Bowditch (1890, 1891) and Sargent (1893) were the first to provide practical percentile charts for height and weight; they concerned only children of school age and above. Boas' norms of 1898 gave standard deviations as well as means, but no simple way in which they could be used. Baldwin (1916) took the data from the schools mentioned above and presented charts for height, weight and vital capacity from 5½ to 18 years, giving only the means (cross-sectionally calculated). The heights and weights were considerably above Boas' means, as might be expected, since the schools concerned were middle class private ones. Later, in 1923 Baldwin and Thomas D. Wood, of the Life Extension Institute (a relative, perhaps, of the Walter D. Wood who was the doctor at Horace Mann), who had also worked up the Horace Mann School results about 1916 (see Gray and Gray, 1917), presented data on '74,000 boys and 55,000 girls from twelve schools in the Eastern and middle USA who had at least five repeated annual or semi-annual measurements' (Baldwin, 1925). These appeared as graphs on cards (revised 1927) and as tables of means of height, weight, and weight-for-height for each half year of age (Baldwin and Wood, 1923; Baldwin, 1924, 1925). These Baldwin–Wood tables were used as standards in North American practice for many years.

Baldwin used his summer teaching at Johns Hopkins University to establish norms for pubertal changes by collecting data on 3,600 boys taking part in athletics in Baltimore, and on a further 1,317 boys in various parts of Maryland. The criterion of pubescence was the appearance of pubic hair. Baldwin emphasized the very wide range of ages at which this occurred: 9½ to 15½ in the country boys and 10 to 18 in the city boys, with modes at

$13\frac{1}{2}$ and 14. A similar study was made of 1,241 girls, but in them the criteria of puberty were sadly muddled up, enlargement of breasts, first menstrual period and presence of axillary hair all being used, it seems indifferently (Baldwin, 1916).

In 1917 Baldwin moved to Iowa City to be director of a new institute established in the University of Iowa. This was the Iowa Child Welfare Research Station (later called the Institute of Child Behavior and Development). It was founded through the vision and quite extraordinary tenacity of one woman, Mrs Cora Bussey Hills (1858–1924), of Des Moines. Her plan, first proposed in 1901, was rejected by the State Agricultural College, four successive presidents of the State University and, the first time the Bill came up, by the State Legislature (University of Iowa, 1933; see note 12.2). On the second occasion that the Bill establishing the Research Station was considered, in April 1917, it was passed. The station prospered and grew and became a major centre for research in child development from the 1930s to the 1960s (see University of Iowa Institute of Child Behavior and Development, 1967). In 1970, however, it was disbanded as a degree-granting institution and the faculty was for the most part dispersed.

As soon as he was appointed, Baldwin set up a longitudinal study programme, with separate groups of infants, pre-school children and schoolchildren enrolled. The study was what we would now call mixed longitudinal, that is, one with children entering and leaving on a continuous basis (as opposed to a cohort study where all children enter during a limited period and then are followed as a group). Baldwin introduced a method of measuring height suitable for use in homes and schools. A paper scale, printed under his supervision, was fixed to the wall. The child stood with his head against it, and the measurer picked up a solid wooden right-angled triangle and brought it down onto the child's head. One face of the triangle was against the wall, the second rested on the head, and the third, the face of the hypotenuse, was shaped so it could be held in the measurer's hand. Baldwin also designed a measuring board for infants, with fixed heel- and sliding head-pieces, an instrument that was universally used until the introduction of the Harpenden digital read-out instruments in the 1950s. The anthropometric laboratory of the Station measured height or length, sitting height; span; upper arm, forearm and calf lengths, shoulder and hip widths; chest depth, breadth and circumference; and six head and face dimensions. Weight was measured, together with vital capacity, and strengths of forearm, wrist and elbow. At the same time the laboratory offered a service to school medical officers and parents, in 'recording and evaluating the semiannual physical measurements of the growth of their boys and girls' (see note 12.3).

While this material was accumulating Baldwin made a cross-sectional study of 140,000 Iowa children aged from birth to 6.0 years. This was part of the nationwide survey made under the auspices of the Federal Children's Bureau, and coordinated by R. M. Woodberry (1912; and see Chapter 10, p.241). It seems that Iowa, goaded by Baldwin, responded better than any other state, and indeed 'the larger portion of her children were weighed and measured' (Baldwin, 1921, p. 58). Only 40,000 measurements were in the end used and Baldwin presented these, amongst much other material, in his classic monograph of 1921, the first volume of the long series of University of Iowa Studies in Child Welfare (see note 12.4).

In the same monograph Baldwin reanalysed the longitudinal data he had previously collected, mostly from the Horace Mann School, the Chicago schools, and the Friends' Schools in Baltimore and Washington, DC. He had records available of 400 individuals, many of them related, followed from 6 to 17, with measurements of sitting height, chest circumference and strength as well as of height and weight. A selection of these records, those of twenty boys and eight girls, were published as graphs in his book. Baldwin analysed the curves in relation to histories of disease and concluded, perhaps incautiously, that 'growth in height is affected by the formation and removal of adenoids' (p. 63), and, with more reason, that 'prolonged disease retards growth in stature'. He gave tables of the intercorrelations between the different measurements at each year of age, and between the same measurements taken at ages 6 and 12, and at ages 10 and 16. He set much store by this particular chapter, which broke new ground in growth studies, and had it personally vetted by Karl Pearson (who added a lot of coefficients of variation). One thing Baldwin's work scotched absolutely: 'Because boys and girls show a wider range of individual distribution and because boys and girls differ more in growth from each other at the adolescent period, it has been concluded that this is a period of irregular growth', he wrote (1921, p. 137); 'The consecutive intercorrelations of the various physical traits show this to be unfounded' (so, of course, did the simple curves themselves). Baldwin also gave a graph of the mean height velocities of groups of girls who had their menarches at ages 11, 12, 13 and 14, demonstrating the relation of menarche to age at peak velocity. This was an exact duplicate of Pagliani's work of 1879; but Baldwin seemed to be unaware of this.

In anybody else such a small ignorance of history would scarcely be surprising, but in Baldwin it amazes. The chief glory of Baldwin's book begins when the analysis of his own data ends. Under the title 'Historical orientation' he gives a magnificent review of previous studies of growth, classified into sections such as 'Theoretical discussions', 'Growth formulae' (Quetelet, Liharzik, Raudnitz, Zeising, Wiener, Robertson), 'National contributions on racial differences', 'General versus individual methods for studying weight and height', 'Climate and season', 'Social status', 'Physiological age'. There follow eighty pages of mean heights and weights of children according to innumerable investigators all over the world. Then comes an annotated bibliography of 911 papers, a source which must be the starting-point for any history of human growth. Without it, the present book would have taken twice as long to write. Indeed if the table of references in this present history serves as adequately for the next historian of growth as Baldwin's has for this one, the writer will be well content (see note 12.5).

Baldwin died of erysipelas in 1928, aged only 53. A complete bibliography of his papers will be found in Bradbury (1931) and Hossfeld (1931). The work he started at Iowa continued unabated. Between 1918 and 1928 some 1,300 radiographs had been taken of the hands and wrists of children from birth to age 17, in an endeavour to develop the original observations of J. W. Pryor, an anatomist at the University of Kentucky (Pryor, 1905, 1906, 1908, 1916, 1923, 1925), and Thomas Morgan Rotch (1849–1914), professor of paediatrics at Harvard 1903–14 (1909, 1910*a, b*; Rotch and Smith, 1910), into a satisfactory scheme for estimating physiological age (see note 12.6). The times of first appearance and of fusion of the epiphyses were studied and the area of the carpals was estimated by planimetry; it was thought this method would have advantages over the 'inspectional' (or atlas) method, which ultimately displaced it (Baldwin, Busby and Garside, 1928; see also Wallis, 1931). Idell Pyle, who later worked on the atlas method with Greulich at Stanford University, was at this time working at Iowa. As to the studies on vital capacity (mostly dating back to the Horace Mann School and the University of Chicago Schools) they were published by Kelly in 1933, and the interest was maintained, Metheny in 1940 publishing a new study of the relation between vital capacity and grip strength in pre-school children.

In 1923 the research on physical growth was made into a separate division of the Station and after Baldwin's death Charles McCloy and later Howard Meredith served, essentially, as directors. Charles McCloy (1886–1959) had been a young intructor in physical education at the Baltimore Athletic League when Baldwin was making his puberty study. He worked in China from 1913 till 1926, returned to take a PhD at Teachers College, Columbia University, and came to Iowa in 1930. He remained primarily a teacher of physical education and is chiefly remembered for introducing the measurement of skinfolds into modern auxology. Kotelmann (1879) had been the first to attempt to measure the amount of subcutaneous fat by applying calipers to a pinched-up fold of tissue (see p. 203) and Richer (1890) in France had followed his example. Oeder (1910, 1911) had revived the method for use in adults to give an index of nutrition, and Neumann (1912), Batkin (1915) and Hille (1923) had applied it again to children, giving age curves from birth to maturity. But the calipers all these used were crude adaptations of the usual instruments of physical anthropometry and it was not possible to obtain a uniform degree of pressure on the fold, a matter obviously of great importance.

In 1929 Raymond Franzen, who was working for the American Child Health Association, designed calipers which incorporated a spring and were more adapted to their particular task (there is a picture in Stolz and Stolz, 1951, p. 28). Though inaccurate, unwieldy and apt to fall to bits, these were used until the 1950s when the Harpenden skinfold caliper displaced them in Europe and the Lange caliper in North America. Whatever caliper is used, however, a valid measurement can only be made at sites where a fold of skin and subcutaneous tissue can be cleanly pulled away from the underlying muscle to form a measurable double fold. Franzen used three sites: on the front and back of the arm (biceps and triceps folds), and on the calf. McCloy found the calf fold to be often impossible to pick up and abandoned it, but he added some folds Batkin had used, at the front of the chest, at the back of the chest (subscapular), at the suprailiac site, and near the umbilicus. In later work the chest-front and umbilical folds were dropped and the four sites of biceps, triceps, subscapular and suprailiac became standard. These have continued to be used in many growth studies throughout the world; other studies have further reduced the number to two, triceps and subscapular, which are the most reliable ones. McCloy also introduced measurements of the widths of elbow and knee, used in adults by Matiegka (1921); these also were adopted in many subsequent growth studies. McCloy's studies (1936, 1938) were aimed at appraising physical status from the point of view of both nutrition (Franzen's concern) and the capacity for muscular activity. McCloy was one of the first to apply factor analysis to body measurements of children at different ages, emerging with factors of 'linear growth', 'cross-sectional growth' and 'fat growth'. He retired in 1954.

The great archive of measurements whose collection Baldwin had initiated was eventually worked up and published by Meredith (1935, for boys) and Boynton (1936, for girls) in what were PhD theses. These monographs had considerable value in that they gave mean cross-sectional curves for skinfolds and other measurements besides the usual height and weight. But the data, which were mixed longitudinal in nature, were treated cross-sectionally. Thus the greater part of the information so carefully collected was thrown away by inefficient statistical technique (see Tanner, 1951). This was not surprising; it took a long time for proper biometrical methods to be introduced into human auxology and indeed it is commonplace even now to see very inefficient analyses being made. All the same Iowa should have been a good place for the introduction of advanced methods: the State Agricultural College at Ames was the home of George Snedecor, the agricultural biometrician, and the place where R. A. Fisher, in the summer of 1931, gave his first course of instruction in the United States. However, as it was, no individual increments were calculated and no true velocity curves given, although the curves of a number of individuals followed over long periods were discussed. Neither author yet quoted Boas' 1930–3 papers.

Howard Meredith (1903–), an Englishman by birth, went to the United States as a young man; after college he took his postgraduate degrees at Iowa and passed practically the whole of his working life there. Neither paediatrician nor psychologist, Meredith was one of the small band of professional human auxologists. He was uncompromising in the accuracy of his anthropometry: the rule at Iowa was that measurements were taken independently by two anthropometrists, with agreement within very tight limits. Meredith's data had the deserved reputation of being the most accurate on the American continent: Tanner once described him as the Tycho Brahe of his subject (see note 12.7). He remained in charge of the physical growth section at Iowa till 1972, contributing a large number of papers to the literature. In the 1940s he set up a pure longitudinal study of some 150 children from age 4 to 18; many of his early papers refer to this (Meredith, 1941, 1944, 1947, 1955; Meredith and Carl, 1946). One of the few studies on the effects of minor childhood illness on growth came from Meredith and his chief associate Virginia Knott. The effects were, in fact, minimal: under the excellent nutritional situation of Iowa City any retardation was rapidly made good (Meredith and Knott, 1962; also Martens and Meredith, 1942).

Later, studies concerned with physical growth came to assume a minor role at the Iowa Research Station (in the Fortieth Anniversary Celebration Symposium all five papers were socio-psychological in content: Iowa Child Welfare Research Station, 1959) and Meredith concentrated increasingly on a study of facial growth he had initiated in 1946 in collaboration with the University of Iowa College of Dentistry. A series of children were examined six-monthly from age 5 to 12, and thereafter annually. Anthropometric measurements, lateral skull radiographs and dental casts were taken at each visit. Some sixty boys and sixty girls were followed to age 12, about half this number to 15, and a few till 17. The study ended in 1960. Meredith and his collaborators produced numerous papers analysing this superb material, and in 1973, on the occasion of his retirement, the Iowa Orthodontic Society republished in book form forty-four of the papers concerning head and face growth that he, Virginia Knott, and his other associates had contributed to the literature between 1949 and 1972 (Meredith and Knott, 1973). The material covers all aspects of face growth, both of single measurements (e.g. Meredith, 1959) and of shape changes (e.g. Meredith, 1960). Meredith stressed above all other considerations the great degree of individual difference shown in the shapes of the growth curves. Thus while most subjects had adolescent spurts in facial dimensions, some apparently did not (e.g. Meredith and Chadha, 1962). The Orthodontic Society book is a major landmark in the study of the growth of the face and head in relation to individual differences and the emerging dentition. Besides this, Meredith produced a series of summaries of the ever-increasing data on growth, continuing this activity well into his retirement (Meredith, 1963, 1970, 1971, 1973, 1978).

The Harvard Growth Study

The second major longitudinal study in North America was set up by W. F. Dearborn. Already accustomed to training his students to take measurements of children simply as a necessary part of their training, in 1922 he initiated a formal, and massive, longitudinal study of schoolchildren (Dearborn and Rothney, 1941). All children entering the public schools of Medford, a town in the Boston Metropolitan area, were measured in their first term, and thereafter at annual intervals as long as they stayed in school. In 1923 children in Revere and Beverly, two other nearby towns, were added, so that in all about 3,600 children were recruited, mostly aged 5 or 6. This is a very large number with which to start a longitudinal study (as opposed to a follow-up study where the children are only seen at intervals of many years). Inevitably there were losses as time went by, but at the end of twelve years, when the study closed, nearly 1,000 children were still being measured. No other longitudinal study attained such numbers until the 1970s. Furthermore, the Harvard Growth Study is exceptional in that all the raw measurement data were eventually published (Dearborn, Rothney and Shuttleworth, 1938), something especially desirable in longitudinal studies since the data take so much time and effort to acquire. The measurements taken were height, ear height, sternal height, sitting height, bi-iliac diameter, chest breadth and depth, head length and breadth, and weight. The number of erupted teeth was recorded, a hand and wrist radiograph taken each year and, in one town only, the date of menarche was recorded. Tests of intelligence and school attainment were also made. Dearborn and Rothney (1941) wrote an excellent general account of the study, and their book includes a telling description of the problems that beset longitudinal work: problems of subjects leaving the town, measuring personnel leaving the study and so forth. Their own study was lavishly organized, with a staff of twenty-one to twenty-six members, 'each with his own special tasks to perform'. Each physical measurement was taken independently by three measurers, each with his own recorder: all had to agree within 1.1 cm for body measurements and 0.5 cm for head measurements (not in fact a very high standard to set). Rules were given for what to do in case of disagreement. As to the decline in numbers of subjects, by age 10 one quarter of the original population was lost, by age 15 one half and by age 17 three quarters. Long-term financial support is always a major problem in longitudinal studies, and the Harvard Growth Study was no exception. From about 1928 onwards the General Education Board of the Rockefeller Foundation took a major interest in this study, as in so many others. A portion of this vast material was briefly analysed by Edwin B. Wilson (1935, 1942), the first professor of biostatistics at the Harvard School of Public Health (whither he had been called in 1922 from a chair in physics at MIT because of a lack of mathematical statisticians with any experience of experimentation: see Box, 1978, p. 312). The Harvard Growth Study, however, is mainly noteworthy for providing the material for the masterly analysis of growth at adolescence by F. K. Shuttleworth.

Frank Shuttleworth (1899–1958) was a psychologist who was born and brought up in Iowa. He obtained his PhD in 1925, in social psychology, at the University of Iowa, where he was a contemporary of Nancy Bayley (see below), though doing work sufficiently different for them to know each other only slightly. He worked in the department of education and the Institute of Human Relations at Yale University from 1928 to 1939; later he was engaged in the counselling of students at the College of the City of New York. Shuttleworth, like Boas, was impressed and dismayed by the lack of understanding amongst psychologists and others of the difference between longitudinal and cross-sectional studies. 'The initial impetus for this study', he wrote, 'was supplied by the many reports which, though based on longitudinal data, have yielded only cross-sectional findings' (1937, p. 180). 'The implications of the so-called 'adolescent spurt' in growth for physical care, diet and mental hygiene have been elaborated in many text-books but upon an insecure foundation of fact derived from cross sectional studies. The suggestion of greater dietary needs during this period is far more forceful than cross-sectional studies would indicate' (1937, p. 189). Already in 1934, the year the Harvard Growth Study ended, Shuttleworth wrote a letter to *Child Develop-*

ment proposing that standards of growth should be in terms of increments, or, as he put it, 'in terms of progress rather than status' (1934, p. 89). Increments showed better the effect of immediate environmental alterations than status could possibly do. Shuttleworth was already some thirty years ahead of his time, and preaching to a quite unconvertible audience. Furthermore, he elaborated in this letter what very recently have come to be called 'conditional standards'. He gives standards for increments of weight during a given year according to the weight at the beginning of that year, and discusses at what ages such standards have an advantage over straight increment standards. This part of his letter was absolutely ignored; conditional standards of this sort, proposed anew by Healy and Goldstein, were not introduced until the late 1970s (see Healy, 1974, Goldstein, 1979; Cameron, 1979).

To clarify the issues he had raised, Shuttleworth set himself to analyse the best longitudinal material available, that of the Harvard Growth Study. A series of classic monographs resulted, a landmark in auxology comparable to the 1930s papers of Franz Boas (see note 12.8).

Shuttleworth began at the beginning. 'The importance of repeated measurements on the same children', he wrote, 'has been urged for so long by so many as to become the first article of faith among students of Child Development. Nevertheless, the multitude of longitudinal studies which have yielded only cross-sectional findings suggest that this faith has been degenerating to the level of pure dogma . . . in longitudinal studies the emphasis should fall on the growth increments rather than on gross dimensions . . . thousands of dollars have been spent on the collection of longitudinal data, hundreds of dollars have been spent on cross-sectional problems, and only dollars have been devoted to the developmental aspects of the data' (Shuttleworth, 1937, p. 9) – a criticism that Tanner (1951), examining in depth the problems of mixed longitudinal data fourteen years later, felt constrained to quote as still entirely applicable. In his first monograph Shuttleworth (1937) took the 315 records of girls with known age at menarche and examined the average growth curves of the groups of individuals having their menarches at 11.5 to 12.0, 12.0 to 12.5, and so on, just as Boas had done. Subsequently he took 'age at maximum growth' (i.e. at maximal growth *rate*, or, as it is often nowadays called, 'peak height velocity') as a measure of an individual's tempo (Shuttleworth, 1939; and see Fig. 10.1–10.5). Putting the growth curves all in terms of maximum growth age (MGA) he showed that each dimension had a characteristic curve, with its peak at a defined position in respect to the other dimensions. Thus the spurt in leg length preceded the spurt in sitting height and was flatter. Those children who spurted early had greater peaks than those who spurted late. On average, children who were relatively tall at age 8 were nearer their spurt than those who were still small; but this did not mean they would end up taller as adults. The correlation of age at menarche with height at age 8 was −0.28, with height at age 12 or 13 −0.54 and height at age 17 or 18 only 0.01 (1937, p. 184). All the analysis was done very simply, and using individual increments, never measurements attained. Yet few of the elaborate and often beautiful curve-fitting procedures done by later generations of auxologists have produced generalizations Shuttleworth failed to make, or proved any of Shuttleworth's generalizations wrong.

Shuttleworth was much interested in standards of growth that took account of tempo and represented the individual reality rather than the statistical artefact. Though he never cast his data in a mould suitable for engendering practical standards, Fig. 12.1, taken from the 1937 monograph, shows how readily he could have done so had the time been ripe for their acceptance. The monograph series was rounded out by an atlas of graphs and pictures illustrating the events of adolescence and stressing, as usual,

Fig. 12.1. Average annual increments in standing height, probable errors of means, and standard deviations of distributions of increments for early and late maturing groups of cases. From Shuttleworth (1937, figure 28); early and late refer to age-at-menarche groups, the first with menarche before age 11.5, the second after age 14.5).

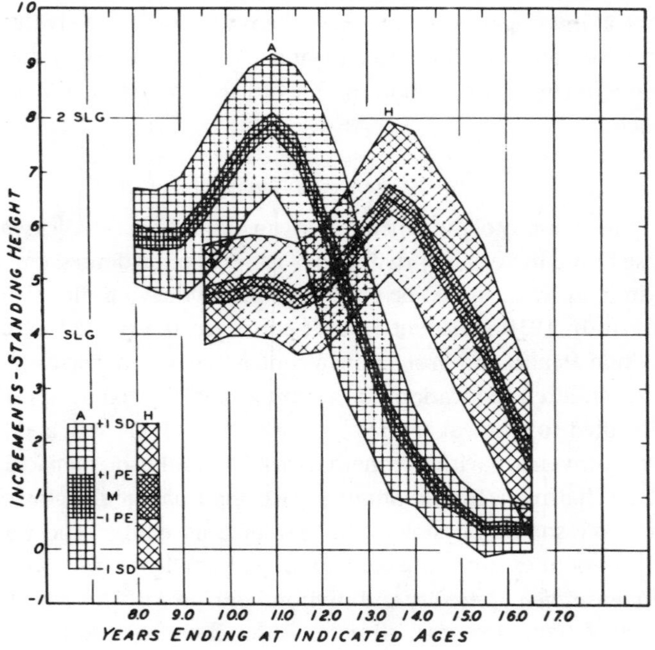

the great difference in tempo between individuals, even of the same sex. Ten years later an extended revision was issued (Shuttleworth, 1938*b*, 1949*a*, *b*): by this time Shuttleworth's work had moulded a whole new generation of auxologists and begun to exert its effect on what was much later to be the practice of adolescent medicine and psychology.

The Harvard School of Public Health longitudinal studies of child health and development

A few years before the Harvard Growth Study ended, its children mature, another graduate school of Harvard University took over, as it were, the baton. The Harvard School of Public Health ran, and runs, postgraduate courses for doctors, nutritionists, nurses and social workers from all over the world. Harold Stuart, the chief of the department of child hygiene (later the department of maternal and child health), saw the creation of conditions for healthy growth of children as a central objective in any programme of public health. Students at the School, then as now, undertook a substantial course in growth and development as a regular part of their training. It was by no means certain, however, what manner of growing and developing should be considered healthy. As we have seen, few children had been followed, even in the anthropometric sense, from birth to maturity; and fewer still had had any physiological investigations made upon them. It was to remedy this that the Center for Research in Child Health and Growth was set up in 1929. Its background includes an important historical occasion, the White House Conference.

The White House Conference

The founding of many of the North American longitudinal studies was in part a response to the increasing privations of the economic depression of 1929–33. Children, at least, should be protected. One endeavour directly reflected this concern. In 1930 President Hoover convened the White House Conference on Child Health and Protection, a conference which discussed medical services to children; education and training; and the handicapped. In the section devoted to medical services, a Committee on Growth and Development was convened, with Kenneth Blackfan, professor of paediatrics at Harvard, as chairman, Harold Stuart as vice-chairman, and Halowell Davis, assistant professor of physiology at Harvard, as editor. The Committee drafted four volumes surveying contemporary knowledge: they were headed *General Considerations* (heredity and environment, prematurity, constitutional types, sleep, fatigue, exercise, climate, seasonal changes, atmospheric pollution, socio-economic factors, immunity, effect of various diseases), *Anatomy and Physiology* (skeleton, muscles, face and dentition, central nervous system, special senses, cardiovascular system, lymphatic and haemopoietic systems, gastrointestinal tract, respiratory system, genito-urinary system and endocrine glands), *Nutrition* (protein requirements, vitamins, iron, energy balance, diet, cooking), and *Appraisal of the Child, Mental and Physical* (intelligence, motor skills, language, personality, skinfolds, muscle, radiography, vital capacity, exercise) (White House Conference, 1932–3). No authors' names are attached to particular sections, as the members of the Committee read and altered (it is said) each other's drafts. The Committee varied somewhat from volume to volume; its membership included Franz Boas, Edith Boyd, Holly Broadbent, Walter Cannon, Charles Davenport, Walter Dearborn, Raymond Franzen, Arnold Gesell, Charles McCloy, Horace Gray, Richard Scammon, Wingate Todd and Edwin Wilson. Blackfan's introduction stressed that they proposed no new standards because 'we simply do not possess the information necessary to state the range of variations which should be regarded as normal in each case' (vol. 1, p. 6). After discussing the contributions of clinical studies and animal experimentation, he went on: 'Another line of approach to problems of growth and development has been recognized only recently for its true worth. That is the type of study which follows an individual over a long period of time . . . certain observations which have recently been completed indicate that the acceleration of growth at puberty is usually much more sudden than the customary group methods had led us to believe. The suddenness of the spurt of growth was masked by the fact that it does not occur at the same age in different individuals, so that the composite curve for the whole group shows only a gradual rise through a period of years' (vol 1, p. 9). Blackfan and his Committee – surely the group of people with the greatest combined knowledge of human auxology ever assembled up to that time, and perhaps, allowing for scientific advance, up to the time of writing – called for more studies and, particularly, where feasible, longitudinal ones.

Harold Stuart (1891–1976) was a paediatrician by training and by nature the gentlest and most courteous of men. He was born in upper New York State, and graduated in medicine at Columbia College of Physicians and Surgeons, where he was drawn to paediatrics by the example of L. Emmett Holt Sr, author of the first American textbook of paediatrics and one of the first American measurers of infants and pre-school children. Stuart went to Harvard in 1921, and from the beginning concerned himself with out-patient care and the long-term follow-up of patients rather than with the care of children with acute illnesses. This preoccupation soon led him to feel the need of a greater knowledge of normal development than was currently available, and in 1928, when he was appointed full-time in the School of Public Health, he was able to start research in this field as well as developing community paediatrics. (Part of this development consisted of a book discussing pregnancy and children's growth, nutrition and diseases, and accidents, that was aimed at parents and grew directly out of the White House Conference (Stuart, 1933.) In many respects Harold Stuart played the same part in North American paediatrics that Alan Moncrieff, a decade later, played in paediatrics in Europe (see Chapter 13). Stuart retired from the School of Public Health in 1954; but in 1965 the children of his longitudinal study, then 30 years old, were recalled for a follow-up. Out of the 134 followed to maturity, by then scattered all over America and beyond, 126 returned. It was a tribute to their guide, mentor and friend.

The Center for Research in Child Health and Growth was opened in 1929, and between 1930 and 1939 Stuart recruited 309 children into a birth-to-maturity normal-child growth study. Mothers registered for delivery at the Boston Lying-In Hospital were contacted during pregnancy, since the effect of the mother's diet on the size of the child was one of Stuart's particular interests. Most of the families were of Irish extraction and belonged to the lower middle class. Some had more than one child in the study. The work of the Center was described in some detail by Stuart in 1939 and again, with hindsight, in 1959, as part of a monograph supplement to *Pediatrics* summarizing the main results of the study (Stuart and Reed, 1959). The earlier description included an illuminating study of six individual children in two families; the later one gave individual height curves from age 1 to 18 of 97 children (Reed and Stuart, 1959). Besides this main study there were subsidiary ones, more intensive but shorter-term, of fetal growth and prematurity, and of school-age children. Out of the original 309 boys and girls of the normal-child group (which, however, included some sets of twins and six children born at low birthweight) 228 continued to age 6 and 134 (67 of each sex) to age 17 or, in most cases, 18 and over. The name 'Center' was dropped following a change of building and the study finally became known as the 'Longitudinal Studies of Child Health and Development of the Harvard School of Public Health'. Much of the finance of the study in the crucial years 1933–42 came from the General Education Board of the Rockefeller Foundation, channelled thence by Lawrence Frank.

The children were seen at birth, 14 days, 3, 6, 9 and 12 months, every six months to 10 years and thereafter annually till 18. Measurements were made of stature (or supine length to age 2), sitting height, shoulder and hip widths, chest breadth, depth and circumference, four head dimensions, and abdominal circumference and weight. Special attention was given to an orthopaedic examination at which radiographs were taken giving lengths of bones in the arm and leg, and widths of chest and heart shadow, in addition to the usual hand and wrist films. The teeth were examined and dental casts made, photographs were taken, a dietary history recorded (see Burke *et al.*, 1959) and a very full account of illness experience obtained from the mother. Finally at some ages there were psychometric tests and observations of behaviour (Slater, 1939).

The illness-experience study was written up by Isabelle Valadian (1927–), Stuart's first assistant and successor, Stuart himself, and Robert Reed (1917–), E. B. Wilson's successor-at-one-remove to the chair of biostatistics (Valadian, Stuart and Reed, 1959, 1961*a, b*). One clear finding emerged: in Boston in the 1930s there was much more illness in the age period 2 to 10 than during adolescence (Valadian *et al.*, 1959). Illness diminished in early adolescence and became still less in late adolescence. Thus in the dispute amongst school doctors of the 1880s Hertel and Key were right (see p. 204). Of course the Boston children were growing up under very different conditions, with sulphur drugs though not yet antibiotics widely available. As to the relation between disease experience and growth, little was demonstrated.

The data on dental development were worked up by Conrad Moorrees (1916–), and his associates at the Forsyth Dental Center (Moorrees, 1959). With Elizabeth Fanning and Edward Hunt, Moorrees described a series of stages for the development of roots, crowns and apices of the teeth as visualized in radiographs, akin to the bone stages of Todd (see below). Standards for dental development, or dental age, were given and the growth of tooth and jaw dimensions plotted on this basis, rather than against chronological age (Moorrees and Reed, 1965). Moorrees analysed material from the Fels Research Institute (see below) as well as the Harvard School, and found that the Fels children were some six months ahead of the Harvard children in development of the posterior mandibular teeth (Moorrees, Fanning and Hunt, 1963). Later Moorrees initiated his own study of dental development, using twins (see Moorrees and Kent, 1978).

Nearly all the North American growth studies took numerous radiographs, and at the Harvard School of Public Health there were X-rays of calf, thigh, foot, forearm, chest and, till age 6, pelvis. No research worker would nowadays take so many radiographs because of the hazard of leukaemia associated with an increase of X-ray dosage above the natural background. But in those days this effect was much less recognized. At present the pendulum of opinion has swung so far in the opposite direction that in the United Kingdom, for example, it is scarcely possible to take even a single hand radiograph except on a hospital patient for diagnostic purposes. Such a radiograph requires an additional 4 millirads dose over background, the same amount that is received during a ten-day holiday in the mountains or in half a dozen aircraft flights (as the amount of natural radiation increases with altitude). The doses used in the studies of the 1930s and 1940s are not accurately known, but must have been of the order of a hundred times higher.

The Harvard radiographs not only gave lengths of femur, tibia and foot, much needed for use in orthopaedic work (Green, Wyatt and Anderson, 1946; Anderson and Green, 1948; Anderson, Blais and Green, 1956). (Anderson, Hwang and Green (1965) later applied the same techniques in a longitudinal study of the growth of children with limbs made asymmetric by poliomyelitis.) The exposures of the calf were made in such a way as to reveal bone, muscle and fat separately. Study of the widths or areas of the separate tissues confirmed the age curves obtained by McCloy for fat using the skinfold calipers, and showed that bone and muscle, not unexpectedly, followed a similar curve to stature (Stuart, Hill and Shaw, 1940; Stuart and Dwinell, 1942; Stuart and Sobel, 1946; Lombard, 1950). The hand radiographs were later analysed by Idell Pyle. They were the first series in which the bone ages of individuals followed the whole way from birth to maturity were presented. They showed that some individuals had velocities consistently a little below the average over the whole period, others consistenlty above. Many reversals from high to low velocity or vice versa occurred between the pre-adolescent and adolescent years, but almost none within those periods. Evidently the factors which controlled bone age advance and created differences between individuals were different before puberty and during it (Pyle, Reed and Stuart, 1959; Pyle *et al.*, 1961).

Though this Harvard study came formally to an end in 1954, many of its participants, like those in other longitudinal studies, seemed loath to forego their annual visits to the research workers they had come to look on as friends and advisers. At the time of writing Valadian continues to study many of the original group, during what it has become fashionable to refer to as 'the middle years'.

The Child Research Council, University of Colorado

Another longitudinal study that was medically orientated was the Child Research Council, situated in the University of Colorado Medical School in Denver. This had its beginnings in a longitudinal programme of research into the detection of disease in children, which had been in existence since 1923. In 1927 it was incorporated as the Council, and in 1930 took on its definitive shape when Alfred Washburn arrived from San Francisco to be its director. A. H. Washburn (1895–1972) was a paediatrician who graduated at Harvard in 1921, practised briefly in Portland, Oregon, and from 1926 to 1930 taught at the University of California Medical School. He was a strong protagonist of the whole-child approach, and stressed on several occasions that 'we are not using human subjects . . . to contribute new knowledge to a specific field, but, instead, all fields of science are being utilised to contribute to our awareness of the unique individuality of each child' (Washburn, 1950, p. 765). Thus the Denver study was specifically orientated towards preventive paediatrics (for its general philosophy see Washburn, 1957). Washburn regarded medicine as a branch of human biology (1955, p. 5: cf. Chapter 13, p. 370) and paediatrics as 'essentially a subdivision of the science of Human Growth' (1955, p. 2). But learning more about the uniqueness of each of the small number of children studied did not readily contribute to knowledge of the science of growth, though it was intensely educative to Washburn's collaborators and students (see Washburn, 1953, for his educational practice). In fact, a lot of information about specific fields did come out of the study; but much of it seemed to carry an almost apologetic air, and with a few notable exceptions modern techniques of data analysis were conspicuous by their absence. Washburn retired in 1960 and was succeeded by Robert McCammon, his deputy for many years. The study finally closed in 1971.

The Child Research Council study is described in a monograph by McCammon (1970) with chapters by his associates. There is a brief history, some results, though only on a cross-sectional basis, and a valuable complete bibliography of Council papers up till 1969. Apart from 78 children inherited from pre-Council days, the children were all recruited before birth, just as at the Harvard School of Public Health. Recruitment was greatest from 1930 to 1939, after which year it was mainly restricted to younger siblings of existing subjects. When recruitment ceased at the end of 1966, there had been 334 children enrolled, coming from 215 families. Two hundred and fifty-six were enrolled before birth and of these 106 (54 boys and 52 girls) remained in the study till they reached maturity. Thus the final number of birth-to-maturity records was slightly less than in the Harvard School of Public Health Study. The social standing of the parents was very different however. While Stuart recruited mothers booked for delivery at a public hospital, Washburn accepted only those attended by physicians privately. Most of the fathers had professional or managerial jobs; in this respect the Denver sample was not unlike that of Iowa.

The examinations were more frequent than in any of the other longitudinal studies: every month till 1 year, then every three months till the end of the adolescent growth spurt, then every year into adulthood. The times when various examinations were introduced into the programme varied, however, so that at no age were as many as 100 boys or 100 girls represented in the 1970 data, except in haematological variables, Washburn's particular interest.

Though the usual (and some less usual) anthropometric measurements were taken – from 1946 to 1957 by Edith Boyd, Scammon's erstwhile assistant – the Child Research Council is perhaps most remarkable for the physiological work done on the children. Blood pressure, electrocardiograms, exercise tolerance, basal metabolic rate, blood protein concentration and the haematological picture were followed for varying periods of time. Reports are in Mugrage and Andresen (1936, 1938), Lewis, Duval and Iliff (1943*a,b, c, d*), Iliff and Lee (1952), Trevorrow *et al.* (1942) and Trevorrow (1957, 1967).

Radiographs were taken in this study, too, in very much the same way as at the Harvard School of Public Health. Lengths of limb bones and widths of bone, muscle and fat in the extremities were measured every six months beginning in 1935, and hand and wrist radiographs were added in 1947 (Maresh, 1943, 1955, 1966; Hansman and Maresh, 1961; Hansman, 1962). Width of the heart was studied (Maresh, 1948), development of the paranasal sinuses followed (Maresh, 1940), and growth of the skull monitored by cephalometric techniques (Nanda, 1955; Bambha, 1961; Bambha and Van-Natta, 1963; Hunter, 1966. Hansman, 1966). Nutrition was investigated (Beal, 1961) and some psychological work was also done; in this study, however, there was rather less emphasis on psychological development than in most of the others.

Little of the work of the Child Research Council appeared in a form which capitalized on its longitudinal nature and the old gibe against longitudinal studies – that they are squirrels burying nuts against the future (or poets waiting for ever for the perfect word) – has undeniably some application. But in at least one field the Child Research Council produced a pioneering paper, using precisely the longitudinality of its data. This was a study of curve-fitting to individual height measurements by Jean Deming (1895–1962), a physician interested in biometry. Deming (1957) fitted Gompertz curves to height measurements of each of twenty-four boys and twenty-four girls over the period of the adolescent growth spurt. She was emphatic about the 'great superiority of longitudinal series of measurements of individuals in contrast to the merged measurements of a group of individuals . . . the great variation in timing of the adolescent cycle . . . and the intensity of growth during the cycle, makes the pattern of the averaged measurements misleading and almost meaningless' (p. 86). The Gompertz curve, originally used to quantify population growth, is a skew logistic (see Fig. 12.2) with the downward leg descending more slowly than the upward leg ascends. Both this curve, and the symmetrical logistic, had been used to fit data on weight growth in animals, in particular by Merrell (1931: see discussion below) and Courtis (1932). But the possibility of quantifying human data in this or a similar way had yet to be realized, and Deming's paper began a whole series of studies, establishing curve-fitting as a method of choice in the analysis of nearly all longitudinal data. In 1957 the method of fit was inefficient, of course, and even so the time it took to fit each individual curve, by hand calculator, was daunting. The results mostly bore out what Shuttleworth and before him Boas had deduced from simpler studies, but they added a precision unknown before. They gave, for example, the correlation coefficients between instanta-

Fig. 12.2. Jean Deming's illustration of the fit of a Gompertz curve to serial height measurements of a single girl. Above, height attained ('distance'); below, rate of growth ('velocity'). Crosses mark the points of inflection of the curve. S.S. represents first breast growth (B2); C., fusion of capitellum of humerus; M., menarche. Reproduced from Deming (1957, p. 89).

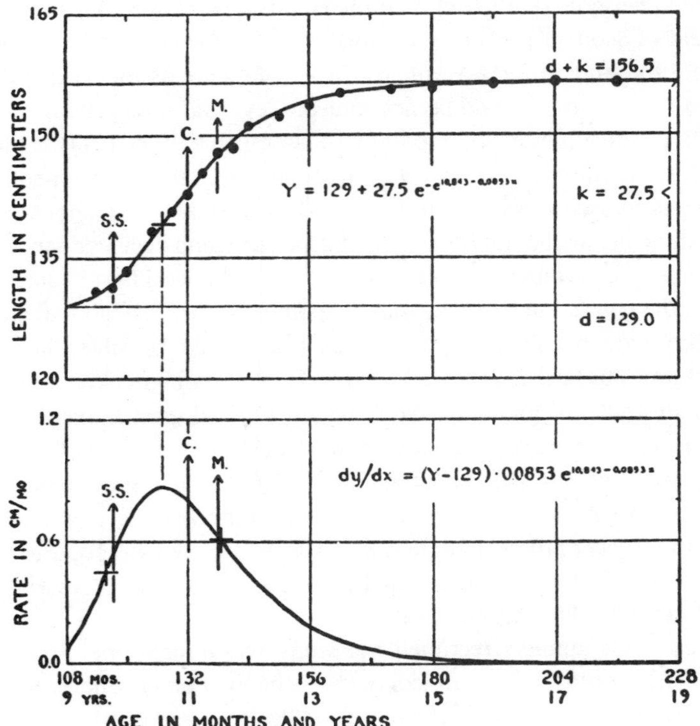

neous peak height velocity and the age at which it occurred, between age at instantaneous peak height velocity and amount of final mature height, between total gain during adolescence and final mature height (0.39 for girls, 0.54 for boys, but a *point de faiblesse* here, for all the logistic curves are uncertain at 'take-off'). In a paper published posthumously (Deming and Washburn, 1963) Deming fitted another curve, that described by Jenss and Bayley (1937: see below), to the length curves of forty boys and forty girls from birth to age 8. This curve is an exponential, but with an added term proportional to age, so that the asymptote is a straight line rising with age, rather than a constant. Again the fit was good, and the differences between individuals were clear-cut.

The Brush Foundation, Western Reserve University

A third medically orientated longitudinal study (before we get back to the educationally orientated ones) started in 1930 in Western Reserve University, Cleveland, under the direction of T. Wingate Todd (1885–1938: for obituaries and bibliography see Krogman, 1939, 1951; Cobb, 1959). Todd, a man of demonic energy, who died of a coronary thrombosis aged 53, was born in Sheffield of Scottish parents, graduated in medicine at Manchester University in 1908 and in 1912 went to Western Reserve as their first full-time professor of anatomy. Todd became one of Cleveland's outstanding citizens. He built up the museum of comparative anatomy that the dean of the Medical School, C. A. Hamann (1868–1930) had founded, until it had a collection scarcely surpassed anywhere in the world. He assembled an almost equally astonishing departmental library. In 1928 a Cleveland businessman, Charles Francis Brush, established a Foundation 'to finance efforts contributing toward the regulation of the increase of population, to the end that children shall be begotten only under conditions which make possible a heritage of mental and physical health, and a favorable environment'. Todd was made chairman and director of the Foundation. He had been a member of the White House Conference and was an editor of *Child Development*, and in 1930 he began a longitudinal study of the physical and mental growth of children who could be considered representatives of just that optimal population that it was the object of the Foundation to foster. The study was supported by the Brush Foundation, and, inevitably, by the Laura Spelman Memorial Fund and the General Education Board.

In terms of numbers recruited the study was a large one. At first only infants of 3 months were enrolled, but subsequently older children were too, especially the older siblings of the infants. Recruitment continued until the study ended in 1942. Thus the data were mixed longitudinal in nature; in all some 515 boys and 484 girls participated, though many for a period of only a few years, and none from infancy to maturity. The selection of families participating was very similar to that in the Child Research Council in Denver: well-off, upper middle class, the majority in professional or managerial jobs. The children were seen at 3, 6, 9 and 12 months and thereafter six-monthly till age 5 and annually from 5 onwards. The investigations were restricted compared with those of Boston or Denver: anthropometry, radiology and psychometric testing only. However the anthropometry was thorough; twenty-five measurements were taken, using a more old-fashioned technique than in other studies, derived from that of the classical physical anthropologists of the nineteenth century.

Many of the anthropometric data were summarized in a well-known monograph by Katherine Simmons (1944; see also Simmons and Todd, 1938). These were still early days in data analysis, and no effort was made to use the mixed longitudinal nature of the data to derive more precise estimates of yearly means, or to estimate velocities and their sampling errors. For the most part the data were simply treated as though cross-sectional. Year-to-year correlations of the same measurement were given, however, and showed how much lower the birth-to-1-year correlations for length and weight were than the later (e.g. age-1-to-age-2) correlations. Simmons and Todd also showed that between birth length and gain in length during the first year there was a negative relation. Twelve years were to pass before the significance of this finding (that there are maternal restraining factors limiting fetal growth, compensated by catch-up growth during the first two years after birth) was fully realized. Simmons calculated the correlation between the results of IQ tests (mostly Stanford–Binet) and stature at each age, finding no relation in the boys, but a low positive correlation in girls from age 3 to 9 and again at 15. William Greulich (1901–), who came from Yale to succeed Todd as director of the Brush Foundation in 1940, had been subjected to the influence of Shuttleworth, and he and Simmons did use the longitudinal element of their data to show the relations between menarcheal age and the adolescent height spurt, precisely as Shuttleworth had done (Simmons and Greulich, 1943).

Todd himself had an over-riding interest in the development of the skeleton and he developed Rotch's system for grading physiological age according to the degree of development of the bones seen in radiographs of the hand and wrist. He saw skeletal maturation as 'the third measure of physical health ... less subject to fluctuations in progress' than the other two of height and weight. 'In robust, healthy children', he wrote, maturation 'is a time-keeper of great reliability . . . the analysis of psychological responses or behaviour patterns is greatly aided by a preliminary assessment of the physical development progress, weight being an indicator of nutrition, stature of health and maturation of constitution' (Todd, 1937, p. 13). Having access to longitudinal series of radiographs on the same individuals, Todd was able to see much finer changes than Rotch. His *Atlas of Skeletal Maturation* came out in 1937, though a first draft was circulated ten years earlier (Simmons, 1944, p. 49; and see Todd, 1930). Having first selected from amongst the radiographs of 7-year-olds, for example, those in which all the bones showed more or less equal degrees of development, he arranged these in order of ascending degrees of maturity and then took the central one to represent a sort of ideal picture of a child aged 7 with what he called an 'evenly matured skeleton'. In Todd's *Atlas* illustrations of this sort were provided for each year of age (or for each six months at certain ages). Girls and boys had separate series of pictures since girls were always ahead of boys. The rater of a radiograph of a child seen in the clinic would match it with one of the standards and thus assign a 'bone age'. If the child's bone age were 7 years and his chronological age 9, then he was two years behind, a late developer. (The exact implications of such a delay at pre-pubertal ages for the future development of the child took a long time to work out and are still not all beyond dispute.) A major problem with Todd's *Atlas* was that it derived from a group of upper middle class children living in very comfortable circumstances. Thus few other children achieved the same degree of advancement of skeletal maturity as the Brush Foundation girls and boys for many decades, and even now most European children average some six months behind. This tends to create confusion (though of course it need not) in a way that is avoided by the later system of assigning a simple maturity score rather than an 'age'.

Todd died before a revision of his *Atlas* was necessary, but one was brought out by Greulich, who in 1944 had moved from Cleveland to be professor of anatomy at Stanford University, in association with Idell Pyle, who had been with the Brush Foundation since 1940 (Greulich and Pyle, 1950). The Greulich and Pyle *Atlas* was more complicated than Todd's in that besides the plates of the whole hand and wrist the successive stages of each individual bone were illustrated, giving the possibility of assigning not only an overall bone age (usually the modal value of all the bones represented) but a measure of the variability of bone age amongst the bones (each of which was assigned an age). This was named the 'red graph method' (Pyle *et al.*, 1948), but has seldom been used in practice. The Greulich–Pyle *Atlas* is still much used clinically, a lasting tribute to Todd's foresight.

Radiographs were also taken of the foot, knee, hip, elbow and shoulder. Standards similar in form to the hand atlas were issued for the knee by Pyle and Normand Hoerr, Todd's successor in the chair of anatomy (Pyle and Hoerr, 1955, 1969); and for the foot and ankle by Hoerr, Pyle and Francis (1962). The last of these authors, C. C. Francis, was associate professor of anatomy and did much of the measuring of the children; he published papers on various aspects of skeletal development and especially on growth of the pituitary fossa (Francis, 1948).

Todd was also at the root of the greatest single advance in the science and practice of orthodontics. B. Holly Broadbent (1894–1977), a young orthodontist, worked as research fellow in Todd's department, and together they adapted the craniostat that Todd was accustomed to use for holding skulls in a standard position to hold, instead, the head of a living subject while precise lateral- and frontal-view radiographs were taken. The first Broadbent Roentgenographic Cephalometer was made in the department of anatomy's machine shop in 1926 (Broadbent, 1931). Broadbent became professor of dentofacial morphology at Western Reserve, and in 1929 set up, still in Todd's department, a longitudinal study of cranio-facial growth, known as the Bolton Study after the family who donated the money which supported it (Broadbent, 1937).

The Bolton Study, made exclusively on children with normal dentition and occlusion, recruited subjects for no less than thirty years. The last new subjects were admitted in 1959, and a little later the study took over the accumulated material of the Brush Foundation Study to become the Bolton–Brush Growth Study Center, located in the School of Dentistry of the now Case Western Reserve University (Broadbent, Broadbent and Golden, 1975). The age range covered by the Bolton Study was effectively 1 to 18 years, in a mixed longitudinal design which permitted the extraction of three pure longitudinal cohorts aged 1 or 2 to 7, 7 to 12 and 12 to 18. An average of fifteen lateral skull radiographs were taken per child (Brodie, 1941, 1953; Lande, 1952). Cross-sectionally considered, between 600 and 2,000 individuals were examined at each year of age, the average being 1,170. Normative values were extracted from this vast archive and published in 1975 by Broadbent, Broadbent and Golden, in a monograph which also contains a brief history of the study (pp. 13–24) and includes portraits of the chief persons concerned. Broadbent radiographs are still a prime instrument in diagnosing and treating disorders of facial growth, just as is Todd's Skeletal Maturity Atlas in diagnosing and treating children with more general growth disorders.

University of California Institute of Child Welfare

It was largely through the agency of Lawrence Frank, yet again, that the University of California established an Institute of Child Welfare. This was in 1926, at a time when the Laura Spelman Memorial Fund was giving large sums to several universities to further the general objective of research on child development. Herbert Stolz (1886–1971) was the Institute's first director.

Stolz spent his early childhood years on his father's cattle ranch in Hawaii, but his father died young and Stolz went to school in New York and California, his mother meanwhile qualifying as a doctor and then entering practice. Stolz studied physiology and medicine at Stanford University, spent 1911–13 as a Rhodes scholar at Oxford, and graduated in 1914.

He was a very 'physical' man. As an undergraduate he sailed as crew in the writer Jack London's ketch *Snark* on the Hawaiian leg of her round-the-world voyage, and he practised all sorts of athletics with conspicuous success. He was a triple blue at Oxford, and a competitor in the long (broad) jump at the Stockholm Olympic Games of 1912. He never practised medicine in the ordinary sense. Instead of entering the hospital as an intern after graduation, he became director of athletics for men at Stanford University, and later, on return from war service, he was appointed first assistant and subsequently director of physical education for the state of California (his predecessor, Clark Hetherington, who appointed him, was an old pupil of Stanley Hall, it almost goes without saying). This position involved health supervision and instruction as well as physical education in the strict sense, and this was the area in which Stolz, in one way or another, spent his whole professional life. He was closely associated with Lawrence Frank in the planning of the Institute of Child Welfare during the period 1924–6. Stolz was not a man to ignore the physical demands and possibilities of the growing body, nor to underestimate its influence on self-perception and on social behaviour. Thus as director of the Institute he laid particular emphasis on the importance of physical development, himself measuring subjects and initiating the theme of the interaction of physical advancement and tempo of growth, particularly at adolescence, with social bearing and personality development that continued to characterise the Berkeley Studies (see Stolz and Stolz, 1944). In 1936, when the initial funding of the Institute lapsed, Stolz returned to the California school system, first as superintendent of guidance to the children of Oakland and later as officer in charge of all the schools for the handicapped in California. He retired in 1956.

Stolz appointed Harold Jones (1894–1960) as director of research, and Jones succeeded Stolz as director of the Institute in general when Stolz returned to school work. Harold Jones was trained as a biologist but took his PhD at Columbia in experimental psychology, with Woodworth. His interest in developmental psychology owed much to his wife, Mary Cover Jones, who was studying infant behaviour when they met, and who served with him on the Institute faculty. But that interest needed little stimulating; Jones' father ran the experimental farm at Massachusetts Agricultural College; and Jones was a man who planted trees, and founded societies, and waited. Frank and Stolz believed firmly in the importance of long-term longitudinal studies and Jones was just the man to implement that belief. All the same, it is perhaps surprising that between them Stolz and Jones started no less than three such studies (see Jones, 1960, for obituary and bibliography).

The Berkeley Growth Study

The first was the Berkeley Growth Study, begun in 1928 and directed by Nancy Bayley, a psychologist, whose contributions to human growth, like those of Frank Shuttleworth, are classic in their penetration and simplicity. Bayley (1899–) took her PhD in 1926 at Iowa, though in the psychology department rather than the Child Welfare Research Station. But she attended, and was much impressed by, Baldwin's seminar on physical growth and kept his 1921 monograph by her as a treasured guide in her early work. After two years teaching academic psychology she was appointed by Harold Jones as research associate at the Institute of Child Welfare, and on the way to her assignment stopped briefly again in Iowa to be taught anthropometry by Idell Pyle. She stayed at the Institute of Child Welfare till 1954. From 1954 to 1964 she worked at the National Institute of Health in Bethesda, but returned in 1964 to what had by then been renamed the Institute of Human Development at Berkeley. She retired in 1968.

The Berkeley Growth Study was a cohort study, initially of thirty-one boys and thirty girls recruited at birth. Nearly two-thirds were successfully followed to maturity, and many still participate in a study of middle age, now run by Dorothy Eichorn. Bayley took twenty-two physical measurements (all herself until the boys reached adolescence and preferred a male technician), as well as photographs, and radiographs of the hand and wrist. There were a number of psychometric tests, whose analysis provided, in Bayley's hands (1955, 1957), some of the most important information on the development of mental performance under good conditions yet gathered.

The children in the study were most of those born in two Berkeley hospitals during a six-month period, subject to the restriction of their parents being of European descent and English-speaking. As a group they were slightly better-off than the average Berkeley citizen, with one-third of the fathers in professional and managerial jobs. The very first paper from the study, rather forgotten in the North American literature, showed the style of scholarship that was to be the norm. The children were measured monthly from 1 to 12 months, then at 15, 18, 24, 30 and 36 months of age. The first analysis (Bayley and Davis, 1935) dealt with this period. The results are set in the context of previous work (mostly North American, though Godin is quoted, for about the first and last time in American literature). The techniques of measurement were mostly derived from the Iowa Child Welfare Research Station; photographs were taken also, as one of the authors' interests was in changes in body build or chubbiness, and how to measure it. (This stimulated a long footnote of editorial comment from Karl Pearson, in whose journal the paper appeared.) For nine measurements tables of means and standard deviations were given on each occasion (calculated cross-sectionally, but the study was almost pure longitudinal and the growth curves of a simple type, so that in this case it mattered little). Correlation coefficients for the same dimensions between different ages were presented. This was the first demonstration that the correlations were much lower between 1 month and 6 months than between 6 months and 1 year; and that the correlations from 1 to 2 years were a little higher than the ones between 6 months and 1 year. Later confirmation was provided by the Brush Foundation study (see above).

In 1940 Bayley reviewed the first ten years of the study, including illness records, temperature, blood pressure and heart rate changes, reflex patterns and the application of the infant scales of motor development and of mental development known by her name (Bayley, 1940). By 1954 the subjects were all mature. More than three quarters remained: to be precise, forty-seven children (from forty-three families) out of the original sixty-one, a high percentage over an eighteen-year-period. Bayley reported the correlations between the children's heights at successive ages and the heights of the parents. Low at 1 month, they rose to stabilize around 2 years (though with some sex differences). Correlations of the children's childhood heights with their own mature heights also were low at 1 month and stabilized at about age 2. This was the first set of correlations published which related infants size to adult size. The results explained the finding relating children's to parents' heights, and extended Bayley's own and Simmons' (1944) results on birth-to-1-year correlations. The explanation appeared two years later, in another paper relating infant to mature measurements (Tanner *et al.*, 1956: see Chapter 13).

Bayley was now in a position to calculate, for the first time, the percentage of his own mature stature that a child had reached at each age. She thus had a valid measure of tempo other than bone age. (The measure 'height age' sometimes used by paediatricians is an *invalid* measure, because children end their growth with different mature heights. An advanced height age, which simply means that a 6-year-old child, for example, is the size of an average 7-year-old, indicates more that the child is going to be a tall adult than anything about his tempo. But percentage of his own mature height is different.) In a very important paper, Bayley used these data to examine again Porter's relationship between height and academic ability. She also had scores giving percentage of IQ at age 21 achieved at different ages, and therefore was able to relate advancement in tempo of physical growth to advancement in tempo of intellectual development. The correlations, unlike those of height and IQ at a given age, were zero or slightly negative (Bayley, 1956*a*, p. 71). The implication was that Boas' explanation of Porter's findings was wrong and that the small correlation between height and IQ would persist into adulthood, as it was later shown to do (see Tanner, 1966, for review).

In the same year Bayley (1956*b*) made the first effort to produce standards for heights which took into account an individual's tempo. For this she used both the Berkeley Growth Study children and the Guidance Study children (see below). All had had skeletal ages assigned to them. Developing Shuttleworth's ideas, Bayley took all those children whose skeletal ages and percentages of mature height indicated a near-average tempo (within ± 1 year of chronological age). The average line for these children is the central one shown in Fig. 12.3. The line immediately above characterizes the growth of children (boys in the figure) with accelerated tempo but average adult height; the line just below, that of children with delayed tempo but average adult height. The two outside lines show the extremes: children with accelerated tempo and high adult height above, and children with delayed tempo and low adult height below. The 'increment curve' represents the velocity of children of average tempo only, and the dots either side of the point of peak velocity represent the peaks for accelerated and delayed children.

This, then, was a radical new departure in the whole approach to standards of growth. It built on Shuttleworth's analysis but sought to be a tool for use rather than a further analytical dissection. It was, perhaps, before its time. It was published, appropriately enough, in the *Journal of Pediatrics*, but at this time paediatricians in general had little understanding of auxology and none whatever of tempo; Boas' and Shuttleworth's papers had appeared in journals unfamiliar to them. In addition it was not clear how these standards could be used to produce a probability statement in the way that conventional centile cross-sectional-type standards could do. They were never much used in practice, not even in the form given in a handbook on *Growth Diagnosis* (Bayer and Bayley, 1959, 1976). But things were moving: a theoreti-

Fig. 12.3. Bayley's growth curves of height-for-age, according to tempo as well as ultimate size. The three central curves of height-for-age represent: accelerated tempo, average mature size; average tempo, average mature size; delayed tempo, average mature size. The increment curve, below, represents the velocity of the average-tempo, average-mature-size group only. The dots ahead of and behind the point of peak velocity represent the peaks for accelerated and delayed children. From Bayley (1956*b*).

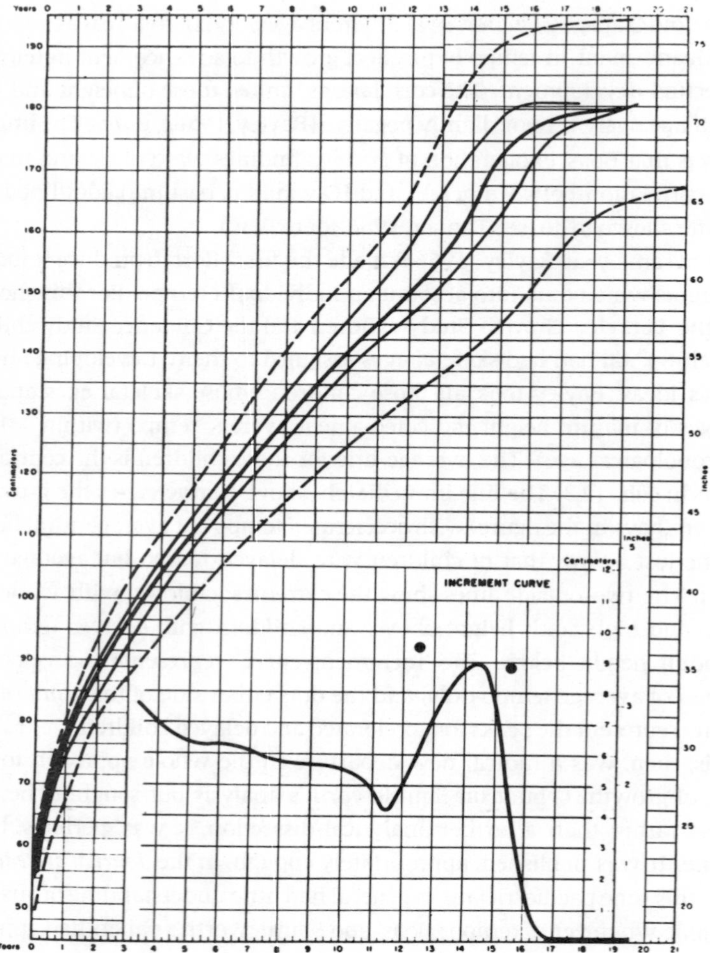

cal article on standards by Tanner (1952) had already appeared in the *Archives of Disease in Childhood* on the other side of the Atlantic. In 1966 Tanner, Whitehouse and Takaishi's much-quoted paper on longitudinal-type standards was published (see Chapter 13). Bayley's seminal paper of ten years earlier lay at the base of it.

Using the same Berkeley Growth Study data, Bayley, with S. R. Pinneau (Bayley and Pinneau, 1952) gave tables for predicting mature height from height and skeletal age at earlier ages. (Bayley (1946) had previously done this using Todd's skeletal ages, but this time the Greulich–Pyle *Atlas* was used.) The same principle again applied. Those children whose growth was known from their skeletal ages to be accelerated were, at a given chronological age, closer to their mature height than those who were delayed in tempo: hence the adjustment of prediction by use of skeletal age. These tables are still extensively used by paediatricians and endocrinologists (see also Bayley, 1962).

The Child Guidance Study

The second of the University of California studies was the Child Guidance Study, started in 1930 and directed by Jean Macfarlane. This was primarily a psychological and sociological study to see if regular guidance averted symptoms of behavioural disturbance; but six physical measurements were taken, together with a test of strength of hand grip and a hand–wrist radiograph. The general programme of the study is given in Macfarlane (1938). Sixty-six boys and seventy girls were followed from birth to 18. Only length and weight were measured before, in general, age 8; the other measurements, taken at six-monthly intervals, began then. The data for all measurements, smoothed to eliminate seasonal fluctuations and to centre them at exact birthdays and half-birthdays, have been given by Tuddenham and Snyder (1954), together with somatotypes and ages at reaching full skeletal maturity. This is an immensely valuable archive, used by a number of later, and sometimes quite unconnected, research workers. In this group peak velocity of height was reached at 11.5 in girls and 13.5 in boys, the earliest recorded to that time (p. 199). Tuddenham and Snyder give intercorrelations between pairs of measurements at each age, and between measurements at one age and the same measurements at other ages; for height and weight these inter-age correlations run from 2 to 18, for the other dimensions from 9 to 18.

Other aspects of the Guidance Study data were analysed by Nicolson and Hanly (1953). At each examination ratings were made on five-point scales (see below) of stages of breast and pubic hair development, and the age of menarche was recorded. Nicolson and Hanley intercorrelated (amongst others) the variables: age at menarche, at breast stage 2, at pubic hair stage 2, at peak height velocity, at bone ages $9\frac{3}{4}$ and $12\frac{3}{4}$, and at 90 per cent mature height. The matrix was factor-analysed to yield a measure of general 'age-at-puberty'.

Faust (1977) analysed the adolescent development of eighty-two girls of the Guidance Study together with fifteen from the Berkeley Growth Study. The average age of menarche in these girls was 12.8 years, S.D. 1.2. Peak height velocity (estimated only as the greatest six-monthly increment) averaged 9.0 cm/yr and centred at 11.7 years. The total gain in height after menarche averaged 7 cm, with a range 1 to 15 cm. Such are some of the findings, which in general are very comparable with those of the later European studies.

The Adolescent Growth Study

The third University of California study was the Adolescent Growth Study. This was started in 1932 and ended in 1939. Some of its subjects were children who were already in the Guidance Study. Some sixty boys and sixty girls were followed for the whole period of seven years; this represented about half of the originally recruited sample. A number of physical measurements were made, including skinfolds every six months. Photographs and hand radiographs were taken, and puberty ratings done. There were records of intellectual and educational achievement. Dr Herbert Stolz took the boys' measurements himself and his and his wife's account of the boys' side of the study, written in their retirement, was something of a landmark in longitudinal work (Stolz and Stolz, 1951.) For one thing there is a complete series of front, side and rear view photographs of one particular boy every six months, showing beautifully the adolescent spurt and the accompanying somatic changes. There are numerous graphic analyses of the growth of individual boys, mostly illustrating the various sorts of individual differences which occur at this time, all illustrated by the relevant photographs. Stolz showed that the spurt in sitting height occurred later than the spurt in leg length, and was on average larger. There is a particularly valuable chapter on the growth of pubic hair and genitalia, with ratings based on photographs, and a description, also based on pictorial appearances, of the pre-pubertal fat increase which disappears as puberty gets under way. The whole book is full of individual examples, given by authors who are masters of their material: it deserves to be better known by auxologists than it is.

Nancy Bayley used these subjects in an analysis of adolescent growth in relation to skeletal maturation (Bayley, 1943*a*) which was a preliminary to her paper on prediction of adult height; and also in a study of the relation of mature body build to early or late maturing (Bayley, 1943*b*).

The Adolescent Growth Study, however, is chiefly remarkable for being deliberately orientated towards the study of physiological changes at puberty. Endocrine tests were too much in their infancy at this time to have been useful, but blood pressure, basal oxygen consumption, heart rate, creatinine excretion and exercise tolerance were followed at six-monthly intervals. Nathan Shock (1906–), a physiologist in the department of paediatrics at Chicago, was recruited to supervise this aspect of the programme, and worked at the Institute from 1932 to 1941, producing a series of papers (Shock, 1941, 1942, 1943, 1944, 1945, 1946*a*, *b*). Most of the data were reported only in terms of chronological age and certainly merit a re-analysis with modern curve-fitting techniques. In one paper, however, Shock analysed the physiological changes by plotting his measurements on a time scale of years before and after menarche, in the manner of Boas and Shuttleworth. Fig. 12.4 is redrawn from this paper, and shows the fall in basal oxygen consumption and resting heart rate, together with the rise in blood pressure, which occur at adolescence. In 1941 Shock transferred his attention to gerontology and at the National Institute for Aging, in Bethesda, Maryland, searched for many years, and in vain, for some measure of physiological aging

Fig. 12.4. Physiological changes at adolescence in relation to age at menarche. Longitudinal data arranged in years before or after menarche. Adolescent height spurt indicated at top of figure. From Tanner (1962); drawn from data of Shock (1946*a*).

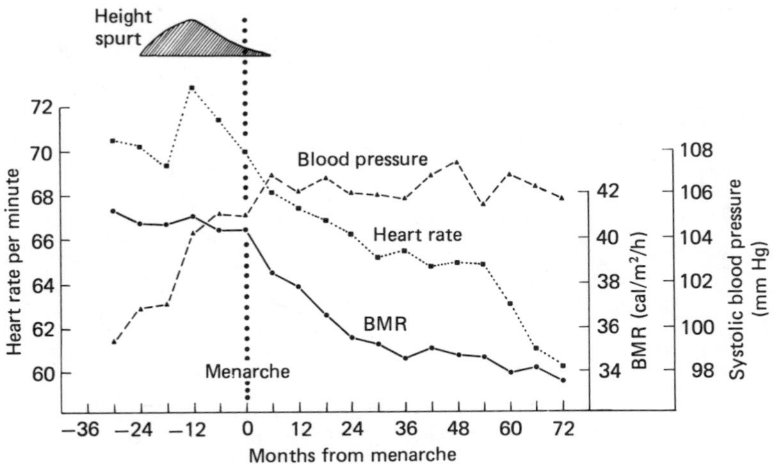

equivalent to the measures of physiological age found so useful by child development specialists and paediatricians. Many of the types of problem in development and gerontology are similar, and a number of research workers have gone from one field to the other.

Strength tests were given to the adolescents of this growth study, and their results reported by Jones himself (Jones, 1949). Again the data were mostly reported in relation to chronological age only, thus muting the spurt, but the results were so clear-cut that even when smoothed out in this way the large male rise in strength at puberty emerged. The tests were done so that the subjects were as far as possible motivated to the maximum; each made three attempts, of which the best was recorded. Fig. 12.5 reproduces some of Jones' data. Jones pointed out that the relative increment in strength in boys at adolescence is much greater than the relative increment in height or weight. He goes into the social implications for adolescents and the guidance implications for counsellors; and, like all his workers in the Institute of Child Welfare, gives individual examples and case histories (another of his publications (Jones, 1943) consists entirely of the case history of a single individual). This material remained the best source of data on strength changes until the Canadian studies (only on boys) of the 1960s (see Chapter 13). Motor perfor-

Fig. 12.5. Changes in strength of arm pull and arm thrust during adolescence. Mixed longitudinal data reported cross-sectionally. From Tanner (1962); drawn from data of Jones (1949).

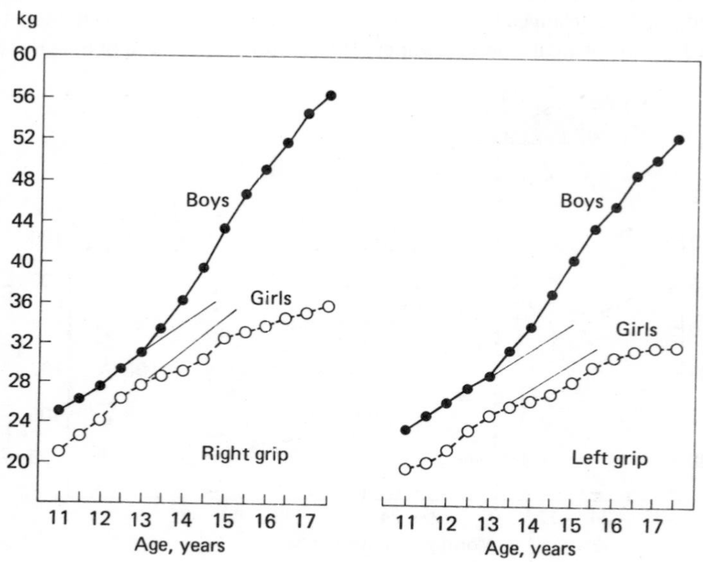

mance tasks, such as running and jumping, and motor coordination as distinct from strength were also studied in these children, though over a shorter period of time. The results were reported by Espenchade (1940, 1947).

An important generalization emerged from these studies. There was no period during the adolescence of these young people when strength or motor performance declined. No evidence could be found of a boy 'outgrowing his strength', as the popular belief has it. The peak of the strength increase, it is true, is probably located slightly later than the peak height velocity and to this extent there might be said to be a period when a boy has not yet the strength to be expected from his increased size (although the strength per centimetre of height increases continually throughout puberty, it increases fastest when the height increase is slowing down). It seems most likely that the old belief has roots in popular folklore and individual psychology (not forgetting sexual psychology) rather than in any true physiological phenomenon.

All the American longitudinal studies studied mental as well as physical development and all fervently hoped that some way of relating the two would be found, some manner of reconstructing 'the whole child'. These hopes have never been realized, but in one perhaps superficial aspect the interaction of bodily development and social function was successfully studied in the Californian subjects. Paul Mussen and Mary Cover Jones (1957, 1958) examined the psychological consequences of early and late physical maturing. Late maturing boys showed more attention-getting behaviour and were rated by their age peers and trained observers as more restless, talkative and bossy. They were less popular and had lower social status. In contrast, the outstanding leaders came from the early maturing group. Projective techniques revealed greater feelings of inadequacy and anxiety in the late maturing and follow-up studies showed that, at least in the society of the United States, the early maturing boys became more sociable and less neurotic.

The Fels Research Institute, Yellow Springs

The largest and longest-lived of all the North American longitudinal studies was founded privately, by the Fels Foundation, and placed on the campus of Antioch College, a small progressive liberal-arts college in Yellow Springs, Ohio. It was very much the creation of one man, Lester Sontag (1901–). Part paediatrician and part psychiatrist, Sontag was not only a man of vision but a brilliant organizer. The study started, in 1929, in a small farmhouse, but twenty years later it had a large eighty-room building, an Institute in itself. The core of the Fels was a longitudinal study of a number of Ohio families, mostly belonging to the farming community. Sontag developed separate departments, not only of physical growth, but of biochemistry, social psychology and psychophysiology. Each department was encouraged to develop its own interests and to obtain grants for research from government and private sources, quite apart from the contribution it made towards the growth study. Thus the Fels was a real Institute of Child Development and indeed, because many of its projects were only distantly related to children, almost a small biomedical research station. Its large and at times distinguished staff made many contributions to the various areas of auxology; but their extreme isolation in a small Ohio town and their consequent lack of any real liaison with any school of medicine or of education prevented their work from having the effect on paediatric and educational practice that a stream of graduate students and a busy undergraduate teaching programme might have produced. The isolation was at first a virtue, for the location secured a group of some 300 families who participated willingly in a conception-to-maturity programme and came back for more with their own children. But heads of departments naturally tended, their names made, to move on to chairs in large universities, and the lack of permanent funding either from a State Legislature or from a university made the future always precarious. In the end, after nearly fifty years, the money ran out and in 1978 the departments were absorbed into the recently founded medical school of the neighbouring university, Wright State.

Sontag (1971) has given a short history of Fels, and the Annual Reports for many years included a full collection of reprints of the staff's papers for the year. There is abundant material there for the future historian of the enterprise. General descriptions of the sample are given in Sontag, Baker and Nelson (1958) and Kagan and Moss (1962). Sontag's own research interest lay largely in fetal behaviour, at a time when practically nobody thought the subject approachable, even if interesting. The physical growth section was run from 1943 to 1950 by Earle Reynolds (1910–), a man who had spent his childhood in a travelling circus, qualified as an anthropologist at Madison, Wisconsin, wrote off-Broadway plays and kept tame racoons in the rafters of his home in Yellow Springs. In a series of papers between 1943 and 1951 (see Tanner, 1962, p. 284 for bibliography) he analysed the development of bone, muscle and fat in the calf as visualized by radiographs (Reynolds, 1944, 1946, 1949, 1950; Reynolds and Grote, 1948) and demonstrated the pre-pubertal sex differences of the pelvis (Reynolds, 1945, 1947). The latter was also a radiographic study, for X-rays were taken at Fels with much the same abandon as at Harvard. Reynolds' calf data, made available to Tanner in a typical gesture, were the basis for the first application of discriminant function analysis in human auxology. Tanner showed, in case anyone should need the demonstration, that a calf X-ray in an adult sufficed to determine in 95 per cent of cases whether its owner was male or female, whereas before puberty misclassification was nearly 40 per cent. Fig. 12.6 shows the graphs resulting from Reynolds' data.

Reynolds also wrote papers detailing the stages of secondary sex character development and the ages at which the Fels girls and boys entered each stage. These descriptions of the pubertal stages (Reynolds and Wines, 1948,

Fig. 12.6. Discrimination of males and females by fat and bone in the calf at ages 7.5, 10.5, 13.5 and in adults. Percentages misclassified are respectively 38, 29, 14 and 5 per cent. From Tanner (1962); data from Reynolds (1949).

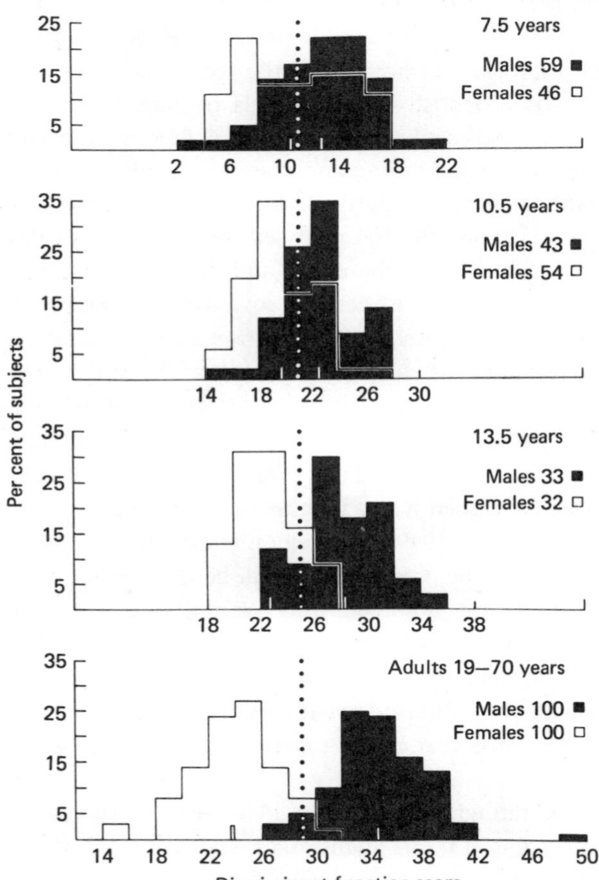

1951), though not by any means original with Reynolds, were in large part the ones adopted by Tanner for illustration in his *Growth At Adolescence* (1955, 1962).

Reynolds had always been a sailor and after he left Fels, apart from helping in the analysis of the effects of atomic bombing on growth, he abandoned scientific research. His boat became an important element in the movement against nuclear bomb tests, and later he was influential in the Greenpeace movement. Eventually he settled in Japan.

He was succeeded at Fels by Stanley Garn (1922–), also an anthropologist, and a graduate of Harvard where he had been a pupil of E. A. Hooton. Garn was (and is) the most prolific writer of papers the field of auxology has yet known and no attempt to list them is made here. He continued Reynolds' work on tissue growth, and particularly growth of fat (Garn, 1955, 1958; Garn and Gorman, 1956; Garn and Haskell, 1960; Garn and Saalburg, 1953; Garn and Young, 1956). In *Methods for Research in Human Growth* (Garn and Shamir, 1958) he gave an account of the techniques then in use at Fels. With A. B. Lewis, a dentist working at Fels, he published a series of papers on tooth eruption and calcification, and with C. G. Rohmann a series dealing with many aspects of the timing and manner of appearance of ossification centres seen in radiographs. A particularly interesting paper (Garn, Rohmann and Blumenthal, 1966) showed that sex differences in ossification were much greater in some areas, the elbow and the knee for example, than in others. Thus besides the girls' general advancement compared with boys there was a sex–bone interaction. Garn's chief interest later centred on the influence of malnutrition on growth and on the genetics of maturation processes (see Garn and Bailey, 1978). In 1968 Garn left Fels for the Centre of Human Growth and Development at Michigan University, and his place was taken by Alex Roche (1921–), an Australian anatomist who had previously been concerned with a longitudinal study of growth in Melbourne (Roche and Sunderland, 1959). His major interests lay in skeletal maturation and its application in the prediction of adult height (see Roche, 1978). When the Fels Institute was absorbed into Wright State University Roche's department was incorporated as the Fels Section of Physical Growth and Genetics: with it went the care of the still-continuing longitudinal study. The total number of children followed from birth to maturity by this time was around 165 (see Thissen *et al.*, 1976), a few more than in the other North American studies.

Lester Sontag retired from Fels in 1970. He was succeeded by Frank Falkner (see Chapter 13), an Englishman who had been Alan Moncrieff's assistant and had set up both the English and French teams of the International Children's Centre longitudinal studies. Falkner left England for the United States in 1956, and became professor of paediatrics in the University of Louisville, Kentucky, where he initiated the twin study that still continues (see Chapter 13). Always primarily a paediatrician, he endeavoured to bring the Fels Research Institute into closer connection with local medical schools, and soon after its absorption into Wright State University he returned to medicine as professor of maternal and child health at the University of Michigan School of Public Health.

Other North American studies 1930–1970

William Greulich

It must not be thought that all North American auxology was confined to these major centres of longitudinal work, though they certainly set the pace and indeed transformed the whole field. There were scientists like William Greulich (1901–), who contributed a number of important papers besides the famous *Atlas of Skeletal Development*. Greulich worked at Yale University at the time when Shuttleworth was there and in 1938 the two were joint authors, with others, of a handbook of methods for the study of adolescent children, with a preface, needless to say, by Lawrence Frank and financial support from the General Education Board (Greulich *et al.*, 1938). Greulich also published the first serious attempt to investigate the relation between somatic and endocrine changes during adolescence, R. I. Dorfman and H. R. Catchpole being the biochemists concerned (Greulich *et al.*, 1942). However, the endocrine methods were rudimentary at that time; it took twenty years before they became usefully applicable, and thirty before the problems Greulich was interested in were really able to be approached. In the same vein Greulich published a longitudinal study of the female pelvis at puberty (Greulich and Thoms, 1944). Then, after he had gone, via Western Reserve, to Stanford University, he wrote two very influential papers. One showed that growth was more disturbed in boys than in girls by the environmental shock of the Hiroshima and Nagasaki radiation (Greulich, Crismon and Turner, 1953): subsequent work demonstrated that this was a regular feature in nearly all environmental stresses. Another (Greulich, 1957) became very well known in the study of the secular trend. Greulich showed that Japanese children born and brought up in California grew to be taller than Japanese children born and brought in Japan. Later, he documented the catching-up of the native-born-and-reared Japanese, so that twenty years after the first study was done there was no longer any difference between the two groups (Greulich, 1976).

Chicago

At the University of Chicago Laboratory Schools the longitudinal study that Baldwin had used continued throughout the 1920s and 1930s, though in rather a desultory manner. H. G. Richey (1937) used its accumulated records of height and weight to analyse growth in groups of girls with different ages of menarche and of boys with different ages of appearance of axillary hair. The study was conceived on very similar lines to Baldwin's; it lacked the penetration of the exactly simultaneous study of Shuttleworth.

The Laboratory Schools data were also used for the first factor analysis of body measurements at successive ages, by F. A. Mullen. L. L. Thurstone and Karl Holzinger both worked at the university, and there was a thriving school of factor analysis. Mullen's (1940) paper was little remarked at the time of publication but in retrospect can be seen as marking the beginning of a tradition of this type of work, and indeed of multivariate analysis in auxology in general.

Philadelphia

In 1940 W. M. Krogman (1903–), a Chicago-trained physical anthropologist who worked with Todd at the Brush Foundation in 1928–9, joined the Laboratory Schools study and revivified it for a while before leaving for Philadelphia in 1948 to become professor of physical anthropology in the University of Pennsylvania Graduate School of Medicine. As outgoing and friendly as he was physically large, Krogman exercised a great influence on auxology in North America in the 1940s and 1950s. He had an especial interest in orthodontics (see Krogman, 1973), which he acquired from contact with Broadbent in Todd's laboratory, and in Philadelphia ran a longitudinal study located in the Children's Hospital of the University of Pennsylvania (Krogman, 1970). (It subsequently became the *Philadelphia* (later *W. M. Krogman*) *Center for Research in Child Growth.*) One of his postgraduate anthropology students there, Frank Johnston (1931–), worked up much of the data of the period 1947–59 (Johnston, 1963, 1964*a*, *b*) and Robert Malina (1937–), a student of Johnston's, based a second study there in 1968–9 (Malina, 1966, 1970, 1972). This Center yielded some of the first reliable comparisons between the growth of Americans of European and African descent ('Whites' and 'Blacks'). Johnston himself became professor of anthropology in the University of Pennsylvania and trained a number of human biologists interested in auxology. Like Meredith, Krogman was a great bibliographer and his collection of studies of growth published in *Tabulae Biologicae* (Krogman, 1941*a*) is a very useful one. So also are his *Bibliography of Human Morphology* (Krogman, 1941*b*), his handbook on measurement of height and weight (1948), his syllabus on human growth (1950), and his appraisal of growth studies made between 1950 and 1955 (1955).

Gray and Ayres

Chicago was also the base for an early and particularly notable cross-sectional study of children in nine private schools : the University of Chicago Laboratory School, the Frances Parker School, and the Latin School for Girls, in Chicago; the City and Country School in New York City; Groton School in Massachusetts; and four others. This was made in 1930–1 by a scholarly paediatrician, Horace Gray (1887–1965 : see note 12.9), who had been a member of the White House Conference, and J. G. Ayres, both at that time working at the Behavior Research Fund and the Institute for Juvenile Research in Chicago. In all 3,110 boys and 1,473 girls were measured, many by Gray and Ayres themselves; they included a number of pre-school age. Besides height and weight, sitting height, shoulder and hip breadths, and chest width, depth and circumference, thigh circumference and seven head and face measurements were taken. Not only means but scales giving standard deviation scores were presented. *Growth in Private School Children* (Gray and Ayres, 1931) exerted a considerable influence on the longitudinal work beginning at about the time it was published. It was thorough; had an excellent bibliography with considerable reference to foreign papers; and used, and explained, biometric methods. Gray moved to Stanford University before the book was published, and continued his measurements there. In 1941 he reported birth-to-maturity measurements of his own three children, giving the raw data for fifteen measurements. Characteristically he compared the growth curve of his own son with those of Wiener's and of Guttman's boys (see above).

Minnesota

In the University of Minnesota Institute of Child Welfare there was a short-lived longitudinal study in the 1920s. Edith Boyd, later of the Child Research Council, did much of the measuring, publishing a paper on the error of measurement (Boyd, 1929) and a small and valuable monograph on growth of surface area (Boyd, 1935). For the calculation of surface areas she used data provided by Carroll Palmer (see below), who had previously been on the staff of the Minnesota Institute and while there had measured 1,000 boys and girls in the Home for Children in Mooseheart, Illinois (see Palmer, 1932).

Kaiser Foundation

Another, later, longitudinal study was made in the Berkeley–Oakland area of San Francisco in connection with routine health check-ups given as part of an insurance plan by the *Kaiser Foundation*. J. Wingerd, biometrician to the study, published with his colleagues papers relating height, weight and skeletal maturity at 2, 5 and 9 years to race, size of parents and social class (Wingerd, 1970; Wingerd, Solomon and Schoen, 1971; Wingerd and Schoen, 1974; Wingerd, Peritz and Sproul, 1974).

Newton, Mass.

In Newton, Massachusetts, a well-off suburb of Boston, a short-term longitudinal study of girls was organized by Leona Zacharias, a biologist attached to MIT. In 1965 all girls aged 9 and 10 in the schools were asked to participate and 781 (62 per cent of total) agreed. Each kept a diary for the study of age of menarche and the characteristics of the first few periods. Heights and weights were taken regularly by the parents. Menarche occurred on average at 12.8 years with a range from 9.1 to 17.7. In this middle class White group there was no relation of age at menarche with number of siblings or with father's occupation (Zacharias, Rand and Wurtman, 1976).

Charles Davenport

Another auxologist, more of the generation of Boas than of Bayley, was Charles B. Davenport (1866–1944). Davenport was a biologist by training, geneticist by profession and eugenicist by avocation. He was the author of one of the first North American texts on statistical methods (Davenport, 1899, 1904, 1914) and wrote both theoretical and practical papers on growth. Brailsford Robertson (see Chapter 11) had postulated three autocatalytic growth cycles, the second causing an acceleration to be seen at about age 6. Davenport (1926), who had accumulated longitudinal records himself on a series of children in a Brooklyn orphanage, thought the six-year acceleration seen in mass data was simply a product of sampling, 6 being the age at which the sampling was radically changed by the fact of children thereafter becoming generally available in schools. In individual curves he could find no evidence of Robertson's second cycle. He proposed, rather, that there were two great growth periods in man: one in fetal life extending to the years directly after birth, the other in adolescence. He maintained that (by analogy with frogs) these two cycles would be under different hormonal control since different parts of the body were preferentially stimulated in either cycle (for instance sitting height at adolescence, head size in fetal life). This was before the discovery of growth hormone; but anterior pituitary extracts, said Davenport, caused great growth of the mouse, and the pituitary gland might well be responsible for the adolescent growth spurt (as Davenport called it, and for the first time: see note 12.10).

In 1931 Davenport published a paper on 'individual versus mass studies in child growth', in which he pointed out again the 'time-spreading' effect of averaging growth curves irrespective of tempo (see Fig. 10.2, p. 238) and in 1934 he repeated this, together with a general critique of the formulation of 'relative growth' then much in vogue amongst zoologists. Davenport organized one of the most extensive studies of the reliability of a single anthropometrist ever undertaken (the unfortunate test-worker was Morris Steggerda, who measured a series of women fifty times each: Davenport, Steggerda and Drager, 1934). Davenport himself measured children at Letchworth Village Colony for the mentally handicapped and babies in Myrtle McGraw's Columbia University Child Development Clinic. He was one of the first to demonstrate the occurrence of an adolescent growth spurt in the nose (Davenport, 1939) and analysed the spurt in various skull dimensions (Davenport, 1940). His final paper (1944) was a monograph on the growth of the limb segments.

Gaston Backman

Davenport was saluted as 'my most serious predecessor' by a scholarly Swedish anatomist whose work may perhaps best be described at this juncture. Gaston Backman (1883–1964) graduated in medicine at Uppsala in 1913, taught at Riga from 1920 to 1928 and then, after a time in Stockholm, became professor of anatomy at Lund in 1933 (see Backman, 1964). He followed Robertson in proposing that the mammalian growth curve was divisible into three phases; but his phases were nearer Davenport's than Robertson's. In man the 'primordial phase' operated from shortly after conception to age 2, with its maximum velocity around the middle of fetal life. The 'infantile phase' extended from birth or a little before till about 12 with its maximum velocity at age 2, and the 'juvenile phase' began around 7 or 8 and reached its maximum velocity at the adolescent spurt (Backman, 1934). Backman fitted the sum of these curves both to the means of large series of data and to individuals, notably Montbeillard's and Lange's sons, and some of Bowditch's purported relatives ('Frank' and 'Ned', 'John' and 'Arthur'). The relative contribution to height of the three cycles at various ages were, at age 2 : 86, 13 and 1 per cent; at age 5 : 68, 30 and 1 per cent; at age 12 : 52, 39 and 9 per cent; at age 20 : 42, 36 and 22 per cent. Backman (1925, 1931, 1939*a*) fitted curves of the sort $\ln h = a + b \ln t + c \ln^2 t$ to each of the phases. In his 1934 monograph he pointed out, with great clarity, the 'time-spreading' effect of differences of tempo on the velocity curve at adolescence. He seems to have come to this realization independently, for Galton is the only author he quotes as a predecessor in this respect. Curiously, he does not mention Boas (1930), though Boas (1933) figures in his bibliography. The latter paper would almost certainly would have come to him after his own manuscript was written. In fact he gives an identical diagram to Boas' and Shuttleworth's. Backman wrote several other long and detailed monographs, on comparative growth (1938), on organic time (1939*b*) and on the secular trend (1948).

Margaret Merrell

At the same period as Davenport's and Backman's theoretical papers and Boas' second phase of auxological work, there was an important paper published on the individualizing-versus-generalizing theme by Margaret Merrell (1931). Suppose one had fitted a mathematical equation to longitudinal data on a number of individuals. Then each individual would have a value for each of the various constants of the equation, three for example in the case of the simple equation $W = a + bt + c \log t$ (W, weight; t, age). Merrell showed mathematically that the means of all these constants, say $\bar{X}_a$, $\bar{X}_b$, $\bar{X}_c$ would not in general be the same as the constants $\bar{a}$, $\bar{b}$, $\bar{c}$ obtained by fitting a curve to the mean values (at each age) of all the individuals. In certain simple curves the two would coincide, but in all non-linear formulations the mean-constant curve, as she called it ($W = \bar{X}_a + \bar{X}_b t + \bar{X}_c \log t$) would be different from the curve of the means.

Raymond Pearl

Merrell's paper, like so many on the theoretical as well as practical aspects of human auxology, appeared in the journal *Human Biology*. Merrell worked in the School of Hygiene and Public Health of Johns Hopkins University where Raymond Pearl (1879–1940), the founder of the journal, was head of the department of biometry and vital statistics from 1918 to 1925. Pearl was a biologist who took his PhD at the University of Michigan with a thesis on the behaviour of flatworms; but a year with Karl Pearson in 1905–6 turned him into a biometrician of force and persuasiveness. He went to Hopkins in 1918, combining his chair of biometry and vital statistics in the school of hygiene and later (1929–30) in an Institute of Biological Research of his own, with a professorship of biology in the undergraduate medical school (1923–40). Pearl was a true follower of the tradition of Galton and Pearson and played a large part in carrying that tradition to the United States. He was, and without doubt will for ever remain, the only person to be president, practically in successive years, of both the American Society for Physical Anthropology and the American Statistical Society (see Jennings, 1943). His *Introduction to Medical Biometry and Statistics* (1923, 1930, 1940) was a classic of its period.

Many of the curve-fitting papers appeared in *Human Biology*. Deming's papers that we have discussed above were there, and the paper by R. M. Jenss and Nancy Bayley (1937) also. Earl Count's papers on fitting a succession of three curves to cover the whole growth period rather in the manner of Robertson were there too (Count, 1942, 1943). Count chose a logistic curve to fit growth at adolescence, but simpler curves for the two earlier periods. He made the logistic ride, so to speak, on the top of the previous ones, adding to them – a technique revived much later by Thissen and his co-workers (Thissen *et al.*, 1976: see review in Marubini, 1978). However Count ignored Merrell's warning and fitted curves to mean values rather than to measurements of individuals.

Carroll Palmer

Another worker who published in *Human Biology* was Carroll Palmer (1903–1972). Palmer left Minnesota (see above) in 1929 to join the department of biostatistics at Johns Hopkins University School of Hygiene and Public Health (see note 12.11). Thence he went into the United States Public Health Service. Between 1930 and 1935 he published a number of important papers on growth, mostly on the measurements of height and weight accumulated in a mixed longitudinal survey of the children in Hagerstown, Maryland, undertaken by the US government from 1921 to 1927 and again in 1933–4. Palmer analysed diurnal changes in height and weight (1930), seasonal changes in weight (1933*a*), the correlation between siblings when at the same age (1934*b*), and the effect of the economic depression of 1929–33 on the weights of children (finding surprisingly little: 1933*b*, *c*, 1934*a*, 1935). His last two auxological papers (Palmer and Reed, 1935; Palmer, Kawakami and Reed, 1937) concerned more theoretical aspects of his data: the relation between size attained and the initiation of the adolescent spurt.

Another worker in the United States Public Health Service, a contemporary of Palmer, wrote a book, *Environment and Growth*, which is much less well known than it deserves to be. The author was Sanders (1934), a statistician otherwise unknown to the growth literature. Sanders gives a valuable bibliography of about 1,200 titles, many of which appear neither in Baldwin (1921) nor the present book.

Palmer's co-author in the last two papers quoted was Lowell Reed, Pearl's successor as professor of vital statistics at Hopkins, and the father of Robert Reed, Wilson's successor at Harvard and the analyser of the Harvard School of Public Health archive. These personal relationships perhaps emphasize, if emphasis is needed, what small groups of people have been responsible for progress in human auxology; in the nineteenth century primarily in Germany, in North America in the middle third of the twentieth century and again in Europe in the third quarter of that century. There have been no university departments of auxology teaching undergraduates, and very few established even as research enterprises. Auxologists have come from and worked in many different disciplines. And, as we have seen, many of the greatest advances have been made by psychologists and anthropologists, particularly in the period we have just discussed.

13

Human biology and the study of growth disorders: the European longitudinal growth studies

The title of this chapter, like that of the preceding one, somewhat oversimplifies. Nevertheless, it directs attention to the salient characteristic of the European longitudinal studies compared with the North American ones. Though in most of the European studies there was an element of psychometry and some study of behaviour, their primary orientation was towards medicine; nearly all were located in medical departments and depended for their existence on the anticipation of medical benefits.

The Aberdeen study

The earliest of the twentieth-century European studies was a highly personal one made by Alexander Low (1868–1950), professor of anatomy in the University of Aberdeen, Scotland. Between 1923 and 1927 Low took no fewer than twenty-one measurements on 900 newborn babies (Low, 1950). He subsequently remeasured sixty-five of the boys and fifty-nine of the girls at each birthday up to age 5. This longitudinal material was never analysed; and it was only published after Low's death, when his successor, R. D. Lockhart, to his surprise found the records in his office. Lockhart thought it useful to make the raw data of so unusual an archive generally available and had them printed in a pamphlet (Low, 1952). The pamphlet happened to be sent to J. M. Tanner for review. As these children were by then adult, Tanner saw the opportunity of obtaining, for the first time, correlations between growth in the pre-school years and eventual size (which Nancy Bayley was in fact calculating at just that moment). Accordingly, with the help of Lockhart, J. D. MacKenzie, a lecturer in Lockhart's department, and Professor Low's secretary, who emerged from retirement to confront her grown-up infants, a total of forty-two men and thirty-eight women were traced and in 1953 remeasured, aged 25 to 30. R. H. Whitehouse was the measurer and M. J. R. Healy the biometrician concerned (Tanner *et al.*, 1956; and see note 13.1). The subjects were of the manual working class, mostly fisherman-families, growing up in the difficult circumstances of the 1920s. Curves of the form $y = a + bt + c\log(t + 0.75)$, where y was the measurement and t the age in years, were fitted to each child's 0 to 5 year data to help in editing out erroneous values. For each measurement, tables of correlations between the values at each age were given, and also the correlations of values at each age with the mature adult value. Correlations between birth measurements and mature measurements were only of the order of 0.25; by age 1 the correlations had risen to between 0.4 and 0.7, depending on the dimension, and by age 2 to values unchanged from then till 5. Using a multiple regression technique, the successive amounts of the adult variability explained by the measurements at birth; at birth and age 1 combined; birth

Fig. 13.1. The percentage of variance of adult measurements explained by measurements of the same individuals at birth; birth and 1 year combined; birth, 1 and 2 years combined; and so on. Data on forty-two men and thirty-eight women in Aberdeen. Reproduced from Tanner *et al.* (1956).

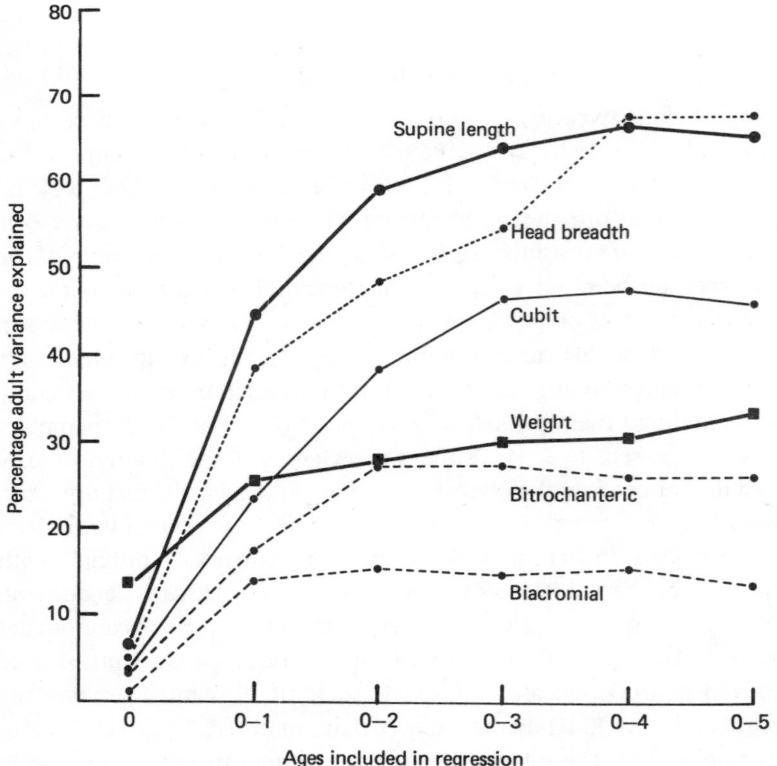

age 1 and age 2 combined; and so on, were calculated: they are illustrated in Fig. 13.1. The authors concluded that 'towards the end of his uterine existence [the child] may be deflected very considerably from [his genetically programmed] curve by the characteristics, both genetical and environmental, of his mother, and after birth he takes a little time to get back onto it, somewhat after the manner of a growing animal who has passed through a period of not too severe malnutrition' (Tanner *et al.*, 1956, p. 378).

Besides this study, there was also during the 1920s a mixed longitudinal study of the growth of schoolchildren, mainly Welsh, by R. M. Fleming (1933), an anthropologist. Some 2,000 boys and 2,000 girls were measured, but one third of them on one occasion only. The chief emphasis was on head measurements. The results give useful means of heights and weights for Welsh children aged 5 to 16 in the 1920s.

The Oxford Child Health Survey

In 1943 John Ryle (1899–1950), the professor of medicine at Cambridge, moved to Oxford to start what was a new academic venture in the United Kingdom. With the aid of the Nuffield Foundation (soon after to found chairs in child health also) an Institute of Social Medicine was set up, with Ryle as the first director. Ryle (1948) was a strong supporter of the concept of 'positive health', much in vogue in the 1940s (partly as a result of the activities of the Peckham Health Centre immediately before World War II: Pearse and Williamson, 1931; Williamson and Pearse, 1938). The first thing he established in his department was the Oxford Child Health Survey, a longitudinal study of healthy children from birth to age 5. Between 1944 and 1947 some 470 children were recruited, mostly through mothers seen in welfare clinics, hence starting at 1 to 3 months old (see Thwaites, 1950). A general account of the survey was given by Alice Stewart, Ryle's deputy and successor, and W. T. Russell, the survey's statistician (Stewart and Russell, 1952). The children were examined four times in the first year and every six months thereafter. Anthropometric measurements were taken, and also radiographs of hand and wrist, knee and chest with some also of hip and skull; illness experience and social changes were carefully recorded. At age 5 some 375 children remained, and the survey closed.

Most of the auxological analysis of these data was done by Roy Acheson, (1921–), later professor of community medicine at the University of Cambridge, and David Hewitt, Russell's successor as statistician. Eventually the raw data were lodged in the department of growth and development of the University of London Institute of Child Health, where they remain available at the time of writing. Acheson and Hewitt (1954*a*) and Acheson, Kemp and Parfitt (1955) gave the basic description of the growth of the children and compared it with the growth of the Brush Foundation children; Tanner, Whitehouse and Takaishi (1966) used the data as a partial basis in the pre-school age for their British Standards for height and weight. The most interesting results of the Oxford survey from an auxological point of view, however, came from the radiographic work. Hewitt (1957, 1958) described the growth curves for widths of bone, muscle and fat in the calf, much after the manner of Harold Stuart's group at Harvard, except that he laid the chief emphasis on the degree of resemblance between brothers and sisters. Acheson and Archer (1959) described the growth of the pituitary fossa.

Acheson's main contribution lay in the development of a new system for assessing skeletal maturation. Instead of assigning a bone age, either to the whole hand or to individual bones, as in the Todd and Greulich–Pyle systems, Acheson adopted the much simpler strategy of regarding each bone separately, determining how many different stages of its appearance could be reliably distinguished, and scoring these stages 1, 2, 3, etc. The scores for all the bones were added up and the resulting sum was called the 'maturity score'. This score could be used entirely like a physical measurement such as height; if desired it could be turned into an equivalent bone age by reference to a standardizing group of children, bone age 5.0 being assigned to a radiograph having the score of the average standard 5.0-year-old (Acheson, 1954, 1966). Acheson applied this method also to the radiographs of the Brush Foundation collection, covering all ages, and found that the maturity scores, unlike skeletal ages, showed adolescent growth spurts just like those in height (Acheson, 1957; Hewitt and Acheson, 1961*a*, *b*). Though this system was never much used clinically, the idea behind it influenced greatly the development of the Tanner–Whitehouse system (see below). Tanner and Whitehouse began taking radiographs in the Harpenden Growth Study in 1950; they were dissatisfied with the Greulich–Pyle system for exactly the same reasons as Acheson, feeling the transformation to 'age' made more for confusion than clarification, and were casting around for a simpler system when Acheson visited them with the manuscript of his paper. In the event the Tanner–Whitehouse system was constructed with a more complex system of scores, but in principle the method was Acheson's.

The Oxford Child Health Survey also published interesting data on its principal concern, the incidence and effects of disease in childhood (Hewitt and Stewart, 1952; Hewitt, Westropp and Acheson, 1955). Acheson (1960) contributed an essay to the Society for the Study of Human Biology Symposium on *Human Growth* (Tanner, 1960) in which he discussed the relation between growth and disease, putting forward the view that illness generally slowed down bone maturation less than bone growth. If this were so it might imply that illness caused ultimate stunting; but later evidence showed this was by no means always the case, so probably in the catch-up phase after the end of the illness bone growth proceeds faster than bone maturation. Acheson's paper is one of the first of the modern ones in which the term 'compensatory growth' is mentioned (see below) (though Robertson (1915*b*) had used the same term for the rapid recovery of weight by the infant during the first week after birth).

The Harpenden Growth Study

The Harpenden Growth Study, a mixed longitudinal study in a Children's Home in the small town of Harpenden some thirty miles from London, began in 1948 and ended in 1971. It started through an approach early in 1948 by E. R. Bransby, a nutritionist at the Ministry of Health, to J. M. Tanner (1920–), then in the department of human anatomy at Oxford. During the war the National Children's Home at Harpenden had collaborated with the Ministry in studies of the adequacy of the food provided under the rationing scheme. Bransby's association with this study had convinced him of the importance of continuing it as a long-term longitudinal study of child growth. Few people in England were interested in human growth, however, and Bransby had to find somebody to run his study. Tanner, who had graduated in medicine from London in 1945 (see note 13.2) and had recently begun to lecture on growth at Oxford was the obvious, indeed only, candidate.

That same year Tanner was invited by the Viking Fund (later the Wenner-Gren Foundation for Anthropological Research) to participate in one of its regular summer seminars in physical anthropology in New York City. After it was over he travelled throughout the United States, visiting all the major longitudinal studies, comparing their methods and learning about their problems (see Tanner, 1948). Thus the first long-term longitudinal study set up in England drew heavily on the experience of North American studies; and the rest of the European studies of physical growth were themselves largely derived from the experience at the Harpenden study.

In the autumn of 1948 Tanner moved back to London and shifted from teaching anatomy to teaching physiology, at St Thomas's Hospital Medical School. He began negotiations with the Ministry for a permanent assistant to help him run their growth study. The Ministry agreed just at the moment when R. H. Whitehouse (1911–) left the Royal Army Medical Corps, in which he had served as a professional soldier since 1938. Whitehouse was innocent of any academic training, but had handled without difficulty the job of Regimental Sergeant-Major at the Royal Army Medical Corps training school; the Harpenden Growth Study seemed scarcely more exacting. Thus began a partnership which lasted till Whitehouse's retirement in 1976.

Tanner and Whitehouse ran the Harpenden Growth Study from the Sherrington School of Physiology at St Thomas's from 1948 to the end of 1955, and then from the department of growth and development at the Institute of Child Health, also of the University of London, where they moved in 1956.

The Harpenden Growth Study has never been written up as a whole in a monograph. There are a number of papers dealing exclusively with its data (Marshall and Tanner, 1969, 1970; Marubini, Resele, Tanner and Whitehouse, 1972; Tanner, Whitehouse, Marubini and Resele, 1976) but for the most part the material was used in conjunction with that from other studies to investigate specific problems. The techniques used are described in Tanner's *Growth at Adolescence* (1962, appendix). Height, sitting height, hip, shoulder, elbow and knee widths, circumferences of upper arms, calf and thigh, foot length, weight and four skinfolds were taken. Every measurement on every child on every occasion was done by Whitehouse, a robust man.

Faced with this labour Whitehouse designed a new range of anthropometric instruments of greater accuracy and convenience than the conventional ones, which had indeed remained virtually unchanged for nearly a century. The Harpenden anthropometer moved on a rack and pinion, giving both a smooth motion with easy control and a digital read-out through a counter attached to the moving pinion. A similar principle was incorporated in the height-measuring machine. For the skinfolds an instrument used for measuring the thickness of leather was adapted by adding a spring, so placed that as the opening between the jaws of the calipers increased, the increased tension in the spring was exactly countered by a decrease in the distance of the spring from the fulcrum, thus keeping constant the turning moment on the jaws. The commissioning of this instrument, the Harpenden skinfold caliper, was organized around a thorough test of the reliability of measurement of skinfolds, both by the same person on different occasions and by different persons. The test was designed by Healy (see below) and is a model of its sort (Edwards *et al.*, 1955). Fig. 13.2 shows Whitehouse measuring height using the Harpenden stadiometer, and subscapular skinfold using the Harpenden skinfold caliper (in a Mark 2 version). In the Harpenden study, and most subsequent European studies, height and sitting height were measured with the subject stretching up to his full height, verbally encour-

Fig. 13.2. (a) Measurement of height using the Harpenden stadiometer. (b) Measurement of subscapular skinfold using the Harpenden skinfold caliper. The measurer is R. H. Whitehouse.

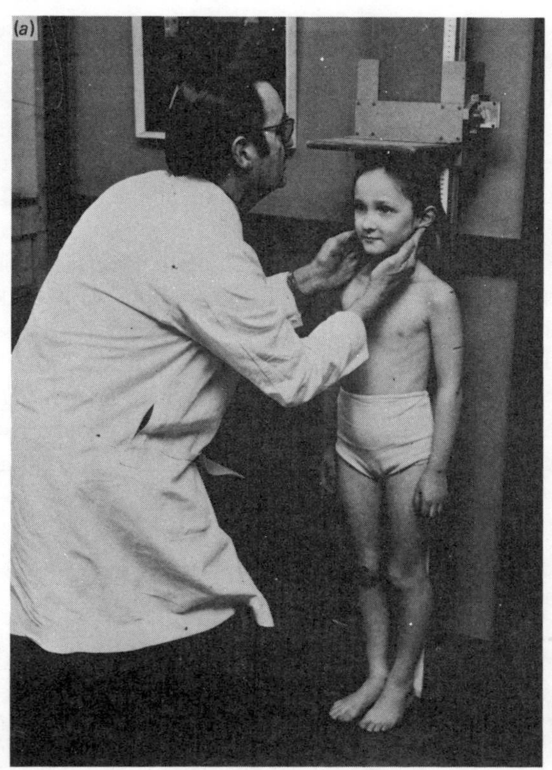

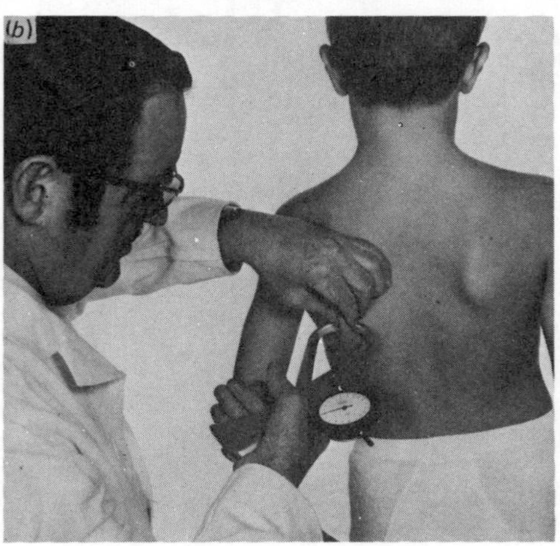

aged by the measurer and assisted by the measurer applying upward pressure under the ears. This diminishes considerably the decrease in height which occurs during the day as the child gets tired, or simply as he gets bored by waiting his turn. The other anthropometric techniques used are more fully described by Cameron (1978).

Photogrammetric pictures were taken, that is, pictures so made that enlargements remain dimensionally accurate to the extent that measurements can be taken off them. The technique is similar to that used in aerial mapping. The child stood motionless on a turntable placed 10 metres from the camera, and was posed in the standard position recommended for somatotyping (Dupertuis and Tanner, 1950). The turntable was then turned (electrically, in later models) to give side and back views. An example is shown in Fig. 13.3. All the pictures were positioned and taken by Tanner. Radiographs were taken of arm, calf and, for a time, thigh, delineating bone, muscle and fat as in the Harvard School of Public Health study. A different pose was adopted, however, that allowed constant magnification whatever the size and age of the child. Clinical examinations were

Fig. 13.3. Photogrammetric picture of child in Harpenden Growth Study pose.

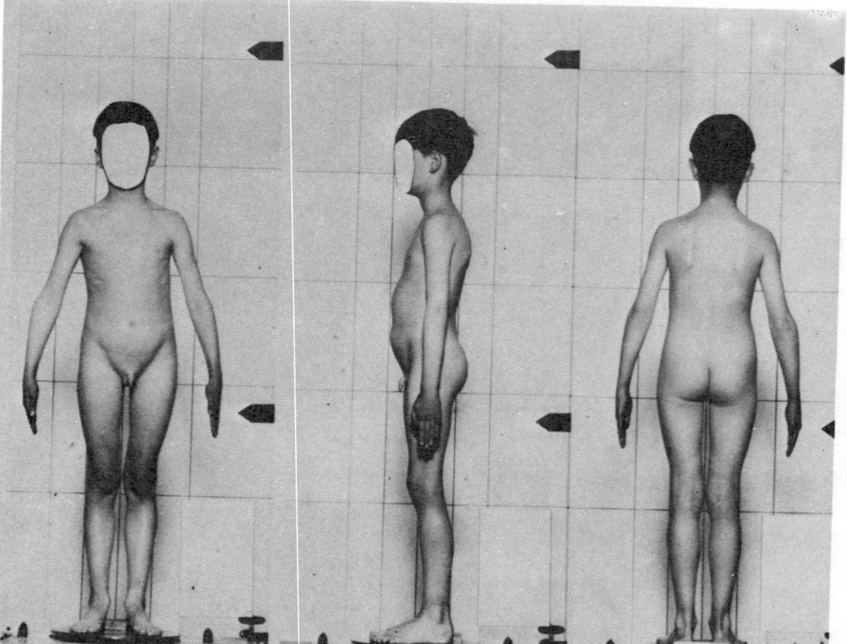

carried out in the usual manner, mostly by Cecile Asher (see Asher, 1975, for postural studies). For a period orthodontic examinations were done. No psychological work was undertaken, nor any physiological research, after an early disappointment over the accuracy of 24-hour urine collections (bottle-sharing and beer-substitution being excessive).

The children were seen six-monthly, or three-monthly during puberty, always within ± 3 weeks of the target date, since the measuring team visited the Home for two consecutive days each month, one for boys, the other for girls. Nearly all the children were over 3 years old on entry, and the majority were over 5 years old. Some were orphans reared in the baby section of the Home since birth; others, an increasing proportion as time went by, were taken into the Home as the result of family breakdown. Most were from families of manual workers or the lower middle class. At the Home they lived in mixed-sex family groups, attended regular schools in the town and had the care and facilities associated with the best boarding schools. They left at varying ages; efforts were made to see them annually thereafter till they had ceased growing. During the whole course of the study some 450 boys and 260 girls were examined; of these eighty-five boys and forty-eight girls were followed for ten years or more. An idea of the general orientation of the study can be obtained from Tanner's Convocation Lecture at the annual convention of the personnel and supporters of the National Children's Home, a United Kingdom charity (Tanner, 1958*a*).

One advantage of the Harpenden study over most others was the frequency of examinations during puberty. Thus the data were appropriate for use in testing the relative efficiencies of the logistic and Gompertz curves in fitting serial measurements of various anthropometric dimensions (Marubini *et al.*, 1972). Both curves fitted well, with the symmetrical logistic slightly the better. In a later paper (Tanner, Whitehouse, Marubini and Resele, 1976) logistic curves were fitted to the data of fifty-five boys and thirty-five girls, mean-constant curves plotted, and the relation between many characteristics of the curves of height, sitting height, and shoulder and hip widths examined. Though the ages at which each measurement reached its peak velocity were closely correlated, the total adolescent gains in the four measurements were to a large degree independent, implying a considerable increase in differentiation of shape between persons during the spurt. The interest in curve-fitting was carried still further by Michael Preece (1944–), a paediatrician with a postgraduate degree in statistical method, who joined Tanner's department at the Institute of Child Health in 1974. Together with M. J. Baines, a mathematician, he described a family of curves which provide what seems at the time of writing the best and most economical fit to measurements of height and other dimensions from about age 2 to maturity (Preece and Baines, 1978). In Fig. 13.4 is shown the fit of the curve to height measurements of a Harpenden Growth Study child. The family of curves shared the differential equation.

$$\frac{dh}{dt} = s(t)\cdot(h_1 - h)$$

where h is height or other measurement, h_1 is the mature height and $s(t)$ a function of time. In the simplest model, which is the one fitted in Fig. 13.4, this function is specified by

$$\frac{ds}{dt} = (s_1 - s)(s - s_0)$$

where s_1 and s_0 are rate constants. The integral form of this equation is the curve in the figure.

Marshall and Tanner (1969, 1970) used the 228 boys and 192 girls of the study who had been followed throughout puberty to define the range of variation seen in the development of pubic hair, genitalia and breasts. The pictorial standards for ratings, on five-point scales, of each of these characteristics, developed from those of previous authors, were given in Tanner (1962) and have since been widely used. One of the results of this work was to show that children differ even in the speed with which they complete each particular developmental sequence, for example of breasts, as well as in the age at which each sequence starts. Some sequences are closely related (menarche and the height spurt for example); others less so.

The radiological data on bone, muscle and fat growth were used mainly to study changes at puberty. Individual growth curves were lined up in terms of years before and after peak height velocity, in the manner of Shuttleworth. In this way it was shown that the peak of muscle increase coincided with the peak of sitting height velocity. In limb fat the velocity curves had the converse shape: at puberty the velocity fell, to the extent that in boys, though not in girls, there was a short-lived actual loss (i.e. negative velocity) coincident with peak height velocity (Tanner, 1965). Boys had a relatively greater spurt in the area of compact, cortical bone than girls (Gryfe *et al.*, 1971).

Another use made of the Harpenden data was in the construction of individual-type standards for distance and velocity of height and weight, the Harpenden material being used to provide the *shape* of the curves, while extensive cross-sectional data gave the *amplitude*: this work is further discussed in Chapter 14. In the same way, when opportunity arose to develop British standards for skeletal maturity, based on large numbers of hand radiographs of a representative sample of children in Scotland (see p. 361), the Harpenden longitudinal series of radiographs were used to define the *stages* of development; the standards were then derived from the extensive cross-sectional data. This fusion of longitudinal and cross-sectional data, particularly in the construction of standards suitable for the clinical following of individual children, was a constant theme of Tanner and his colleagues.

Such an endeavour required considerable biometrical expertise, and

Fig. 13.4. Fit of Preece–Baines Model I curve to height measurements of a Harpenden Growth Study girl. From Preece and Baines (1978).

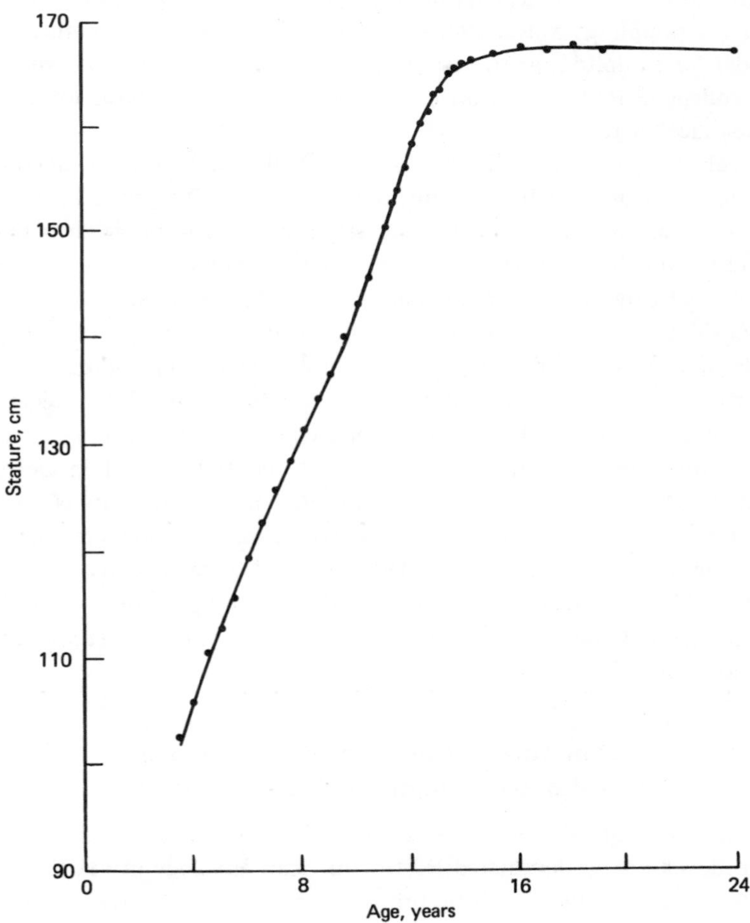

this was supplied to Tanner's group by M. J. R. Healy (1923–), a close collaborator from 1950 onwards. Rothamsted Agricultural Research Station, made famous as the place in which R. A. Fisher developed most of modern biometry and experimental design, is situated in Harpenden, about fifteen minutes' walk from the Children's Home. Though Fisher himself had left Rothamsted for Cambridge by the time the Harpenden Growth Study began, his successor Frank Yates was there, and it was natural for Tanner, with his own biometrical background, to seek help from so celebrated and convenient a source. The immediate fruit of this visit was a paper outlining the methods necessary for the efficient extraction of information from mixed longitudinal data (Tanner, 1951). Boas and Shuttleworth had outlined the problem, and Wilson (1942) had dealt with it in more mathematical terms. But only with the appearance of Yates' text on sampling (Yates, 1949) and the development of a mathematical model for sampling on successive occasions with partial replacement, by his colleague H. D. Patterson (1950), was there a solid basis for a useful auxological method.

Healy was a junior colleague of Frank Yates; he had the distinction of having been brought up in a doctor's family, and his prime and lasting interest was in helping medical research workers improve their understanding and handling of their often extremely awkward data (see Healy, 1978*b*, *c*). Healy's interest in the manipulation of growth data brought about a considerable transformation in human auxology; his papers form the basis of a modern understanding of reliability, standards and multivariate interrelationships (Healy, 1952, 1958, 1974, 1978*a*). One of his protégés was Harvey Goldstein (1939–), a statistician who worked in Tanner's department from 1964 to 1971, subsequently at the National Child Development Study (see Chapter 14) and from 1977 at the University of London Institute of Education. Goldstein, who wrote the first monograph specifically on auxological biometry (1979), played a particularly prominent part in the biometrical education of the doctors and psychologists working in the International Children's Centre longitudinal studies. To these studies we now turn.

The International Children's Centre Coordinated Longitudinal Growth Studies

Just as the longitudinal studies arising, for the most part, in a context of child development dominated North American auxology in the 1940s and 1950s so growth studies arising primarily in a medical context dominated European work in the 1960s and 1970s. And just as the American studies had their fairy godmother in Lawrence Frank, so the European studies found their champion in Natalie Masse, a French paediatrician who worked in the International Children's Centre in Paris.

The prime mover at the beginning, however, was Alan Moncrieff (1901–1971), professor of child health at the Institute of Child Health of the University of London, the postgraduate school of preventive and therapeutic paediatrics attached to The Hospital for Sick Children, Great Ormond Street. Moncrieff was a man of quite exceptional vision. With the help of the Nuffield Foundation he founded the Institute of Child Health in 1946 and served as its director till 1964; he greatly developed the midwifery services both nationally and internationally, and in 1946 played a prominent part in setting up the National Perinatal Mortality Survey which later gave rise to the National Child Development Study (see Chapter 14). It was Moncrieff who more than any other put preventive and social paediatrics on the European map.

In 1947 James Spence (1892–1954), the professor of child health at Newcastle-upon-Tyne, who was also very prominent in the establishment of social paediatrics, began an extensive follow-up type survey of 1,000 families in Newcastle (see Chapter 14). Partly to complement this study, and partly, perhaps, because of his contacts with Ryle at Oxford, Moncrieff proposed to start an intensive, longitudinal, birth-to-maturity study in the Institute of Child Health. This was in 1949. Paediatricians and educational psychologists would share the responsibility equally; the children would be examined by both, and in this way integration of the disciplines, that still-sought grail, would be assured. The Institute of Education willingly agreed, and after a brief pilot study a cohort of 224 children was recruited from a more or less random sample of women attending ante-natal clinics in the West Central area of London during the years 1951–4. Facilities were minimal, however, and in the pilot study doctors in The Hospital for Sick Children were used to measure the babies. Thus for the second time (see note 13.2) a delegation arrived at Tanner's laboratory to demand why it was that babies shrank instead of growing. Tanner had the answer in Whitehouse, and recommended that a full-time professional be assigned to the study. Thus Frank Falkner (1918–), a young paediatrician recently returned from a year in Cincinatti and awaiting appointment as Moncrieff's medical assistant, was put in charge of the project. Though he was soon to leave this particular study, Falkner remained as a central figure in the life of the International Children's Centre Studies, coordinating the work of the several teams, and securing their safe passage through the vicissitudes inseparable from all longitudinal studies.

Shortly after Falkner joined Moncrieff, R. Debré (1883–1978), the professor of paediatrics in Paris and a friend of Moncrieff, decided that France, too, should have such a study (see note 13.3). Sensibly, the two men agreed that exactly the same base-line methods should be used in both studies, alike on the somatic and psychological sides, so that eventually the results could be, if not actually pooled, at least used as some sort of replicates. In 1953 Falkner himself was despatched for a year to Paris to set up the French work. Debré was chairman of, amongst other things, the International Children's Centre (ICC), an organization founded in 1950 and supported partly by the French government and partly by UNICEF. The Centre had a strong department for postgraduate education in social paediatrics, particularly as applied to developing countries, and studies of growth fitted in nicely with its programme. Natalie Masse (1919–1975) was appointed director of medical activities of the Centre in 1954 and the French growth study was placed under her wing for administrative and teaching purposes, though it was actually conducted in a hospital in a different area of Paris.

Thus began the ICC Coordinated Longitudinal Growth Studies. The professors of paediatrics at Stockholm and Zurich, A. Wallgren and G. Fanconi, became interested and set up their exactly parallel cohorts; Marcel Graffar, professor of social medicine at Brussels, did the same; and three out-stations joined the group: one in Dakar, West Africa, under Jean Sénécal, a paediatric nutritionist, the second in Kampala, East Africa, under R. F. A. Dean, and the third in Louisville, Kentucky, where by this time Falkner himself had gone to be professor of paediatrics. Every two years the members of all the teams met for a few days, at each of the European laboratories in turn, to discuss initially their methods and later their results (see e.g. Falkner *et al.*, 1958). In this way the methods were kept standardized, some uniformity of analysis was achieved, and a considerable process of mutual education took place. As time went by, other, similar, longitudinal studies sprang up in Europe and their members joined the ICC meetings as guests of the teams. In this way longitudinal studies in Helsinki, Prague, Wroclaw, and Istanbul shared many of the methods and objectives of the ICC teams.

Masse continued to provide the organizational backing until her early death from cancer in 1975; Falkner remained as coordinator of the teams on the somatic side, and Colin Hindley, professor of child development at the Institute of Education in London, and psychologist with the London team from the beginning, coordinated the psychological side. The proceedings of these meetings (International Children's Centre, 1954–76) give a good idea of the type of work undertaken; and a very valuable and complete bibliography of all publications by members of the teams from 1951 to 1976 was published by the International Children's Centre in 1977. A general description of the studies was given in a monograph edited by Falkner (1960). A seminar on 'Growth of the normal child during the first three years of life', held in 1960 at Zurich, brought together members of the North American studies as well as the European ones; the proceedings (Merminod, 1962) include discussions of most of the problems seen at that time to be of paramount importance.

Another, more complete and authoritative book, was published in 1966 under Falkner's editorship. Entitled *Human Development*, it consisted of twenty-nine essays, by several of the most distinguished experts in the United States, England, France and Denmark, concerning both physical and psychological growth. It provides a very clear conspectus of the state of the field in the 1960s.

On the somatic side of the ICC studies, the base-line investigations, that is, those which were undertaken by every team, consisted of anthropometric measurements based on the Harpenden Growth Study techniques, ratings of pubertal characteristics similarly based, radiographs of the hand and wrist, and a medical examination. Each team was at liberty to add any other investigations it wanted; most, but not all, took photogrammetric photographs. A social history was taken and a psychological examination made. Children were seen at 1, 3, 6, 9, 12, 18 and 24 months and thereafter at yearly intervals, except that in some studies six-monthly visits were arranged during puberty (see Falkner, 1955).

The ICC London study

The London member of this group of studies began with 224 children (Moore, Hindley and Falkner, 1954), of whom about 40 per cent (forty-three boys and forty-five girls) were successfully followed until their growth was completed (a point taken to be reached when the increase in height was less than 1 cm over a period of one year). In 1956 Falkner went to the United States to join the department of paediatrics in Louisville, Kentucky, and Tanner moved from St Thomas's Hospital to the Institute of Child Health. Whitehouse went with him. Thus the ICC London study and the Harpenden Growth Study were joined under one roof. Moncrieff firmly believed that auxology was the basic science of paediatrics, and soon a full department of growth and development was set up in the Institute. A large grant from the Nuffield Foundation established it in the form of a sort of micro-Fels Research Institute, with sections devoted to experimental auxology and to child ethology, the latter under N. Blurton-Jones. After some years a regular Chair of child health and growth was established in the University, thus giving to human auxology the permanent academic home it lacked elsewhere.

In 1962 W. A. Marshall (1929–), a physician with a PhD in experimental reproductive physiology, joined the department. He remained with the ICC study, doing much of the measuring, until he was appointed professor of human biology at Loughborough University of Technology in 1977. Besides his studies with Tanner on puberty in the children of the Harpenden Growth Study (see above; also Marshall, 1974) Marshall began the analysis of the enormous archive of photographs that had been built up of Harpenden and ICC children, and of children suffering from growth disorders (Marshall and Ahmed, 1976). He made studies of seasonal variation in height growth (Marshall, 1971, 1975) and published an important paper showing how prediction of the age at which menarche would occur in a presently pre-menarcheal girl could be improved by knowledge of her skeletal age (Marshall and de Limongi, 1976). This paper made clear the crucial distinction between *prediction* (of the time of a future event) and *estimation*. The two had become confused through the careless use of the word 'prediction' in association with the statistical method of regression. The distinction lies in the distribution of errors. The most likely age at which a pre-menarcheal girl now aged 12.2 will have menarche is, say, 13.0 (the mean for the population.) The distribution of possible ages around this most likely (that is, predicted) value is already asymmetrical, for some girls in the population will have menarche before 12.2, and our particular girl is not amongst them. The error of prediction from, say, her present skeletal maturity, or her height and weight is not ± 2 years (the 2 S.D. population range) but less. Further, if a second girl is aged 14.2 and we are required to predict her age of menarche, the most likely time is tomorrow and the error distribution is even narrower. Much growth work concerns such time-entrained, or censored, events. Marshall and de Limongi showed that the additional knowledge of the girl's skeletal maturity under most circumstances improved the prediction of when menarche would occur.

Harvey Goldstein was also a member of the London team for several years and remained statistical mentor to the whole ICC group. He was largely responsible for one of the few papers in which data from all the teams were actually combined. This was a restudy of the problem attacked by Bayley in 1954 and by Tanner and Israelsohn in 1963, both with distressingly small numbers: the correlation between heights of children at increasing ages and the heights of their parents. The combined data from London, Paris, Brussels, Stockholm and Zurich produced 518 mothers and 243 fathers actually measured by the teams. The correlations, like Bayley's, increased sharply from 1 month to 2 years but from then on remained sensibly constant (at about 0.4 with heights of mothers and fathers singly, and 0.5 with the average of parents' heights) until the beginning of adolescence (during which they fell because of tempo effects). Tanner, Goldstein and Whitehouse (1970) produced regression standards of children's heights at ages from 2.0 to 9.0 years on average parental height, providing a practical way for the clinician to allow for the parents' heights when judging the normality or otherwise of a small or large child. No significant differences were found in these ICC data between the four correlations mother–daughter, mother–son, father–daughter and father–son. Tanner and Israelsohn (1963), using the London data only, gave the changes in these correlations for a whole series of body measurements.

The London group also developed a new method for the assessment of skeletal maturity from radiographs of the hands and wrists. This sprang from the circumstances of being given by the Ministry of Health a large sample of radiographs, taken on some 900 boys and 900 girls in the schools of an urban and a rural area in southern Scotland (in a Ministerial effort to reassure Parliament about the benign effects of compulsory fluoridation of water supplies). This representative sample contrasted with the optimally circumstanced sample that generated the Greulich–Pyle standards. The Acheson method was adopted, though Healy and Goldstein developed a more sophisticated weighting technique whereby the stages of each of the bones were scored not simply 1, 2, 3, etc. but by means of weights chosen so that the overall maturity score represented that estimate for which the concurrence, or agreement, of all the bones was greatest. The longitudinal samples of the Harpenden Growth Study and ICC London study were used in the delineation of successive stages which could be reliably distinguished, and the subjects of these studies plus those of the Oxford Child Health Survey were used as the standardizing group in the pre-school years. The system was circulated in a preliminary form amongst the ICC teams from 1962 onwards, and eventually published in 1975, together with a table of prediction of adult heights resulting from its use (Tanner, Whitehouse, Marshall, Healy and Goldstein, 1975). The London study closed in 1974.

The Paris study

A total of 497 children were recruited into the Paris study between 1953 and 1955 but only 119 of them were still available for measurement at age 16. The study closed in 1975. Accounts are given in Roy-Pernot (1959), Sempé, Tutin and Masse (1964), Roy-Pernot *et al.* (1972) and Roy-Pernot, Sempé and Filliozat (1976).

The Zurich study

In Zurich 412 children were recruited and 110 girls and 112 boys remained at maturity, an impressive figure compared with London. This was the largest birth-to-maturity cohort that has yet been followed. The investigation was led by Andrea Prader (1919–), who succeeded Fanconi as professor of paediatrics at Zurich. Prader's main clinical interest was in paediatric endocrinology but he rapidly became a proficient auxologist, always with an eye to using the normal to illuminate what was happening to his patients. It was he who, with Masse, was chiefly responsible for turning the ICC longitudinal studies in the direction of useful clinical application, and for bringing Tanner back from pure research to the practice of medicine. For Prader the Zurich Growth Centre was an interesting adjunct to the real life of hospital practice; for Tanner hospital practice was a useful and enjoyable interlude in the real life of biological research. The two combined their skills in a paper whose rather inelegant terminology passed into the general literature (Prader, Tanner and von Harnack, 1963). Prader was much impressed with the ability of children to accelerate back towards their normal growth curve after some force that had been restraining their growth had been removed. He and a German colleague, von Harnack, collected examples and Tanner helped in the analysis. The problem was what to call the phenomenon. Acheson (1960) had talked of 'compensatory growth' but that term had been pre-empted by zoologists to refer to two quite different phenomena: the regrowth of an amphibian limb after its removal, and the excess growth of a remaining kidney or testis, for example, when the other of the pair had been removed. Prader and his colleagues decided on the term 'catch-up growth', this being a literal translation (in German term *aufholen,* to catch up with, as in a race; for history of catch-up growth see Tanner, 1979). Tanner (1963*b*) later outlined a hypothetical physiological model for how catch-up was controlled, in a lecture at the Institute of Child Welfare in Berkeley, commemorating Harold Jones. He regarded catch-up as simply an example of the general phenomenon of homeorhesis or canalization, described by C. H. Waddington in his highly influential book *The Strategy of the Genes* (Waddington, 1957).

The Zurich team published an important analysis of growth in the first four years after birth (Heierli, 1960), and later, when the children in the study had passed through puberty, were amongst the first to use smoothing spline functions to estimate individual growth curves (Largo *et al.*, 1978). They were fortunate in having a permanent anthropometrist in M. Willisegger, who measured all the children from age 8 onwards, making six-monthly measurements throughout puberty. Their results were in general similar to those obtained by Deming and later authors: peak height velocity occurred on average at 13.9 years in boys and 12.2 in girls with 'instantaneous' intensities of 9.0 cm/yr and 7.1 cm/yr respectively (1978, p. 428; and see Fig. 13.5). The Zurich authors made the new point that the degree to which the velocity rose from the pre-spurt minimum to the peak spurt maximum was not, like the absolute value of the peak velocity, related to the age at which the peak occurred. The basal velocity dropped with age, so late maturers took off from a lower level: they gained on average as much as early maturers, even though reaching a lower peak. The point was of importance in the elucidation of the endocrine mechanisms of the spurt.

The Zurich group also published a paper on the growth of the testes (Zachmann *et al.*, 1974), something which had not been studied properly in the Harpenden Growth Study or in the North American investigations. The size of the testis was estimated by tactile comparison with a string of wooden models of increasing size known as the Prader–Zachmann orchidometer,

Fig. 13.5. Mean height velocity curves, peak height velocity (PHV) centred, of 112 boys and 110 girls in the Zurich longitudinal study. Individual curves estimated by smoothing spline functions. From Largo *et al.* (1978).

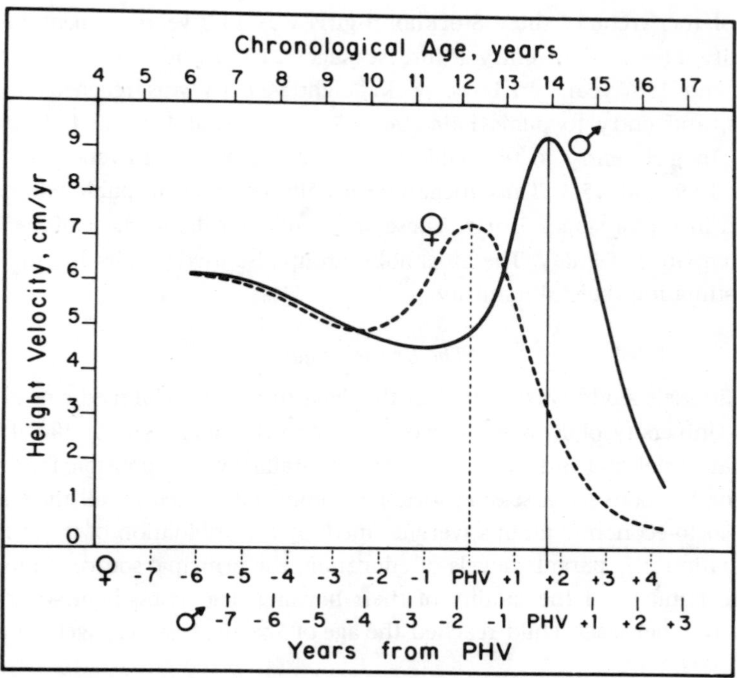

now used by paediatric endocrinologists throughout the world. In the study of their hand–wrist radiographs also, the Zurich team imported a new technique. Bonnard (1968) measured the thickness of the external bony cortex of the metacarpals compared with the internal medulla, and showed how it changed with age.

The Stockholm study

The Stockholm study began, like the Zurich study, in 1955, and was directed by Petter Karlberg (1919–), Wallgren's first assistant and later professor of paediatrics at Göteborg. Out of an initial recruitment of 122 boys and 90 girls no less than 80 per cent (100 boys and 73 girls) were followed to age 18. The main results of the study were published as a supplement to volume 258 of *Acta paediatrica Scandinavica* (Karlberg *et al.*, 1976; Taranger, 1976; Taranger *et al.*, 1976*a, b, c, d*). Taranger and his associates used the probit technique to obtain mean ages for all the usual stages of pubertal development, using log post-conception age (i.e. log $[t + 0.75]$ years). This technique gives average ages of *transition* from one stage to another, i.e. the age of very first appearance of a stage, to be differentiated from the condition of being-in-a-stage which is what is observed in clinical work (a point of difference not always made clear in the literature). The average age of menarche of these Stockholm girls was 13.0 years, of peak height velocity 11.9, and of entry to breast stages 2, 3, 4 and 5 ages 11.0, 11.8, 13.1 and 15.6 years. In boys, peak height velocity was reached at 14.1 years, and entry to pubic hair stages 2 to 5 was at 12.5, 13.4, 14.1 and 15.5. In girls entry to the pubic hair stages occurred on average at 11.5, 12.0, 12.9 and 15.2. Thus menarche usually occurred in pubic hair stage 4 and in breast stages 3 or 4. These are similar results to those of the Harpenden Growth Study. The Stockholm group also used probits as a method for estimating skeletal maturity.

The Brussels study

The Brussels study was located in the department of social medicine at the Free University of Brussels, directed by Marcel Graffar (see Graffar, 1959; Graffar, Asiel and Emery-Hauzeur, 1962). Graffar was responsible for devising the methods for assessing socio-economic status used by all the teams. Five socio-economic groups were defined, by a combination of the father's occupation, the parents' levels of education, the principal source of income of the family and the quality of their housing. The assessment was done initially when each child reached the age of 3 years. In a very well-worked paper incorporating the first joint analysis by all the ICC teams, Graffar and Corbier (1966) showed that socio-economic group differences in height were already present in these urban children at age 1 and increased through age 3 to age 5. They were, however, much smaller than differences between equivalent groups of twenty years before. The boys, as usual, showed larger differences than the girls. In weight the differences were smaller, since, again as usual, in European urban children the lower socio-economic groups had greater weight for height. No significant differences were visible in skeletal maturity up to age 5. Differences in mental test results were also given. We shall meet similar results for height and weight in the survey data discussed in Chapter 14; only in Sweden in the 1960s had social class differences disappeared from urban children's growth (see below).

The Dakar and Kampala studies

The Dakar study provided an interesting counterweight to the studies in European cities, especially in respect of skeletal maturation and skinfolds (Massé and Hunt, 1963; Massé, 1969). At the time it began little was known about differences in growth between Africans and Europeans. The group in Kampala showed that Africans were advanced in skeletal maturation during the first year or two relative to Europeans living in the same surroundings, but thereafter fell behind. Only several years later did studies on well-off Americans of African descent in North America, by Garn, Johnston and others, show that given identity of economic provision Africans remained slightly ahead of Europeans in all aspects of tempo of growth (see summary in Eveleth and Tanner, 1976).

The Louisville study

The Louisville study was different from all the others in that Falkner, its director, had chosen to investigate twins (see Falkner, 1978). There had been studies of the measurements of twins before, but never one in which they had been followed longitudinally from birth. In 1962 Falkner set up an organization which recruited continuously a proportion of all the twins born in Louisville. Their growth was measured in the usual ICC manner. Concurrent psychological studies were made by Steven Vandenberg. Both Falkner and Vandenberg left Louisville before the work came to fruition; but it continued and continues under Ronald Wilson (1933– ; see Wilson, 1974, 1976, 1979). By 1979 over 900 twins had been recruited, with 600 followed to at least 5 years and 400 to 8 years. Wilson showed that at birth the differences in weight and length between monozygotic (MZ) twin pairs were actually greater than the differences between same-sexed dizygotic (DZ) pairs; but as the catch-up after the uterine restriction proceeded so the monozygotic pairs came more and more to resemble each other, while the dizygotic pairs remained unlike. Thus the average difference in length between MZ twins was 1.8 cm at birth and 1.1 cm at age 4; between same-sexed DZ pairs it was 1.6 cm at birth and 3.2 cm at age 4. Much of the reassortment of status had occurred by 1.0 years, when the between-pair correlation for length of MZ twins had risen from 0.62 at birth to 0.86; by age 4.0 it was 0.94.

The Stockholm School of Education Longitudinal Study

One other European study needs to be mentioned here. It was originally little related to the ICC studies, being sponsored by educationists and conceived more, perhaps, in the image of the Harvard School of Education Growth Study. But its director Bengt-Olov Ljung (1927–) and his associates Siv Fischbein and Gunilla Lindgren came to have close connections with the London ICC group. The study concerned a large sample of twins, drawn randomly from the forty largest urban school areas of Sweden. These twins were followed from age 10 to 16 (girls) or 18 (boys). The 323 pairs represented approximately one third of all twins born in Sweden in 1954–5. As well, there was a control group of 740 singletons, classmates to the twins and matched for age and sex. Height and weight were measured twice-yearly by specially trained school nurses, and tests of ability were administered.

The singletons provided a true random sample of the Swedish urban population and were used in assessing the secular trend towards earlier maturity and greater size in Sweden (Ljung, Bergsten-Brucefors and Lindgren, 1974). The study of their measurements gave the very important result that, for the first time in any published data, the usual socio-economic gradient of height had disappeared in Swedish towns (Lindgren, 1976). Lindgren (1978) described the adolescent growth of the singletons in terms of years before and after peak height velocity (which occurred on average at 14.1 years in the boys and 11.9 years on the girls, with menarche at 13.1 years). Fischbein (1977; Fischbein and Nordquist, 1978) reported, for the first time, on the degree of resemblance in both amount and timing of the adolescent spurt in monozygotic twins. A comparison between the growth of twins and that of singletons showed only trifling differences in the case of boys, but a lower height and weight in twin girls than in singleton girls (Ljung, Fischbein and Lindgren, 1977). These Swedish data complement well the Louisville data of Wilson.

Other European studies

The ICC studies, and their meetings, served both to channel existing longitudinal work into their own mould, and to stimulate new work. In Helsinki a study very similar to the ICC ones was made by N. Hallman, L. Backström-Järvinen, R.-L. Kantero and R. Tiisala (Hallman *et al.*, 1966, 1971; see also Tiisala, Kantero, Backström and Hallman, 1969, and Tiisala, Kantero and Tamminen, 1971). In Prague a longitudinal study was carried out by V. Kapalín, a paediatrician, and M. Prokopec, an anthropologist (Kapalín, Kotasková and Prokopec, 1969). Other studies were made in West Germany by W. Hagen, a public health doctor (Hagen 1964, 1966), in Florence by H. Bourterline-Young, a British paediatrician settled there; and in Istanbul by O. Neyzi, a paediatrician. Many of the results of these studies can be traced through the bibliography of the ICC teams' meetings.

In Poland in the 1960s there were two longitudinal studies. In Lublin some ninety boys and ninety girls born in 1964–5 were followed from birth to age 7 by Hanna Chrzastek-Spruch (1968, 1977, 1979*a, b,)*, a paediatrician. The children were seen every month in the first year and every three months in subsequent years, a greater frequency of measurement than in any other series. No curve-fitting studies have yet been reported however. In Wroclaw, Tadeusz Bielicki, director of the Institute of Anthropology, and his associates studied growth during puberty in some 200 boys and 250 girls followed from age 8 to 17 or 18 (Bielicki, 1975, 1976; Bielicki and Walisko, 1976; Walisko and Jedlinska, 1976). The study ran from about 1961 to 1971. As in all recent Polish studies, cross-sectional (see Wolanski, 1979) as well as longitudinal, emphasis was at first laid on the development of the parent–offspring correlations (Bielicki and Welon, 1966; Welon and Bielicki, 1971). Later the sequence of pubertal events in girls was considered in some detail. Menarche occurred on average at 13.1 years. The first of the sequence of peak velocities in the various measurements to be reached was foot length, which was seven months earlier, in both boys and girls, than the next measurement, leg length, and a whole year earlier than hand length. Trunk length peaked last of the measurements, at 12.3 in girls and 14.4 in boys (Bielicki and Welon, 1973; Welon and Bielicki, 1979). In 1967 a study of the growth of twins was also initiated in Wroclaw.

Orthodontically based studies

Orthodontists had naturally always been interested in growth, since their whole specialty depended on programmes for modifying growth or predicting what would happen in the absence of modification. We have seen how the cephalometer was developed in Todd's workshop in the 1920s, and how the medically oriented North American longitudinal studies (Harvard School of Public Health, Denver, Fels) incorporated orthodontic studies in their programmes. The ICC studies did not do this, chiefly for financial reasons. However in Nijmegen from 1971 to 1975 a growth study based at the department of orthodontics was made (Prahl-Andersen, Kowalski and Heydendael, 1979). It was especially distinguished by using a different sampling design from that of the ICC studies. Those had all used one single longitudinal cohort. The long-term studies such as the Fels and Harpenden used a 'mixed longitudinal' design, with children entering and leaving at different ages. Nijmegen used a 'linked mixed longitudinal' design, where six cohorts were studied, each for five years, but each starting at a different age so that by combining the results the age range of 4 to 14 could be covered in the five years (4 to 9, 5 to 10, etc.). In all, 232 boys and 254 girls were followed (Prahl-Andersen and Kowalski, 1973). Birte Prahl-Andersen also introduced a new anthropometer whose design permitted automatic registration of the reading (Prahl-Andersen *et al.*, 1972). The age range covered by this study, though good for some purposes, precluded a study of adolescent changes (van t'Hof, Roede and Kowalski, 1976, 1977). In Utrecht a still shorter-term, two-year study was made (Venrooij-Ysselmuiden and Ipenburg, 1978).

Another orthodontically based study, to a large extent deriving from European experience, was that of Arto Demirjian (1931–) on French-Canadian children at the Montreal Human Growth Research Centre (see Demerjian, Goldstein and Tanner, 1973). Two cohorts were followed, one from 6 to 15 in 1967–76, and the other from 10 to 19. The facial growth of fifty girls of the former cohort, given a cephalometric radiograph each year, was reported by Baughan *et al.*, (1979) in a paper justly critical of much of the previous literature on this subject. Facial measurements showed a small adolescent spurt with the peak on average coincident with that of sitting height.

Amongst the orthodontists who were making studies of growth, Arne Björk (1911–) of the Royal Dental College, Copenhagen, was a pioneer in Europe, though most of his studies concerned patients with orthodontic pathology (see Björk, 1947, 1955, 1963, Björk and Palling, 1955). In Burlington, Toronto, at the Burlington Orthodontic Research Center, a mixed longitudinal study was made from 1952 to 1971 on facial development in 1,258 normal children, examined annually and starting at ages 3, 6, 8, 10 and 12. In all some 48,000 cephalograms, 8,000 radiograms of hands and 16,000 facial photographs were accumulated. This study was initiated by

R. E. Moyers (1919–), then head of orthodontics at Toronto University, and later director of the Center for Human Growth and Development at the University of Michigan (see Moyers and Krogman, 1971). A report on 111 girls followed from age 4 to 14 has been published (Thompson, Popovitch and Anderson, 1976); much more remains to be done.

In Portland, Oregon, another orthodontist, Bhim Savara (1924–), and his associates followed the growth of individual bones of the skull from age 3 to 16 by using trigonometric methods to locate specific points on them from a combination of lateral and antero-posterior radiographs (Singh and Savara, 1966; Tracy and Savara, 1966; Nakamura, Savara and Thomas, 1972).

Low-birthweight children

There have also been numerous longitudinal studies of children who were small at birth (see e.g. Drillien, 1964; Cruise, 1973; Neligan *et al.*, 1976; Brandt, 1978). From them emerged the finding that simply being born early was not in itself deleterious to the child. Children born after less than 37 weeks' gestation are called 'pre-term' children. By the 1970s paediatric handling of such children had advanced to the degree that even a child born as early as 28 weeks and weighing only 1.00 kg could, in the best hospitals, be kept growing in an incubator at the very rapid rate appropriate to its age, and sent home 8 to 10 weeks later at the normal weight for a full-term infant. Thereafter growth was normal. On the other hand babies born at the normal time but with below-normal birthweight were shown not to be able on average to catch up to the size or the mental ability of normal children. The average such baby reaches about the 25th centile of normal height. These are babies who have suffered some intrauterine pathology, either through noxious substances reaching them from the mother, lack of necessary food from the mother or disorder of the placenta or the fetus itself. It is these children, colloquially referred to as 'small for dates', that were proposed above as the model of what occurred in a more widespread way to children of working mothers in the nineteenth century.

The Society for the Study of Human Biology

Soon after the ICC longitudinal studies began, another event of importance in the history of human auxology took place. Growth had traditionally been studied (if never with very high priority) as part of the syllabus of physical anthropology, or biological anthropology as it was called on the European continent. But some aspects of the older physical anthropology, such as its insistence on the immutability of the cephalic index and the importance of the colour of the hair and eyes, were felt to be irksome; and to emphasize that the subject had changed almost beyond recognition a number of its practitioners started calling themselves human biologists. The term first came into prominence when Raymond Pearl, himself a biologist interested in man, named his new journal *Human Biology* (see Chapter 12). But no Society of human biologists existed, and indeed, lacking the support of one, in 1949 the journal nearly collapsed.

In North America there was at least the American Society of Physical Anthropology; but in England no equivalent existed. However, in 1957 the Royal Anthropological Society sponsored a symposium entitled 'The scope of physical anthropology and human population biology and their place in academic studies' (see Roberts and Weiner, 1958), and this resulted in the formation of the Society for the Study of Human Biology in May 1958 (see note 13.4). By 1973 the Society, from the beginning multinational, had 410 members in thirty-one countries, and began to feel the need for a journal of its own. Up till then its members had frequently published in *Human Biology*, of which indeed Tanner, a founder-member of the Society, was European editor. But that was not quite the same as having a journal officially sponsored by the Society. Thus in 1973 the *Annals of Human Biology* was begun. Since then papers on the normal and theoretical aspects of human growth have been found for the most part in that journal, with some still in *Human Biology*, a few in the *American Journal of Physical Anthropology* and the remainder in clinical journals, either of paediatrics or of public health and epidemiology. 'Human biology', wrote P. B. Medawar in his introduction to the first textbook on the subject (Harrison *et al.*, 1964), 'is not so much a discipline as a certain attitude of mind towards the most interesting and important of animals ... it portrays mankind on the canvas that serves for other living things'. Amongst other things he thought a case can be made for thinking our medical education anew and rebuilding it upon a foundation of human biology'. Tanner (1958*b*, 1964) went still further: like Florence Nightingale prescribing Quetelet's social physics for intending Cabinet Ministers, he recommended human biology to the attention of poets and law-givers, educators and all interested in rebuilding a common humanist culture. More immediately a degree course in the subject, then called human sciences, was initiated in 1970 at the University of Oxford, with G. A. Harrison (1927–) as director. Thus human auxology, over the last twenty or thirty years, has come to have its feet planted firmly where it is reasonable they should stand: one in human biology and the other in paediatrics.

The study of growth disorders and the physiology of growth

Much of human auxology grew up in partnership with preventive paediatrics; but therapeutic paediatrics, that is to say the branch of paediatrics dealing with the diagnosis and treatment of children already disordered or ill, came to see the relevance of auxology very late. Doctors working in schools and infant welfare clinics had long recognized that normal growth was the best criterion of good health by the time that paediatric endocrinologists, for example, had realized that growth was also the best criterion of the effectiveness of many of their treatments. Lawson Wilkins, of Johns Hopkins, the great founder of paediatric endocrinology, was not much concerned with growth and he left behind him in North America a generation of pupils of whom, with only a few exceptions, the same was true. Thus the North American longitudinal studies had scarcely any impact on therapeutic paediatrics (the exception as usual being Nancy Bayley who, with a paediatrician, Leona Bayer, published a monograph called *Growth Diagnosis* (Bayer and Bayley, 1959, 1976) that was largely ignored until the boom in auxology of the 1970s caused the publishers to reissue it). Modern methods of auxology came to be applied in the therapeutic field mostly after Prader had joined the ICC study of normal growth and Tanner had returned to clinical medicine to start a Growth Disorder Clinic at The Hospital for Sick Children in London (see e.g. Barr, Schmerling and Prader, 1972).

The London Growth Disorder Clinic began in 1963; the methods developed at Harpenden and in the ICC were applied to children with short stature due to endocrine and other causes. A few years earlier M. Raben in Boston had for the first time successfully treated a child with growth hormone deficiency with human growth hormone, and The Hospital for Sick Children clinic became the major centre from which this treatment in the United Kingdom was controlled (see Tanner *et al.*, 1971). Because of a relative shortage of human growth hormone, which had to be extracted from cadavers, careful control was essential, both of diagnosis and treatment, and eventually a network of growth clinics was established covering the whole of the United Kingdom, each with a trained anthropometrist using modern equipment and techniques. The existence of such clinics soon revealed a great reservoir of children suffering from short stature for a variety of reasons. Some had treatable disorders, some required diagnosis, prognosis and genetic advice, many suffered only from an uncomfortably slow tempo of growth and sought explanation and reassurance. At the same time the longitudinal study of patients revealed a small clinical universe of disorders of the timing and sequence of puberty. This was another field where precise knowledge of the vast but orderly range of normal variation was essential; the morphological changes provided a solid background against which the more transient biochemical responses could be evaluated. Gradually, too, auxology brought a better understanding of events subject to the dimension of time; no longer was it blithely (and incorrectly) assumed, for example, that because a child satisfactorily retained nitrogen over a first week or two of treatment he would grow adequately on the same treatment over the period of a whole year.

Conversely, the range of pathology in patients studied longitudinally with the full technology of modern biochemistry, radiology and histology, brought a new illumination to the basic science of auxology. The investigation of children with endocrine disorders, for example, threw light on the mechanisms controlling the events of puberty in normal children (see e.g. Grumbach, 1978). Longitudinal studies of children with chromosome disorders such as Turner's syndrome also reflected knowledge upon the normal, showing both that without oestrogen secretion no adolescent spurt took place in girls (Brook *et al.*, 1974) and that despite this the height correlation of adult offspring with parents' height was practically fully preserved (Brook *et al.*, 1977).

As for the history of the physiology of growth, it is soon written. We still know lamentably little about the mechanisms by which human growth is so precisely controlled. We do not even know why the velocity of growth gets steadily less from birth to puberty. We do not know what causes a fast tempo, what a slow one. We do not know why the skinfolds reach their maximum towards the end of the first year and then decline till 7 or 8. Most of the history of auxological physiology lies in the future.

The first physiological function to be investigated was stength, by Quetelet and others (see below). The next was 'vital capacity', the amount of air the lungs can hold at maximum inspiration above what remains when the subject exhales forcefully. It was investigated, like chest circumference (see e.g. Liharzik, 1858), because of the light it might shed on the resistance of the lungs to the all-pervading respiratory diseases, especially tuberculosis. Soon after Hutchinson invented the spirometer, the instrument with which vital capacity is measured, Schnepf (1857) in Paris gave minimum and maximum values for boys and girls separately at most years of age from 8 to 19. Students at Amherst College in the United States (age 16 and above) had vital capacity measured, as well as chest circumference, from 1861 onwards (Hitchcock, 1893). Pagliani measured vital capacity in the 1870s both in the Institut Bonafous and the Turin schoolchild studies. Kotelmann too included it in his measurements of Hamburg boys, as did Winfield Hall in the Friends' Schools of Philadelphia. It was part of the routine at the Horace

Mann and the University of Chicago schools, and Baldwin introduced it into the Iowa study. In the 1920s it was still considered of great importance (for general health reasons now that tuberculosis was on the decline) and so influential an educationist as S. A. Courtis wrote that he considered that 'lung capacity is the vital element in growth' (Courtis, 1917). In more recent times a considerable number of studies, mostly cross-sectional, have been done, which have included more complicated tests of lung function as well (see Godfrey 1974*a*, *b*).

Aitken (1887) recognized that at puberty 'a marked impetus was given to the growth of the lungs' (p. 120), and an even greater one to the growth of the heart. Boyd (1861) had given tables of weights of heart, lungs and other organs in subjects aged 7 to 14 and 14 to 20, and Aitken, in the light of these and other data, emphasized how rapid was the change in the size of the heart during puberty. If the changes of puberty were accomplished 'say in one year' then the heart doubled in size in that year. Aitken was in charge of young army recruits and much concerned with their fitness to withstand fatigue and military duties. 'The enormous importance of these changes to the individual', he wrote, 'may be inferred from the fact that the heart may in this one year of puberty development grow three times as much as it did in the preceding year' (p. 116). A modern longitudinal study of the growth of heart and lungs from 5 to 19 years by means of serial radiographs was reported by Simon and his associates in 1972 (Simon *et al.*, 1972). More data on organ growth were gathered by the younger Vierordt for his *Daten und Tabellen* (1906) and there were studies of single organs such as those of Hammar and later Boyd on the thymus (see above; and Tanner, 1962, pp. 176–80). The importance of the endocrine glands in relation to growth was recognized in the late nineteenth century, when the link was made between gigantism (an exceedingly rare condition) and tumours of the pituitary gland and between cretinism and a lack of thyroid. But it was not until the 1940s that bioassays of pituitary hormones became available and a chemical method for assessing androgens, in the shape of 17-ketosteroids in the urine, was established.

Most research on physiological change has centred on either the first few months after birth, or puberty. Both are periods when change is considerable. In infancy the chief research has concerned body composition, the drying-out process. Nearly all has been cross-sectional, yielding general results of much medical importance but throwing a limited light on the growth process itself. However, in the 1970s S. J. Foman and his associates, in the department of pediatrics at Iowa, made longitudinal studies on the relation of food intake to growth in babies from birth to about 5 months (for review see Bergman and Bergman, 1979). This was a modern echo of the work of Camerer, Russow and the paediatricians of a hundred years before. Metabolism was much better quantified, though, and it was possible to calculate the relative amount of nutrition used for growing and for all other activities. Two to three weeks after birth nearly half the daily energy intake was used for growth; by 4 months only one fifth; and by 12 months barely 3 per cent. A good account of the history of research into nutritional intakes of older children is incorporated in Elsie Widdowson's classical report on her own survey done in 1939 (Widdowson, 1947).

In the 1960s and 1970s there was much interest in plotting the changes in composition of the body tissues themselves. In fetal life tissues such as muscle or liver are made up of cells separated by much extracellular fluid. The cells consist of nuclei surrounded by only small amounts of cytoplasm. As growth proceeds the percentage of extracellular fluid drops; and the amount of cytoplasm per cell rises (reviews in Brasel and Gruen, 1978; Brook, 1978).

The difficulty about investigating growth changes of this sort is that the necessary techniques are usually invasive, indirect, or both. There was a period in the 1950s when some medical research workers let enthusiasm outrun judgement and subjected patients, and sometimes volunteer students and others, to procedures which caused pain and even carried some degree of risk. The inevitable revulsion from this brought about a much tighter control of all such 'experiments' in the 1960s. Every hospital or group of hospitals in the industrialized countries was led to appoint an Ethical Committee, which in most cases included persons who were not doctors and which had to pass all projected investigations. As regards children, the stringency of the standards laid down varied from country to country. The measurement of height and the collection of urine specimens were nowhere regarded as significantly painful or risky; the collection of a blood sample was usually passed for genuine volunteers (if any) above age 10 say, but not below. In several countries all radiographs of normal subjects were forbidden, in what was probably an over-reaction not unrelated to other aspects of the public's fear and suspicion of radiation; in other countries a level commensurate with the natural background radiation was permitted. Patients were as well protected as normal volunteers; no investigations could be done on them that they had not expressly and with full understanding volunteered for, except those necessary for their own individual diagnosis and treatment.

These clearly necessary rules inevitably slowed down research in the 1970s on changes in body composition. None of the radiographic studies of bone, muscle and fat widths of the North American longitudinal studies, or of the Harpenden Growth Study, would be permitted today. Probably the same would apply to the studies of Don Cheek (1924–), an Australian paediatrician at Johns Hopkins who took muscle biopsies on volunteer siblings of children being investigated for growth hormone deficiency and other disorders. Such biopsies carry no appreciable risk, but are slightly painful. Cheek (1968, 1975) sought to discern at what age new muscle-cell nuclei stopped appearing, and whether, as he supposed, new nuclei were again produced during the large male adolescent growth spurt in muscle. Similar work has been carried out on fat tissue (see Brook, 1978). The importance of differentiating here between an increase in the number of fat cells and an increase in the size of fat cells as causes of the increase in skinfolds before and after puberty lies in the light it might throw on the mechanics of childhood obesity. At the time of writing the answers to these problems, as regards both tissues, remain somewhat uncertain. A particularly active worker in the field of body composition, exercise and obesity has been Jana Parišková (1931–) in Prague (see e.g. Parišková, 1968, 1973, 1977).

One entirely non-invasive approach to the measurement of muscle mass is through estimating the amount of the naturally occurring radioactive isotope of potassium, ^{40}K, that is present in the body. Sensitive counters are able to do this, without any intrusion on the subject whatever (as Forbes (1978) says, the only hazard is claustrophobia). Potassium is found almost entirely within cells rather than in the body fluids, and much, though not all of it, is in muscle. A particular exponent of this technique has been Gilbert Forbes of Rochester, New York, whose review of changes in body composition discusses the results of this and other techniques in demonstrating the adolescent spurt in 'lean body mass' (body mass minus fat). All these data, however, are cross-sectional (Forbes, 1978).

Growth in strength, allied to growth in muscle, has been investigated on and off since Quetelet's time. Harold Jones' results have already been discussed (see Chapter 12). A carefully controlled longitudinal study was made by Don Bailey (1934–) and his associates at the University of Saskatchewan in Saskatoon. Unfortunately it was of boys only, and it ended at age 16, exactly when the most interesting and controversial relationships between bone, muscle and strength growth appear (Carron and Bailey, 1974).

The effects of muscular exercise on growth, both of muscles and of the skeleton, has also been discussed from the earliest times, but seldom effectively investigated. The difficulty here is a common one in human research: obtaining a proper control group. Children tend not to accept being assigned at random to low-exercise and high-exercise groups; some in the first want to join the second, and some in the second refuse absolutely to stay there. Reviews of the subject are given in Parišková (1973), Rarick (1973), Bailey, Malina and Rasmussen (1978) and Malina (1979).

Shock's longitudinal studies on changes in basal heart rate, blood pressure and oxygen consumption during puberty have been mentioned in Chapter 12. The Denver work on metabolism of Iliff and Lee has also been described. Other similar studies are discussed in Tanner (1962, pp. 165–75). An excellent account of the history of studies of basal metabolism in children is given in Benedict and Talbot's classical monograph (1921). The first studies of carbon dioxide production in children at rest were done as early as 1843. Benedict and Talbot's own work was carried out in a Children's Home where nearly all the children were pre-adolescent. It was the Denver workers who first demonstrated that the basal energy per kilogram of weight or square metre of surface area increased somewhat at puberty, presumably because of the requirement of increased growth.

Knowledge of the endocrine changes during the growing period and their effects began to accumulate in the 1940s, but was very uncertain until the development, in the 1960s, of much more sensitive chemical means of estimating hormone concentrations in blood and urine than had been available before. During the 1960s and 1970s a number of newly discovered hormones were isolated and virtually all became measurable at physiological concentrations. Detailed knowledge then depended on establishing longitudinal studies of this aspect of growth, for exactly the same reasons as applied in the study of morphological changes. The difficulties were and are formidable. It takes brave children and persuasive investigators to obtain blood or urine samples repeatedly over many years. Even then only some hormones can be meaningfully studied by single blood samples; some, like growth hormone, are secreted only in occasional bursts, or pulses, so that a real study requires sampling the blood continuously for twenty-four hours or more. In the late 1960s Charles Faiman and Jeremy Winter in Winnipeg successfully followed blood concentrations of a variety of hormones in fifty-six boys and fifty-eight girls over five years, though only with one sample each year (see Faiman and Winter, 1974; Tanner, 1975; Winter, 1978). Fig. 13.6 is drawn from their data. Tanner and his associates made a study from 1971 to 1978 of 24-hour urinary levels of pituitary and gonadal hormones in about forty boys and thirty girls followed three times a year for seven years: few results have yet been published. At the time of writing there seem to be only two longitudinal studies of normal children's endocrinological development in progress. One, by Michael Preece, is located in a school in Chard, Somerset;

the other is by Pierre Sizonenko and his associates in Geneva. Both use blood samples and study children aged 11 and over.

The second major change in endocrinology that occurred in the late 1960s was the recognition of the importance of receptors, the substances on to which the hormones have to lock before they have any effect. Earlier it was thought, because of the facts of precocious puberty and the response of small children to ingested sex hormones, that changes in receptor sensitivity or concentration played little part in pubertal development. Now such changes are seen as probably very important in the case of some hormones. There is no way known at present in which receptors could be studied longitudinally

Fig. 13.6. Mean values of serum pituitary hormones FSH and LH, and of testosterone, plotted against stage of puberty in mixed longitudinal data on fifty-six normal boys. Redrawn from data of Faiman and Winter (1974).

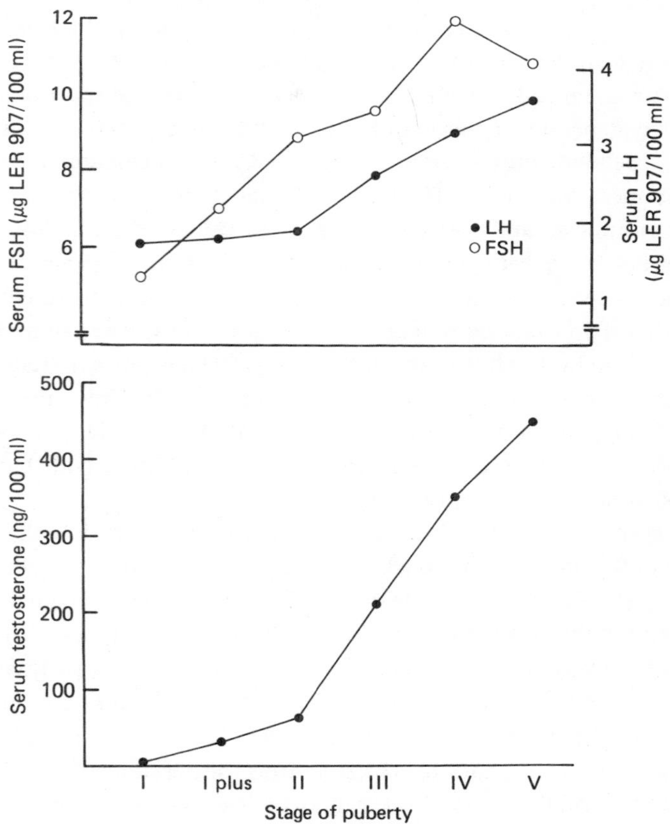

in children, but a solution of this previously ignored problem would now be recognized as a large step forward.

As to the individual hormones: growth hormone, recognized in rodents and dogs in the 1930s, was only certainly shown to be of importance in children in 1958. Twenty years earlier, animal-derived hormone had been given to children with presumed deficiency (because of their smallness unallied with any disproportion) but it had been without effect. At last it was realized that growth hormone, unlike most others, was species-specific in its action. When Raben administered hormone extracted from cadavers it was at once successful in promoting growth. Clinical experience has since shown that the hormone is necessary from birth or very soon after, and remains necessary throughout the whole period of growth. The question arose as to whether it was in fact necessary during the adolescent growth spurt, which was known to require, and perhaps be entirely controlled by, androgenic hormones, that is testosterone and adrenal gland androgens. The Zurich and London groups, simultaneously and using quite different approaches, showed that in fact both hormones had to be present for a normal adolescent spurt to occur; if growth hormone was absent, the spurt was much reduced in amplitude. Furthermore, at least in boys, the spurt of the legs was highly dependent on growth hormone while the spurt of the trunk was mostly dependent on testosterone (Aynsley-Green, Zachmann and Prader, 1976; Tanner, Whitehouse, Hughes and Carter, 1976).

Both androgens and oestrogens appear to cause a turning-down of the curve of height velocity after its adolescent peak; in their absence the curve is extended onwards in time. This onward extension is mainly or entirely of limb rather than trunk growth and gives rise to what was recognized a hundred years ago in the clinical and anthropological literature as the eunuchoid physique. During these hundred years there has been little advance in knowledge (though much uncontrolled speculation) about the relation between differential hormone secretion during growth and ultimate body build; this situation is now beginning to change.

The recent history of endocrinology itself will not be dealt with here: an account of the present state couched in non-technical language will be found in Tanner (1978*a*). A view of the development of knowledge about the endocrinology of puberty and the relation of endocrine and morphological events can be usefully obtained, perhaps, by comparing a series of fairly complete accounts written by the same author, and so with constant bias, between 1955 and 1981 (Tanner, 1955, 1962, 1969, 1975, 1981). The major advance in understanding the mechanism which controls the onset of puberty is further reviewed in Grumbach (1978; see also Grumbach, Grave and Mayer, 1974).

Though it has been clear since the classical transplantation experiments of G. W. Harris and Dora Jacobsohn in 1952 that maturity is registered in the hypothalamic portion of the brain and not in the pituitary gland itself, we are still ignorant of the nature of the metronome which sets the tempo of maturation and of the factors which cause it to run fast or slow in different individuals.

14

National monitoring: population surveys and standards of growth

Monitoring populations and monitoring individuals

At the outset of this chapter we should make clear the distinction between monitoring populations and monitoring individuals. Villermé, Quetelet and Horner were interested in populations; if factory children were small because they were overworked and undernourished then appropriate action should be taken. The action was social or political in nature; the depressed subpopulation had to be brought up to the level of the more favoured groups in the population. Roberts, Bowditch and Galton also dealt in terms of populations; they had become more sophisticated in that they compared the depressed subpopulation with a subpopulation that was privileged, that of the Public (i.e. private) Schools. Godin, on the other hand, and Camerer and the paediatricians, were interested in monitoring individuals. They wanted to know whether a particular schoolboy or a particular infant departed from some measure of expectation. If so, the action taken was individual not social; medical rather than in the sphere of politics.

There has been, and to some extent still is, considerable confusion over this point (see discussion in Tanner, 1978*b*, and Goldstein and Tanner, 1980). We have already seen the difference between charts which show individual growth and charts which show growth for populations: we met the former first as background to the height growth of Friedrich Schiller, the latter first in relation to the height of the average of Marine Society boys. The charts are different because of the differences in tempo amongst individuals making up a population. The population mean at adolescence shows a prolonged increase with its maximum rate of height increment reaching about 7 cm/yr for boys; the individual curve has a sharper increase, the maximum rate reaching 9–10 cm/yr in a boy of average tempo and 11 cm/yr or more in a boy of accelerated tempo with an early spurt. When comparing populations there is, strictly speaking, no need of standards; one group is simply compared with another (or with itself at an earlier time) by means of the usual methods of statistics. Some workers like to have a 'reference population', but when they have adopted it, ostensibly for purely pragmatic reasons, it has nearly always become gradually transmuted into the old Platonic Ideal to which all populations should aspire. Such workers forget there were no separate gene-pools nor even separate environments in Plato's Republic. For judging individuals' growth, on the other hand, standards really are necessary, incorporating some measure of range so that a statement can confidently be made as to the likelihood of a given child being subject to the same influences as the rest of the group, and – the point – to those influences only. In this chapter the history of population surveillance will be considered first, and the history of the construction of standards of growth second.

Population surveys

Quetelet's and Horner's surveys have been discussed already. Roberts, with Galton, endeavoured to survey well-off and poor, and Bowditch called for a nationwide survey in the schools of the United States. In the late nineteenth century several towns in Germany and Russia surveyed good samples of their schoolchildren, and in Denmark and Sweden national school surveys were undertaken. The history of surveys of London schoolchildren has been described by Cameron (1979) and of Oslo schoolchildren by Brundtland, Liestøl and Walløe (1980).

Comprehensive surveys of a country's whole population of children, or even of defined parts of it, are of very recent origin. They have been made usually to define regional differences, differences between socio-economic classes, differences between races in multiracial societies, and secular trends. Some of the national surveys have combined cross-sectional and longitudinal approaches. Thus in the last of the London County Council surveys, undertaken in 1966–7, one-half of the original 29,000 children were remeasured approximately a year after the initial measuring. In this way standards for growth increment were obtained, as well as standards for height at given age (Cameron, 1977). Standards for increment obtained this way, however, do not apply to individuals, since they take no cognizance of tempo. They are appropriate only if we want to compare populations not only with regard to their *mean* increments during a given year of age, but also in respect of the *range* of increments in each population.

Another category of national surveys cannot really be classified as either cross-sectional or longitudinal. The National Child Development Study for example (see below), concerned a true national sample studied at birth. These children were then seen again at ages 7, 11 and 15. This is not what one normally means by a longitudinal study; but it has great advantages over a cross-sectional study in terms of information gathered. Such studies are called 'follow-up studies'. Clearly the line of demarcation between a follow-up study and a longitudinal study is drawn according to the prejudice of the drawer.

The first national population study was done in the United Kingdom, and it was of the follow-up type. A second such study, of a single town only, was started soon after the first, and a third, again nationwide, twelve years later. We begin, therefore, by looking at these United Kingdom studies.

The National Survey of Health and Development (1946)

The first of these follow-up studies began as a simple survey at birth, with little thought of a long-term commitment. The Royal Commission on Population, of 1945, required information on the use made of the maternity services in the UK. A joint committee of the Population Investigation Committee and the Royal College of Obstetricians and Gynaecologists was set up, and conducted a massive survey of the circumstances and costs of all confinements in Great Britain occurring during the first week of March 1946. Alan Moncrieff was a member of this committee, and became director of the new Institute of Child Health of London University just as the study was taking place. It seemed to him and others, including the director of the study, James W.B. Douglas, (1914–), a young physician interested in social medicine, that such a unique national sample of children should not simply be left to disappear again into the general mass. A new committee was formed, the Institute of Child Health and the Society of Medical Officers of Health taking the place of the Royal College of Obstetricians and Gynaecologists. Moncrieff became vice-chairman; David Glass, of the London School of Economics and the Population Investigation Committee, secretary. A sub-sample of the original 13,687 children was drawn, first by excluding twins and illegitimate children, and then by taking, at random, only 25 per cent of the children from families of manual workers, the self-employed and farmers, who comprised four out of the nine socio-economic occupational categories though accounting for 72 per cent of the total (the other categories being professional workers, employers of ten or more people, salaried workers, black-coated wage earners and agricultural workers). Thus a sub-sample of 5,386 children was followed, but with the possibility of regaining the original national sample by suitable weighting.

The first follow-up was made when the children were aged approximately 2.0 years and the second when they were about 4.25. Measurements of height and weight were taken by health visitors in Infant Welfare Centres and in the children's homes; the controls of this early survey were poor however, and half the children were weighed with clothes, half without. In all 4,037 children were measured at age 2 and 4,298 at 4. Rural children were taller than urban ones except in Wales, a result unusual in European studies but obtaining already in England at the beginning of the century and in Scotland in 1919–23 (Paton and Findlay, 1926). The children of professional and salaried fathers averaged 2.3 cm taller than children of the unskilled manual workers at age 2, and 3.5 cm taller at age 4. The prevalence of illness and accidents was studied, the frequency of children being separated from their mothers, and the degree to which the social services were used (Douglas and Blomfield, 1958). A particular effort was made to assess the degree of maternal care and its effect on the amount of illness and on growth: so far as growth was concerned there was no effect at either of the ends of the social scale, but some small effect amongst wage-earners and skilled manual workers.

During the school age the follow-ups concentrated mainly on educational research, and after a period in Edinburgh, Douglas moved the headquarters of the study to the London School of Economics. However, height and weight were measured by school doctors or nurses at ages 7, 11 and 15 and ratings of breast, pubic hair and genitalia development were made. By the age of 15 the sample followed had been reduced to approximately 3,000 children.

The social class difference in height remained sensibly constant. At ages 7 and 11 boys with upper middle class fathers averaged about 2.2 cm taller than those whose fathers were unskilled manual workers; at age 15 the difference had slightly diminished, to 2.0 cm. For girls the differences were a little greater: about 3 cm at ages 7 and 11, falling to 2.5 cm at age 15. Though boys usually show greater reactivity to environmental stresses and hence a larger social class difference, the National Survey of Health and Development is not unique in the converse finding: the rural–urban difference described by Paton and Findlay (1926) in Scotland was also greater in girls than boys. The usual relationship between height and the number of siblings (see Fig. 14.1 below) was present at all ages except in the upper middle class boys and girls; in the well-off, children in large families were as tall as children in small ones.

Age at menarche (as recollected at age 15) was strongly related to number of siblings, being later as the number increased; but it was hardly related at all to father's occupational class. Advancement in boys was also more closely related to number of siblings than to social class. The average age of menarche for all the girls, given in a probit analysis by Miller and his associates (1974, p. 64) was 13.2 years.

Douglas (1964; also Douglas, Ross and Simpson, 1965) made a particular effort to trace the relationships between physical maturing and school progress. At this time there was still a selective examination at age 11 + for all pupils; those who passed went on to a more academic education than those who failed. Douglas showed that early maturers had a considerably greater chance of passing the 11 + examination than late maturers, in that they scored higher at that age on tests of school ability. This held true even within social class, and when equated for number of siblings. As for the relation with height, the taller children scored higher in the tests at all ages, and the low but still significant correlation between height and test score was not reduced when the effects of early or late development, social class and number of siblings were all allowed for. Thus the correlation was not accounted for by co-advancement in physical and mental fields. Not only was Porter's finding again confirmed, but also his prognostication that the relationships persisted into adult life.

The Newcastle-upon-Tyne study of a thousand families

A year after the London-based follow-up survey began, a much more intensive study, more medically oriented, started in Newcastle-upon-Tyne. Newcastle had for many years been an industrially depressed area, with up to 20 per cent unemployed in the 1930s. James Spence (1892–1954), a paediatrician and from 1943 professor of child health there, found at this time a marked difference between the heights and weights of pre-school children in professional families and those attending welfare clinics, and surmised that the greater incidence of disease in the poorer children was largely responsible. In 1936 infant mortality actually reached 96 per 1,000 in Newcastle, compared with 59 per 1,000 for all England and Wales. During World War II the health of the children improved because of the emphasis given to proper nutrition of children and child-bearing mothers, but even so Spence decided in 1947 to study on a long-term basis the prevalence of disease in a thousand families, mostly of the manual working class. In all, 1,142 infants born in May and June 1947 enrolled; 967 remained at the end of the first year and the large number of 763 (70 per cent) after fifteen years. At age 22, 442 subjects were traced and remeasured.

The study was more a cohort follow-up than a longitudinal one. Height and weight were only measured seven times during the fifteen years, and on one of the occasions, at age 3, the measurements were done by the parents. At 5, 9, 13, 14 and 15 years the children were measured in school. Weighing was done in indoor clothing, without jackets or shoes. Health visitors kept a detailed record of illnesses, particularly during the first seven years, and recorded their impressions of maternal concern and efficiency. Tests of mental performance were administered at school. The results were presented in three books (Spence *et al.*, 1954; Miller *et al.*, 1960, 1974). Social class was assessed by father's occupation; the height differential at age 5 between children of the professional class and the unskilled working class was 6 cm, twice that found by Douglas in the National Survey. By age 22 the difference was 4 cm in men, but nil in women (Miller, Billewicz and Thomson, 1972). The mean age of menarche was later than in the National Survey: 13.5 for all girls, with the non-manual class at 13.4, skilled manual 13.5 and semi- and unskilled manual 13.6 years. This then, was the situation in the north-east of England about 1960. As usual, scores in various mental tests at ages 10–14 all related to height, and this was already true of height at age 3. Scores at age 10 to 14 were also related to age at menarche, with early maturers, who had menarche before 12.0, scoring typically about 8 per cent higher than late maturers, who had menarche after 15.0. Children with disturbed behaviour were on average smaller than others.

The major emphasis was placed on the prevalence of various disease patterns. The most common in the pre-school years was, of course, respiratory disease, but the Newcastle team were unimpressed with any relation between frequency of respiratory disease and stature at age 3 (Miller *et al.*, 1960, p. 149). A small group who had severe respiratory infections with more than four episodes of bronchitis and pneumonia before age 3 had a lower height than others, but lesser degrees of infection seemed to be without effect. Miller summed up his conclusions about growth and infection in Newcastle by saying that 'a common factor is more likely to explain both an increased frequency of severe illness and small stature rather than that one should account for the other, and the most likely common factor is inadequate nutrition'. As regards this, Court and his colleagues give a description that any family doctor serving a depressed urban area will instantly recognize: 'Some children seemed to live almost entirely on bread and jam, rarely having cooked meals. Some parents never set a proper meal and the children stood while helping themselves to the bread or other scraps of food which for ever littered the table. When feeding at home was unsatisfactory, as soon as the children could walk they foraged amongst the neighbours' (Miller *et al.*, 1960 p. 44). Despite their standards being 'not exacting' they found low levels of personal cleanliness in 7 per cent of families and of domestic cleanliness in 12 per cent. A century after Charles Roberts' description of the conditions in the factory towns there were still some pockets untouched by change and prosperity.

The National Child Development Study

During the period 1946–58 extensive social changes took place in the United Kingdom, and the National Health Service was established. By 1958, therefore, it seemed time to make another survey of the maternity services on the same lines as the one done twelve years earlier. Once again Alan Moncrieff was influential; the circumstances of the births of all children during one week in March 1958 were recorded in a nationwide Perinatal Mortality Survey (see Butler and Bonham, 1963). Again efforts were made to follow this cohort of approximately 17,000 children, but lack of funds prevented any follow-up until age 7. In 1964, however, the National Child Development Study (NCDS) was set up and follow-ups were conducted at ages 7, 11 and 15. At age 7, 92 per cent of the original sample were traced although not all were measured within the appropriate time limit. Much of the study concerned educational variables (Pringle, Butler and Davie, 1966); the major results on growth were presented in a highly cogent paper by Goldstein (1971), who started work with the NCDS while still at the Institute of Child Health and later became its full-time statistician.

Fig. 14.1. Height at 7.0 years of British children (sexes combined) according to number of children in the family and socio-economic level (1–5). Data from the National Child Development Study (1965), given in Goldstein (1971); figure from Tanner (1978*a*).

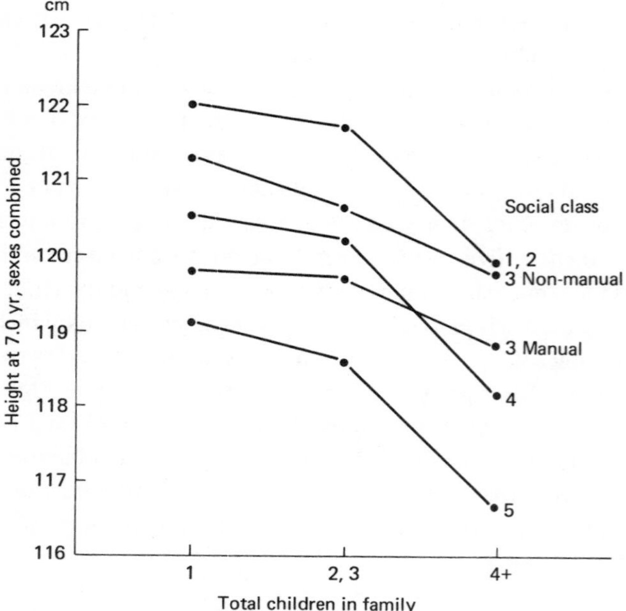

Goldstein analysed the heights of the 13,127 children who were measured during April to December 1965. The size of the sample and its truly national character makes this study a landmark in auxology. The associations between height at 7 and the factors birthweight, length of gestation, mother's parity, age and height, whether she smoked during pregnancy, number of younger siblings and social class according to father's occupation were computed separately and in combination. Children in professional and managerial families averaged 3.3 cm taller than those whose fathers were unskilled manual workers; boys and girls did not differ in this respect. This was a very similar result to that obtained by the National Survey of Health and Development twelve years previously: thus disappointing. When all other variables were allowed for, the length of gestation was without effect on height at 7, but the weight at birth remained associated, with a gain on 2 cm of height per kilogram of birthweight. Children with many siblings were smaller than those with few, and this held good in all social classes. This aspect of the analysis is shown in Fig. 14.1. Smoking more than ten cigarettes a day during pregnancy was associated with a diminution of 1 cm of height, even when all the other factors were allowed for. We can get an idea of the additive effects of all these factors if we take as '50th centile' a child born with an average birth weight as a second baby to a non-smoking mother married to a junior-grade manager: then a baby weighing half a kilogram less, born as a fourth child to a smoking mother married to a semi-skilled manual worker would on average be at about the 30th centile, even supposing the two mothers' heights to be the same.

A non-technical account of the NCDS 7-year results is given in Davie, Butler and Goldstein (1972). Regional differences, already shown in the 1946 study, were present; on average children in England were 0.5 cm taller than those in Wales and 1.1 cm taller than those in Scotland. For children in the same social class these differences disappeared when mother's height was taken into account. Analyses of the 11-year and 15-year data have not appeared at the time of writing.

National Study of Health and Growth and Pre-school Child Growth Survey

In 1972 the British Department of Health and Social Security set up two studies, one monitoring growth in children from birth to age 5, the other growth from age 5 to 11. Until 1968 milk was provided at school free of charge to all British schoolchildren; in 1968 this supply was restricted to those under 11, and in 1971 to those under 7, with exceptions in the case of children of poor parents or those who were medically certified as being in need of it. The reduction caused some discussion, and the government wished to monitor any possible effects on health and growth.

The survey of pre-school children was carried out by a group under John Waterlow, professor of nutrition at the London School of Hygiene and Tropical Medicine. No attempt was made to obtain a national sample, since it was felt that particularly at these ages measurement was difficult and should only be done by properly trained observers. Instead, six geographical areas were chosen, said by demographic opinion to be between them as characteristic as possible of the country at large; two, however, defaulted. Of the remaining four one was rural; the others represented small and large towns. Within each area a random sample of children was recruited, based on the local birth notification register. Measuring teams, all centrally trained, were sent to each area. The study was made up of a longitudinal cohort of some 3,500 children measured twice-yearly from age 3 months to age 5, and a cross-sectional element of some 4,000 children divided into equal groups of 2- , 3- and 4-year-olds. Thus the sample was a considerable one. Height (or length), weight, sitting height (or crown–rump length), arm and head circumferences and triceps and subscapular skinfolds were taken and the usual records kept of health, father's occupation, number of siblings and so forth. Heights of the parents were also measured. No reports are yet publicly available; it is known, however, that in the cross-sectional element social class differences were slightly less than in the previous surveys, differences between the top and the bottom groups, out of three, averaging about 1 cm at ages 2 and 3 and 1.2 cm at age 4. At age 4 the whole group of children averaged about 0.5 cm taller than those of the British standards of 1965 (Tanner, Whitehouse and Takaishi, 1966).

The study of children in primary schools (aged 5 to 11) was somewhat differently planned. Twenty-eight areas in the UK were selected by stratified random sampling with proportionately more areas including less well-off social groups. In all 9,815 children were measured. The study was mixed longitudinal in design, children of all ages from 5 to 11 entering in the first year (1972) and those below 11 being measured annually till they reached that age. Measurements were done by local nurses, trained and supervised by three fieldworkers from the department of community medicine of St Thomas's Hospital, London, where the study was based. Despite this, the standard deviation of the differences between the measurements of height by the fifty nurses and the three fieldworkers was 0.4 cm, about twice that of a single well-trained measurer.

The results for height, weight and triceps skinfold obtained in the first year, and hence cross-sectional, were reported by Rona and Altman (1977), and the relations with social class and other factors by Rona, Swan and Altman (1978). The latter analysis used the same techniques as Goldstein, and produced generally similar results. In England, however, the effect of number of siblings did not appear in the better-off (non-manual occupations) group until there were five or more children in the family. In Scotland the effect was present in all occupational groups. One new and important finding emerged. Children whose fathers were unemployed had lower heights, by some 2 cm, than those whose fathers, though in the same occupational category, were in employment. A deficit of 2 cm cannot, of course, be brought about by sudden undernutrition; the slower growth rate had to have begun in most cases well before the time at which the father became unemployed. A more complex analysis showed that the deficit was not accounted for by the effect of the number of siblings in the family. A further study (Rona and Florey, 1980) related height to history of symptoms of respiratory disease. The results were very much the same as in Newcastle; except for asthmatic children, who were small, shortness was only marginally associated with respiratory symptoms, and then only in the poorly-off families with large numbers of children.

The Dutch national surveys

The first straight cross-sectional national survey was undertaken in Holland in 1955. Approximately 17,000 children and young people were measured, aged 1 to 25, in a sample selected to be representative of the whole country (de Wijn and de Haas, 1960). Ten years later the survey, this time including infants, was repeated (van Wieringen, Wafelbakker, Verbrugge and de Haas, 1971; van Wieringen 1972). Fifty-four thousand subjects were measured this time. Pre-school children were about 1 cm taller than they had been ten years earlier, schoolchildren about 2 cm taller and young adults 3 cm taller. Social class differences were still present, and already visible at 1 year of age. The difference amounted to between 2 and 3 cm at ages 4 to 19. There were also regional differences, children in the north of the country being some 3 to 4 cm taller than those in the south. There were no urban-rural differences within social class.

A separate study was made of the development of pubertal characteristics. Seven school medical officers examined some 4,000 boys and 2,500 girls aged 8 to 9 in various parts of Holland and rated the development of breasts, pubic hair and genitalia on the usual five-point scales. (The monograph of van Wieringen and his colleagues (1971) gives excellent colour-photograph standards for each of the stages.) In addition they measured the size of the testes by comparison with a string of models. The mean ages of transition from each stage to the next were estimated by probits, this being the first time this had been done in a population survey. Menarche occurred on average at 13.4 years, as contrasted with 13.7 years in 1955. Entries to breast stages were at 11.0, 12.1, 13.4 and 15.2 years; to pubic hair stages in girls at 11.3, 12.2, 13.3 and 14.9 years. (Menarche therefore fell most usually in breast and pubic hair stages four.) In boys genitalia stages began an average at 11.0, 13.2, 14.2 and 15.9, and pubic hair stages at 11.8, 13.5, 14.4 and 16.0.

The Dutch famine follow-up study

One very particular follow-up study of much importance was done in Holland in 1970. It concerned the ultimate effects of a short but very sharp period of famine experienced by the people of the central part of Holland between October 1944 and May 1945, as a consequence of the war. In Rotterdam, Amsterdam and Leiden the infants whose mothers suffered the starvation during the last three months of their pregnancy averaged 9 per cent below the norm in birthweight and 2.5 per cent below in birthlength. The mothers' own weights averaged 10 per cent below the norm. (Babies exposed only in the earlier part of fetal life were unaffected.) At age 19 the males of this low-birthweight group entered military service and their heights and weights were measured, as well as some aspects of their mental abilities. The material was collated by Stein, Susser, Saenger and Marolla (1975), a group working at Columbia University. They found that complete recovery had occurred; the young men who had been undernourished in fetal life were no different in height, weight or mental ability from those who had not been undernourished. In the continual debate about the possible effects of prenatal malnutrition – such as we have discussed in respect of the nineteenth century in Europe – the result was important. So long as *post-natal* nutrition was adequate, and so long as we are considering fetal undernutrition not accompanied by exposure to noxious substances, then the effect on mature height was nil.

The Cuban national survey

In 1972 a nationwide survey was made in Cuba, as a joint undertaking of the Children's Institute and Ministry of Health in Cuba, and the department of growth and development of the Institute of Child Health in London. Some 50,000 children were measured. The methods adopted were described in detail by Jordan *et al.* (1975). The number of subjects was chosen so that the standard error of the 3rd height centile of the charts to be constructed (an error which is about double the error of the mean) was about 0.3 cm. Goldstein, who was responsible for the design of the study, introduced the idea of making the number of children sampled at each age proportional to the rate of growth at that age. Thus for each sex 3,600 infants aged 0 to 1 were measured, 2,000 children aged 1 to 2, 1,100 aged $6\frac{1}{2}$ to $7\frac{1}{2}$, and subsequently, until puberty when the numbers again increased.

Multistage sampling was used: within each of eight provinces, rural and urban parts were separated and each divided into areas containing some 1,200 to 2,000 households. A random sample of areas was then taken. Each area had been divided by the census into sectors containing about 120 households, and within each of the selected areas a random selection of two sectors was made. The sectors having been located, every child living in them was listed and from the list a random selection of the designated number of children was made. The procedure is illustrated in Fig. 14.2.

There were nine measuring teams, but all the measurers had been trained by a single anthropometrist, R. H. Whitehouse. Fifteen anthropometric measurements were taken, and hand–wrist radiographs were done in a 10 per cent sample. Every few weeks quality-control sessions were undertaken at which the teams met and each measured the same individuals. Stringent editorial checks were made on all data before the final statistics were issued

Fig. 14.2. Illustration of three-stage sampling adopted in Cuban National Survey of 1972. Oriente province, rural areas. The 34 (of 254) selected areas are shown stippled Within each selected area 2 sectors (out of an average of 12) were selected, shown in black. Reproduced from Jordan *et al.* (1975).

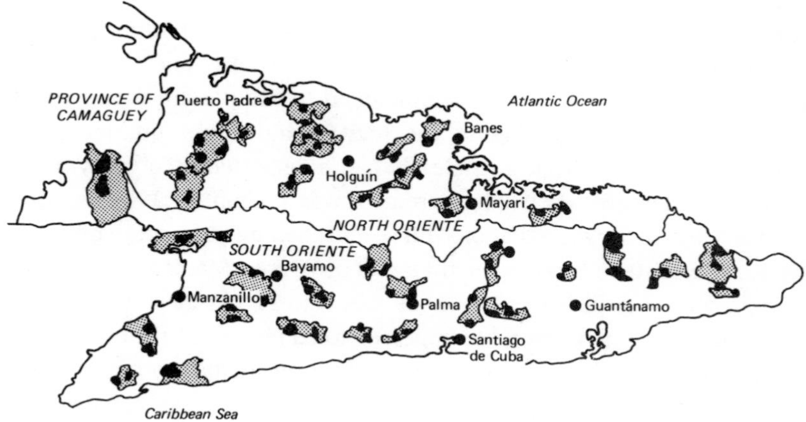

(Jordan, 1979). It can be seen that the technique of a modern survey is relatively complicated, and there are many precautions of sampling and method to be observed.

The United States national survey

Holland and Cuba both had the advantage of being geographically small countries, with relatively small populations. The larger the country the more complex the task of national monitoring. However, the National Center for Health Statistics of the United States surveyed a representative sample of some 7,100 children aged 6 to 11 years in 1963–5, and of some 6,800 youths aged 12–17 in 1966–70, 2,100 of whom had been subjects of the first sample (The Health Examination Survey : see Hamill, Johnston and Grams, 1970; Hamill, Johnston and Lemeshow, 1973*a*). In 1969 this was extended by a new surveillance system known as the Health and Nutrition Examination Survey (HANES). This aimed to survey a national probability sample of 30,000 persons aged 1–74 years, and in 1971–4 some 5,500 children aged 1–17 were measured. The whole, plus some data from 0 to 3 years from the Fels Research Institute, formed the basis of national population height, weight and pre-pubertal weight-for-height charts (Hamill *et al.*, 1977). Nineteen other measurements beside height and weight were made, and hand–wrist radiographs were taken. Reports were issued comparing European-descended, 'White', and African-descended, 'Black', Americans (Hamill, Johnston and Lemeshow, 1973*b*), and rich and poor (Hamill, Johnston and Lemeshow, 1972). Others dealt with skinfold thickness (Johnston, Hamill and Lemeshow, 1972, 1974), skeletal maturity (Roche, Roberts and Hamill, 1974, 1975, 1976, 1978) and vital capacity (Hamill, Palmer and Drizd, 1977).

Children who came from rich families were taller than children from poor ones, with trends that were about the same for girls and boys and for Whites and Blacks. The difference amounted to about 3 cm at ages 6 to 11 between children in families in the highest and lowest quarters of the income distribution. Black girls clearly had a faster tempo of growth than White girls: their greatest height increment was at about 11.25 years, which was 0.5 years earlier than that of Whites, and their average age of menarche was 12.5 years, 0.3 years earlier than that of Whites (MacMahon, 1973). In consequence Black girls were taller at ages 6 to 13, but thereafter little different. No such tempo difference appeared to be present in the boys, but the sample of Black children was not large and the age of maximum velocity of the boys was not well determined. The White boys' greatest increment was centred at 13.25 years, a surprising 0.75 years earlier than in north European boys, although the American White girls were only 0.25 years earlier than their European counterparts.

The skeletal maturity investigators used the Greulich–Pyle technique of giving a bone age to each bone separately and obtaining an average. The national sample was found to be less advanced in bone maturity at a given age than the 1930s sample of the Brush Foundation from which the standards were derived. The Americans of European origin had bone ages very similar to their contemporary European counterparts, whereas Americans of African origin were considerably advanced over their counterparts in Africa or Jamaica. African-descended boys were in advance of European-descended boys at most ages from 6 to 10, and African-descended girls were in advance at nearly all ages during childhood, the differences being generally of the order of two to three months. There were only very small differences between urban and rural children, geographical regions or family-income groups.

In the original reports dealing with the adolescent period (Hamill *et al.*, 1973*a*, p. 21) the authors were perfectly clear about the difference between population standards and standards for clinical use. But when the charts appeared, Pagliani and Shuttleworth, Boas and Bayley were thrown to the winds. The distinction was forgotten, and a commercial firm issued free to paediatricians huge numbers of charts, overprinted with advertising matter, with the clear implication that they were the appropriate instrument for following the growth and possible treatment of individuals. It would not have been difficult to use the Health Examination Survey results in conjunction with velocity data to produce standards for clinical use which would have incorporated a consideration of tempo. To such standards we are about to turn.

Other population surveys

Before we do so, we should indicate, at least, that in the period from about 1960 on there have been many less extensive or less detailed cross-sectional surveys. They can be traced through the book *Worldwide Variation in Human Growth* (Eveleth and Tanner, 1976), a compendium of the results of 'all serious studies of child growth in the world in the period 1960–74' (p. 2) made under the auspices of the Human Adaptability Section of the International Biological Programme, a programme sponsored by the International Council of Scientific Unions that ran from 1964 to 1974. In particular there have been surveys in New South Wales, Australia (Jones, Hemphill and Myers, 1973; Jones and Hemphill, 1974), in New Zealand (1971), in the Caribbean (see papers of Ashcroft *et al.* cited in Eveleth and Tanner, 1976) and in India (see Indian Council of Medical Research and papers by Agarwal *et al.* and Banik *et al.* cited in Eveleth and Tanner), as well as in Poland (Wolanski *et al., ibid.*), Turkey (Neyzi *et al., ibid.*), Hong Kong (Chang, *ibid.*), Switzerland (Heimendinger, *ibid.*) and Finland (Backström-Järvinen, 1964; Backström-Järvinen and Kantero, 1971).

Standards for clinical use

We have given some glimpses of the history of clinical standards in the preceding chapters. Always the question was the same: what is the probability that this particular child belongs to the motley army of normal children, of so many shapes and sizes; and what the probability that he belongs to another battalion altogether, the company of the sick? The two probabilities are not necessarily mutually exclusive; the distributions might overlap. The formal model for this state of affairs was not easily made out until statistical theory had clarified questions of sampling and problems of distribution. The first clear statement of the models underlying clinical standards seems not to have appeared till 1952 (Tanner, 1952).

Galton's invention of percentiles in 1875 was the essential step, and percentiles, or centiles for short, have been at the basis of most standards ever since. This is because their use leads to a simple and clear statement of the chances of a given child belonging to the normal group. Bowditch employed them in formulating standards for schoolchildren in 1891, following Galton (1885) in using the 5th, 10th, 20th, . . . , 80th, 90th and 95th centiles in his charts. Sargent used them in 1893, Beyer in 1895; and Porter in his charts of 1922–3 added the 1st and 99th centiles to his 'grades'. When Stuart in Boston and Meredith in Iowa promulgated their very influential standards in 1946 their charts gave the 10th, 25th, 50th, 75th and 90th centiles, from 5 to 18 years (see note 14.1). An alternative system used standard deviations. When a distribution is Gaussian (as is the case for height at most ages) the standard deviation scores are equivalent to the centiles, but when it is not (as in the case of weight) then the standard deviations no longer lead to a simple probability-of-normal statement: indeed, they are no longer an appropriate statistic at all. Even when the distribution *is* Gaussian the problem of communication arises: a parent who is told his child's height is at the 10th centile understands readily: 10 per cent of normal healthy children are shorter than his child. But the equivalent statement that the child's height lies 1.28 standard deviations below the mean is incomprehensible, and has to be translated by a doctor equipped with both tables and understanding. Hence centile standards have normally been preferred. Standard deviations were given, however, by Boas in 1898 (though no charts) and in 1922 Robertson published charts giving −2, −1, 0, +1, +2 S.D. lines for weight from birth to 1 year, and for height from 7.0 to 14.0 years. Gray and Ayres (1931) used standard deviations, as also did Sontag and Reynolds (1945) at the Fels Research Institute, sometimes with a mathematical transformation (e.g. log weight instead of weight) to turn an awkwardly distributed measurement into Gaussian form.

These are standards for height or weight at given age; they answer the question about whether the *size* of the child is normal. But a more important and immediate question is: what is the probability that this child is growing at a normal *rate*? Rate, or velocity, as D'Arcy Thompson so cogently argued, reflects what is happening *now*; size, or, as we might call it, distance travelled, reflects too much the events of the past. Hence for clinical use velocity standards are desirable, as well as distance ones.

Shuttleworth (1934) was really the first to appreciate this (see above). Palmer and his associates published annual increment figures for height and weight from 6 to 14, with standard deviations (e.g. Palmer, Kawakami and Reed, 1937). Simmons and Todd (1938) gave annual increments with their variability from 3 months to 13 years for a number of measurements; so also did Robinow (1942) at Fels, Meredith and Meredith (1950) at Iowa and Hammond (1955) in England. However, the serious advocacy of velocity standards started with Tanner's theoretical paper of 1952. In this paper Tanner described the basic model on which standards are founded, considered the problems of centiles and standard deviations, gave graphs of velocity standards from age 1 to 12 and discussed also regression standards such as weight-for-height at given age (now called 'conditional standards') and standards based on discriminant functions. He analysed the underlying principles on which the Wetzel grid, a then very popular weight-for-height standard, was constructed (Wetzel, 1941, 1943) and the much less well known but more securely founded weight-for-height 'auxogram' of Correnti (1948). Tanner prudently stopped his velocity centiles just before puberty, realizing that to cope with the problems of the adolescent spurt 'some slight complications will have to be introduced if the precision of the standards is to be maintained' (1952, p. 10). The true nature of what was necessary seems only to have dawned on him several years later, perhaps because of rereading Boas' papers for his contribution to the Boas centennial volume (Tanner, 1959). If the average, or 50th centile individual is to grow at all times along the 50th centile standard curve, as seems desirable, then that curve must show the average *individual* velocity, not the increment of the population. This average individual will have an average tempo of growth; hence his peak velocity during the adolescent growth spurt will be at the average age of peaking in the population considered. The 50th centile will be that centile that characterizes, both for distance and for velocity, the individual who has average *tempo* throughout his growth as well as average size. Eventually standards incorporating this notion were published by Tanner, Whitehouse and Takaishi in 1966 (see note 14.2). The shape of the curves in such standards had to be obtained from longitudinal data, handled in Boas' way, that is, individually aligned on peak height velocity at puberty (in the case of height). But such longitudinal data were insufficient in volume to provide good estimates of the *amplitude* of the standards, that is of the positions of the centiles (and particularly the important outer centiles). Superimposed on the Harpenden Growth Study shape, therefore, were the London County Council cross-sectional values for height and weight, obtained from 1,000 boys and 1,000 girls at each age from 5 to 16.

In terms of velocity the conventional cross-sectional type of approach implies calculating increments each year irrespective of tempo, that is, without any alignment by peak height velocity. In the 1966 paper Tanner and his associates gave this system of lines (the 50th velocity being the one shown in the population chart so often used above) as well as the new peak-height-centred lines. This tended to cause confusion between the 'population' and 'individual' sets of lines (see, for example, Lowrey, 1973, p. 84) and in 1976 Tanner and Whitehouse published a new set of standard charts incorporating the same data, but leaving out altogether the 'population' lines. This enabled velocities for early and later maturers also to be indicated, together with the limits for normal variation in tempo, at least at adolescence. These are the charts used throughout this book for plotting the growth of individuals.

15

Coda: Causes and effects of studies of growth

Finally, now that we have outlined the history, and some of the results, of the major studies of growth, and sketched, however roughly, the main actors in the story, it is time to consider why such studies were made, and what effects their results produced. More especially it would be interesting to see how they contributed, if they contributed at all, to our present ideas about society or our present practice of children's care.

There seem to have been three distinguishable, though not always separable, impulses behind the studies: the social, the medical and the intellectual. We shall consider each in turn.

Social impulses

The social impulses have by no means always been benign. If the auxologist requires yet further examples of man's inhumanity to man, he has but to look behind him. Chaussier initiated measurement of the newborn to enable mothers who committed infanticide to be more surely arraigned. The Marine Society, who left us the oldest known series of children's heights, measured the boys it recruited so they could be better identified and punished if they deserted. Not till the nineteenth century did growth studies serve a humanitarian purpose. From the time of Villermé, returned from the horrors of the war in Spain, to the last report of the British Association Anthropometric Committee in 1883, heights and weights of children were measured primarily to provide ammunition in the continual battle for social reform. The location of the studies followed rather precisely the course of development of the Industrial Revolution; in England first, where first the satanic mills sucked in their victims from the countryside, and first the Factory Acts forced the employers to move at least a little towards the light. In England and Scotland, though not in continental Europe, country children have always grown up faster than children in the towns, despite the well-documented poverty in many rural areas in the nineteenth century. And, of course, children working in factories have always been smaller than children not so working; not, it was thought by Roberts, through any cause of the work itself, but because a child working in a factory was a symptom of a family's poverty, ill-housing, undernourishment and often ignorance. With the Factory Acts the symptom was removed, but the disease persisted. In England in 1910 the children were only a little taller than the factory boys of fifty years before. In Scotland in 1920 the rural children were taller than the urban ones by as much as children in private schools were taller than children in factories in the England of 1870.

Even in the 1970s, in almost every country that supplied statistics, children of non-manual workers (the professional class, managers, intellectual workers – the terminology follows the political persona of the country considered) were taller at all ages than children of the manual working class. In England the gap lessened only slightly between 1950 and 1970 despite what were felt to be far-reaching and significant social reforms.

This, then, has been and is one of the main uses of studies of growth: to monitor social conditions and in particular to pick out those groups in a population whose children betray the signs of neglect. The neglect that the population cross-sectional type of survey reveals is neglect by the state: neglect of education, neglect of amenity, neglect of work, neglect of money. Neglect by the parents is revealed by individual monitoring, to which we turn in a moment. Recent Swedish and Norwegian surveys have shown that in a genetically fairly homogeneous population social class differences in children's growth can indeed be eliminated. Perhaps this is as good a criterion as any of the classless state, for the care of children cuts across all other preferences of life-style, and growth measures very precisely the excellence of such care. If everyone in a population had the same amount of money to spend and the same access to services, and yet there were regional, occupational, linguistic, or educational groups whose children grew worse than the rest, then surely the criterion would not be met.

There are, however, two qualifications which must be entered. First, that we are discussing population *differences* in rates of growth, not absolute rates. There is no reason to think that swiftness of development is a good in itself (there may even be some reason to believe the opposite). It is only because increasing rapidity of growth has gone hand in hand with improvement of the quality of life (at least as seen by a consumerist society) in the industrialized countries that rapidity of growth tends to be equated with social advance. Equally healthy growth could probably be achieved at a slower rate given the necessary nutritional and social customs. Such slower growth might or might not be arranged to lead to the maximum attainable adult size. It is not certain that growth up to the maximum size permitted by the genes is essential for the full expression of a person's intellectual and cultural potential. (Small people, of course, may be as gifted as large ones, but this is not the same thing; we are talking of people not small because of their genes, but because of deliberate social custom restricting the attainment of maximum size.) On a limited Earth with a scarcity of energy available to its inhabitants, we must suppose that nature, at least, prefers small people to big ones. Already people in rich countries limit their energy intake so as to prevent the accumulation of fat. The desirable absolute growth rate is something which the coming generations of human biologists and biological humanists will have continually to debate.

The second very necessary qualification relates to genetic differences between populations. Growth represents the result of a continuous interaction between the genotype and the changing environment; an interaction which is sometimes more complex than simple (see Tanner, 1978*a*, p. 117). Thus population differences in growth are only indicative of the environmental situation if the gene-pools (that is, the whole distribution of genes in the population) of the subpopulations are the same, at least in respect of genes concerned in growth. Fortunately, height differences between populations growing up under similar, and good, environmental circumstances seem to be relatively minor. (Sitting height or leg length differences are very much greater.) However, the Japanese, Chinese and Indo-Malay populations, with some of their descendants on the American continent, are clearly less tall than Europeans or Africans even when growing up under excellent conditions. The secular trend in Japan was spectacular during the period 1950–75 but now has practically ceased. The mean height is at about the British 25th centile, the difference being entirely attributable to length of legs. Thus it would be wrong necessarily to attribute the smaller stature of a subpopulation of Japanese children growing up in, say, London to the environmental deprivation which one would unhesitatingly diagnose in a subpopulation of African children attaining only a similar size. Thus an over-simple view must be avoided; there are subpopulations with different gene-pools and different proclivities for growth (see, for examples, Eveleth and Tanner, 1976). But this serves only to complicate somewhat, and not to negate, the use of growth studies as population monitoring devices.

Medical impulses

The medical impulse to growth studies is, in a sense, the reverse side of the same coin. Just as cross-sectional studies of populations aim to monitor provision for communities, so longitudinal studies of individuals aim to monitor provision, or lack of it, *within* the population, in the individual family. The German paediatricians of the 1870s and the German school doctors of the same period, led the way. Part of the concern overlapped with that of social reform: factory mothers' babies weighed little and fared badly. But even in the middle class the mortality was appalling, and it was widely, and probably correctly, believed that doing those things that conferred a good growth rate on a child also conferred a high degree of resistance to the numerous infections to which the child was subject.

For the schoolchild the concern was similar, but with the important added worry about over-pressure of brain work. It seems that the brain was looked on in much the same way as the muscles. Exercise caused fatigue; over-exercise, exhaustion. And just as it was beleved that overdeveloping the muscles caused stunting of stature, so it was thought that overburdening the brain caused disorders of reproduction, if not of growth. This opinion has regressed, and the surveillance of schoolchildren has declined in this century. More recently there are signs of its revival. This seems to come chiefly from the teachers, who now have to deal far more with adolescents than they did prior to, say, 1920. The psychological effects of the differences in tempo between individuals are nowadays a regular part of the life of the secondary school – hence a demand for knowledge about growth, and for medical help if possible. It was this type of demand which led to the studies of Porter and Bowditch and Boas. It contributed, too, to the great explosion of growth studies in the United States in the 1930s, though that also had a source in the belief that correct upbringing in the early, pre-school years had a beneficial effect lasting for the rest of life.

The second generation of longitudinal growth studies, that is the European ones, were the first to be in part oriented towards overtly therapeutic medicine. Except for the practitioners of infant care, paediatricians realized only recently the usefulness of growth as a guide to restored health. Centile standards, the first step in any individual monitoring system, only became generally available in the 1940s through Stuart and Meredith and the editor of Nelson's textbook of paediatrics. Standards suitable for following individuals appeared only in the 1960s, at about the same time that the paper on catch-up growth appeared in one of the most widely read paediatric journals. Every paediatrician must have seen catch-up occur, but like measurements of height itself, until it was charted its potential information remained unrealized. The decade of the 1970s has been the period of explosion of paediatric interest in growth, with a profusion of textbooks, chapters in textbooks, a book entitled *Scientific Foundations of Paediatrics* (Davis and Dobbing, 1974) largely devoted to growth of organ systems, and a full-scale three-volume handbook, by some seventy authors from a dozen countries, called simply *Human Growth* (Falkner and Tanner, 1978–9).

Orthodontists were much ahead of paediatricians in their interest in growth. Particularly in North America, and particularly after the development of the cephalostat, orthodontic research took growth as its major subject. The question was whether one would make things worse rather than better by treatment, and only an accurate way of predicting outcome in the absence of treatment could answer that question. Hence orthodontists have been strongly concerned with prediction of adult face shape from earlier shape. They have also been concerned with predicting when the adolescent spurt will take place. This is because they wish to apply their treatments during the period of most rapid growth, since it is believed that the most rapid results will be obtained at that time.

Thus medicine, in the therapeutic rather than the epidemiological sense, has influenced, and been influenced by, growth studies only recently. But nowadays, with the development of paediatric endocrinology as a specialty, the increasing concern of paediatricians about puberty, and the rise of specialists in adolescent medicine, research on growth is seen as of considerable benefit to medicine. Numerous paediatricians have established special clinics for growth disorders and from these, so often concerned with reassuring the small, delayed child as to his ultimate height and development, comes a demand for more accurate prediction and for ways of influencing tempo, if not size.

Intellectual impulses

Lastly, there has been the still small voice of curiosity. Very few growth studies were made in the spirit of pure inquiry: Buffon's was one, Wiener's another. It is an element in recent longitudinal studies, particularly in the fitting of curves to measurements, the analysis of sibling relationships, the investigation of hormonal changes. Since Darwin, the impulse has centred around the light that growth studies of man can shed on the mechanism and history of evolution. Adult form is altered during the course of evolution through the alteration of growth rates in various parts of the body; the foot matures exceptionally early in man, reaching its nearly-adult status considerably earlier than the hand or the legs or the trunk. That was both necessary for, and conducive to, the exercise of the upright posture early in childhood. Comparative longitudinal growth studies of other primates, which in the last twenty years have become established in a few places, are the foil against which, in this context, human growth is set. The problems of form are presently very intractable, both chemically and mathematically. But, as D'Arcy Thompson, no less than Aristotle, saw so clearly, it is in them that the intellectual challenge offered by studies of growth has its sharpest point.

Finally, there is the contribution the study of growth may have made to our ideas about society. Probably it was in part responsible for the change in how the adult world regarded children that took place between the middle of the seventeenth and the end of the nineteenth century. While children were regarded as small adults it was natural for them to be worked, not reared. (It was a constant complaint of the founders of the Iowa Child Welfare Research Station that far more attention was devoted to the scientific rearing of hogs than of children.) But when children's growth began to be studied, that forced a different view on the observer, the view of a developing organism. Child development moved into the foreground, child work output receded. The process of course was circular, growth studies being both caused by, and stimulating, the new view. (And even should the process change again, it will surely be because work may be seen as beneficial to the child's own development in society, a very different view from that held by the workhouses of the 1700s.)

Growth has also contributed in some small, exemplary, way to a modern appraisal of man's relation with society. Earlier, there have been extreme views, both of men created unequal, immutably occupying niches in society fixed by their birth and nature; and of men, born as *tabulae rasae,* able if suitably trained to become anything, to fulfil every desire. The modern biologist has no use for such views; for him everything is a product of the interaction of the genotype with the continuously changing environment. Nowhere is this better exemplified than in the findings of growth studies down the centuries. But the biologist's view is both complex and general; to achieve a clearer picture we need to discern the details within the generalization. Here also lies an impulse to the study of growth. Together with the rest of human biology, growth studies seek to define man's place in nature, and nature's place in man.

REFERENCES

Aaboe, A. (1974). Scientific astronomy in antiquity. In: *The Place of Astronomy in the Ancient World*, ed. F. R. Hodson, pp. 21–42. London: Oxford University Press.

Acheson, R. M. (1954). A method of assessing skeletal maturity from radiographs: a report from the Oxford Child Health Survey. *Journal of Anatomy* (London), **88**, 498–508.

Acheson, R. M. (1957). The Oxford method of assessing skeletal maturity. *Clinical Orthopaedics*, **10**, 19–39.

Acheson, R. M. (1960). Effects of nutrition and disease on human growth. In: *Human Growth*, ed J. M. Tanner, Symposium for the Study of Human Biology, **3**, pp. 79–92. Oxford: Pergamon Press.

Acheson, R. M. (1966). Maturation of the skeleton. In: *Human Development*, ed. F. Falkner, pp. 465–502. Philadelphia: W. B. Saunders.

Acheson, R. M. and Archer, M. (1959). Radiological studies of the growth of the pituitary fossa in man. *Journal of Anatomy* (London), **93**, 52–67.

Acheson, R. M. and Hewitt, D. (1954*a*). Oxford Child Health Survey. Stature and skeletal maturation in the pre-school child. *British Journal of Preventive Medicine*, **8**, 59–65.

Acheson, R. M. and Hewitt, D. (1954*b*). Physical development in the English and the American pre-school child. A comparison between the findings in the Oxford and the Brush Foundation Surveys. *Human Biology*, **26**, 343–55.

Acheson, R. M., Kemp, F. H. and Parfit, J. (1955). Height, weight and skeletal maturity in the first five years of life. *Lancet*, **i**, 691–2.

Ackerknecht, E. H. (1952). Villermé and Quetelet. *Bulletin of the History of Medicine*, **26**, 317–29.

Aetios of Amida. *Tetrabiblon*, Trans. J. V. Ricci. Philadelphia: Blakiston (1950).

Aitken, W. (1862). *On the Growth of the Recruit and Young Soldier, with a View to a Judicious Selection of 'Growing Lads' for the Army and a Regulated System of Training for Recruits*. London: Griffin, Bohn.

Aitken, W. (1887). *On The Growth of the Recruit and Young Soldier, with a View to a Judicious Selection of 'Growing Lads' for the Army and a Regulated System of Training for Recruits*, 2nd edn. London: Macmillan.

Alberti, L. B. *On Painting and On Sculpture. The Latin texts of 'De Pictura' and 'De Statua' edited with translations . . . by C. Grayson*. London: Phaidon (1972).

Albertus Magnus, pseudo — (1615). *De secretis mulierum libellus, scholiis auctus et amendis repurgatus*. Strassburg: Zetzner.

Ammon, O. (1893). *Die natürliche Auslese beim Menschen*. Jena: Fischer.

Ammon, O. (1894). Die Körpergrösse der Wehrpflichtigen in Grossherzogtum Baden in den Jahren 1840–1864. *Beiträge zur Statistik der Grossherzogtum Baden*, N.S., part 5, (Cited in Baldwin (1921); not seen.)

Amundsen, D. W. and Diers, C. J. (1969). The age of menarche in classical Greece and Rome. *Human Biology*, **41**, 125–32.

Amundsen, D. W. and Diers, C. J. (1973). The age of menarche in medieval Europe. *Human Biology*, **45**, 363–9.

Anderson, M., Blais, M. and Green, W. T. (1956). Growth of the normal foot during childhood and adolescence. Length of the foot and interrelations of foot, stature and lower extremity as seen in serial records of children between 1–18 years of age. *American Journal of Physical Anthropology*, N.S., **14**, 287–308.

Anderson, M. and Green, W. T. (1948). Lengths of the femur and tibia. Norms derived from orthoroentgenograms of children from five years of age until epiphyseal closure. *American Journal of Diseases of Children*, **75**, 279–90.

Anderson, M., Hwang, S. C. and Green, W. T. (1965). Growth of the normal trunk in boys and girls during the second decade of life; related to age, maturity and ossification of the iliac epiphyses. *Journal of Bone and Joint Surgery*, **47**, 1554–64.

Angerstein, W. (1865). *Die Massverhältnisse des menschlichen Körpers und das Wachstum der Knaben dargestellt und erläutert.* Cologne: Selbsverlag.

Anon. (1918). A physical census and its lesson. *British Medical Journal*, **ii**, 348–9.

Aristotle. *Historia animalium*, trans. D'Arcy Thompson. Oxford: Clarendon Press (1910).

Aristotle. *De partibus animalium*, trans. W. Ogle. Oxford: Clarendon Press (1911).

Aristotle. *De generatione et corruptione*, trans. H. H. Joachim. Oxford: Clarendon Press (1922).

Aristotle. *Politics*, trans. E. Barker. Oxford: Clarendon Press (1948).

Aristotle. *Generation of Animals*, with English trans. by A. L. Peck, revised edn Loeb Classical Library. London: Heinemann (1963).

Aristotle. *Historia animalium* (3 vol.), with English trans. by A. L. Peck. Loeb Classical Library, London: Heinemann (vol. 1, 1965; vol. 2, 1970; vol. 3, unpublished).

Aristotle. *Parts of Animals*, with English trans. by A. L. Peck. Loeb Classical Library. London: Heinemann (1968).

Arnald of Villanova (1585). *Opera omnia cum Nicolai Tavrelli . . . annotationibus.* Basle: Waldkirch.

Ascher, — (1912). Ueber Schülerfürsorge. *Zeitschrift für Medizinalbeamte*, **25**, 79–89.

Asher, C. (1975). *Postural Variations in Childhood.* London: Butterworth.

Astruc, P. (1932). Louis-René Villermé. Médecin-sociologue (1782–1863). *Progrès médical*, **9**, 49–54.

Astruc, P. (1933). Villermé, Louis-René, 1782–1863. In: *Biographies médicales*, vol. **7**, pp. 225–44. Paris: Baillière.

Auden, G. A. (1910). Heights and weights of Birmingham school children in relation to infant mortality. *School Hygiene*, **1**, 290–1.

Audran, G. (1683). *Les proportions du corps humain mesurées sur les plus belles figures de l'antiquité.* Paris: Gerard Audran.

Aynsley-Green, A., Zachmann, M. and Prader, A. (1976). Interrelation of the therapeutic effects of growth hormone and testosterone on growth in hypopituitarism. *Journal of Pediatrics*, **89**, 992–9.

Babbage, C. (1851). *The Exposition of 1851*, 2nd edn. with additions. London: Murray. (Reprinted 1968 by Frank Cass Ltd, London.)

Backman, G. (1925). Über generalle Wachstumsgesetze beim Menschen. *Acta Universitatis Latviensis Riga.* **12**, 315–65.

Backman, G. (1931). Das Wachstumsproblem. *Ergebnisse der Physiologie*, **33**, 883–973.

Backman, G. (1934). Das Wachstum der Körperlänge des Menschen. *Kunglicke Svenska Vetenskapsakademiens Handlingar*, series 3, **14** (1), 145pp.

Backman, G. (1938). Wachstumszyklen und phylogenetische Entwicklung. *Lunds Universitets Årsskrift*, N.S., Part 2, **34** (5), 142pp.

Backman, G. (1939*a*).Methodik der theoretischen Wiedergabe beobachteter Wachstumsserien. *Lunds Universitets Årsskrift*, N.S., Part 2, **35**(8), 20pp.

Backman, G. (1939*b*). Die organische Zeit. *Lunds Universitets Årsskrift*, N.S., Part 2, **35** (7), 72pp.

Backman, G. (1948). Die beschleunigte Entwicklung der Jugend. Verfrühte Menarche, verspätete Menopause, verlängerte Lebensdauer. *Acta anatomica,* **4**, 421–80.

[Backman, G. V.] (1964). Obituary. *Kunglicke Fysiografiska Sällskapets i Lund, Årsbok 1964,* pp. 63–70. Lund: Gleerup.

Backström-Järvinen, L. (1964). Heights and weights of Finnish children and young adults. *Annales paediatriae Fenniae, Supplement 23,* 116pp.

Backström-Järvinen, L. and Kantero, R. (1971). Cross-sectional studies of height and weight of Finnish children aged from birth to 20 years. *Acta paediatrica Scandinavica, Supplement 220,* 9–12.

Bacon, F. (1620). *Instauratio magna. Novum organum,* part 2. London: Bill and Norton.

Bacon, F. (1629). *Sylva Sylvarum, or A Natural Historie,* 2nd edn. London: William Lee. (1st edn., 1626).

Baelz, E. von (1880–8). Die körperliche Eigenschaften der Japaner. *Mittheilungen der deutschen Gesellschaft für Natur und Voelkerkunde Ostasiens,* **3**, 330–59 (1880–4); **4**, 35–103 (1884–8).

Bailey, D. A., Malina, R. M. and Rasmussen, R. L. (1978). The influence of exercise, physical activity and athletic performance on the dynamics of human growth. In: *Human Growth,* ed. F. Falkner and J. M. Tanner, vol. 2, pp. 475–505. New York: Plenum.

Baldwin, B. T. (1914). *Physical Growth and School Progress: A Study in Experimental Education.* US Bureau of Education Publication No. 10. Washington: USBE. 212pp.

Baldwin, B. T. (1916). A measuring scale for physical growth and physiological age. In: *Fifteenth Yearbook of the National Society for the Study of Education,* part 1, pp. 11–23. Bloomington, Illinois: Public School Publishing Co.

Baldwin, B. T. (1921). The physical growth of children from birth to maturity. *University of Iowa Studies in Child Welfare,* **1**(1), 411pp.

Baldwin, B. T. (1924). The use and abuse of weight–height–age tables as indexes of health and nutrition. *Journal of the American Medical Association,* **82**, 1–4.

Baldwin, B. T. (1925). Weight–height–age standards in metric units for American-born children. *American Journal of Physical Anthropology,* **8**, 1–10.

Baldwin, B. T., Busby, L. M. and Garside, H. V. (1928). Anatomic growth of children: a study of some bones of the hand, wrist and lower forearm by means of roentgenograms. *University of Iowa Studies in Child Welfare,* **4**(1), 88pp.

Baldwin, B. T. and Wood, T. D. (1923). *Weight–Height–Age Tables. Tables for Boys and Girls of School Age.* New York: American Child Health Association. 11pp.

Bambha, J. K. (1961). Longitudinal cephalometric roentgenographic study of face and cranium in relation to body height. *Journal of the American Dental Association,* **63**, 776–99.

Bambha, J. K. and Van-Natta, P. (1963). Longitudinal study of facial growth in relation to skeletal maturation during adolescence. *American Journal of Orthodontics,* **49**, 481–93.

Barker, E. (trans.) (1948). *Aristotle: Politics.* Oxford: Clarendon Press.

Barr, D. G. D., Schmerling, D. H. and Prader, A. (1972). Catch-up growth in malnutrition: studies in celiac disease after institution of gluten-free diet. *Pediatric Research,* **6**, 521–7.

Bartholomeus Anglicus, *De proprietatis rerum.* As *Batman uppon Bartholome,* trans. S. Batman London: East (1582).

Bartholomew the Englishman, see Bartholomeus Anglicus.

Batkin, S. (1915). Die Dicke des Fettpolsters bei gesunden und kranken Kindern. *Jahrbuch für Kinderheilkunde,* **82**, 103–22.

Batman, S. (trans.) (1582). *Batman uppon Bartholome.* London: East.

Baudelocque, J. L. (1790). *A System of Midwifery,* trans. J. Heath. London: J. L. Baudelocque. (First French edn., 1781.)

Baudier, F. (1952). Principaux aspects de la vie et de l'oeuvre de Buffon. In: *Buffon,* ed. Muséum National d'histoire Naturelle. Paris: Muséum National d'Histoire Naturelle.

Baughan, B., Demerjian, A., Levesque, G. Y. and Lapalme-Chaput, L. (1979). The pattern of facial growth before and during puberty, as shown by French-Canadian girls. *Annals of Human Biology*, **6**, 59–76.

Bayer, L. M. and Bayley, N. (1959). *Growth Diagnosis*. Chicago: University of Chicago Press.

Bayer, L. M. and Bayley, N. (1976). *Growth Diagnosis*, 2nd edn. Chicago: University of Chicago Press.

Bayley, N. (1940). *Studies in the Development of Young Children*. Berkeley: University of California Press. 45pp.

Bayley, N. (1943*a*). Skeletal maturing in adolescence as a basis for determining percentage of completed growth. *Child Development*, **14**, 1–46.

Bayley, N. (1943*b*). Size and body build of adolescents in relation to rate of skeletal maturing. *Child Development*, **14**, 47–90.

Bayley, N. (1946). Tables for predicting adult height from skeletal age and present height. *Journal of Pediatrics*, **28**, 49–64.

Bayley, N. (1954). Some increasing parent–child similarities during the growth of children. *Journal of Educational Psychology*, **45**, 1–21.

Bayley, N. (1955). On the growth of intelligence. *American Psychologist*, **10**, 805–18.

Bayley, N. (1956*a*). Individual patterns of development. *Child Development*, **27**, 45–74.

Bayley, N. (1956*b*). Growth curves of height and weight by age for boys and girls, scaled according to physical maturity. *Journal of Pediatrics*, **48**, 187–94.

Bayley, N. (1957). Data on the growth of intelligence between 16 and 21 years as measured by the Wechsler–Bellevue Scale. *Journal of Genetic Psychology*, **90**, 3–15.

Bayley, N. (1962). The accurate prediction of growth and adult height. *Modern Problems in Paediatrics*, **7**, 234–55.

Bayley, N. and Davis, F. C. (1935). Growth changes in bodily size and proportions during the first three years: a developmental study of 61 children by repeated measurements. *Biometrika*, **27**, 26–87.

Bayley, N. and Pinneau, S. R. (1952). Tables for predicting adult height from skeletal age: revised for use with the Greulich–Pyle hand standards. *Journal of Pediatrics*, **40**, 423–41. (Erratum corrected in *Journal of Pediatrics*, **41**, 371.)

Beal, V. A. (1961). Dietary intake of individuals followed through infancy and childhood. *American Journal of Public Health*, **51**, 1107–17.

Bean, R. B. (1914). The stature and eruption of the permanent teeth of American, German-American and Filipino children. Deductions from the measurements and examinations of 1445 public school children in Ann Arbor, Michigan, and 776 in Manila, PI. *American Journal of Anatomy*, **17**, 113–60.

Beddoe, J. (1885). *The Races of Britain. A Contribution to the Anthropology of Western Europe*. London: Hutchinson.

Bell, E. T. (1945). *The Development of Mathematics*, 2nd edn. New York: Macmillan.

Bellier, P. (1598). *Les œuvres de Philon Juif mises de grec en françois*. Paris: Michèl Sounives.

Beneckendorf, K. F. (1787–9). *Karakterzüge aus dem Leben König Friedrich Wilhelms* (9 vols.) (Cited by Ergang (1941); not seen.)

Benedict, F. G. and Talbot, F. B. (1921). *Metabolism and Growth from Birth to Puberty*. Carnegie Institute Publication No. 302. Washington: Carnegie Institute. 213pp.

Beneke, F. W. (1878). *Die anatomischen Grundlagen der Constitutionsanatomie*. Marburg: Elwert.

Benesch, O. (1966). *German Painting from Dürer to Holbein*. Geneva: Skira.

Benestad, G. (1914). Die Gewichtsverhältnisse reifer norwegischer Neugeborener in der ersten 12 Tagen nach der Geburt. *Archiv für Gynäkologie*, **101**, 292–350.

Berdot le Fils, M. (1774). *De l'art d'accoucher*, vol. 1. Basle: Imhof.

Bergman, R. L. and Bergman, K. E. (1979). Nutrition and growth in infancy. In: *Human Growth*, ed. F. Falkner and J. M. Tanner, vol. 3, pp. 331–60. New York: Plenum.

Bergmüller, J. G. (1723). *Anthropometria, sive statura hominis a nativitate ad consummatum aetatis incrementum ad dimensionum et proportionum regulas discriminata. Oder: Statur des*

Menschen von der Geburt an nach seinem Wachsthum und verschieden Alter . . . Augsburg: J. J. Lotter.

Bergsten-Brucefors, A. (1976). A note on the accuracy of recalled age at menarche. *Annals of Human Biology*, **3**, 71–3.

Berman, A. (1965). Romantic Hygeia: J. J. Virey (1775–1846), pharmacist and philosopher of Nature. *Bulletin of the History of Medicine*, **39**, 134–42.

Berridge, V. (1979). Opium in the fens in nineteenth century England. *Journal of the History of Medicine and Allied Sciences*, **34**, 293–313.

Berriman, A. E. (1953). *Historical Metrology*. New York: Greenwood Press.

Berry, F. M. D. (1904). On the physical examination of 1580 girls from the elementary schools of London. *British Medical Journal*, **i**, 1248–9.

Beyer, H. G. (1895). The growth of US naval cadets. *Proceedings of the United States Naval Institute*, **21**, 297–333.

Bidloo, G. (1685). *Anatomia humani corporis*. Amsterdam: Someren.

Bielicki, T. (1975). Interrelationships between various measures of maturation rate in girls during adolescence. *Studies in Physical Anthropology*, **1**, 51–64.

Bielicki, T. (1976). On the relationships between maturation rate and maximum velocity of growth during adolescence. *Studies in Physical Anthropology*, **3**, 79–84.

Bielicki, T. and Walisko, A. (1976). Wroclaw Growth Study. I. Females. *Studies in Physical Anthropology*, **2**, 53–81.

Bielicki, T. and Welon, Z. (1966). Parent–child height correlations at ages eight to 12 years in children from Wroclaw, Poland. *Human Biology*, **38**, 167–74.

Bielicki, T. and Welon, Z. (1973). The sequence of growth velocity peaks of principal body measurements in girls. *Materialy i Prace Antropologiczne*, **80**, 3–10.

Björk, A. (1947). The face in profile. *Svensk Tandlakare-Tijdskrift*, **40**, *Supplement 513*, 180pp.

Björk, A. (1955). Cranial base development. A follow-up X-ray study of the individual variation in growth occurring between the ages of 12 and 20 years and its relation to brain case and face development. *American Journal of Orthodontics*, **41**, 198–225.

Björk, A. (1963). Variations in the growth pattern of the human mandible: longitudinal radiographic study by the implant method. *Journal of Dental Research*, **42**, 400–11.

Björk, A. and Palling, M. (1955). Adolescent age changes in sagittal jaw relation, alveolar prognathy, and incisal inclination. *Acta odontologica Scandinavica*, **12**, 201–32.

Blaug, M. (1964). The poor law report re-examined. *Journal of Economic History*, **24**, 229–45.

Boas, F. (1892*a*). The growth of children. *Science*, (original series), **19**, 256–57, 281–2.

Boas, F. (1892*b*). The growth of children. *Science*, (original series), **20**, 351–2.

Boas, F. (1894). The correlation of anatomical or physiological measurements. *American Anthropologist*, **7**, 313–24.

Boas, F. (1895*a*). On Dr William Townsend Porter's investigation of the growth of the school children of St Louis. *Science*, **1**, 225–30.

Boas, F. (1895*b*). The growth of first-born children. *Science*, **1**, 402–4.

Boas, F. (1897). The growth of children. *Science*, **5**, 570–3.

Boas, F. (1898). The growth of Toronto children. In: *Report of US Commissioner of Education for 1896–7*, pp. 1541–99. Washington: Department of Education.

Boas, F. (1904). The history of anthropology. *Science*, **20**, 513–24.

Boas, F. (1912*a*). The growth of children. *Science*, **36**, 815–18.

Boas, F. (1912*b*). *Changes in Bodily Form of Descendants of Immigrants*. New York: Columbia University Press.

Boas, F. (1912*c*). Changes in the bodily form of descendants of immigrants. *American Anthropologist*, N.S., **14**, 530–62.

Boas, F. (1922). Review of R. M. Woodbury's 'Statures and weights of children under six years of age'. *American Journal of Physical Anthropology*, **5**, 279–82.

Boas, F. (1928). *Materials for the Study of Inheritance in Man.* New York: Columbia University Press.

Boas, F. (1930). Observations on the growth of children. *Science,* **72**, 44–8.

Boas, F. (1932). Studies in growth [I]. *Human Biology,* **4**, 307–50.

Boas, F. (1933). Studies in growth: II. *Human Biology,* **5**, 429–44.

Boas, F. (1935*a*). Studies in growth: III. *Human Biology,* **7**, 303–18.

Boas, F. (1935*b*). The tempo of growth of fraternities. *Proceedings of the National Academy of Sciences, USA,* **21**, 413–18. (Reprinted in Boas (1940).)

Boas, F. (1935*c*). Conditions controlling the tempo of development and decay. *Association of Life Insurance Medical Directors of America,* **22**, 212–23. (Reprinted in Boas (1940).)

Boas, F. (1936). Effects of American environment on immigrants and their descendants. *Science,* **84**, 522–5.

Boas, F. (1940). *Race, Language and Culture.* New York: Macmillan.

Boas, F. (1941). The relation between physical and mental development. *Science,* **93**, 339–42.

Boas, F. and Wissler, C. (1906). Statistics of growth. In: *Report of US Commissioner of Education for 1904,* pp. 25–132. Washington: Department of Education.

Boerhaave, H. (1708). *Institutiones medicae in usus annuae exercitationes domesticos.* Leyden: van der Linden.

Boerhaave, H. (1709). *Aphorismi de cognoscendis et curandis morbis in usum doctrinae domesticae.* Leyden: van der Linden.

Boerhaave, H. (1732). *Elementa chemiae, quae anniversario labore docuit in publicis privatisque scholis Hermannus Boerhaave.* Leyden: Isaacus Severinus.

Boismont, A. Brière de (1841). De la menstruation. Faire connaître l'influence que cette fonction exerce sur les maladies et celle qu'elle en reçoit. *Mémoires de l'Académie Royale de Médecine,* **9**, 104–233.

Bonnard, G. D. (1968). Cortical thickness and diaphysial diameter of the metacarpal bones from the age of three months to 11 years. *Helvetica paediatrica acta,* **23**, 445–63.

Bonnier, G. (1900). Croissance. In: *Dictionnaire de physiologie,* ed. C. Richet, vol. 4, pp. 488–548. Paris: Alcan.

Borelli, G. A. (1680–1). *De moto animalium . . . opum posthumum: pars prima* (1680), *pars altera* (1681). Rome.

Boudin, M. (1863). Histoire médicale du recrutement des armées et de quelques autres institutions militaires chez divers peuples anciens et modernes. *Annales d'hygiène publique,* 2nd series, **20**, 5–82.

Boulton, P. (1876). Some anthropometrical observations. *British Medical Journal,* **i**, 280–2.

Boulton, P. (1880). On the physical development of children, or, the bearing of anthropometry to hygiene. *Lancet,* **ii**, 610–12.

Bourdier, F. (1952). Principaux aspects de la vie et de l'œuvre de Buffon. In: *Buffon,* ed. L. Bertin *et al.,* pp. 15–86. Paris: Muséum National d'Histoire Naturelle.

Bowditch, H. P. (1872). Comparative rate of growth in the two sexes. *Boston Medical and Surgical Journal,* N.S., **10**, 434; Old Series, **87**.

Bowditch, H. P. (1874–5). Letters to Edward Jarvis. Unpublished: in Countway Library, Boston, Mass.

Bowditch, H. P. (1877). The growth of children. In: *8th Annual Report of the State Board of Health of Massachusetts,* pp. 275–327. Boston: Wright.

Bowditch, H. P. (1879). Growth of children. In: *10th Annual Report of the State Board of Health of Massachusetts,* pp. 33–62. Boston: Wright.

Bowditch, H. P. (1881). The relation between growth and disease. *Transactions of the American Medical Association,* **32**, 371–7.

Bowditch, H. P. (1890). The physique of women in Massachusetts. In: *21st Annual Report of the State Board of Health of Massachusetts,* pp. 287–304. Boston: Wright and Potter.

Bowditch, H. P. (1891). The growth of children studied by Galton's percentile grades. In: *22nd Annual Report of the State Board of Health of Massachusetts,* pp. 479–525. Boston: Wright and Potter.

Bowditch, M. (1958). Henry Pickering Bowditch: an intimate memoir. *Physiologist,* **1**, 7–11.

Box, J. F. (1978). *R. A. Fisher: The Life of a Scientist.* New York: Wiley.

Boyd, E. (1929). The experimental error inherent in measuring the growing human body. *American Journal of Physical Anthropology,* **13**, 389–432.

Boyd, E. (1935). *The Growth of the Surface Area of the Human Body.* Minneapolis: University of Minnesota Press.

Boyd, E. (1936). Weight of the thymus and its component parts and number of Hassall corpuscles in health and disease. *American Journal of Diseases of Children,* **51**, 313–35.

Boyd, E. (1953). R. E. Scammon: obituary notice. *Anatomical Record,* **116**, 259–62.

Boyd, R. (1861). Weight of internal organs. *Philosophical Transactions of the Royal Society of London,* **151**, 241–62.

Boyd-Orr, J. (1936). *Food, Health and Income.* London: Macmillan. (2nd edn, 1937.)

Boynton, B. (1936). The physical growth of girls. A study of the rhythm of physical growth from anthropometric measurements on girls between birth and eighteen years. *University of Iowa Studies in Child Development,* **12**(4), 105pp.

Brabrook, E. W. (1904–5). On the recording of observations on physical conditions in schools. *Journal of the Royal Sanitary Institute,* **26**, 895–902.

Bradbury, D. E. (1931). A review of the published writings of Bird Thomas Baldwin. *Psychological Bulletin,* **28**, 257–76.

Brandt, I. (1978). Growth dynamics of low-birth-weight infants with emphasis on the perinatal period. In: *Human Growth,* ed. F. Falkner and J. M. Tanner, vol. 2, pp. 557–618. New York: Plenum.

Bransby, E. R. and Gelling, J. W. (1946). Variations and the effect of weather on the growth of children. *Medical Officer,* **75**, 213–17.

Brasel, J. A. and Gruen, R. K. (1978). Cellular growth: brain, liver, muscle and lung. In *Human Growth,* ed. F. Falkner and J. M. Tanner, vol. 2, pp. 3–19. New York: Plenum.

Braunfels, S. (1973). Von Mikrokosmos zum Meter. In: *Der vermessene Mensch: Anthropometrie in Kunst und Wissenschaft,* ed. S. Braunfels *et al.,* pp. 43–73. Munich: Mors.

Brebiss, J. G. (1705). *De mensium muliebrium fluxu.* Halle: Henckel.

Breslau, —. (1860). Über die Veränderung im Gewichte der Neugebornen. In: *Denkschrift der medizinisch-chirugische Gesellschafft des Kantons Zurich zur Feier des Fünfzigsten Stifungstages,* pp. 111–18. Zurich: Zürcher and Furrer.

Breslau, —. (1862). Neue Ergebnisse aus Schadelmessungen an Neugebornen. *Wiener medizinisches Wochenschrift,* **12**, 785–7.

Bridges, J. H. and Holmes, T. (1873). Report to the Local Government Board on proposed changes in hours and ages of employment in textile factories. *Parliamentary Papers: Accounts and papers (17),* **55**, 803–64.

British Association for the Advancement of Science (1834). *Report of the British Association for the Advancement of Science Meeting held at Cambridge in 1833.* London: John Murray.

British Association Anthropometric Committee (1878). In: *Report of the British Association for the Advancement of Science Meeting, 1878,* pp. 152–209. London: BAAS. (Document.)

British Association Anthropometric Committee (1879). In: *Report of the British Association for the Advancement of Science Meeting, 1879,* pp. 175–209. London: BAAS. (Document.)

British Association Anthropometric Committee (1880). In: *Report of the British Association for the Advancement of Science Meeting, 1880,* pp. 120–59. London: BAAS. (Document.)

British Association Anthropometric Committee (1881). In: *Report of the British Association for the Advancement of Science Meeting, 1881,* pp. 225–72. London: BAAS. (Document.)

British Association Anthropometric Committee (1882). In: *Report of the British Association for the Advancement of Science Meeting, 1882,* pp. 278–80. London: BAAS. (Document.)

British Association Anthropometric Committee (1883). In: *Report of the British Association for the Advancement of Science Meeting, 1883*, pp. 1–54. London: BAAS. (Document.)

Broadbent, B. H. (1931). A new X-ray technique and its application to orthodontia. *Angle Orthodontist*, **1**, 45–66.

Broadbent, B. H. (1937). The face of the normal child. *Angle Orthodontist*, **7**, 183–208.

Broadbent, B. H. Sr, Broadbent, B. H. Jr and Golden, W. H. (1975). *Bolton Standards of Dentofacial Development Growth.* St Louis: Mosby.

Brock, J. (trans.) (1916). *Galen: On the Natural Faculties.* London: Heinemann.

Brodie, A. G. (1941). On the growth pattern of the human head from the third month to the eighth year of life. *American Journal of Anatomy*, **68**, 209–62.

Brodie, A. G. (1953). Late growth changes in the human face. *Angle Orthodontist*, **23**, 146–57.

Brody, S. (1945). *Bioenergetics and Growth.* New York: Reinhold.

Brook, C. G. D. (1978). Cellular growth: adipose tissue. In: *Human Growth*, ed. F. Falkner and J. M. Tanner, vol. 2, pp. 21–33. New York: Plenum.

Brook, C. G. D., Gasser, T., Werder, E. A., Prader, A. and Vanderschueren-Lodewykx, M. (1977). Height correlations between parents and mature offspring in normal subjects and in subjects with Turner's and Klinefelter's and other syndromes. *Annals of Human Biology*, **4**, 17–22.

Brook, C. G. D., Mürset, G., Zachmann, M. and Prader, A. (1974). Growth in children with 45 XO Turner's Syndrome. *Archives of Disease in Childhood*, **49**, 789–95.

Bruce-Murray, H. (1924). *Child Life Investigations. The Effect of Maternal Social Conditions and Nutrition upon Birth Weight and Birth Length.* Special Report Series of the Medical Research Council No. 81. London: HMSO. 34pp.

Brudevoll, J. E., Liestøl, K. and Walløe, L. (1979). Menarcheal age in Oslo during the last 140 years. *Annals of Human Biology*, **6**, 407–16.

Bruford, W. H. (1962). *Culture and society in classical Weimar. 1775–1806.* London: Cambridge University Press.

Brundtland, G. H., Liestøl, K. and Walløe, L. (1980). Height, weight and menarcheal age of Oslo school children during the last 60 years. *Annals of Human Biology*, **7**, 307–22.

Brundtland, G. H. and Walløe, L. (1976). Menarcheal age in Norway in the 19th century: a re-evaluation of the historical sources. *Annals of Human Biology*, **3**, 363–74.

Brunet, P. (1925). Guéneau de Montbeillard. *Mémoires de l'Académie des Sciences, Arts et Belles-Lettres de Dijon*, pp. 125–31.

Brunniche, A. (1866). Ein Beitrag zur Beurteilung der Körperentwicklung der Kinder. *Journal für Kinderkrankheiten*, **47**, 1–28.

Bücking, J. (1968). *Kultur und Gesellschaft in Tirol um 1600. Des Hippolytus Guarinonius' 'Grewel der Verwüstung menschlichen Geschlechts'. (1610) als Kulturgeschichtliche Quelle des frühen 17 Jahrhunderts.* Historische Studien No. 401. Lubeck and Hamburg: Matthiesen.

Buffon, G. L. L. de. (1749–1804). *Histoire naturelle, générale et particulière avec la description du Cabinet de Roi* (44 vols.). Paris: Imprimerie Royale. (Vols. 2 and 3 issued in 1749.)

Buffon, G. L. L. de (1780). *Epoques de la nature.* Paris: Imprimerie Royale.

Buffon, G. L. L. de. (1812). *Histoire naturelle*, trans. W. Smellie, vol. 3, p. 134. London: Cadell and Davies.

Buffon, G. L. L. de. (1836). *Œuvres complètes avec les suppléments, augmenté de la classification de G. Cuvier*, vol. 4, p. 70. Paris: Dumenil.

Burdach, K. F. (1838). *Die Physiologie als Erfahrungswissenschaft*, 2nd edn (3 vols.). Leipzig: Voss.

Burk, F. (1898). Growth of children in height and weight. *American Journal of Psychology*, **9**, 253–326.

Burke, B. S., Reed, R. B., Van den Berg, A. S. and Stuart, H. C. (1959). Longitudinal studies of child health and development. II. Caloric and protein intakes of children between 1 and 18 years of age. *Pediatrics*, **24**, 922–40.

Burnett, J. (1966). *Plenty and Want. A Social History of Diet in England from 1815 to the Present day.* London: Nelson.

Burschell, F. (1968). *Schiller.* Reinbek: Rowohlt.

Büsch, O. (1962). *Militärsystem und Sozialleben in alten Preussen.* Berlin: de Gruyter.

Bussemaker, U. C. and Daremberg, C. V. (trans.) (1851–76). *Oeuvres d'Oribase.* Paris: Imprimerie nationale.

Butler, N. R. and Bonham, D. G. (1963). *Perinatal Mortality.* Edinburgh: Livingstone.

Caelius Aurelianus. *Gynaecia*, ed. M. F. and I. E. Drabkin. *Bulletin of the History of Medicine. Supplement 13* (1951) .

Caesar J. *The Conquest of Gaul*, trans. S. A. Handford, Harmondsworth: Penguin Books (1951).

Calderini, G. (1875). *Le dimensioni del feto negli ultimi tre mesi della gravidanza.* Turin: Loescher.

Camerer, [W.] (1880). Versuche über den Stoffwechsel angestellt mit 5 Kindern in Alten von 2 – 11 Jahren. *Zeitschrift für Biologie*, **16**, 24–41.

Camerer, [W.] (1882). Gewichtszunahme von 21 Kindern im ersten Lebensjahr. *Jahrbuch für Kinderheilkunde*, **18**, 254–64.

Camerer, W. (1893). Untersuchungen über Massenwachstum und Längenwachstum der Kinder. *Jahrbuch für Kinderheilkunde*, **36**, 249–93.

Camerer, W. (1901). Das Gewichts- und Längenwachstum des Menschen, insbesondere im 1 Lebenjahr. *Jahrbuch für Kinderheilkunde*, **53**, 381–446.

[Camerer, W.] (1910). Obituary. *Jahrbuch für Kinderheilkunde*, **71**, 651–4.

Camerer, W., Jr (1906). Gewichts- und Längenwachstum des Kindes. In: *Handbuch für Kinderheilkunde*, ed. M. Pfaundler and A. Schlossman. Vol. 1, pp. 385–. (English trans., Camerer (1908).)

Camerer, W., (Jr) (1908). Children's growth in weight and height. In: *The Diseases of Children* (4 vols.), ed. M. Pfaundler and A. Schlossman, English trans. ed. H. L. K. Shaw and L. L. Fetra, vol. 1, pp. 409–24. Philadelphia: Lippincott.

Cameron, N. (1977). 'An analysis of the growth of London schoolchildren: the London County Council's 1966–1967 growth survey'. PhD thesis, University of London.

Cameron, N. (1978). The methods of auxological anthropometry. In: *Human Growth*, ed. F. Falkner and J. M. Tanner, vol. 2, pp. 35–90. New York: Plenum.

Cameron, N. (1979). The growth of London schoolchildren 1904–1966: an analysis of secular trend and intra-county variation. *Annals of Human Biology*, **6**, 505–25.

Campbell, D. A., (1967). *Greek Lyric Poetry.* London: Macmillan.

Campbell, J. (1862). On the appearance of the menses in Siamese women. *Edinburgh Medical Journal*, **8**, 233–6.

Camper, P. (1794). *On the connexion between the science of Anatomy and the Arts of Drawing, Painting and Statuary* (2 vols.). London: Dilly.

Cannon, W. B. (1924). Henry Pickering Bowditch. *Biographical Memoirs of the National Academy of Sciences, USA*, **17**, 183–96. Washington: Government Printing Office.

Carlier, G. (1892). Recherches anthropométriques sur la croissance. Influence de l'hygiène et des exercices physiques. *Mémoires de la Société d'Anthropologie de Paris*, **4**, 265–346.

Carlson, A. J. (1949). William Townsend Porter, 1862–1949. *Science*, **110**, 111–12.

Carron, A. V. and Bailey, D. A. (1974). Strength development in boys from 10 through 16 years. *Monographs of the Society for Research in Child Development*, **39** (4), 37 pp.

Carstädt, F. (1888). Über das Wachstum der Knaben vom 6 bis zum 16 Lebensjahre. *Zeitschrift für Schulgesundheitspflege*, **1**, 65–9.

Celsus. *De Medicina*, trans. W. G. Spencer. Loeb Classical Library. London: Heinemann (1938).

Chadwick. E. (1842). *Report on the Sanitary Condition of the Labouring Population of Great Britain from the Poor Law Commissioners.* London: Clowes.

Chadwick, J. and Mann, W. N. (1950). *The Medical Works of Hippocrates.* Oxford: Blackwell Scientific Publications.

Charleton, W. (1680). *Enquiries into Human Nature, in VI Anatomic Praelections.* London: White.

Cheek, D. B. (1968). Muscle cell growth in normal children. In: *Human Growth*, ed. D. B. Cheek, pp. 337–51. Philadelphia: Lea and Febiger.

Cheek, D. B. (1975). Growth and body composition. In: *Fetal and Postnatal Cellular Growth*, ed. D. B. Cheek, pp. 389–408. New York: Wiley.

Christopher, W. S. (1900). Measurements of Chicago children. *Journal of the American Medical Association*, **35**, 618–23, 683–7.

Chrzastek-Şpruch, H. (1968). Longitudinal study on the physical development in Lublin infants. *Prace i Materialy Naukowe IMD*, **11**, 65–104.

Chrzastek-Spruch, H. (1977). Some genetic problems in physical growth and development – a longitudinal study on children aged 0–7 years. *Acta geneticae medicae et gemellologiae*, **26**, 205–20.

Chrzastek-Spruch, H. (1979*a*). Genetic control of some somatic traits in children as assessed by longitudinal studies. *Studies in Human Ecology*, **3**, 27–51.

Chrzastek-Spruch, H.(1979*b*). Longitudinal studies of environmental influences on child growth. *Studies in Human Ecology*, **3**, 179–202.

City of Edinburgh Charity Organization Society (1906). *Report on the Physical Condition of 1400 School Children in the City together with Some Account of their Homes and Surroundings.* London: King.

Clark, W. E. Le Gros, and Medawar, P. B. (eds.) (1945). *Essays on Growth and Form presented to D'Arcy Wentworth Thompson.* Oxford: Clarendon Press.

Clarke, E. H. (1873). *Sex in Education, or a Fair Chance for Girls.* Boston: Osgood. (5th edn, 1875.)

Clarke, E. H. (1874). *The Building of a Brain.* Boston: Osgood.

Clarke, J. (1786). Observations on some causes of the excess of the mortality of males above that of females. *Philosophical Transactions of the Royal Society of London*, **76**, 349–64.

Clever, W. (1590). *The Flower of Phisicke.* London: Roger Ward.

Cnopf, J. (1872). Über die Wichtigkeit der Anwendung der Waage in der Kinderpraxis, *Journal für kinderkrankheiten*, **58**, 219–34.

Cobb, W. M. (1959). Thomas Wingate Todd, 1885–1938. *Journal of the National Medical Association*, **51**, 233–46.

Cohn, N. (1957). *The Pursuit of the Millennium.* London: Secker and Warburg.

Collard, A. (1934). Goethe et Quetelet. Leurs relations de 1829 à 1832. *Isis*, **20**, 426–35.

Collins, R. (1835). *A Practical Treatise on Midwifery, containing the Result of 16,654 Births occurring in the Dublin Hospital during a Period of Seven Years, commencing November 1826.* London: Longman Rees.

Collins, R. (1849). *A Short Sketch of the Life and Writings of the late Joseph Clarke, Esq. MD.* London: Longman, Brown, Green and Longman.

Combe, J. (1896). Körperlange und Wachstum der Volkschulkinder in Lausanne. *Zeitschrift für Schulgesundheitspflege*, **9**, 569–89.

Cone, T. E. (1961). De pondere infantum recens natorum. The history of weighing the new-born infant. *Pediatrics*, **28**, 490–8.

Cone, T. E. and Khoshbin, S. (1978). Botticelli demonstrates the Babinski reflex more than 400 years before Babinski. *American Journal of Diseases of Children*, **132**, 188.

Conring, H. (1666). *De habitus corporum germanicorum antiqui ac novi causis.* Helmstaedt: Muller.

Cook, T. A. (1914). *The Curves of Life: An Account of Spiral Formations and their Application to Growth in Nature, to Science and to Art.* London: Constable.

Correnti, V. (1948). On the correlation between weight and height in human growth: evaluation of the variations through use of the auxogram. *Rivista di antropologia*, **36**, 120–51. (English trans. in *Yearbook of Physical Anthropology* (1948–9), **5**, 259–79.)

Count, E. W. (1942). A quantitative analysis of growth in certain human skull dimensions. *Human Biology*, **14**, 143–65.

Count, E. W. (1943). Growth patterns of the human physique: an approach to kinetic anthropometry: I. *Human Biology*, **15**, 1–32.

Courtis, S. A. (1917). Measurement of the relation between physical and mental growth. *American Physical Education Review*, **22**, 464–81.

Courtis, S. A. (1932). *The Measurement of Growth*. Ann Arbor, Michigan: Blumfield and Blumfield.

Cowell, J. W. (1833). See Parliamentary Papers (1833).

Crampton, C. W. (1908). Physiological age, a fundamental principle. *American Physical Education Review*, **13**, (3–6). (Reprinted in *Child Development* (1944), **15**, 1–12.)

Creutz, R. (1938). Die Medizin in Speculum Maius des Vincentius von Beauvais. *Sudhoffs Archiv für Geschichte der Medizin und der Naturwissenschaften*, **31**, 297–313.

Cruise, M. O. (1973). A longitudinal study of the growth of low birth weight infants. I. Velocity and distance growth, birth to 3 years. *Pediatrics*, **51**, 620–8.

Crum, F. S. (1916). Anthropometric statistics of children, ages 6 to 48 months. *Quarterly Publication of the American Statistical Association*, **15**, 332–6.

Culpeper, N. (1675). *A Directory for Midwives* [2nd edn]. London: Sawbridge.

Curry, W. C. (1925). Chaucer's Doctor of Physik. *Philological Quarterly*, **4**, 1–24.

Daffner, F. (1884*a*). Vergleichende Untersuchungen über die Entwicklung der Körpergrösse und des Kopfumfanges. *Archiv für Anthropologie*, **15**, 37–44.

Daffner, F. (1884*b*). Über Grösse, Gewicht, Kopf und Brustumfang beim männlichen Individuum von 13 bis 22 Lebensjahre nebst vergleichender Angabe einige Kopfmasse. *Archiv für Anthropologie*, **15**, 121–6.

Daffner, F. (1897). *Das Wachstum des Menschen. Anthropologische Studien*, 1st edn. Leipzig: Englemann. 129 pp.

Daffner, F. (1902). *Das Wachstum des Menschen. Anthropologische Studien*, 2nd edn. Leipzig: Englemann. 475pp.

Dally, E. (1879). Croissance. In: *Dictionnaire encyclopédique des sciences médicales*, series 1, vol. 23, pp. 372–490. Paris: Asselin, Masson.

Damon, A. and Bajema, C. J. (1974). Age at menarche: accuracy of recall after 39 years. *Human Biology*, **46**, 381–4.

Damon, A., Damon, S. T., Reed, R. B. and Valadian, I. (1969). Age at menarche of mothers and daughters, with a note on accuracy of recall. *Human Biology*, **41**, 161–75.

Danson, J. T. (1862). Statistical observations relative to the growth of the human body (males) in height and weight, from 18 to 30 years of age, as illustrated by the records of the Borough Gaol of Liverpool. *Journal of the Statistical Society*, **25**, 20–6.

Davenport, C. B. (1899). *Statistical Methods with Special Reference to Biological Variation*. New York: Wiley. (2nd edn, 1904; 3rd edn, 1914.)

Davenport, C. B. (1926). Human metamorphosis. *American Journal of Physical Anthropology*, **9**, 205–32.

Davenport, C. B. (1931). Individual versus mass studies in child growth. *Proceedings of the American Philosophical Society*, **70**, 381–9.

Davenport, C. B. (1934). Critique of curves of growth and of relative growth. *Cold Spring Harbor Symposia on Quantitative Biology*, **2**, 203–8.

Davenport, C. B. (1939). Postnatal development of the human outer nose. *Proceedings of the American Philosophical Society*, **80**, 175–356.

Davenport, C. B. (1940). Postnatal development of the head. *Proceedings of the American Philosophical Society*, **83**, 1–215.

Davenport, C. B. and Love, A. G. (1920). Report on defects found by draft boards in USA conscripts. *Science Monthly*, Jan., 5–25; Feb., 125–41.

Davenport, C. B., Steggerda, M. and Drager, W. (1934). Critical examination of physical anthropometry of the living. *Proceedings of the American Academy of Arts and Sciences*, **69**, 265–84.

Davie, R., Butler, N. and Goldstein, H. (1972). *From Birth to Seven.* London: Longman.

Davis, J. A. and Dobbing, J. (eds.) (1974). *Scientific Foundations of Paediatrics.* London: Heinemann.

Daw, S. F. (1970). Age of boys' puberty in Leipzig 1727–49 as indicated by voice breaking in J. S. Bach's choir members. *Human Biology,* **42**, 87–9.

Dearborn, W. F. and Rothney, J. W. M. (1941). *Predicting the Child's Development.* Cambridge, Mass.: Sci-Art Publishers.

Dearborn, W. F., Rothney, J. W. M. and Shuttleworth, F. K. (1938). Data on the growth of public school children. *Monographs of the Society for Research in Child Development,* **3**(1), 90pp.

De Giovanni, A. (1904–9). *Morfologia del Corpo Umano* (3 vols.), 2nd edn. Milan: Hoepli.

Deming, J. (1957). Application of the Gompertz curve to the observed pattern of growth in length of 48 individual boys and girls during the adolescent cycle of growth. *Human Biology,* **29**, 83–122.

Deming, J . and Washburn, H. H. (1963). Application of the Jenss curve to the observed pattern of growth during the first eight years of life in 40 boys and 40 girls. *Human Biology,* **35**, 484–506.

Demirjian, A., Goldstein, H. and Tanner, J. M. (1973). A new system of dental age assessment. *Human Biology,* **45**, 211–27.

Demol, J. (1731). *Dissertatio phisiologico-mechanica de naturali catamaeniorum fluxu.* Montpelier: Martel.

Denman, T. (1788). *Introduction to the Practice of Midwifery.* London: Johnson.

Deslandres, M. and Michelin, A. (1938). *Il y a cent ans. Etat physique et morale des ouvriers au temps du libéralism. Témoinage de Villermé.* Paris: Editions Spes.

Diderot, D. and D'Alembert, J. L. R. (1751–65). *Encyclopédie, ou dictionnaire raisonné des sciences, des arts et des métiers . . .* (17 vols.). Paris: Briasson, David, Le Breton and Durand.

Diderot, D. and D'Alembert, J. L . R. (1777). *Encyclopédie, ou dictionnaire raisonné des sciences, des arts et des métiers,* new edn. Geneva: Pellet.

Dietz, J. F. G. (1757). *De temporum in graviditate et partu aestimatione.* Göttingen: Schulz.

Dodds, E. R. (1963). *Proclus: The Elements of Theology. A Revised Text with Translation, Introduction and Commentary.* Oxford: Clarendon Press.

Dörrer, A., Grass, F., Sauser, G. and Schadelbauer, K. (eds.) (1954). *Hippolytus Guarinonius (1571–1654). Zur 300 Wiederkehr seines Todestages.* Schlern-Schriften No. 126. Innsbruck: Universitätsverlag Wagner.

Douglas, J. W. B. (1964). *The Home and the School.* London: MacGibbon and Kee.

Douglas, J. W. B. and Blomfield, J. M. (1958). *Children under Five.* London: Allen and Unwin.

Douglas, J. W. B., Ross, J. M. and Simpson, H. R. (1965). The relation between height and measured educational ability in school children of the same social class, family size and stage of sexual development. *Human Biology,* **37**, 178–86.

Douglas, J. W. B. and Simpson, H. (1964). Height in relation to puberty, family size and social class. *Milbank Memorial Fund Quarterly,* **42**, 20–35.

Drabkin, M. F. and Drabkin, I. E. (eds.) (1951). Caelius Aurelianus: Gynaecia. *Bulletin of the History of Medicine, Supplement 13.*

Drillien, C. M. (1964). *The Growth and Development of the Prematurely Born Infant.* Edinburgh: Livingstone.

Ducros, A. and Pasquet, P. (1978). Evolution de l'âge d'apparition des premières règles (ménarche) en France. *Biométrie humaine,* **13**, 35–43.

Duffy, C. (1974). *The Army of Frederick the Great.* Newton Abbot: David and Charles.

Duffy, J. (1968). Mental strain and 'overpressure' in the schools: a 19th century viewpoint. *Journal of the History of Medicine and Allied Sciences,* **23**, 63–79.

Duncan, J. M. (1864–5). On the weight and length of the newly-born child in relation to the mother's age. *Edinburgh Medical Journal,* **10**, 497–502.

Duncan, J. M. (1866). *Fecundity, Fertility, Sterility and Allied Topics.* Edinburgh: A. and C. Black.

Dupertuis, C. W. and Tanner, J. M. (1950). The pose of the subject for photogrammetric anthropometry, with special reference to somatotyping. *American Journal of Physical Anthropology, N. S.*, **8**, 27–42.

Dürer, A. (1525). *Underweysung der Messung mit dem Zirkel im Richtscheyt, in Linien ebnen unnd gantzen Corporen durch Albrecht Dürer zu samen getzoge und zu nutz alle Kunstlieb habenden mit zu gehörigen Figuren in truck gebrocht.* Nürnberg.

Dürer, A. (1528). [*Proportionslehre*] *Hierin sind begriffen vier Bücher von menschlicher Proportion, durch Albrechten Dürer von Nuremberg erfunden und beschriben zu nutz allen denen so zu dieser kunstlieb tragen.* Nuremberg: Jeronymum Formschneyder.

Dürer, A. (1532). *Albertus Durerus . . . versus e germanica lingua in latinam . . . adeo exacte. Quatuor his suarum institutionum geometricarum libris, lineas, superficies et solida corpora tractavit.* Paris: Wechel.

Dürer, A. (1557). *De symmetria partium humanorum corporum libri quatuor, a germanica lingua in latinam versi.* Paris: Perier.

Dürer, A. (1591). *Della symmetria de i corpi humani, libri quattro*, trans. G. P. Gallucci Salodiano. Venice: Nicolini.

Dürer, A. (1613). *Les quatres livres d'Albert Dürer traduits par Loys Meigret Lionnois.* Arnhem: Jeansz.

Edgerton, S. Y., Jr (1975). *The Renaissance Rediscovery of Linear Perspective.* New York: Basic Books.

Edwards, D. A. W., Hammond, W. H., Healy, M. J. R., Tanner, J. M. and Whitehouse, R. H. (1955). Design and accuracy of calipers for measuring subcutaneous tissue thickness. *British Journal of Nutrition*, **9**, 133–43.

Ehrenberg, V. (1973). *From Solon to Socrates*, 2nd edn. London: Methuen.

Elderton, E. M. (1914). Height and weight of school children in Glasgow. *Biometrika*, **10**, 288–339.

Elsholtz, J. S. (1654). *Anthropometria. Accessit doctrina naevorum.* Padua: Pasquati.

Elsholtz, J. S. (1672). *Anthropometria*, editio post Patavinam altera. Stade: E. Gohlium.

Emmett, T. A. (1879). *The Principles and Practice of Gynaecology.* Philadelphia: Lea.

Emmison, F. G. (1976). *Elizabethan Life: Home, Work and Land.* Council Record Office, Publication No. 69. Chelmsford: Essex County Council.

Engelmann, G. J. (1891). The health of the American girl, as imperilled by the social conditions of the day. *Southern Surgical and Gynecological Transactions*, **3**, 1–29.

Engelmann, G. J. (1901*a*). Rapport du développement mental ou développement fonctionnel chez la jeune fille américaine. *Annales de gynécologie et d'obstétrique*, **55**, 30–43.

Engelmann, G. J. (1901*b*). Age of the first menstruation on the North American continent. *Transactions of the American Gynecological Society*, **26**, 77–110.

[Engelmann, G. J.] (1904). Obituary. *Transactions of the American Gynecological Society*, **29**, 485–8.

Ergang, R. (1941). *The Potsdam Führer. Frederick William I, Father of Prussian Militarism.* New York: Columbia University Press.

Erismann, F. (1888). Schulhygiene auf der Jubiläumsausstellung der Gesellschaft für Beförderung der Arbeitsamkeit in Moskau. *Zeitschrift für Schulgesundheitspflege*, **1**, 347–73, 393–419.

Erismann, F. (1889). Untersuchungen über die körperliche Entwicklung der Arbeiterbevölkerung in Zentralrussland. *Archiv für soziale Gesetzgebung in Statistik* (Tübingen), **1**, 98–135, 429–84.

Ernst, L. H. and Meumann, E. (1906). *Das Schulkind in seiner körperlichen und geistigen Entwicklung. I. Anthropologisch-psychologische Untersuchungen an Züricher Schulkindern nebst einer Zusammenstellung der Resultate der wichtigsten Untersuchungen an Schulkindern in andern Ländern, von Dr. phil. Lucy Hoesch Ernst.* Leipzig: Nemnich.

Ersch, J. S. and Graber, J. G. (eds.) (1834). *Allgemeine Encyclopädie der Wissenschaften und Künste in alphabetischen Folge.* Hospitäler, Abschnitt 3: Findlungshäuser und Waisenhäuser, Leipzig: Gleditsch and Brockhaus.

Espenschade, A. (1940). Motor performance in adolescence, including the study of relationships with measures of physical growth and maturity. *Monographs of the Society for Research in Child Development,* **5**, 126pp.

' penschade, A. (1947). Development of motor coordination in boys and girls. *Research Quarterly of the American Association for Health,* **18**, 30–44.

Evans, R. J. W. (1973). *Rudolf II and his World. A Study in Intellectual History 1576–1612.* Oxford: Clarendon Press.

Eveleth, P. B. and Tanner, J. M., (1976). *Worldwide Variation in Human Growth.* IBP Synthesis Series 8. Cambridge University Press.

Evitsky, E. (1881). On the growth of children during the first year, and on the nutritive conditions of early childhood. *New York Medical Journal,* **33**, 172–89.

Faiman, C. and Winter, J. S. D. (1974). Gonadotropins and sex hormone patterns in puberty: clinical data. In: *Control of the Onset of Puberty,* ed. M. M. Grumbach, G. D. Grave and F. E. Mayer, pp. 32–55. New York: Wiley.

Falkner, F. (1955). *A Baseline of Investigations for Longitudinal Growth Studies in the Child.* Paris: Centre International de l'Enfance.

Falkner, F. (ed.) (1960). Child development: an international method of study. *Modern Problems in Paediatrics,* **5**, 237pp.

Falkner, F. (ed.) (1966). *Human Development: by 29 authorities.* Philadelphia: W. B. Saunders.

Falkner, F. (1978). Implication for growth in human twins. In: *Human Growth,* ed. F. Falkner and J. M. Tanner, vol. 1, pp. 397–413. New York: Plenum.

Falkner, F., Pernot-Roy, M. P., Habich, H., Sénécal, J. and Massé, G. (1958). Some international comparisons of physical growth in the two first years of life. *Courrier du Centre International de l'Enfance,* **8**, 1–11.

Falkner, F. and Tanner, J. M. (eds.) (1978–9). *Human Growth* (3 vols.). New York: Plenum.

Farquaharson, R. (1876). On overwork. *Lancet,* **i**, 9–10.

Fasbender, H. (1878). Mutter- und Kindeskörper; das Becken des lebenden Neugeborenen. *Zeitschrift für Geburtshülfe und Gynäkologie,* **3**, *278–304.*

Fassman, D. (1735). *Leben und Thaten . . . Friedrich Wilhelms.* Hamburg: [no publisher given]. [Author's name not on title page.]

Faust, M. S. (1977). Somatic development of adolescent girls. *Monographs of the Society for Research in Child Development,* **42** (1), 90pp.

Faye, F. C. and Vogt, H. (1866). Statisticke Resultater støttede til 3000 paa Fødselsstiftelsen i Christiania undersøgte Svangre of Fødende samt Børn. *Norsk Magazin for Laegevidenskaben,* **20**, 1–39, 193–219, 289–312, 393–415.

Faye, L. and Hald, J. (1896). Undersøgelser om Sundhetstilstanden ved norske høiere Gutte og Pigeskoler samt faellesskoler udförte i 1891 og 1892. Med un résumé en français. *Kongeriget Norges 45 ordentlige Stortags Forhandlinger i Aaret 1896.* (Cited in Kiel (1939); not seen.)

Fergus, W. (1884). Over-pressure in schools. *Lancet,* **ii**, 616.

[Fergus, W.] (1887). Obituary notice *Lancet,* **i**, 105.

Fergus, W. and Rodwell, G. F. (1874). On a series of measurements for statistical purposes recently made at Marlborough College. *Journal of the Anthropological Institute,* **4**, 126–30.

Fetter, V., Suchý, J. and Prokopec, M. (1966). Nová anthropologická norma vývoje mladeze v čssr. *Časopis lékařu českých,* **105**, 1323–4.

Fischbein, S. (1977). Intra-pair similarity in physical growth of monozygotic and of dizygotic twins during puberty. *Annals of Human Biology,* **4**, 417–30.

Fischbein, S. and Nordquist, T. (1978). Profile comparisons of physical growth for monozygotic and dizygotic twin pairs. *Annals of Human Biology,* **5**, 321–8.

Fischer, A. (1927). Hippolyt Guarinonius, ein deutscher Vorkämpfer für Gesundheitsrecht und Gesundheitspflicht (1571–1654). *Sozial hygienisches Mitteilungen*, **11**, 105–14.

Fitt, A. B. (1941). *Seasonal Influence on Growth Function and Inheritance*. Wellington, NZ: New Zealand Council for Educational Research.

Fleischmann, L. (1877). Über Ernährung und Körperwägungen der Neugeborenen und Säuglingen. *Wiener Klinik*, **3**, 145–94.

Fleming, R. M. (1933). *A Study of Growth and Development: Observations in Successive Years on the Same Children: With a Statistical Analysis by W. J. Martin*. Medical Research Council Special Report Series No. 190. London: MRC. 85pp.

Flourens, P. (1860). *Des manuscripts de Buffon*. Paris: Garnier.

[Fontenu, Abbé, L. F. de] (1727) Sur les accroisemens et décroissmens alternatifs du corps humain. *Histoire de l'Académie Royale des Sciences*, [1725], p. 16.

Forbes, G. B. (1978). Body composition in adolescence. In: *Human Growth*, ed. F. Falkner and J. M. Tanner, vol. 2, pp. 239–72. New York: Plenum.

Forbes, J. D. (1836). Experiments on the weight, height and strength of men at different ages. *Report of the British Association for the Advancement of Science Sixth Meeting: Notes and Abstracts*, pp. 38–9. London: BAAS. (Document.)

Forbes, [J. D.] (1837). On the result of experiments on the weight, height and strength of above 8000 individuals. *Proceedings of the Royal Society of Edinburgh*, 16 Jan. 1837, vol. 1, pp. 160–1 [but there is an error of pagination so pages 148–64 are duplicated; this volume covers 1832 to 1844].

Ford, E. H. R. (1958). Growth in height of ten siblings. *Human Biology*, **30**, 107–19.

Forrest, D. W. (1974). *Francis Galton: The Life and Work of a Victorian Genius*. London: Elek.

Förster, F. C. (ed.) (1834–5). *Friedrich Wilhelm I: König von Preussen* (5 vols.). Potsdam: Riegel.

Foster, M. (1901). *Lectures on the History of Physiology during the Sixteenth, Seventeenth and Eighteenth Centuries*. London: Cambridge University Press. (Reprinted, 1924.)

Fowler, H. N. (trans.) (1977). *Plato: Phaedo*. Loeb Classical Library, London: Heinemann.

Fox, A. L. (1877). Measurements taken of the officers and men of the 2nd Royal Surrey Militia according to the general instructions drawn up by the Anthropometric Committee of the British Association. *Journal of the Anthropological Institute*, **6**, 443–57.

Francis, C. C. (1948). Growth of the human pituitary fossa. *Human Biology*, **20**, 1–20.

Francke, A. H. (1787). *The Footsteps of Divine Providence*. London: Justins.

Frank, L. K. (1935). The problem of child development. *Child Development*, **6**, 7–18.

Frank, L. K. (1943). Research in child psychology: history and prospect. In: *Child Behavior and Development*, ed. R. G. Barker, J. S. Kounin and H. F. Wright, pp. 1–16. New York: McGraw-Hill.

Frank, L. K. (1962). The beginnings of child development and family life education in the twentieth century. *Merrill-Palmer Quarterly of Behavior and Development*, **8**, 207–27.

Frankenhauser, —. (1859). Ueber einige Verhältnisse die Einfluss auf die stärkere oder schwächere Entwicklung der Frucht während der Schwangerschaft haben. *Monatschrift für Gebürtskunde und Frauenkrankheiten*, **13**, 170–9.

Franzen, R. (1929). *Physical Measures of Growth and Nutrition*. School Health Research Monographs No. 2. New York: American Child Health Association. 138pp.

Freeman, K. (1926). *The Work and Life of Solon, with a Translation of his Poems*. Cardiff: University of Wales Press Board.

Freind, J. (1703). *Emmenologia*. London: Bennet.

Freind, J. (1729). *Emmenologia*, trans. T. Dale. London: Cox.

French, P. J. (1972). *John Dee: The World of an Elizabethan Magus*. London: Routledge and Kegan Paul.

Friedenthal, H. (1914). *Allgemeine und spezielle Physiologie des Menschenwachstum*. Berlin: Springer. (Also in *Ergebnisse der inneren Medizin und Kinderheilkunde* (1912–13), **8**, 254–99; **9**, 504–30; **11**, 685–753.)

Friedlander, M. (1815). *De l'éducation physique de l'homme.* Paris: Treuttel and Würtz.

Friend, G. E. (1935). *The Schoolboy: A Study of his Nutrition, Physical Development and Health.* Cambridge: Heffer.

Gadol, J. (1969). *Leon Battista Alberti: Universal Man of the Early Renaissance.* Chicago: University of Chicago Press.

Galen. *On the Natural Faculties,* trans. J. Brock. London: Heinemann (1916).

Galen. *De sanitate tuenda: A Translation of Galen's Hygiene by R. M. Green.* Springfield, Illinois: Thomas (1951).

Galen. *De usu partium: On the Usefulness of the Parts of the Body. Translated from the Greek with an Introduction and Commentary by M. T. May.* Ithaca: Cornell University Press (1968).

Galen. *De usu partium: Libri XVII ad codicum fidem recensuit Georgius Helmreich.* Bibliotheca scriptorium Graecorum et Romanorum Teubneriana. (Reprinted 1968, by Hakkete, Amsterdam.)

Galton, F. (1873–4). Proposal to apply for anthropological statistics from schools. *Journal of the Anthropological Institute,* **3**, 308–11.

Galton, F. (1874). Notes on the Marlborough School statistics. *Journal of the Anthropological Institute,* **4**, 130–5.

Galton, F. (1875–6). On the height and weight of boys aged 14 years in town and country Public Schools. *Journal of the Anthropological Institute,* **5**, 174–81.

Galton, F. (1883). *Inquiries into Human Faculty and its Development.* London: Macmillan.

Galton, F. (1884–5). Some results of the Anthropometric Laboratory. *Journal of the Anthropological Institute,* **14**, 275–88.

Galton, F. (1885). Anthropometric percentiles. *Nature* (London), **31**, 223–5.

Galton, F. (1885–6). Hereditary stature. *Journal of the Anthropological Institute,* **14**, 488–99.

Galton, F. (1886). Family likeness in stature. *Proceedings of the Royal Society of London,* **40**, 42–73.

Galton, F. (1889). *Natural Inheritance.* London: Macmillan.

Gandevia, B. (1977*a*). *Tears often Shed: Child Health and Welfare in Australia from 1788.* Oxford: Pergamon Press.

Gandevia, B. (1977*b*). A comparison of the heights of boys transported to Australia from England, Scotland and Ireland, c. 1840, with later British and Australian developments. *Australian Paediatric Journal,* **13**, 91–7.

Garn, S. M. (1955). Relative fat patterning: an individual characteristic. *Human Biology,* **27**, 75–89.

Garn, S. M. (1958). Fat, body size and growth in the newborn. *Human Biology,* **30**, 265–80.

Garn, S. M. and Bailey, S. M. (1978). The genetics of maturational processes. In: *Human Growth,* ed. F. Falkner and J. M. Tanner, vol. 1, pp. 307–32. New York: Plenum.

Garn, S. M. and Gorman, E. L. (1956). Comparison of pinch-caliper and teleroentgenogrammetric measurements of subcutaneous fat. *Human Biology,* **28**, 407–13.

Garn, S. M. and Haskell, J. A. (1960). Fat thickness and development status in childhood and adolescence. *American Journal of Diseases of Children,* **99**, 746–51.

Garn, S. M., Rohmann, C. G. and Blumenthal, T. (1966). Ossification sequence polymorphism and sexual dimorphism in skeletal development. *American Journal of Physical Anthropology,* **24**, 101–16.

Garn, S. M. and Saalberg, J. H. (1953). Sex and age differences in the composition of the adult leg. *Human Biology,* **25**, 144–53.

Garn, S. M. and Shamir, Z. (1958). *Methods for Research in Human Growth.* Springfield, Illinois: Thomas.

Garn, S. M. and Young, R. W. (1956). Concurrent fat loss and fat gain. *American Journal of Physical Anthropology, N. S.,* **14**, 497–504.

Garnier, F. (1973). L'iconographie de l'enfant au moyen âge. *Annales de démographie historique,* [no vol. no.], 135–6.

Gassner, U. K. (1862). Über die Veränderungen des Körpergewichtes bei Schwangeren, Gebärenden, und Wöcherinnen. *Monatsschrift für Gebürtskunde und Frauenkrankheiten*, **19**, 1–67.

Geissler, A. (1892). Messungen von Schulkindern in Gohlis-Leipzig. *Zeitschrift für Schulgesundheitspflege*, **5**, 249–53.

Geissler, A. and Uhlitzsch, R. (1888). Die Grössenverhältnisse der Schulkinder im Schulinspektionsbezirk Freiberg. *Zeitschrift des Königlichen Sächsischen Statistiken Bureau's*, **34**, 28–40.

George, M. D. (1965). *London Life in the Eighteenth Century*. New York: Capricorn.

Ghiberti, L. *I commentari. A cura di Ottavio Morisani*. Naples: Ricciardi (1947).

Ghyka, M. C. (1927). *Esthetique: des proportions dans la nature et dans les arts*. Paris: Gallimard.

Giles, A. E. (1901*a*). Primary amenorrhoea. *Clinical Journal*, **17**, 225–33.

Giles, A. E. (1901*b*). The factors which lead to variations in the age of puberty and the clinical characters of menstruation. *Medical Chronicle*, **34** (**1** of 4th series), 161–79, 254–64.

Godfrey, S. (1974*a*). Growth and development of the respiratory system: functional development. In: *Scientific Foundations of Paediatrics*, ed. J. Davis and J. Dobbing, pp. 254–71. London: Heinemann.

Godfrey, S. (1974*b*). The growth and development of the cardiopulmonary responses to exercise. In: *Scientific Foundations of Paediatrics*, ed. J. Davis and J. Dobbing, pp. 271–80. London: Heinemann.

Godin, P. (1903). *Recherches anthropométriques sur la croissance des diverses parties du corps. Détermination de l'adolescent type aux differents âges pubertaires d'après 36,000 mensurations sur 100 sujets suivis individuellement de 13 à 18 ans*. Paris: Maloine.

Godin, P. (1913). *La croissance pendant l'âge scolaire: applications éducatives*. Neuchatel: Delachaux and Niestlé.

Godin, P. (1914). Lois de croissance. *Journal of the Royal Anthropological Institute*, **44**, 295–301.

Godin, P. (1919*a*). *Manuel d'anthropologie pédagogique: basée sur l'anatomo-physiologie de la croissance méthode auxologique*. Paris: Delachaux and Niestlé.

Godin, P. (1919*b*). La méthode auxologique. *Le médecin français*, 15 Mar. (Cited in Godin (1935); not seen.)

Godin, P. (1920). *Growth During School Age; Its Application to Education*, trans. S. L. Eby. Boston: Badger.

Godin, P. (1921). Mon enseignement à Genève 1912–1924. *Le médecin français*. (Cited in Godin (1935); not seen.)

Godin, P. (1935). *Recherches anthropométriques sur la croissance des diverses parties du corps*, 3nd edn. Paris: Legrand.

Goldschmidt, P. (1910). *Berlin in Geschichte und Gegenwart*. Berlin: Springer.

Goldschmidt, W. (ed.) (1959). *The Anthropology of Franz Boas*. San Francisco: Chandler. (Also in *American Anthropologist*, **61**.)

Goldstein, H. (1971). Factors influencing the height of seven-year-old children: results from the National Child Development Study. *Human Biology*, **43**, 92–111.

Goldstein, H. (1979). *The Design and Analysis of Longitudinal Studies*. London: Academic Press.

Goldstein, H. and Tanner, J. M. (1980). Ecological considerations in the creation and use of child growth standards. *Lancet*, **i**, 582–5.

Gombrich, E. H. (1972). *Symbolic Images: Studies in the Art of the Renaissance*. London: Phaidon.

Gordon, B. L. (1960). *Medieval and Renaissance Medicine*. London: Owen.

Gould, B. A. (1877). Letter to H. P. Bowditch, dated 4 August 1877. MS. in Countway Library, Boston, Mass.

Gould, B. A. (1869). *Investigations in the Military and Anthropological Statistics of American Soldiers*. New York/Cambridge, Mass.: US Sanitary Commission/Riverside Press.

Graffar, M. (1959). Les recherches du Centre d'Etude de la Croissance de Bruxelles: exposé introductif. *Acta paediatrica Belgica,* **13**, 293–304.

Graffar, M. (1962). Influence du milieu social sur la croissance. *Modern Problems in Paediatrics,* **7**, 159–70.

Graffar, M., Asiel, M. and Emery-Hauzeur, C. (1962). La crossance de l'enfant normal jusqu' à trois ans: analyse statistique des données relatives au poids et à la taille. *Acta paediatrica Belgica,* **16**, 5–23.

Graffar, M. and Corbier, J. (1966). Contribution à l'étude de l'influence des conditions socio-économiques sur la croissance et le développement. *Courrier du Centre International de l'Enfance,* **16**, 1–25. (English translation in *Early Child Development and Care* (1972), **1**, 141–79.)

Granger, F. (ed.) (1931–4). *Vitruvius on Architecture,* edited from the Harleian manuscript (2 vols.) London : Heinemann.

Grass, F. (1954). Hippolytus Guarinonius: ein Vorkämpfer für deutsche Volksgesundheit im 17. Jahrhundert. In: *Hippolytus Guarinonius,* ed. A. Dörrer *et al.,* pp. 41–90. Schlern-Schriften No. 126. Innsbruck: Universitätsverlag.

Graupner, H. (1904). Wachstumgesetze der Körperlange nach Untersuchung von 57,000 Dresdener Schulkinder. *Berichte der erst International Kongress für Schulhygiene,* **1**, 421–5. (Cited in Baldwin (1921); not seen.)

Gray, A. H. (1909). The laws of individual growth. *Proceedings of the Royal Philosophical Society of Glasgow,* **40**, 139–55.

Gray, H. (1921). Ideal tables for size and weight of private school boys. *American Journal of Diseases of Children,* **22**, 272–83.

Gray, H. (1941). Individual growth rates from birth to maturity for 15 physical traits. *Human Biology,* **13**, 306–33.

Gray, H. (1948). Prediction of adult stature. *Child Development,* **19**, 167–75.

Gray, H. and Ayres, J. G. (1931). *Growth in Private School Children.* Chicago: University of Chicago Press.

Gray, H. and Gray, K. M. (1917). Normal weight. *Boston Medical and Surgical Journal,* **177**, 894–9.

Gray, H. and Jacomb, W. J. (1921). Size and weight in one hundred and thirty-six boarding school boys (Groton). *American Journal of Diseases of Children,* **22**, 259–71.

Grayeff, F. (1974). *Aristotle and his School.* London: Duckworth.

Grayson, C. (1972). *Leon Battista Alberti: on Painting and on Sculpture. The Latin texts of 'De Pictura' and 'De Statua', edited with translations, introduction and notes.* London: Phaidon.

Green, P. (1973). *A Concise History of Ancient Greece to the Close of the Classical Era.* London: Thames and Hudson.

Green, R. M. (trans.) (1951). *De sanitate tuenda: A Translation of Galen's Hygiene.* Springfield, Illinois: Thomas.

Green, W. T., Wyatt, G. M. and Anderson, M. (1946). Orthoroentgenography as a method of measuring the bones of the lower extremities. *Journal of Bone and Joint Surgery,* **28**, 60–5.

Greenwood, A. (1915). *The Health and Physique of School Children.* Ratan Tata Foundation. London: London School of Economics.

Greulich, W. W. (1957). A comparison of the physical growth and development of American-born and native Japanese children. *American Journal of Physical Anthropology,* N.S., **15**, 489–515.

Greulich, W. W. (1976). Some secular changes in the growth of American-born and native Japanese children. *American Journal of Physical Anthropology,* **45**, 553–68.

Greulich, W. W., Crismon, C. S. and Turner, M. L. (1953). The physical growth and development of children who survived the atomic bombing of Hiroshima or Nagasaki. *Journal of Pediatrics,* **43**, 121–45.

Greulich, W. W., Dorfman, R. I., Catchpole, H. R., Solomon, C. I. and Culotta, C. S. (1942).

Somatic and endocrine studies of pubertal and adolescent boys. *Monographs of the Society for Research in Child Development*, **7** (3), 85pp.

Greulich, W. W., Day, H. G., Lachman, S. E., Wolfe, J. B. and Shuttleworth, F. K. (1938). A handbook of methods for the study of adolescent children. *Monographs of the Society for Research in Child Development*, **3** (2), 406pp.

Greulich, W. W. and Pyle, S. I. (1950). *Radiographic Atlas of Skeletal Development of the Hand and Wrist*. Stanford, California: Stanford University Press. 190 pp. (2nd edn, 1959, 256pp.)

Greulich, W. W. and Thoms, H. (1944). The growth and development of the pelvis of individual girls before, during and after puberty. *Yale Journal of Biology and Medicine*, **17**, 91–7.

Griese, E. and Hagen, B. von (1958). *Geschichte der medizinischen Facultät der Friedrich-Schiller-Universität Jena*. Jena: Fischer.

Grinder, R. E. (1969). The concept of adolescence in the genetic psychology of G. Stanley Hall. *Child Development*, **40**, 355–69.

Grumbach, M. M. (1978). The central nervous system and the control of the onset of puberty. In: *Human Growth*, ed. F. Falkner and J. M. Tanner, vol. 2, pp. 215–38. New York: Plenum.

Grumbach, M. M., Grave, G. D. and Mayer, F. E. (eds.) (1974). *Control of the Onset of Puberty*. New York: Wiley.

Grüsdeff, —. (1894). [No title: Referat über die Verhandlungen auf der gynäkologischen Sektion des V Kongresses Russischer Ärzte zum Andenken an Pirogoff zu St Petersburg am 28 und 31 December 1893.] *Centralblatt für Gynäkologie*, **18**, 568.

Gryfe, C. I., Exton-Smith, A. V., Payne, P. R. and Wheeler, E. F. (1971). Pattern of development of bone in children and adolescence. *Lancet*, **i**, 523–6.

Guarinoni, H. (1610). *Die Grewel der Verwüstung menschlichen Geschlechts* . . . Ingolstadt: Angermayr.

[Guarinoni, H.] (1879, 1910). Biography. *Allgemeine deutsche Biographie* (1879), **10**, 83–4; (1910), **55**, 889.

Guillot, N. (1852). Klinische Bemerkungen über Ammen und Säuglinge. *Journal für Kinderkrankheiten*, **19**, 113–25.

Guttman, M. (1915). Einige Beispiele individueller körperlicher Entwicklung. *Zeitschrift für Kinderheilkunde*, **13**, 248–56.

Guy, W. A. (1845). On menstruation. *Medical Times*, **12**, 363–4.

Haase, —. (1875). Entbindings-Anstalt: Jahresbericht für 1875. *Charité-Annalen*, **2**, 669–96.

Hagen, W. (1964). *Wachstum und Entwicklung von Schulkindern im Bild*. Munich: Barth.

Hagen, W. (1966). Wachstum und Entwicklung von Schulkindern: Bericht über die Langsschnittuntersuchungen in Deutschland. *Deutsche medizinische Wochenschrift*, **91**, 1490–7.

Hall, G. S. (1904). *Adolescence: Its Psychology and its Relations to Physiology, Anthropology, Sociology, Sex, Crime, Religion and Education* (2 vols.). New York: Appleton.

[Hall, G. S.] (1925). In memoriam. *Publications of Clark University Library*, **7** (6), 135pp.

Hall, W. S. (1896). Changes in the proportions of the human body during the period of growth. *Journal of the Anthropological Institute*, **25**, 21–46.

Halle-Waisenhaus (1799). *Beschreibung des Hallischen Waisenhauses und der übrigen damit Verbundenen franckischen Stiftung nebst der Geschichte ihres ersten Jahrhunderts*. Halle: Waisenhaus.

Haller, A. von (1752). *Disputationes anatomicae selectae*, vol. **6**, p. 785. Göttingen. (Translated in Cone (1961).)

Haller, A. von (1757–66). *Elementa Physiologiae corporis humani* (8 vols.) Lausanne: Bousquet and Haak.

Haller, A. von (1766–79). *Bibliotheca medicinae practicae* (3 vols.) Bern: Haller.

Haller, A. von (1774–7). *Bibliotheca anatomica* (2 vols.). Zurich: Orell, Gessner and Fuessli.

Haller, A. von. (1778). *Elementa physiologiae corporis humani* (8 vols.), 2nd edn. Lausanne: Pott.

Hallman, N., Backström-Järvinen, L., Kantero, R.-L. and Tiisala, R. (1966). Finnish centre for study in child growth and development. *Annales paediatriae Fenniae*, **12**, 8–12.

Hallman, N., Backström, L., Kantero, R.-L. and Tiisala, R. (1971). Studies on growth of Finnish children from birth to ten years. *Acta paediatrica Scandinavica, Supplement 220*, 48pp.

Hamberger, G. E. (1751). *Physiologia medica*. Jena: Güth.

Hamill, P. V. V., Drizd, T. A., Johnson, C. L., Reed, R. R. and Roche, A. F. (1977). *NCHS Growth Curves for Children from Birth to 18 years: United States*. US Department of Health, Education and Welfare Publication No. (PHS) 78–1690; Vital and Health Statistics Series 11, No. 165, 74pp. Hyattsville, Maryland: USDHEW.

Hamill, P. V. V., Johnston, F. E. and Grams, W. (1970). *Height and Weight of Children: United States*. US Department of Health, Education and Welfare, Public Health Service Publication No. 1000 – Series 11, No. 104, 49pp. Rockville, Maryland: USDHEW;

Hamill, P. V. V., Johnston, F. E. and Lemeshow, S. (1972). *Height and Weight of Children: Socioeconomic Status: United States*. US Department of Health, Education and Welfare Publication No. (HSM) 73–1601; Vital and Health Statistics Series 11, No. 119, 87pp. Rockville, Maryland: USDHEW.

Hamill, P. V. V., Johnston, F. E. and Lemeshow, S. (1973*a*). *Height and Weight of Youths 12–17 Years: United States*. US Department of Health, Education and Welfare Publication No. (HSM) 73–1606; Vital and Health Statistics Series 11, No. 124, 81pp. Rockville, Maryland: USDHEW.

Hamill, P. V. V., Johnston, F. E. and Lemeshow, S. (1973*b*). *Body Weight, Stature and Sitting Height: White and Negro Youths 12–17 Years*. US Department of Health, Education and Welfare Publication No. (HRA) 74–1608; Vital and Health Statistics Series 11, No. 126, 34pp. Rockville, Maryland: USDHEW.

Hamill, P. V. V., Palmer, A. and Drizd, T. A. (1977). *Forced Vital Capacity of Children, 6–11 years: United States*. US Department of Health, Education and Welfare Publication No. (PHS) 78–1691. Vital and Health Statistics, Series 11, No. 164, 30 pp. Hyatsville, Maryland: USDHEW.

Hamilton, A. (1781). *A Treatise on Midwifery*. London: Murray.

Hammar, J. A. (1906). Über Gewicht, Involution und Persistenz der Thymus in Postfötalleben des Menschen. *Archiv für Anatomie und Entwicklungsgeschichte*, vol. for 1906 [no number] *(supplement)*, 91–182.

Hammar, J. A. (1926). Die Menschenthymus in Gesundheit und Krankheit. I. Das normale Organ. *Zeitschrift für mikroskopisch-anatomische Forschung*, **6** (*Supplement*), 570pp.

Hammond, W. H. (1955). Body measurements of pre-school children. *British Journal of Preventive and Social Medicine*, **9**, 152–8.

Hamy, E. T. (1872). Proportions des bras aux differents âges de la vie. *Bulletin de la Société d'Anthropologie de Paris*, **7**, 495–513.

Hanks, L. (1966). *Buffon avant 'L'Histoire Naturelle'*. Paris: Presses Universitaires.

Hannover, A. (1865). *Undersøgelser angaaende menstruationen i dens normale tilstand og i dens forhold til sygdomme*. Copenhagen: Reitzel. 94pp.

Hannover, A. (1869). *Les rapports de la menstruation en Danemark et l'époque, en général, de la première menstruation chez les differents peuples*. (Given in Quetelet (1869*a*).)

Hansman, C. F. (1962). Appearance and fusion of ossification centers in the human skeleton. *American Journal of Roentgenology*, **88**, 476–82.

Hansman, C. F. (1966). Growth of interorbital distance and skull thickness as observed in roentgenographic measurements. *Radiology*, **86**, 87–96.

Hansman, C. F. and Maresh, M. M. (1961). A longitudinal study of skeletal maturation. *American Journal of Diseases of Children*, **101**, 305–21.

Hanway, J. (1762). *Serious Consideration on the Salutary Design of the Act of Parliament for a*

Regular, Uniform Register of the Parish-Poor in all the Parishes within the Bills of Mortality. (Cited in Pinchbeck and Hewitt (1969–73), vol. 1, p. 177.)

Hargenvilliers, A. A. (1817). *Recherches et considerations sur la formation et le recrutement de l'armée française.* Paris : Firmin-Didot. (See Hargenvilliers (1937).)

Hargenvilliers, A. A. (1937). *Compte general de la conscription: publié d'après le manuscrit original . . .* Paris: Recueil Sirey.

Harris, G. W. and Jacobsohn, D. (1952). Functional grafts of the anterior pituitary gland. *Proceedings of the Royal Society of London, Series B,* **139**, 263–76.

Harris, J. A., Jackson, C. M., Patterson, D. G. and Scammon, R. E. (1930). *The Measurement of Man.* Minneapolis: University of Minnesota.

Harrison, G. A., Weiner, J. S., Tanner, J. M. and Barnicott, N. A. (1964). *Human Biology: An Introduction to Human Evolution, Variation, Growth and Ecology.* Oxford: Clarendon Press. (2nd edn, 1977.)

Hartmann, W. (1970). 'Beobachtungen zur Acceleration des Langenwachstums in der Zweiten Hälfte des 18 Jahrhunderts'. Thesis, University of Frankfurt am Main.

Hasse, E. (1891). *Beiträge zur Geschichte und Statistik des Volkschulwesens von Gohlis.* Leipzig: Dunder and Humbert. (Cited in Baldwin (1921); not seen. Tables in Schmidt (1892–3).)

Hastings, W. W. (1900). Anthropometry studies in Nebraska. *American Physical Education Review,* **5**, 53–66.

Hatfield, H. C. (1943). *Winckelmann and his German Critics 1755–1781. A Prelude to the Classical Age.* New York: King's Crown Press.

Havers, C. (1729). *Osteologia nova,* 2nd edn. London: Innys. (1st edn, 1691.)

Hawkins, E. (1899). *Physical Examination of School Boys.* Medical Officers of Schools Association. (Cited in Friend (1935); not seen.)

Hay, D. R. (1851). *The Geometric Beauty of the Human Figure Defined. To which is Prefixed a System of Aesthetic Proportion.* Edinburgh: Blackwood.

Healy, M. J. R. (1952). Some statistical aspects of anthropometry. *Journal of the Royal Statistical Society, Series B,* **14**, 164–84.

Healy, M. J. R. (1958). Variation within individuals in human biology. *Human Biology,* **30**, 210–18.

Healy, M. J. R. (1962). The effect of age-grouping on the distribution of a measurement affected by growth. *American Journal of Physical Anthropology,* **20**, 49–50.

Healy, M. J. R. (1974). Notes on the statistics of growth standards. *Annals of Human Biology,* **1**, 41–6.

Healy, M. J. R. (1978*a*). Statistics of growth standards. In: *Human Growth,* ed. F. Falkner and J. M. Tanner, vol. 1, pp. 169–81. New York: Plenum.

Healy, M. J. R. (1978*b*). Is statistics a science? *Journal of the Royal Statistical Society, Series A,* **141**, 385–93.

Healy, M. J. R. (1978*c*). Truth and consequences in medical research. *Lancet,* **ii**, 1300–1.

Hecker, C. (1864). *Klinik der Geburtskunde,* vol. 2. Leipzig: Engelmann.

Hecker, C. (1865). Über Gewicht und Länge der neugeborenen Kinder im Verhältniss zum Alter der Mutter. *Monatsschrift für Gebürtskunde und Frauenkrankheiten,* **26**, 348–63.

Hecker, C. (1866). Über das Gewicht des Fötus und seiner Anhänge in den verschiedenen Monaten der Schwangerschaft. *Monatsschrift für Gebürtskunde und Frauenkrankheiten,* **27**, 286–99.

Hecker, C. and Buhl, L. (1861). *Klinik der Geburtskunde,* vol. 1. Leipzig: Engelmann.

Heierli, E. (1960). Longitudinale Wachstumstudien: Resultate von Länge, Gewicht und Kopfumfang in den ersten vier Lebensjahren. *Helvetica paediatrica acta,* **15**, 311–35.

Heim, R. (1952). Préface à Buffon. In: *Buffon,* ed. Muséum National d'Histoire Naturelle. Paris: Muséum National d'Histoire Naturelle.

Henderson, J. (1975). *The Maculate Muse.* New Haven and London: Yale University Press.

Herskovits, M. J. (1953). *Franz Boas: The Science of Man in the Making.* New York: Scribner.

Hertel, A. (1885). *Overpressure in High Schools in Denmark*, trans. C. Godfrey Sörenson, intro. J. Crichton-Browne. London: Macmillan.

Hertel, A. (1888). Neuere Untersuchungen über den allgemeinen Gesundheitszustand der Schüler und Schülerinnen. *Zeitschrift für Schulgesundheitspflege*, **1**, 167–83, 201–15.

Hewitt, D. (1957). Some familial correlations in height, weight and skeletal maturity. *Annals of Human Genetics*, **22**, 26–35.

Hewitt, D. (1958). Sib resemblance in bone, muscle and fat measurements of the human calf. *Annals of Human Genetics*, **22**, 213–21.

Hewitt, D. and Acheson, R. M. (1961*a*). Some aspects of skeletal development through adolescence. I. Variations in the rate and pattern of skeletal maturation at puberty. *American Journal of Physical Anthropology*, **19**, 321–31.

Hewitt, D. and Acheson, R. M. (1961*b*). Some aspects of skeletal development through adolescence. II. The inter-relationship between skeletal maturation and growth at puberty. *American Journal of Physical Anthropology*, **19**, 333–44.

Hewitt, D. and Stewart, A. M. (1952). The Oxford Child Health Survey: a study of the influence of social and genetic factors on infant weight. *Human Biology*, **24**, 308–19.

Hewitt, D., Westropp, C. K. and Acheson, R. M. (1955). Oxford Child Health Survey: effect of childish ailments on skeletal development. *British Journal of Preventive and Social Medicine*, **9**, 179–86.

Hewitt, M. (1958). *Wives and Mothers in Victorian Industry*. London: Rockliff.

Heyn, A. (1920). Über Menstruation, Haarfärbung and Libido und ihre gegenseitigen Beziehungen, *Zeitschrift für Geburtshilfe und Gynäkologie*, **82**, 136 :52.

Higman, B. W. (1979). Growth in Afro-Caribbean slave populations. *American Journal of Physical Anthropology*, **50**, 373–85.

Hille, G. (1923). Die Fettpolsterdicke bei der Beurteilung des Ernährungszustandes von Kindern. *Archiv für Kinderheilkunde*, **73**, 134–9.

Hiller, F. (1973). Mass und Freiheit: Anthropometrie in der griechisch-römischen Antike. In: *Der vermessene Mensch: Anthropometrie in Kunst und Wissenschaft*, ed. S. Braunfels *et al.*, pp. 33–41. Munich: Moos.

Hilts, V. L. (1973). Statistics and social science. In: *Foundations of Scientific Method: The Nineteenth Century*, ed. R. N. Gieve and R. S. Westfall, pp. 206–33. Bloomington: Indiana University Press.

Hippocrates. *The Medical Works of Hippocrates*, trans. J. Chadwick and W. N. Mann. Oxford: Blackwell Scientific Publications (1950).

Hippocrates. Airs, waters and places. In: *Works* (4 vols.), with English trans. by W. H. S. Jones, vol. 1, pp. 71–137. Loeb Classical Library. London/Cambridge, Mass.: Heinemann/ Harvard University Press (1972).

His, W. (1874). *Unser Körperform und das physiologische Problem ihrer Entstehung*. Leipzig: Vogel.

Hitchcock, E. (1893). Anthropometric statistics of Amherst College. *Quarterly Publication of the American Statistical Association*, N. S., **3**, 588–99.

Hoerr, N. L., Pyle, S. I. and Francis, C. C. (1962). *Radiographic Atlas of Skeletal Development of the Foot and Ankle: A Standard of Reference*. Springfield, Illinois: Thomas.

Hollerius, J. (1572). *De morbis internis*. Paris. [Thus cited in Müller-Hess (1938); perhaps *De morborum internorum*, Venice, Angelerius.]

Hollerius, J. (1611). *De morbis internis*. Paris: Perier.

Holt, L. E. (1897). *The Diseases of Infancy and Childhood*. New York: Appleton.

[Horner, L.] (1837*a*). Practical application of physiological facts. *Penny Magazine*, **6**, 270–2 (editorial).

Horner, L. (1837*b*). Letter to Mr Senior of May 23rd, 1837. In: *Letters on the Factory Act as it Affects Cotton Manufacture . . . by N. W. Senior . . . to which is Appended a Letter to Mr Senior from Leonard Horner Esq.*, pp. 30–42. London: Fellowes.

Horner, L. (1840). *On the Employment of Children in Factories and other Works in the United Kingdom and in some Foreign Countries.* London: Longman.

Horner, L. (1842). Appendix II in Quetelet, A. (1870). *A Treatise on Man . . .*, trans. R. Knox. Edinburgh: Chambers.

Hossfeld, M. (1931). Bibliography of the published writings of Bird Thomas Baldwin. *Psychological Bulletin*, **28**, 269–76.

HOUILLER, see Hollerius.

Hultkräntz, J. V. (1927). *Über die Zunahme der körpergrösse in Schweden in den Jahren 1840–1926. Nova Acta Reg. Soc. Scient. Upsaliensis*, extraordinary volume.

Hunter, C. (1966). Correlation of facial growth with body height and skeletal maturation at adolescence. *Angle Orthodontics*, **36**, 44–54.

IAMBLICHUS. *De mysteriis Aegyptorum, Chaldaerum, Assyriorum. Proclus in Platonicum Alcibiadem de anima atque demone. Idem de sacrificio et magia* . . . trans. M. Ficino. London: Tornaesius (1549).

Iliff, A. and Lee, V. A. (1952). Pulse rate, respiratory rate, and body temperature of children between two months and 18 years of age. *Child Development*, **23**, 237–45.

Ingerslev, E. (1876). Birth measurements of infants. *Obstetrical Journal of Great Britain and Ireland*, **3**, 705–15, 777–99.

International Children's Centre (1954–76). *Proceedings of the Meetings of the Coordinated Growth Teams (Comptes rendus des Réunions des Equipes Chargées des Etudes sur la Croissance et le Développement de l'Enfant Normal), vols. 1–13.* (Paris, 1954; Stockholm, 1955; London, 1956; Brussels, 1958; Zurich, 1960; London, 1962; Paris, 1964; Stockholm, 1966; Brussels, 1968; Davos, 1970; London, 1972; Paris, 1974; Rennes, 1976.) Paris: International Children's Centre.

International Children's Centre (1977). *Growth and Development of the Child: International Children's Centre Coordinated Studies Publications and Index 1951–1976.* Paris: International Children's Centre.

Iowa Child Welfare Research Station (1959). The fortieth anniversary. *Monographs of the Society for Research in Child Development*, **24**, (5), 78pp.

Ireland, M. W., Love, A. G. and Davenport, C. B. (1919). *Physical Examination of the First Million Draft Recruits.* War Department: Office of the Surgeon General. Bulletin No. 11. Washington: Government Printing Office.

Issmer, E. (1887). Zwei Hauptmerkmale der reife Neugeborenen und deren physiologische Schwankungen. *Archiv für Gynäkologie*, **30**, 277–315.

Iversen, E. (1955). *Canon and proportions in Egyptian Art.* London: Sidgwick and Jackson.

Jackson, C. M. (1915). Morphogenesis. In: *Morris' Human Anatomy*, 5th edn, ed. C. M. Jackson, pp. 7–25. London: Churchill.

Jacob, R. (1938). Heights and weights in a girls' public school. *Nature* (London), **142**, 436–7.

Jampert, C. F. (1754). *De causis incrementum corporis animalis limitantes.* Halle: Fürstenia.

Janes, M. D. (1975). Physical and psychological growth and development. *Environmental Child Health*, **121**, 26–30.

Jennings, H. S. (1943). Biographical memoir of Raymond Pearl 1879–1940. *Biographical Memoirs of the National Academy of Sciences, USA*, **22**, 295–347.

Jenss, R. M. and Bayley, N. (1937). A mathematical method for studying the growth of a child. *Human Biology*, **9**, 556–63.

Joachim, H. H. (trans.) (1922). Aristotle: *De generatione et corruptione.* Oxford; Clarendon Press.

John, V. (1898). *Quetelet bei Goethe.* Jena: Fischer.

Johnston, F. E. (1963). Skeletal age and its prediction in Philadelphia children. *Human Biology*, **35**, 192–201.

Johnston, F. E. (1964*a*). The relationship of certain growth variables to chronological and skeletal age. *Human Biology,* **36**, 16–27.

Johnston, F. E. (1964*b*). Individual variation in the rate of skeletal maturation between five and 18 years. *Child Development,* **35**, 75–80.

Johnston, F. E., Hamill, P. V. V. and Lemeshow, S. (1972). *Skinfold Thickness of Children 6–11 years. United States.* US Department of Health Education and Welfare, Publication No. (HSM) 73–1602; Vital and Health Statistics Series 11, No. 120, 60 pp. Rockville, Maryland: USDHEW.

Johnston, F. E., Hamill, P. V. V. and Lemeshow, S. (1974). *Skinfold Thickness of Youths 12–17 Years: United States.* US Department of Health, Education and Welfare, Publication No. (HRA) 74–1614; Vital and Health Statistics Series 11, No. 132, 68 pp. Rockville, Maryland: USDHEW.

Jones, D. L. and Hemphill, W. (1974). *Height, Weight and other Physical Characteristics of New South Wales Children,* part 2, *Children under Five Years of Age.* New South Wales: Department of Health.

Jones, D. L., Hemphill, W. and Myers, E. S. A. (1973). *Height, Weight and Other Characteristics of New South Wales children,* part 1, *Children Aged Five Years and Over.* New South Wales: Department of Health.

Jones, H. E. (1943). *Development in Adolescence.* New York: Appleton-Century.

Jones, H. E. (1949). *Motor Performance and Growth. A Developmental Study of Static Dynamometric Strength.* Berkeley: University of California Press.

[Jones, H. E.] (1960). Harold Ellis Jones, 1894–1960: obituary. *Child Development,* **31**, 593–608. (Obituary and bibliography, by R. N. Sanford, D. H. Eichorn and M. P. Honzik.)

Jones, W. H. S. (trans.) (1972). Airs, waters and places. In: *Hippocrates: Works,* vol. 1, pp. 71–137. Loeb Classical Library. London/Cambridge, Mass. Heinemann/Harvard University Press.

Jordan, J. R. (1979). *Desarrollo Humano en Cuba.* Havana: Editorial Cientifico-Técnica.

Jordan, J. R., Ruben, M., Hernandez, J., Bebelagua, A., Tanner, J. M. and Goldstein, H. (1975). The 1972 Cuban national child growth study as an example of population health monitoring: design and methods. *Annals of Human Biology,* **2**, 153–71.

Kagan, J. and Moss, H. (1962). *Birth to Maturity: A Study in Psychological Development.* New York: Wiley.

Kaiser, W. and Krosch, K.-H. (1965). Zur Geschichte der Medizinischen Fakultät der Universität Halle in 18 Jahrhundert: VIII, IX, X. *Wissenschaftlichen Zeitschrift der Martin-Luther-Universität, Halle-Wittenberg. Mathematisch-Naturwissenschaftlichen Reihe,* **14**, 284, 388–9.

Kaiser, W. and Krosch, K.-H. (1966). Zur Geschichte der Medizinischen Fakultät der Universität Halle im 18 Jahrhundert. XVI. Die Desputationen und Doktoranden der Jahre 1700–49. *Wissenschaftlichen Zeitschrift der Martin-Luther-Universität, Halle-Wittenberg. Mathematisch-Naturwissenschaftlichen Reihe,* **15**, 1011–24.

Kaiser, W. and Krosch, K.-H. (1967). Wissenschaftsbeziehungen Halle-Russland aus medizinhistorischer Sicht (18 Jahrhundert). *Wissenschaftliche Beiträge der Martin-Luther-Universität Halle-Wittenberg,* 71pp.

Kaiser, W. and Piechocki, W. (1969). Anfänge des geburtshilflich-gynäkologischen Unterrichts an der Universität Halle. In: *In Memoriam Herman Boerhaave (1688–1738),* ed. W. Kaiser and C. Beierlein, pp. 125–33. Halle: Martin-Luther University.

Kajava, Y. (1927). Mitteilungen über die Körpergrösse des Finnischen Mannes Ende des 18 und Anfang des 19 Jahrhunderts. *Annales Academiae Scientarum Fennicae, Series A,* **25** (5), 56pp.

Kapalín, V., Kotasková, J. and Prokopec, M. (1969). *Tělesný a Duševni vyvoj současná Generace Našich Děti* (with English summary). Prague: Academia.

Karlberg, P., Taranger, J., Engström, I., Karlberg, J., Landstrom, T., Lichtenstein, H., Lindstrom, B. and Svennberg-Redegren, I. (1976). The somatic development of children in a Swedish urban community: a prospective longitudinal study. I. Physical growth from birth to 16 years and longitudinal outcome of the study during the same period. *Acta paediatrica Scandinavica*, **288**, *(Supplement)*, 7–76.

Kay, T. (1904–5). Tables showing height, weight, mental capacity, condition of nutrition teeth etc. of Glasgow schoolchildren. *Journal of the Royal Sanitary Institute*, **26**, 907–13.

Keill, J. (1708). *An Account of Animal Secretion.* London: Straham. (4th edn, 1738).

Kelly, H. G. (1933). A study of individual differences in breathing capacity in relation to some physical characteristics. *University of Iowa Studies in Child Welfare*, **7**(5), 59pp.

Kelly, H. J. (1937). Anatomic age and its relation to stature. *University of Iowa Studies in Child Welfare*, **12**(5), 38pp.

Kennedy, W. (1933). The menarche and menstrual type: notes on 10,000 case records. *Journal of Obstetrics and Gynaecology of the British Empire*, **40**, 792–804.

Kerckring, T. (1672). An account of what hath been of late observed by Dr Kerkringius concerning eggs ... *Philosophical Transactions of the Royal Society of London*, **7**(81), 4018.

Key, A. (1885). *Läroverkskomitens underdaniga utlåtande och förslag angående organisationen af rikets allmänna läroverk och dermed samenanhängende frågor. Bilaga E. Redogöele den Hygieniska Undersökningen.* (2 vols.: vol. 1, text; vol. 2, tables and figures.) Stockholm: Kongl. Boktryckeriet.

Key, A. (1889). *Schulhygienische Untesuchungen.* Leipzig: Burgerstein. (Translation of Key (1885).)

Key, A. (1891). Die Pubertätsentwicklung und das Verhältniss derselben zu den Krankheitserscheinungen der Schuljugend. In: *Verhandlungen des 10 Internationale Medizinischen Congresses, Berlin, 1890*, ed. Redactions-Comité, vol. 1, pp. 66–129. Berlin: Hirschwald.

Kiil, V. (1939). Stature and growth of Norwegian men during the past two hundred years. *Skrifter utgitt av det norske Videnscaps-Academi i Oslo I. Mat-Nat. Klasse*, **6**.

Kilian, J. G. C. (1765). Herr Joh. George Bergmiller, Historien-Mahler. In: *Neue Bibliothek der schönen Wissenschaft und der freyen Künste*, vol. 1, pp. 156–9. Leipzig: Dyck.

Klapisch, C. (1973). L'enfance en Toscane au début de XVe siècle. *Annales de démographie historique*, **10**, 99–122.

Kosmowski, W. (1895). Über Gewicht und Wuchs der Kinder der Armen in Warschau. *Jahrbuch für Kinderheilkunde*, **39**, 70–6.

Kotelmann, K. (1879). Die Körperverhaltnisse der Gelehrtenschüler des Johanneums Hamburg. *Zeitschrift des Königlich Preussischen Statistischen Bureaus* (Berlin), **19**, 1–16.

Kramer, R. (1978). *Maria Montessori.* Oxford: Blackwell.

Krieger, E. (1869). *Die Menstruation.* Berlin: Hirschwald.

Krogman, W. M. (1939). Contributions of T. Wingate Todd to anatomy and physical anthropology. *American Journal of Physical Anthropology*, **25**, 145–86.

Krogman, W. M. (1941*a*). *Growth of Man. Tabulae Biologicae*, vol. 20. The Hague: Dr W. Junk.

Krogman, W. M. (1941*b*). *Bibliography of Human Morphology, 1914–1939.* Chicago: University of Chicago Press.

Krogman, W. M. (1948). A handbook of the measurement and interpretation of height and weight in the growing child. *Monographs of the Society for Research in Child Development*, **13**(3), 68pp.

Krogman, W. M. (1950). The physical growth of the child: syllabus. *Yearbook of Physical Anthropology*, **5**, 280–99.

Krogman, W. M. (1951). T. Wingate Todd: catalyst in growth research. *American Journal of Orthodontics*, **37**, 679–87.

Krogman, W. M. (1955). Physical growth of children: an appraisal of studies 1950–1955. *Monographs of the Society for Research in Child Development*, **20**(1), 91pp.

Krogman, W. M. (1970). Growth of the head, face, trunk and limbs in Philadelphia white and Negro children of elementary and high school age. *Monographs of the Society for Research in Child Development*, **35**(3), 80pp.

Krogman, W. M. (1973). Forty years of growth research and orthodontics. *American Journal of Orthodontics*, **63**, 357–65.

Lande, M. J. (1952). Growth behavior of the human bony facial profile as revealed by serial cephalometric roentgenology. *Angle Orthodontist*, **22**, 78–90.

Landis, E. M. (1949). William Townsend Porter 1862–1949. *American Journal of Physiology*, **158**, v–vii.

Landsberger, –. (1888). Das Wachstum in Alter der Schulpflicht. *Archiv für Anthropologie*, **17**, 229–65.

Landus, B. (1556). *De incremento*. Venice: Balthassar Constantine.

Lange, E. von (1896). *Die normale Körpergrösse des Menschen von der Geburt bis zum 25. Lebensjahre. Nebst Erläuterungen über Wesen und Zweck der Skala-Messtabelle zum Gebrauche in Familie, Schule und Erziehungs-Anstalten*. Munich : Lehmann. 38pp.

Lange, E. von (1903). Die Gesetzmässigkeit im Längenwachstum des Menschen. *Jahrbuch für Kinderheilkunde*, **57**, 261–324.

Lee, D. (trans.) (1971). Plato: *Timaeus and Critias*. Harmondsworth: Penguin Books.

Largo, R. H., Gasser, T., Prader, A., Stuetzle, W. and Huber, P. J. (1978). Analysis of the adolescent growth spurt using smoothing spline functions. *Annals of Human Biology*, **5**, 421–34.

Lehmann, J. W. H. (1841). Bemerkungen bei Gelegenheit der Abhandlung von Quetelet: Über den Menschen und die Gesetze seiner Entwickelung. *Jahrbuch für 1836 etc.*, ed. H. C. Schumacher, *Jahrbuch für 1841*, pp. 137–219. Stuttgart: Cotta.

Lehmann, J. H. W. [in error for J. W. H.] (1843*a*). Versuch den Wachstum junger Menschen männlichen Geschlechts, nach Höhe und Statur auf mathematische Gesetze zurückzuführen. *Magazin für die gesamte Heilkunde* (Berlin), **60**, 3–95.

Lehmann, J. W. H. (1843*b*). Über den Menschen und die Gesetze seiner Entwickelung. *Jahrbuch für 1836 etc.*, ed. H. C. Schumacher. *Jahrbuch für 1843*, pp. 146–230. Stuttgart: Cotta.

Lehmann, J. W. H. (1844). Über die mathematische Theorie des menschlichen Wachsthumes. *Amtliche Berichte über die Versammlungen deutschen Naturforscher und Aertze 1843, Gratz*, **21**, 299–302.

Lemnius, L. (1582). *De habitu et constitutione corporis quam Graeci Κρασιν, triviales complexionem vocant*, vol. 2, 2nd edn. Erfurt: Mechlerus. (1st edn, 1561, Antwerp.)

Lemnius, L. (1633). *The Touchstone of Complexions*. London: Michael Sparke.

Lenner, A. (1944). Das Menarchealter. Eine Untersuchung über der Einfluss verschiedener Faktoren auf des Menarchealter. *Acta obstetricia et gynecologica Scandinavica*, **24**, 113–64.

Levinson, A. (1925). Liharzik and his law of human growth. *Bulletin of the Society of Medical History* (Chicago), **3**, 479–87.

Levret, A., (1761). *L'art des accouchemens démontré par des principes de physique et de mécanique*, 2nd edn. Paris: Leprieur.

Levy, P. M. G. (1974). Quetelet, le poête et le statisticien. In: *Adolphe Quetelet 1796–1874*, ed. pp. 44–56. Brussels: Conseil Supérieur de Statistique.

Levy, S. L. (1970). *Nassau W. Senior 1790–1864*. Newton Abbot: David and Charles.

Lewes, G. H. (1864). *Aristotle: A Chapter from the History of Science, including Analyses of Aristotle's Scientific Writings*. London: Smith, Elder and Co.

Lewis, R. C., Duval, A. M. and Iliff, A. (1943*a*). The basal metabolism of normal boys and girls from two to 12 years old, inclusive. Report of a further study. *American Journal of Diseases of Children*, **65**, 834–44.

Lewis, R. C., Duval, A. M. and Iliff, A. (1943*b*). Effect of adolescence on basal metabolism of normal children. *American Journal of Diseases of Children*, **66**, 396–403.

Lewis, R. C., Duval, A. M. and Iliff, A. (1943*c*). Standards for the basal metabolism of children from two to 15 years of age, inclusive. *Journal of Pediatrics*, **23**, 1–18.

Lewis, R. C., Duval, A. M. and Iliff, A. (1943*d*). Basal metabolism of normal children from 13 to 15 years old, inclusive. *American Journal of Diseases of Children*, **65**, 845–57.

Lichtensteger, G. (1746). *Die aus der Arithmetik und Geometrie heraus geholten menschlichen Gründe zur Proportion an das Licht gestellet*. Nürnberg: Fleischmann.

Liharzik, F. (1858). *Das Gesetz des menschlichen Wachstumes und der unter der Norm zurückgebliebene Brustkorb als die erste und wichtigste Ursache der Rachitis, Scrophulose und Tuberculose*. Vienna: Gerold's Sohn.

Liharzik, F. (1862*a*). *Der Bau und das Wachstum der Menschen, die Proportionslehre aller menschlichen Körpertheile für jedes Alter und für die beide Geschlechter*. Vienna: State Printing Office.

Liharzik, F. (1862*b*). *The Law of Increase and the Structure of Man*. Vienna: State Printing Office.

Liharzik, F. (1865). *Das Quadrat die Grundlage aller Proportionalität in der Natur und das Quadrat aus der Zahl sieben die Uridee des menschlichen Körperbaues*. Vienna: Herzfeld and Bauer.

Lindgren, G. (1976). Height, weight and menarche in Swedish urban school children in relation to socioeconomic and regional factors. *Annals of Human Biology*, **3**, 501–28.

Lindgren, G. (1978). Growth of schoolchildren with early, average and late ages of peak height velocity. *Annals of Human Biology*, **5**, 253–68.

Lindner, J. C. (1713). *De sanguine menstruo*. Jena: Krebs.

Linforth, I. M. (1919). *Solon the Athenian*. University of California Publications in Classical Philology vol. 6. Berkeley, California: University of California.

Littré E. (1853). *Oeuvres complètes d'Hippocrate*. Paris: Baillière.

Livi, R. (1896). *Antropometria militare, parte 1, Dati antropologici ed etnologici*. Rome: Giornale Medico del Regio Esercito.

Livi, R. (1905) *Antropometria militare, parte 2. Dati demografici e biologici*. Rome: Giornale Medico del Regio Esercito.

Livson, N. and McNeill, D. (1962). The accuracy of recalled menarche. *Human Biology*, **34**, 218–21.

Livy, J. (1885). On the periods of eruption of the teeth as a test of age. *British Medical Journal*, **ii**, 241–4.

Ljung, B.-O., Bergsten-Brucefors, A. and Lindgren, G. (1974). The secular trend in physical growth in Sweden. *Annals of Human Biology*, **1**, 245–56.

Ljung, B.-O., Fischbein, S. and Lindgren, G. (1977). A comparison of growth in twins and singleton controls of matched age followed longitudinally from 10 to 18 years. *Annals of Human Biology*, **4**, 405–15.

Lomax, E. (1973). The uses and abuses of opiates in 19th century England. *Bulletin of the History of Medicine*, **47**, 167–76.

Lomazzo, G. P. (1591). *Idea del tempio della pittura*. Milan: Ponto.

Lombard, O. M. (1950). Breadth of bone and muscle by age and sex in childhood. Studies based on measurements derived from several roentgenograms of the calf of the leg. *Child Development*, **21**, 229–39.

London County Council (1905). *Annual Report of the Medical Officer, No. 1010*. London: London County Council.

Longmate, N. (1978). *The Hungry Mills*. London: Temple Smith.

Lorey, C. (1888). Über Gewicht und Mass normalentwickelten Kinder in den ersten Lebensjahren. *Jahrbuch für Kinderheilkunde*, **27**, 339–40.

Lowrey, G. H. (1973). *Growth and Development of Children*, 6th edn. Chicago: Yearbook Publishers.

Low, A. (1950). Measurements of infants at birth. *Annals of Eugenics*, **15**, 210–18.

Low, A. (1952). *Growth of Children*. Aberdeen: Aberdeen University Press.

Lyell, K. M. (1890). *Memoir of Leonard Horner FRS, FGS, consisting of Letters to his family and from some of his Friends* (2 vols.). London: Women's Printing Society.

McCammon, R. W. (1970). *Human Growth and Development.* Springfield, Illinois: Thomas.

McCarthy, J. M. (1976). *Humanistic Emphases in the Educational Thought of Vincent of Beauvais.* Leiden: Brill.

McCleary, G. F. (1933). *The Early History of the Infant Welfare Movement.* London: Lewis.

McCleary, G. F. (1945). *The Development of British Maternity and Child Welfare Services.* London: National Association of Maternity and Child Welfare Centres and for the Prevention of Infant Mortality.

McCloy, C. H. (1936). Appraising physical status: the selection of measurements. *University of Iowa Studies in Child Welfare,* **12**(2), 126pp.

McCloy, C. H. (1938). Appraising physical status: methods and norms. *University of Iowa Studies in Child Welfare,* **15**(2), 260pp.

MacDonald, A. (1899). Experimental study of children. In: *Report of the Commissioner for Education for the year 1897–1898,* pp. 985–1204, 1281–390. Washington: Department of Education.

Macfarlane, J. W. (1938). Studies in child guidance. I. Methodology of data collection and organization. *Monographs of the Society for Research in Child Development,* **3**(6), 254pp.

McGregor, A. S. M. (1908). The physique of Glasgow children: an enquiry into the physical condition of children admitted to the City of Glasgow Fever Hospital, Belvidere during the years 1907–8. *Proceedings of the Royal Philosophical Society* (Glasgow), **40**, 156–76.

Mackeprang, E. P. (1907–11). De Vaernepfligtiges Legenshöjde i Danmark; *Meddelelser om Danmarks Antropologi,* **1**, 11–68.

MacMahon, B. (1973). *Age at Menarche: United States.* Department of Health, Education and Welfare Publication No. (HRA)74–1615; Vital and Health Statistics Series 11, No. 133, 29pp. Rockville, Maryland: USDHEW.

Macrobius, Ambrosius, Theodosius, *Commentary on the Dream of Scipio,* trans. with introduction and notes by W. H. Stahl. New York: Columbia University. (1952).

Mailly, E. (1875). Essai sur la vie et les ouvrages de L. A. H. Quetelet. *Annuaire de l'Académie Royale des Sciences, des Lettres et des Beaux-Arts de Belgique,* **41**, 109–297. (Also Brussels: Hayez.)

Malcolm, L. (1978). Protein-energy malnutrition and growth. In: *Human Growth,* ed. F. Falkner and J. M. Tanner, vol. 3, pp. 361–72. New York: Plenum.

Malina, R. M. (1966). Patterns of development in skinfolds of Negro and White Philadelphia children. *Human Biology,* **38**, 89–103.

Malina, R. M. (1970). Skeletal maturation studied longitudinally over one year in American Whites and Negroes 6 through 13 years of age. *Human Biology,* **42**, 377–90.

Malina, R. M. (1972). Weight, height and limb circumferences in American Negro and White children: longitudinal observations over a one year period. *Environmental and Child Health,* **13**, 280–3.

Malina, R. M. (1979). The effects of exercise on specific tissues, dimensions and functions during growth. *Studies in Physical Anthropology,* **5**, 21–52.

Malling-Hansen, R. (1883). *Über Periodizität in Gewicht der Kinder an täglichen Wägungen.* Copenhagen: Jørgensen. 35pp.

Malling-Hansen, R. (1886). *Perioden im Gewicht der Kinder und in der Sonnenwarme.* Copenhagen: Tryde. 286pp.

Malmio, H. R. (1919), Über das alter der Menarche in Finnland. Eine statistische Studie. *Acta Societatis Medicorum Fennicae 'Duodecim',* **1**(2), 146pp.

Mancini, C. and Fravega, G. (1963). Il libro di Arnaldo da Villanova sul modo di conservare la gioventu e ritardare la vecchiaia. *Scientia veterum,* Genoa, **38**, 57pp.

Mandeville, B. de (1723). *An Essay on Charity and Charity Schools.* London: Tonson. (4th edn, 1725.)

Mann, H. (1840). Letter to L. Horner In: *On the Employment of Children in Factories and Other Works in the United Kingdom and in some Foreign Countries*, by L. Horner, pp. 107ff. London: Longman.

Mansfeld, J. (1971). *The Pseudo-Hippocratic Tract Π ερι ἑβδομαδων Chapters I–II and Greek Philosophy*. Assen: Van Gorcum.

Maresh, M. M. (1940). Paranasal sinuses from birth to late adolescence. I. Size of the paranasal sinuses as observed in routine postero-anterior roentgenograms. *American Journal of Diseases of Children*, **60**, 55–78.

Maresh, M. M. (1943). Growth of major long bones in healthy children. A preliminary report on successive roentgenograms of the extremities from early infancy to 12 years of age. *American Journal of Diseases of Children*, **66**, 227–57.

Maresh, M. M. (1948). Growth of the heart related to bodily growth during childhood and adolescence. *Pediatrics*, **2**, 382–404.

Maresh, M. M. (1955). Linear growth of long bones of extremities from infancy through adolescence. *American Journal of Diseases of Children*, **89**, 725–42.

Maresh, M. M. (1961). Bone, muscle and fat measurements. Longitudinal measurements of the bone, muscle and fat widths from roentgenograms of the extremities during the first six years of life. *Pediatrics*, **28**, 971–84.

Maresh, M. M. (1966). Changes in tissue widths during growth. *American Journal of Diseases of Children*, **111**, 142–55.

Marina, G. (1896). *Ricerche antropologiche ed etnografiche su ragazzi*. Turin: Fratelli Bocca.

Marinello, G. (1574). *Le medicine partenenti alle infermita delle donne*. Venice: Valgriso.

Marshall, D. (1926). *The English Poor in the Eighteenth Century*. London: Routledge.

Marshall, W. A. (1971). Evaluation of growth rate in height over periods of less than a year. *Archives of Disease in Childhood*, **46**, 414–20.

Marshall, W. A. (1974). Inter-relationships of skeletal maturation, sexual development and somatic growth in man. *Annals of Human Biology*, **1**, 29–40.

Marshall, W. A. (1975). The relationship of variation of children's growth rates to seasonal climatic variations. *Annals of Human Biology*, **2**, 243–50.

Marshall, W. A. and Ahmed, L. (1976). Variation in upper arm length and forearm length in normal British girls: photogrammetric standards. *Annals of Human Biology*, **3**, 61–70.

Marshall, W. A. and de Limongi, Y. (1976). Skeletal maturity and the prediction of age at menarche. *Annals of Human Biology*, **3**, 235–43.

Marshall, W. A. and Tanner, J. M. (1969). Variation in the pattern of pubertal changes in girls. *Archives of Disease in Childhood*, **44**, 291–303.

Marshall, W. A. and Tanner, J. M. (1970). Variation in the pattern of pubertal changes in boys. *Archives of Disease in Childhood*, **45**, 13–23.

Martens, E. J. and Meredith, H. V. (1942). Illness history and physical growth. I. Correlation in junior primary children followed from fall to spring. *American Journal of Diseases of Children*, **64**, 618–30.

Martin, L. (1951). Notice sur une lettre écrite par Florence Nightingale à Adolphe Quetelet en 1872. *Revue de l'Association des Médicins sortis de l'Université Libre de Bruxelles*, **7**, 3–11.

Martinet, J. F. (1777–9). *Katechismus der Natur* (4 vols.). Amsterdam: J. Allart.

Marubini, E. (1978). Mathematical handling of long-term longitudinal data. In: *Human Growth*, ed. F. Falkner and J. M. Tanner, vol. 1, pp. 209–25. New York: Plenum.

Marubini, E., Resele, L. F., Tanner, J. M. and Whitehouse, R. H. (1972). The fit of Gompertz and logistic curves to longitudinal data during adolescence on height, sitting height and biacromial diameter in boys and girls of the Harpenden Growth Study. *Human Biology*, **44**, 511–24.

Masaracchia, A. (1958). *Solone*. Florence: Nuova Italia Editrice.

Massé, G. (1969). Croissance et développement de l'enfant à Dakar. *Biométrie humaine*, **4**, 13–23.

Massé, G. and Hunt, E. E. (1963). Skeletal maturation of the hand and wrist in West African children. *Human Biology*, **35**, 1–25.

Matiegka, H. (1898). Über die Beziehung von Körperbeschaffenheit und geistiger Tätigkeit bei Schulkindern. *Mitteilungen der Anthropologischen Gesellschaft in Wien*, **27**, 122–6.

Matiegka, J. (1921). The testing of physical efficiency. *American Journal of Physical Anthropology*, **4**, 223–30.

May, M. T. (trans.) (1968). *Galen: De usu partium*. Ithaca: Cornell University Press.

Mattingley, H. and Handford, S. A. (trans.) (1948). Tacitus: *The 'Agricola' and the 'Germania'*. Harmondsworth: Penguin Books.

Maygrier, J. P. (1819). Menstruation. In: *Dictionnaire des sciences médicales*, **32**, 375–96.

Medawar, P. B. (1945). Size, shape and age. In: *Essays on Growth and Form presented to D'Arcy Wentworth Thompson*, ed. W. E. le Gros Clark and P. B. Medawar, pp. 157–87. Oxford: Clarendon, Press.

Meredith, H. V. (1935). The rhythm of physical growth: a study of 18 measurements on Iowa City White males ranging in age between birth and 18 years. *University of Iowa Studies in Child Welfare*, **11**(3), 128pp.

Meredith, H. V. (1939). Length of head and neck, trunk, and lower extremities on Iowa City children aged 7 to 17 years. *Child Development*, **10**, 129–44.

Meredith, H. V. (1941). Stature and weight of children of the United States, with reference to the influence of racial, regional, socioeconomic and secular factors. *American Journal of Diseases of Children*, **62**, 909–32.

Meredith, H. V. (1944). Human foot length from embryo to adult. *Human Biology*, **16**, 207–82.

Meredith, H. V. (1947). Length of upper extremities in *Homo sapiens* from birth through adolescence. *Growth*, **11**, 1–50.

Meredith, H. V. (1955). A longitudinal study of change in size and form of the lower limbs on North American white schoolboys. *Growth*, **19**, 89–106.

Meredith, H. V. (1959). Change in a dimension of the frontal bone during childhood and adolescence. *Anatomical Record*, **134**, 769–80.

Meredith, H. V. (1960). Changes in form of the head and face during childhood. *Growth*, **24**, 215–64.

Meredith, H. V. (1963). Change in the stature and body weight of North American boys during the last 80 years. *Advances in Child Development and Behaviour*, **1**, 69–107.

Meredith, H. V. (1970). Body weight at birth of viable human infants: a worldwide comparative treatise. *Human Biology*, **42**, 217–64.

Meredith, H. V. (1971). Growth in body size: a compendium of findings on contemporary children living in different parts of the world. *Advances in Child Development and Behaviour*, **6**, 154–238.

Meredith, H. V. (1973). Gingival emergence of human deciduous teeth: a synoptic report. *Journal of Tropical Paediatrics and Environmental Child Health*, **19**, 195–9.

Meredith, H. V. (1978). Secular change in sitting height and lower limb height of children, youths and young adults of Afro-Black, European and Japanese ancestry. *Growth*, **42**, 37–42.

Meredith, H. V. and Carl, L. J. (1946). Individual growth in hip width: a study covering the age period from 5–9 years based upon seriatim data for 55 non-pathologic white children. *Child Development*, **17**, 157–72.

Meredith, H. V. and Chadha, J. M. (1962). A roentgenographic study of change in head height during childhood and adolescence. *Human Biology*, **34**, 299–319.

Meredith, H. V. and Knott, V. B. (1962). Illness history and physical growth. III. Comparative anatomic status and rate of change for school children in different long-term health categories. *American Journal of Diseases of Children*, **103**, 146–51.

Meredith, H. V., Knott, V. B. and collaborators (1973). *Childhood Changes of Head, Face and Dentition. A Collection of Research Reports*. Iowa City: Iowa Orthodontic Society.

Meredith, H. V. and Meredith, E. M. (1950). Annual increment norms for 10 measures of physical growth on children 4 to 8 years of age. *Child Development*, **21**, 141–7.

Merminod, A. (ed.) (1962). The growth of the normal child during the first three years of life. The proceedings of a seminar organised by the International Children's Centre at Zurich, October 1960. *Modern Problems in Pediatrics*, **7**, 256pp.

Merrell, M. (1931). The relationship of individual growth to average growth. *Human Biology*, **3**, 37–70.

Mertins, E. (1968). *Die Militärschule zu Potsdam: ein Gedenkbuch*. Berlin: Mertins.

Metheny, E. (1940). Breathing capacity and grip strength of preschool children. *University of Iowa Studies in Child Welfare*, **18**, 207pp.

MICHAEL THE SCOTSMAN, *see* Michael Scotus.

Michael Scotus (1615). *De secretis naturae opusculum*. Strassburg: Zerzner.

Militairwaisenhaus (1874). *Das Königliche Potsdamsche Grosse Militair-Waisenhaus in den Jahren von 1824 bis 1874*. Berlin: Königlichen Geheimen Ober-Hofbuchdruckerei.

Miller, F. J. W., Billewicz, W. Z. and Thomson, A. M. (1972). Growth from birth to adult life of 442 Newcastle-upon-Tyne children. *British Journal of Preventive and Social Medicine*, **26**, 224–30.

Miller, F. J. W., Court, S. D. M., Knox, E. G. and Brandon, S. (1974). *The School Years in Newcastle-upon-Tyne, 1952–1962*. London: Oxford University Press.

Miller, F. J. W., Court, S. D. M., Walton, W. S. and Knox, E. G. (1960). *Growing Up in Newcastle-upon-Tyne*. London: Oxford University Press.

Minot, C. S. (1892). *Human Embryology*. New York: Wood.

Montessori, M. (1913). *Pedagogical Anthropology*, trans. F. T. Cooper. London: Heinemann.

Monti, A. (1897–9). Das Wachstum des Kindes von der Geburt bis Einschliesslich der Pubertät. In: *Kinderheilkunde in Einzeldarstellungen*, vol. 1, pp. 541–70. Berlin and Vienna: Urban and Schwartzkopf.

Moore, T., Hindley, C. B. and Falkner, F. (1954). A longitudinal research in child development and some of its problems. *British Medical Journal*, **ii**, 1132–7.

Moorrees, C. F. A. (1959). *The Dentition of the Growing Child. A Longitudinal Study of Dental Development between Three and 18 years of Age*. Cambridge, Mass.: Harvard University Press.

Moorrees, C. F. A., Fanning, E. A. and Hunt, E. E. (1963). Age variation of formation stages for ten permanent teeth. *Journal of Dental Research*, **42**, 1490–502.

Moorrees, C. F. A. and Kent, R. L., Jr (1978). A step function model using tooth counts to assess the developmental timing of the dentition. *Annals of Human Biology*, **5**, 55–68.

Moorrees, C. F. A. and Reed, R. B. (1965). Changes in dental arch dimensions expressed on the basis of tooth eruption as a measure of biologic age. *Journal of Dental Research*, **44**, 129–41, 161–73.

Morant, G. M. (1950). Secular changes in the heights of British people. *Proceedings of the Royal Society of London, Series B*, **137**, 443–52.

Morgenstern, S. J. (1973). *Über Friedrich Wilhelm I*. Braunschweig: Auns.

Moyers, R. E. and Krogman, W. M. (eds.) (1971). *Cranio-facial Growth in Man*. Oxford and New York: Pergamon Press.

Mugrage, E. R. and Andresen, M. I. (1936). Values for red blood cells of average infants and children. *American Journal of Diseases of Children*, **51**, 775–91.

Mugrage, E. R. and Andresen, M. I. (1938). Red blood cell values in adolescence. *American Journal of Diseases of Children*, **56**, 997–1003.

Mullen, F. A. (1940). Factors in the growth of girls. *Child Development*, **11**, 27–42.

Müller, H. W. (1973). Der Kanon in der ägyptischen Kunst. In: *Der vermessene Mensch: Anthropometrie in Kunst und Wissenschaft*, pp. 9–31. Munich: Moos.

Müller-Hess, H. G., (1938). Die Lehre von der Menstruation von Beginn der Neuzeit bis zur Begründung der Zellenlehre. *Abhandlung zur Geschichte der Medizin und der Naturwissenschaften*, **27**, 1–102.

Mumford, A. A. (1910). Imperfections in physical growth and hindrances to social development. *Child Study (Journal of the Child Study Society, London)*, **3**, 85–98.

Mumford, A. A. (1927). *Healthy Growth: A Study of the Relation between the Mental and Physical Development of Adolescent Boys in a Public Day School.* London: Oxford University Press.

Munster, L. (1966). La ricerca delle proporzioni del corpo umano dai tempi antichi all'anthropometria di Giovanni Sigismondo Elsholtius. *Atti e memorie della Accademia di Storia dell'Arte Sanitaria*, **32**, 142–6.

Murat, —. (1816). Foetus. *Dictionnaire des sciences médicales*, **16**, 49–80.

Murphy, E. W. (1845). A report of the obstetric practice of University College Hospital London. *Dublin Journal of Medical Science*, **26**, 177–229.

Mussen, P. H. and Jones, M. C. (1957). Self-conceptions, motivations and interpersonal attitudes of late- and early-maturing boys. *Child Development*, **28**, 243–56.

Mussen, P. H. and Jones, M. C. (1958). The behaviour-inferred motivations of late- and early-maturing boys. *Child Development*, **29**, 61–7.

Nagler, G. K. (1835). Bergmüller, J. G. In: *Neue allgemeines Kunstler-Lexicon*, vol. 1, pp. 439, 440. Munich: Fleischmann.

Nakamura, S., Savara, B. and Thomas, D. (1972). Norms of size and annual increments of the sphenoid bone from four to 16 years. *Angle Orthodontics*, **42**, 35–43.

Nanda, R. S. (1955). The rates of growth of several facial components measured from serial cephalometric roentgenograms. *American Journal of Orthodontics*, **41**, 658–73.

Needham, J. (1934). *A History of Embryology.* London: Cambridge University Press.

Neligan, G. A., Kolvin, I., Scott, D. M. and Garside, R. F. (1976). Born too soon or born too small. *Clinics in Development Medicine*, **61**, 101 pp.

Neumann, H. (1912). Die Dicke des Fettpolsters bei Kindern. *Jahrbuch für Kinderheilkunde*, **75**, 481–8.

Newman, G. (1906). *Infant Mortality: A Social Problem.* London: Methuen.

New Zealand Department of Health (1971). *Physical Development of New Zealand Schoolchildren, 1969.* Health Services Research Unit Special Report No. 38. Wellington: Department of Health.

Nicolai, F. (1786). *Beschreibung der Königlichen Residenzstädte Berlin und Potsdam aller daselbst befindlicher Merkwürdigkeiten, und der umliegenden Gegend*, 3rd edn (3 vols.). Berlin: Nicolai.

Nicomachus, Gerasinus. *Introductionis arithmeticae, libri II*; ed. R. Hoche, p. 66. Leipzig: Teubner (1866). (Also as *Introduction to Arithmetic*, trans. M artin Luther D'Ooge. New York: Macmillan (1926).)

Nicolson, A. B. and Hanley, C. (1953). Indices of physiological maturity: derivation and interrelationships. *Child Development*, **24**, 3–38.

Norman, H. B. (1939). Public school and secondary school boys. A comparison of their physique. *Lancet*, **ii**, 442–5.

Oakland, City of (1893). Physical development of Oakland children. *Annual Report of the Public Schools of the City of Oakland for the Year ending June 30th 1893*, pp. 38–44. Oakland, Calif.: Board of Education.

Oeder, G. (1910). Die Fettpolsterdicke als Index des Ernährungszustandes bei Erwachsene. *Medizinische Klinik*, **6**, 657–62.

Oeder, G. (1911). Fettpolsterdicke und Fettpolstermessung. *Fortschritte der Medizin*, **29**, 961–72.

Ogle, W. (trans.). (1911) *Aristotle: De partibus animalium.* Oxford: Clarendon Press.

O'Neil, W. (1965). *Proclus: Alcibiades I, A Translation and Commentary.* The Hague: Nijhoff.

Oppers, V. M. (1966). The secular trend in growth and maturation in the Netherlands. *Tijdschift voor Sociale Geneeskunde*, **44**, 539–48.

Oribasius, Sardianus. *Œuvres d'Oribase*, trans. U. C. Bussemaker and C. V. Daremberg. Paris: Imprimerie nationale (1851–76).

Osiander, F. C. (1795). Resultate von Beobachtungen und Nachrichten über die erste Erscheinung des Monatliches. *Denkwürdigkeiten für das Heilkunde und Geburtshilfe*, **2**, 380–8.

Paccioli, L. (1509). *De divina proportione.* Venice: Paganini.

Pagliani, L. (1875–6). Sopra alcuni fattori dello sviluppo umano; richerche anthropometriche. *Atti della Reale Accademia di Scienze di Torino,* **11**, 694–760.

Pagliani, L. (1877). I fattori della statura umana. *Archivio di statistica,* **4**, 92–120.

Pagliani, L. (1879). Lo sviluppo umano per età, sesso, condizione sociale ed etnica studiato nel peso, statura, circonferenza toracica, capacita vitale e forza muscolare. *Giornale della Società Italiana d'Igiene,* **1**, 357–76, 453–91, 589–610.

Pagliani, L. (1925). *Lo sviluppo dell'organismo umano nell'infanzia, puerizia, adolescenza, puberta e giovinezza, con deduzioni igieniche, pedagogiche e sociali.* Turin: Paravia.

Palmer, C. E. (1930). Diurnal variations of height and weight in the human body during growth. *Anatomical Record,* **45**, 234–5.

Palmer, C. E. (1932). The relationship of erect body length to supine body length. *Human Biology,* **4**, 262–71.

Palmer, C. E. (1933*a*). Seasonal variation of average growth in weight of elementary schoolchildren. *Public Health Reports* (Washington), **48**, 211–33.

Palmer, C. E. (1933*b*). Variations of growth in weight of elementary schoolchildren. *Public Health Reports* (Washington), **48**, 993–1005.

Palmer, C. E. (1933*c*). Growth and the economic depression. A study of the weight of elementary schoolchildren in 1921–27 and in 1933. *Public Health Reports* (Washington), **48**, 1277–92.

Palmer, C. E. (1934*a*). Further studies on growth and the economic depression. A comparison of weight and weight increments of elementary schoolchildren in 1921–27 and in 1933–34. *Public Health Reports* (Washington), **49**, 1453–69.

Palmer, C. E. (1934*b*). Age changes in physical resemblance of siblings. *Child Development,* **5**, 351–60.

Palmer, C. E. (1935). Height and weight of children of the depression poor. *Public Health Reports* (Washington), **50**, 1107–13.

Palmer, C. E. and Ciocco, A. (1946). Physical growth and development. In: *Mitchell-Nelson's Textbook of Pediatrics,* 4th edn, ed. W. E. Nelson, pp. 13–40. Philadelphia: W. B. Saunders.

Palmer, C. E., Kawakamo, R. and Reed, L. J. (1937). Anthropometric studies of individual growth. II. Age, weight and rate of growth in weight: elementary school children. *Child Development,* **8**, 47–61.

Palmer, C. E. and Reed, L. J. (1935). Anthropometric studies of individual growth. I. Age, height and growth in height: elementary school children. *Human Biology,* **7**, 319–34.

Panofsky, E. (1955). The history of the theory of human proportions. In: *Meaning in the Visual Arts,* ed. E. Panofsky, pp. 55–107. New York: Doubleday.

Panofsky, E. (1968). *Idea: A Concept in Art Theory,* trans. J. J. S. Peake. New York: Harper and Row. (First published Leipzig, 1924; 2nd edn 1960, Berlin.)

Paparella, S. (1573). *De calido libri III.* Perugia: Brixiani.

Pariset, E. (1836). Eloge de Chaussier. *Mémoires de l'Académie Royale de Médicine,* **5**, 1–40.

Parišková, J. (1968). Longitudinal study of the development of body composition and body build in boys of various physical activity. *Human Biology,* **40**, 212–25.

Parišková, J. (1973). Body composition and exercise during growth and development. In: *Physical Activity: Growth and Development,* ed. G. L. Rarick, pp. 97–124. New York: Academic Press.

Parišková, J. (1977). *Body Fat and Physical Fitness,* (trans. K. Osankova). The Hague: Nijhoff.

Parliamentary Papers (1833). Reports from Commissioners (4), vol. 20, *First Report . . . into the Employment of Children in Factories.* London: HMSO.

Parliamentary Papers (1873). Accounts and Papers (17), vol. 55, *Report to the Local Government Board on Proposed Changes in Hours and Age of Employment in Textile Factories,* pp. 803–64. London: HMSO.

Parliamentary Papers (1876). Reports from Commissioners (15), vol. 29, *Report from Commis-*

sioners on the Working of the Factory and Workshops Act, with a View to their Consolidation and Amendment. London: HMSO.

Parliamentary Papers (1904). vol. 32, *Report of the Inter-Departmental Committee on Physical Deterioration.* Cd. 2175, 2210, 2186. London: HMSO.

Paton, D. N. and Findlay, L. (1926). *Poverty, Nutrition and Growth: Studies of Child Life in Cities and Rural Districts of Scotland.* Medical Research Council Special Report Series No. 101. London: HMSO. 333pp.

Patterson, H. D. (1950). Sampling on successive occasions with partial replacement of units. *Journal of the Royal Statistical Society, Series B,* **12**, 241–55.

PAULUS AEGINATA. *The Seven Books of Paulus Aeginata,* (3 vols.), trans. F. Adams. London: Sydenham Society (1844).

Pavisius, J. J. (1559). *De accretione.* Ulm: Marc Antonicus.

Pearce, E. H. (1901). *Annals of Christ's Hospital.* London: Methuen.

Pearl, R. (1923). *Introduction to Medical Biometry and Statistics.* Philadelphia: W. B. Saunders. (2nd edn, 1930; 3rd edn, 1940.)

Pearse, I. H. and Williamson, G. S. (1931). *The Case for Action.* London: Faber and Faber.

Pearson, K. (1900). Data for the problem of evolution in Man. III. On the magnitude of certain coefficients of correlation in man. *Proceedings of the Royal Society of London,* **66**, 23–32.

Pearson, K. (1914–30). *The Life, Letters and Labours of Francis Galton* (3 vols.) Cambridge University Press.

Peck, A. L. (trans.) (1963). *Aristotle: Parts of Animals.* Loeb Classical Library. London: Heinemann.

Peck, A. L. (trans.) (1965–). *Aristotle: Historia Animalium* (2 vols.). Loeb Classical Library. London: Heinemann.

Peck, A. L. (trans.) (1968). *Aristotle: Generation of Animals,* revised edn Loeb Classical Library. London: Heinemann.

Peckham, A. W. (1882). The growth of children. In: *Report of the State Board of Health, Wisconsin, for 1881,* vol. 6, pp. 28–73. Madison: Atwood. (Also *ibid* vol. 7, pp. 185–8.)

Peiper, E. (1912). Körperliche Entwicklung der Schuljugend in Pommern. *Archiv für soziale Hygiene,* **7**, 109–37.

Perier, J. A. N. (1867). Le docteur Boudin. Notice historique sur sa vie et ses travaux. *Recueil des mémoires de médicine, de chirurgie et de pharmacie militaires,* 3rd series, **19**, 39pp.

Pfannkuch, W. (1872). Über die Körperform der Neugeborenen. *Archiv für Gynäkologie,* **4**, 297–310.

Pfeiffer-Belli, W. (ed.) (1950). *Goethes Gespräche.* Zurich: Artemis.

PHILO JUDAEUS. De septenario. In: *Omnia quae extant opera.* Paris: Lutetia (1440). [Solon's poem is on p. 24.]

Pinchbeck, I. and Hewitt, M. (1969–73). *Children in English Society* (2 vols.). London: Routledge and Kegan Paul.

Pitcairn, A. (1718). *The Philosophical and Mathematical Elements of Physick.* London: Bell.

PLATO. *Timaeus and Critias,* trans. with introduction and appendix on Atlantis by D. Lee. Harmondsworth: Penguin Books (1971).

PLATO. *Phaedo,* with English trans. by H. N. Fowler, pp. 193–204. Loeb Classical Library. London: Heinemann (1977).

Ploss, H. H., Bartels, M. and Bartels, P. (1935). *Woman,* 11th edn, trans. E. J. Dingwall. London: Heinemann.

Polinière, M. (1820). Puberté. *Dictionnaire des sciences médicales,* **46**, 32–64.

Porter, W. T. (1893*a*). The physical basis of precocity and dullness. *Transactions of the Academy of Science of St Louis,* **6**, 161–81.

Porter, W. T. (1893*b*). The relation between the growth of children and their deviation from the physical type of their sex and age. *Transactions of the Academy of Science of St Louis*, **6**, 233–50.

Porter, W. T. (1893*c*). On the application to individual school children of the mean values derived from anthropological measurements by the generalizing method. *Quarterly Publications of the American Statistical Association*, N.S., **3**, 576–87.

Porter, W. T. (1894). The growth of St Louis children. *Transactions of the Academy of Science of St Louis*, **6**, 263–380.

Porter, W. T. (1920). The seasonal variation in the growth of Boston school children. *American Journal of Physiology*, **52**, 121–31.

Porter, W. T. (1922). The relative growth of individual Boston school boys. *American Journal of Physiology*, **61**, 311–25.

Porter, W. T. (1923). Percentile charts of the height and weight of Boston school children. *Boston Medical and Surgical Journal*, **188**, 639–44.

Portmann, M. P. (1962). Louis-René Villermé (1782–1863) und sein Werk über die Lage der französischen Textilarbeiter im 1830–1840. *Praxis*, **51**, 721–6.

Post, J. B. (1971). Ages at menarche and menopause: some medieval authorities. *Population Studies*, **25**, 83–7.

Potsdamsches Waisenhaus (1824). *Geschichte des Königlichen Potsdamschen Waisenhauses von seiner Entstehung bis auf die jetztige Zeit*. Berlin: Mittler.

Potsdamsches Grosses Waisenhaus (1924). *Festschrift zur Zweihundertjahrfeier, 1724–1924*. Potsdam: Edmund Stein.

Potsdamsches Grosses Waisenhaus (1930). *Erziehungsheim und Schule*. Potsdam: Fernruf.

Power, D. A. (reviser) (1930). *Plarr's Lives of the Fellows of the Royal College of Surgeons of England* (2 vols.). Bristol: Wright.

Prader, A., Tanner, J. M. and Harnack, G. A. von (1963). Catch-up growth following illness or starvation. *Journal of Pediatrics*, **62**, 646–59.

Prahl-Andersen, B. and Kowalski, C. J. (1973). A mixed longitudinal, interdisciplinary study of the growth and development of Dutch children. *Growth*, **37**, 281–95.

Prahl-Andersen, B., Kowalski, C. J. and Heydendael, P. H. J. m. (eds.) (1979). *A Mixed Longitudinal Interdisciplinary Study of Growth and Development*. New York: Academic Press.

Prahl-Andersen, B., Pollman, A. J., Raaben, D. J. and Peters, K. A. (1972). Automated anthropometry. *American Journal of Physical Anthropology*, **37**, 151–4.

Preece, M. A. and Baines, M. J. (1978). A new family of mathematical models describing the human growth curve. *Annals of Human Biology*, **5**, 1–24.

Pringle, M. L. K., Butler, N. R. and Davie, R. (1966). *11,000 Seven-Year-Olds*. London: Longman.

Pryor, J. W. (1905). Development of bones of the hand as shown by the X-ray method. *Bulletin of the State College of Kentucky*, **2**(5), 3–45, Lexington: Press of Transylvania Company.

Pryor, J. W. (1906). Ossification of the epiphyses of the hand: X-ray. method. *Bulletin of the State College of Kentucky*, **3**(4), 36pp. (Cited in Pryor (1925); not seen.)

Pryor, J. W. (1908). The chronology and order of ossification of the bones of the human carpus. X-ray method. *Bulletin of the State College of Kentucky*, New Series **1**(2), 24pp. Lexington: Press of Transylvania Company.

Pryor, J. W. (1916). Some observations of the ossification of the bones of the hand. *Bulletin of the State College of Kentucky*, New Series **8**(11), 1–67. Lexington: University Press.

Pryor, J. W. (1923). Differences in the time of development of centers of ossification in the male and female skeleton. *Anatomical Record*, **25**, 257–74.

Pryor, J. W. (1925). Time of ossification of the bones of the hand of the male and female and union of epiphyses with diaphyses. *American Journal of Physical Anthropology*, **8**, 401–10.

Pyle, S. I. and Hoerr, N. L. (1955). *Radiographic Atlas of Skeletal Development of the Knee*. Springfield, Illinois: Thomas.

Pyle, S. I. and Hoerr, N. L. (1969). *A Radiographic Standard of Reference for the Growing Knee.* Springfield, Illinois: Thomas.

Pyle, S. I., Mann, A., Dreizen, S., Kelly, H. J., Macey, I. G. and Spies, T. D. (1948). A substitute for skeletal age (Todd) for clinical use: the red graph method. *Journal of Pediatrics,* **32**, 125–36.

Pyle, S. I., Reed, R. B. and Stuart, H. C. (1959). Longitudinal studies of child health and development, series II: patterns of skeletal development in the hand. *Pediatrics,* **24**, 886–903.

Pyle, S. I., Stuart, H. C., Cornoni, J. and Reed, R. B. (1961). Onsets, completions, and spans of the osseous stage of development in representative bone growth centers of the extremities. *Monographs of the Society for Research in Child Development,* **26**(1), 126pp.

Quabeck, C. J. (1752). *De insolito corporis augmento frequenti morborum futurorum signo.* Halle: Hendel.

Quetelet, A. (1830). Sur la taille moyenne de l'homme dans les villes et dans les campagnes, et sur l'âge ou la croissance est complètement achevée. *Annales d'hygiène publique.* **3**. 24–6.

Quetelet, A. (1831). Recherches sur la loi de croissance de l'homme. *Annales d'hygiène publique,* **6**, 89–113.

Quetelet, A. (1833). Recherches sur le poids de l'homme aux différents âges. *Mémoires de l'Academie Royale de Bruxelles,* N.S., **7**, 1–44.

Quetelet, (L.) A. (J.) (1835). *Sur l'homme et le développement de ses facultés. Essai sur physique sociale* (2 vols.). Paris: Bachelier.

Quetelet, A. (1839). Ueber den Menschen und die Gesetze seiner Entwickelung. *Jahrbuch für 1836 etc.,* ed. H. C. Schumacher. *Jahrbuch für 1839,* pp. 180–97. Stuttgart: Cotta.

Quetelet, A. (1842). *A Treatise on Man and the Development of his Faculties,* trans. R. Knox. Edinburgh: Chambers.

Quetelet, A. (1846). *Lettres à SAR le duc régnant de Saxe-Coburg et Gotha, sur la théorie des probabilités appliquée aux sciences morales et politiques.* Brussels: Hayez.

Quetelet, A. (1848). *Du système social et des lois qui le régissent.* Paris: Guillaumin.

Quetelet, A. (1849). *Letters Addressed to HRH the Grand Duke of Saxe-Coburg and Gotha on the Theory of Probabilities as Applied to the Moral and Political Sciences,* trans. Olinthus Gregory Downes. London: Layton.

Quetelet, A. (1869*a*). *Physique sociale. Essai sur le développement des facultés de l'homme* (2 vols.). Brussels: Muquardt.

Quetelet, A. (1869*b*). Sur le tome XIXè des *Annales de l'Observatoire royal de Bruxelles* et sur le tome seconde de la nouvelle édition de la *Physique Sociale. Bulletin de l'Academie Royale des Sciences, des Lettres et des Beaux-Arts de Belgique,* 2nd series, **28**, 149–68.

Quetelet, A. (1870). *Anthropométrie, ou mésure des différentes facultés de l'homme.* Brussels: Muquardt.

Raciborski, A. (1844). *De la puberté et de l'âge critique chez la femme.* Paris: Baillière.

Raciborski, A. (1868). *Traité de la menstruation.* Paris: Baillière.

Raleigh, W. (1614). *The History of the World* (5 vols.). London: Burre.

Rall, G. F. (1669). *De generatione animalium* . . . Stettin: Hoffner.

Ranke, O. (1905). Beiträge zur Frage des kindlichen Wachstums. *Archiv für Anthropologie,* **3**, 161–80.

Rapp, L. (1903). *Hippolytus Guarinoni, Stiftartz in Hall.* Brixen: Weger.

Rarick, G. L. (ed.) (1973). *Physical Activity: Human Growth and Development.* New York: Academic Press.

Raudnitz, R. W. (1892). Über Lebensbücher und das Massenwachstum der Säuglinge. *Prager medizinische Wochenschrift,* **17**, 66–9, 82–5.

Ravn, N. E. (1850). Menstruationens Physiologi. In: —. Fenger, Sjette or syvende Halvaarsberetning fra det Kgl. mediciniske Selskabs Statistiske Comite, *Bibliothek for Laeger 3 die raikhe,* 7, 2–17.

Rea, J. D. (1925). Jaques on the microcosm. *Philological Quarterly,* **4**, 345–50.

Rea, J. D. (1928–9). Coleridge's intimations of immortality from Proclus. *Modern Philology*, **26**, 201–13.

Reed, R. B. and Stuart, H. C. (1959). Longitudinal studies of child health and development, series II: patterns of growth in height and weight from birth to 18 years of age. *Pediatrics*, **24**, 904–21.

Regnier, R. (1860). *Des maladies de croissance*. Paris.

Retzius, G. (1904). Zur Kenntnis der Entwicklung der Körperformen des Menschen während der fötalen Lebensstufen. *Biologischen Untersuchungen von Prof. Dr Gustav Retzius*, N.S., **11**, 33–76.

Reynolds, E. L. (1944). Differential tissue growth in the leg during childhood. *Child Development*, **15**, 181–205.

Reynolds, E. L. (1945). The bony pelvic girdle in early infancy. *American Journal of Physical Anthropology*, N.S., **3**, 321–54.

Reynolds, E. L. (1946). Sexual maturation and the growth of fat, muscle and bone in girls. *Child Development*, **17**, 121–44.

Reynolds, E. L. (1947). The bony pelvis in prepubertal childhood. *American Journal of Physical Anthropology*, N.S., **5**, 165–200.

Reynolds, E. L. (1949). The fat–bone index as a sex differentiating character in man. *Human Biology*, **21**, 199–204. (Reprinted with individual measurement data in *Yearbook of Physical Anthropology*, **5**, 249–58.)

Reynolds, E. L. (1950). The distribution of subcutaneous fat in childhood and adolescence. *Monographs of the Society for Research in Child Development*, **15**,(2), 189pp.

Reynolds, E. L. and Grote, P. (1948). Sex differences in the distribution of tissue components in the human leg from birth to maturity. *Anatomical Record*, **102**, 45–53.

Reynolds, E. L. and Wines, J. V. (1948). Individual differences in physical changes associated with adolescence in girls. *American Journal of Diseases of Children*, **75**, 329–50.

Reynolds, E. L. and Wines, J. V. (1951). Physical changes associated with adolescence in boys. *American Journal of Diseases of Children*, **82**, 529–47.

Reynolds, L. D. and Wilson, N. G. (1974). *Scribes and Scholars. A Guide to the Transmission of Greek and Latin Literature*, 2nd edn. Oxford: Clarendon Press.

Ricci, J. V. (1950). *Aetios of Amida: Gynecology and Obstetrics of the Sixth Century*, trans. from the Latin of Cornarius, 1542. Philadelphia: Blakiston.

Richards, O. W. and Kavanagh, A. J. (1945). The analysis of growing form. In: *Essays on Growth and Form presented to D'Arcy Wentworth Thompson*, ed. W. E. Le Gros Clark and P. B. Medawar, pp. 188–230. Oxford: Clarendon Press.

Richer, P. (1890). Du rôle de la graisse dans la conformation extérieure du corps humain. *Nouvelle iconographie de la Sulpêtrière*, **3**, 30–5.

Richer, P. (1893). *Canon des proportions du corps humain* (3rd edn. 1919). Paris: Delagrave. 95pp.

Richey, H. G. (1937). The relation of accelerated and retarded puberty to the height and weight of school children. *Monographs of the Society for Research in Child Development*, **2**(1), 67pp.

Riedle, J. J. (1834). Die Ergebnisse der militär-Conscriptionen in Beziehung auf körperliche Beschaffenheit der Conscriptionspflichtigen. In: *Württembergishce Jahrbücher für 1833*, pp. 369–393. Stuttgart and Tübingen : Memminger.

Riehl, B. (1898). *Die Kunst an der Brennerstrasse*. Leipzig. Breitkopf and Härtl.

Rietz, E. (1904). Das Wachstum Berliner Schulkinder während der Schuljahre. *Archiv für Anthropologie*, **1**, 30–42.

Rietz, E. (1906). Körperentwicklung und geistige Begabung. *Zeitschrift für Schulgesundheitspflege*, **19**, 65–98.

Rigden, W. (1869). On the age at which menstruation commences. *Transactions of the Obstetric Society of London*, **11**, 243.

Roberton, J. (1830). An inquiry respecting the age of puberty in women. *North of England Medical and Surgical Journal*, **1**, 69–85, 179–91.
Roberton, J. (1832). An enquiry into the natural history of the menstrual function. *Edinburgh Medical and Surgical Journal*, **38**, 227–54.
Roberton, J. (1842). On the period of puberty in Negro women. *Edinburgh Medical and Surgical Journal*, **58**, 112–20.
Roberton, J. (1844). On the alleged influence of climate on female puberty in Greece. *Edinburgh Medical and Surgical Journal*, **62**, 1–11.
Roberton, J. (1845*a*). On the period of puberty in Esquimaux women. *Edinburgh Medical and Surgical Journal*, **63**, 57–65.
Roberton, J. (1845*b*). On the period of puberty in Hindu women. *Edinburgh Medical and Surgical Journal*, **64**, 156–69, 257–64, 423–9; **66**, 36–64.
Roberton, J. (1846). On the age of puberty in the island of Madeira. *Edinburgh Medical and Surgical Journal*, **66**, 281–5.
Roberton, J. (1848). On the period of puberty in the Negro. *Edinburgh Medical and Surgical Journal*, **69**, 69–77.
Roberton, J. (1851). *Essays and Noted on the Physiology and Diseases of Women and on Practical Midwifery*. London: Churchill.
[Roberton, J.] (1876). Obituary notice. *British Medical Journal*, **ii**, 385.
Roberts, C. (1872–4). Flat foot. *St George's Hospital Reports*, **7**, 211–16.
Roberts, C. (1874–6). The physical development and the proportions of the human body. *St George's Hospital Reports*, **8**, 1–48.
Roberts, C. (1876). The physical requirements of factory children. *Journal of the Statistical Society*, **39**, 681–733.
Roberts, C. (1878). *A Manual of Anthropometry*. London: Churchill.
Roberts, C. (1890). Relative growth of boys and girls. *Nature* (London), **42**, 390.
Roberts, C. (1895). Memorandum on the medical inspection of, and physical education in, secondary schools. In: *Report of the Royal Commission on Secondary Education in England*, pp. 352–74. London: HMSO.
[Roberts, C.] (1902). Obituary notice. *British Medical Journal*, **i**, 181.
[Roberts, C.] (1930). Entry in: *Plarr's Lives of the Fellows of the Royal College of Surgeons of England*, revised D'Arcy Power, vol. 2, p. 231. Bristol: Wright.
Roberts, D. F., Rozner, L. M. and Swan, A. V. (1971). Age at menarche, physique and environment in industrial north-east England. *Acta paediatrica Scandinavica*, **60**, 158–64.
Roberts, D. F. and Weiner, J. S. (eds.) (1958). *The Scope of Physical Anthropology and Its Place in Academic Studies*. Oxford: Society for the Study of Human Biology.
Robertson, T. B. (1908). On the normal rate of growth of an individual and its biological significance. *Archiv für Entwicklungsmechanik der Organismen*, **25**, 581–614.
Robertson, T. B. (1915*a*). Studies on the growth of man. I. The pre- and post-natal growth of infants. *American Journal of Physiology*, **37**, 1–42.
Robertson, T. B. (1915*b*). Studies on the growth of man. II. The postnatal loss of weight in infants and the compensatory overgrowth which succeeds it. *American Journal of Physiology*, **37**, 74–85.
Robertson, T. B. (1916). Studies on the growth of man. III. The growth of British infants during the first year succeeding birth. *American Journal of Physiology*, **41**, 535–46.
Robertson, T. B. (1922). Criteria of normality in the growth of children. *Medical Journal of Australia*, **i**, 570–6.
Robertson, T. B. (1923). Growth and development. In: *Pediatrics*, ed. I. A. Abt, vol. 1, pp. 445–519. Philadelphia: W. B. Saunders.
[Robertson, T. B.] (1932). Obituary and bibliography (by H. R. Marston). *Australian Journal of Experimental Biology and Medical Science*, **9**, 1–21.
Robinow, M. (1942). The variability of weight and height increments from birth to six years. *Child Development*, **13**, 159–64.

Roche, A. F. (1978). Bone growth and maturation. In: *Human Growth*, ed. F. Falkner and J. M. Tanner, vol. 2, pp. 317–55. New York: Plenum.

Roche, A. F., Roberts, J. and Hamill, P. V. V. (1974). *Skeletal Maturity of Children 6–11 years, United States.* US Department of Health, Education and Welfare Publication No. (HRA)75–1622; National Center for Health Statistics Series 11, No. 140, 62pp. Rockville, Maryland: USDHEW.

Roche, A. F., Roberts, J. and Hamill, P. V. V. (1975). *Skeletal Maturity of Children 6–11 years. Racial, Geographic Area and Socioeconomic Differentials, United States.* US Department of Health, Education and Welfare Publication No. (HRA)76–1631; National Center for Health Statistics Series 11, No. 149, 81pp. Rockville, Maryland: USDHEW.

Roche, A. F., Roberts, J. and Hamill, P. V. V. (1976). *Skeletal Maturity of Youths 12–17 Years, United States.* Department of Health, Education and Welfare Publication No. (HRA)77–1642; National Center for Health Statistics Series 11, No. 160, 90pp. Rockville, Maryland: USDHEW.

Roche, A. F., Roberts, J. and Hamill, P. V. V. (1978). *Skeletal Maturity of Youths 12–17 Years: Racial, Geographic Area and Socioeconomic Differentials.* Department of Health, Education and Welfare Publication No. (PHS) 79–1654; National Center for Health Statistics Series 11, No. 167, 98pp. Hyatsville, Maryland: USDHEW.

Roche, A. F. and Sunderland, S. (1959). Melbourne University Child Growth Study. *Medical Journal of Australia*, **i**, 559–62.

Roederer, J. G. (1754). De pondere et longitudine infantum recens natorum. *Commentarii Societatis Regiae Scientarum Göttingensis*, **3**, 410–24.

Roederer, J. G. (1763). *Opuscula medica: de temporum in graviditate et partu aestimatione.* Göttingen: Bossigelius.

Rona, R. J. and Altman, D. G. (1977). National Survey of Health and Growth: standards of attained height, weight and triceps skinfold in English children 5 to 11 years old. *Annals of Human Biology*, **4**, 501–23.

Rona, R. J. and Florey, C. du V. (1980). National study of health and growth: respiratory symptoms and height in primary school children. *International Journal of Epidemiology*, **9**, 35–43.

Rona, R. J., Swan, A. V. and Altman, D. G. (1978). Social factors and height of primary school children in England and Scotland. *Journal of Epidemiology and Community Health*, **32**, 147–54.

Rooff, M. (1957). *Voluntary Societies and Social Policy.* London: Routledge and Kegan Paul.

Roscher, W. H. (1919). Die Hippokratische Schrift von der Siebenzahl und ihr Verhältnis zum Altpythagoreismus. *Berichte über die Verhandlungen der Sächsischen Akademie der Wissenschaften zu Leipzig*, **71**(5), 114pp.

Rosen, G. (1955). Problems in the application of statistical analysis to questions of health 1700–1800. *Bulletin of the History of Medicine*, **29**, 27–45.

Rossi-Doria, T. (1908). Über das Alter der ersten Menstruation in Italien und über ein Verhältniss, welches zwischen demselben und der Entwicklung des Beckens besteht. *Archiv für Gynäkologie*, **86**, 505–41.

Rössle, R. and Böning, H. (1924). Das Wachstum der Schulkinder. Ein Beitrag zur pathologischen Physiologie des Wachstums. Nebst einem Anhang über das Wachstum einiger innere Organe beim Kinde. *Veröffentlichungen aus dem Gebiet der Kriegs- und Konstitutionspathologie*, **4**, 1–72.

Rotch, T. M. (1909). A study of the development of the bones in childhood by the Roentgen method, with the view of establishing a developmental index for the grading of and the protection of early life. *Transactions of the Association of American Physicians*, **24**, 603–24.

Rotch, T. M. (1910*a*). Roentgenray methods applied to the grading of early life. *American Physical Education Review*, **15**, 396–420.

Rotch, T. M., (1910*b*). A comparison in boys and girls of height, weight and epiphyseal development. *Transactions of the American Pediatric Society*, **22**, 36–8.

Rotch, T. M. and Smith, H. W. (1910). A study of the development of the epiphyses of the hand and wrist for the purposes of classifying the cadets at Annapolis. *Transactions of the Association of American Physicians*, **25**, 200–10.

Rowntree, B. Seebohm (1913). *Poverty: A Study of Town Life*, 2nd edn. London: Nelson.

Royal Statistical Society (1934). *Annals of the Royal Statistical Society 1834–1934*. London: Royal Statistical Society.

Roy-Pernot, M.-P. (1959). Etude longitudinale sur la croissance de l'enfant. Résultats de quelques mésures somatiques effectives pendant les trois premiers années de la vie. *Archives françaises de pédiatrie*, **16**, 202–14.

Roy-Pernot, M.-P., Sempé, M. and Filliozat, A.-M. (1976). Rapport d'activité terminal de l'équipe française. Supplement to: *Compte rendu de la 13ième Réunion des Equipes Chargées des Etudes sur la Croissance et le Développement de l'Enfant Normal*. Paris: Centre International de l'Enfance.

Roy-Pernot, M.-P., Sempé, M., Orssaud, E. and Pedron, G. (1972). Evolution clinique de la puberté de la fille. *Archives françaises de pédiatrie*, **29**, 155–68.

Rufus of Ephesus. *De vesicae renumque morbis. De purgantibus medicamentis. De partibus corporis humani. Nunc iterum typis mandavit Gulielmas Clinch*. London: Clarke (1726).

Russow, A. (1880–1). Vergleichende Beobachtungen über den Einfluss der Ernährung mit der Brust und der künstlichen Ernährung auf das Gewicht und den Wuchs (Länge) der Kinder. *Jahrbuch für Kinderheilkunde*, **16**, 86–131.

Ruysch, F. (1701). *Thesaurus anatomicus*. Amsterdam: Wolters.

Ryle, J. A. (1948). *Changing Disciplines: Lectures on the History, Method and Motives of Social Pathology*. London: Oxford University Press.

Sack, N. (1893). Über die körperliche Entwicklung der Knaben in der Mittelschulen Moskau's. *Zeitschrift für Schulgesundheitspflege*, **6**, 649–63.

Sainte-Hilaire, I. G. (1832–7). *Histoire générale et particulière des anomalies de l'organisation chez l'homme et les animaux* (3 vols.). Paris: Baillière.

Sanders, B. S. (1934). *Environment and Growth*. Baltimore: Warwick and York.

Sargent, D. A. (1887). The physical proportions of the typical man. *Scribner's Magazine*, **2**, 3–17.

Sargent, D. A. (1889). The physical development of women. *Scribner's Magazine*, **5**, 172–85.

Sargent, D. A. (1893). *Anthropometric Charts for Different Ages, Male and Female, Ranging from 10 to 26 years of age*. Cambridge, Mass.: Sargent. [MS charts.]

Sarton, G. (1927–48). *Introduction to the History of Science* (3 vols.). Baltimore : Williams and Wilkins.

Sarton, G. (1952–9). *A History of Science* (2 vols.). Cambridge, Mass.: Harvard University Press.

Sarton, G. (1957). *Six Wings: Men of Science in the Renaissance*. Indiana: Indiana University Press.

Scammon, R. E. (1923). A summary of the anatomy of the infant and child. In: *Pediatrics*, ed. I. A. Abt, pp. 257–444. Philadelphia: W. B. Saunders.

Scammon, R. E. (1927*a*). The first seriatim study of human growth. *American Journal of Physical Anthropology*, **10**, 329–36.

Scammon, R. E. (1927*b*). The literature on the growth and physical development of the fetus, infant, and child: a quantitative summary. *Anatomical Record*, **35**, 241–67.

Scammon, R. E. (1930). The measurement of the body in childhood. In: *The Measurement of Man*, by J. A. Harris, C. M. Jackson, D. G. Patterson and R. E. Scammon, pp. 171–215. Minneapolis: University of Minnesota.

Scammon, R. E. and Calkins, L. A. (1929). *The Development and Growth of the External Dimensions of the Human Body in the Fetal Period*. Minneapolis: University of Minnesota.

Schadow, J.-G. (1834–5). *Polyclète, ou théorie des mésures de l'homme* (2 vols.). Berlin: Sachse.

Schaeffer, R. (1908). Über das Alter des Menstruationsbeginns. *Archiv für Gynäkologie*, **84**, 657–86.

Scheiber, S. H. (1881). Untersuchungen den mittleren Wuchs in Urgarn. *Archiv für Anthropologie*, **13**, 233–67.

Schiötz, C. (1923). Physical development of children and young people during the age of 7 to 18–20 years. An investigation of 28,700 pupils of Public (elementary) and Higher (secondary) schools in Christiania. *Skrifter utgitt av det Norske Videnskaps-Akademi*, **4**, 54pp.

Schlichting, F. X. (1880). Statistisches über den Eintritt der ersten Menstruation und über Schwangerschaftsdauer. *Archiv für Gynäkologie*, **16**, 203–32.

Schmidt, E. (1892–3). Die Körpergrösse und das Gewicht der Schulkinder des Kreises Saalfeld (Herzogthum Memingen). *Archiv für Anthropologie*, **21**, 385–434.

Schmidt, F. A. and Lessenich, H. H. (1903). Über die Beziehung zwischen körperlicher Entwicklung und Schulerfolg. *Zeitschrift für Schulgesundheitspflege*, **16**, 1–17.

Schmidt, M. C. P. (1825). *Geschichte und Topographie der Königlichen Preussischen Residenzstadt Potsdam*. Potsdam: Riegel.

Schmidt-Monnard, K. (1892). Über den Einfluss des Militärdienstes der Väter auf die körperliche Entwicklung ihrer Nachkommenschaft. *Jahrbuch für Kinderheilkunde*, **33**, 327–50.

Schmidt-Monnard, K. (1895). Über den Einfluss der Jahreszeit und der Schule auf das Wachstum der Kinder. *Jahrbuch für Kinderheilkunde*, **40**, 84–107.

Schmidt-Monnard, K. (1897). Die Chronische Kränklichheit in unseren Mittleren und Höheren Schulen. *Zeitschrift für Schulgesundheitspflege*, **11**, 593–615; **12**, 666–85.

Schmidt-Monnard, K. (1900). Über den Werth von Körpermassen zur Beutheilung des Körperzustandes von Kindern. *Korrespondenzblatt für Anthropologie*, **31**, 130–3.

Schmidt-Monnard, K. (1901). Über den Werth von Körpermassen zur Beurteilung des Körperzustandes bei Kindern. *Jahrbuch für Kinderheilkunde*, **53**, 50–8.

Schnepf, —. (1857). Influence de l'âge sur la capacité vitale du poumon. *Gazette médicale de Paris*, **12**, 331–5, 386–92.

Schott, L. (1974*a*). Zur Kritik an der Phasenlehre der ontogenetischen Entwicklung in Kindes- und Jugendalter. *Biologisches Rundschau*, **12**, 341–5.

Schott, L. (1974*b*). Carl Heinrich Stratz (1858–1924), ein Beitrag aus Anlass der 50 Wiederkehr seines Todestages. *Mitteilungen der Sektion Anthropologie der Biologischen Gesellschaft der DDR*, **30**, 43–51.

Schulz, A. (1962). *Winckelmann und seine Welt*. Berlin: Akademie-Verlag.

Schurig, M. (1729). *Parthenologia historico-medica, hoc est virginitatis consideratio, qua ad eam pertinens pubertates et menstruatio, item varia de insolitis mensium viis, nec non de partium genitalium miliebrium pro virginitatis custodia*. Dresden: Hekel.

Schuster, N. H. (1968). English doctors in Russia in the early 19th century. *Proceedings of the Royal Society of Medicine*, **61**, 185–90.

Scouller, R. E. (1966). *The Armies of Queen Anne*. Oxford: Clarendon Press.

Seggel, —. (1904). Über das Verhältnis von Schädel- und Gehirnentwicklung zum Längenwachstum des Körpers. *Archiv für Anthropologie*, **1**, 1–25.

Sempé, M., Tutin, C. and Masse, N.-P. (1964). La croissance de l'enfant de 0 à 7 ans. (Mésures pratiquées sur des enfants de la région parisienne de 1953–1962.) *Archives françaises de pédiatrie*, **21**, 111–34.

[Senac, J.,] (1825). Biography. *Dictionnaire des sciences médicales: biographie médicale*, **7**, 197–8.

Sharp, J. (1671). *The Midwives Book, or the Whole Art of Midwifery Discovered*. London: Miller.

Sharpe, W. D. (1964). Isidore of Seville: the medical writings. An English translation, with an introductory commentary. *Transactions of the American Philosophical Society*, N.S., **54**(2), 5–75.

Sherrington, C. (1946). *The Endeavour of Jean Fernel*. London: Cambridge University Press.

Shock, N. W. (1941). Age changes and sex differences in alveolar CO_2 tension. *American Journal of Physiology*, **133**, 610–16.

Shock, N. W. (1942). Standard values for basal oxygen consumption in adolescents. *American Journal of Diseases of Children*, **64**, 19–32.

Shock, N. W. (1943). The effect of menarche on basal physiological function in girls. *American Journal of Physiology*, **139**, 288–92.

Shock, N. W. (1944). Basal blood pressure and pulse rate in adolescents. *American Journal of Diseases of Children*, **68**, 16–22.

Shock, N. W. (1945). Creatine excretion in adolescents. *Child Development*, **16**, 167–80.

Shock, N. W. (1946*a*). Some physiological aspects of adolescence. *Texas Reports on Biology and Medicine*, **4**, 368–86.

Shock, N. W. (1946*b*). Physiological responses of adolescents to exercise. *Texas Reports on Biology and Medicine*, **4**, 368–86.

Shuttleworth, F. K. (1934). Standards of development in terms of increments. *Child Development*, **5**, 89–91.

Shuttleworth, F. K. (1937). Sexual maturation and the physical growth of girls aged six to 19. *Monographs of the Society for Research in Child Development*, **2**(5), 253pp.

Shuttleworth, F. K. (1938*a*). Sexual maturation and the skeletal growth of girls aged six to 19. *Monographs of the Society for Research in Child Development*, **3**(5), 56pp.

Shuttleworth, F. K. (1938*b*). The adolescent period: a graphic and pictorial analysis. *Monographs of the Society for Research in Child Development*, **3**(3), 246pp.

Shuttleworth, F. K. (1939). The physical and mental growth of girls and boys age six to 19 in relation to age at maximum growth. *Monographs of the Society for Research in Child Development*, **4**(3), 291pp.

Shuttleworth, F. K. (1949*a*). The adolescent period: a graphic analysis. *Monographs of the Society for Research in Child Development*, **14** (1), 453 Figs. [167pp].

Shuttleworth, F. K. (1949*b*). The adolescent period: a pictorial analysis. *Monographs of the Society for Research in Child Development*, **14** (2), 69pp.

Siebold, E. von (1860). Über die Gewichts und Längenverhältnisse der neugeborenen Kinder, über die Verminderung ihres Gewichts in den ersten Tagen und die Zunahme desselben in der ersten Wochen nach der Geburt. *Monatsschrift für Geburtskunde*, **15**, 337–54.

Siebold, E. G. C. de (1891). *Essai d'une histoire de l'obstétricie*, trans. F. J. Herrgott. Paris: Steinheil.

Sigerist, H. E. (1961). *A History of Medicine*, vol. 2, *Early Greek, Hindu and Persian Medicine*. New York: Oxford University Press.

Simmons, K. (1944). The Brush Foundation study of child growth and development. II. Physical growth and development. *Monographs of the Society for Research in Child Development*, **9** (1), 87pp.

Simmons, K. and Greulich, W. W. (1943). Menarcheal age and the height, weight and skeletal age of girls age 7 to 17 years. *Journal of Pediatrics*, **22**, 518–48.

Simmons, K. and Todd, T. W. (1938). Growth of well children: analysis of stature and weight, 3 months to 13 years. *Growth*, **2**, 93–134.

Simon, G., Reid, L., Tanner, J. M., Goldstein, H. and Benjamin, B. (1972). Growth of radiologically determined heart diameter, lung width and lung length from 5–19 years, with standards for clinical use. *Archives of Disease in Childhood*, **47**, 373–81.

Simon, M. D. (1959). Body configuration and school readiness. *Child Development*, **30**, 493–512.

Simpson, J. Y. (1844). Memoir on the sex of the child as a cause of difficulty and danger in human parturition. *Edinburgh Medical and Surgical Journal*, **62**, 387–439.

Singh, I. and Savara, B. (1966). Norms of size and annual increments of seven anatomical measures of maxillae in girls from three to 16 years of age. *Angle Orthodontics*, **36**, 312–24.

Sinibaldus, J. B. (1642). *Geneanthropeiae sive de hominis generatione decateuchon*. Rome: Francisus Caballus.

Slater, E. (1939). Studies from the Center for Research in Child Health and Development,

Harvard School of Public Health. II. Types, levels and irregularities of response to a nursery school situation of 40 children observed with special reference to the home environment. *Monographs of the Society for Research in Child Development*, **4** (2), 148pp.

Smedley, F. W. (1902). Child study in Chicago. In: *Annual Report of the Department of the Interior*, pp. 1095–138. (Cited in Baldwin (1921); not seen.)

Sneath, P. H. A. (1967). Trend-surface analysis of transformation grids. *Journal of Zoology* (London), **151**, 65–122.

Snell, B. (1953). *Discovery of the Mind.* Cambridge, Mass.: Harvard University Press.

Sontag, L. W. (1971). The history of longitudinal research: implications for the future. *Child Development*, **42**, 987–1002.

Sontag, L. W., Baker, C. and Nelson, V. M. (1958). Mental growth and personality development: a longitudinal study. *Monographs of the Society for Research in Child Development*, **23**(2), 143pp.

Sontag, L. W. and Reynolds, E. L. (1945). The Fels composite sheet. *Journal of Pediatrics*, **26**, 327–35.

Soranus *Gynaeciorum*. In: *Corpus medicorum graecorum*, vol. 4, ed. J. Ilberg. Berlin: Teubner (1927).

Soranus. *De signis fracturarum*. In: *Corpus medicorum graecorum*, vol. 4, ed. J. Ilberg. Berlin: Teubner (1927).

Soranus. *Gynecology*, trans. O. Temkin. Baltimore: Johns Hopkins Press (1950).

Soulié, E. and Bartholémy, E. de (1868). *Journal de Jean Heroard sur l'enfance et la jeunesse de Louis XIII (1601–1628) extrait des manuscrits originaux* (2 vols.). Paris: Firmin Didot.

Southgate, D. A. T. (1978). Fetal measurements. In: *Human Growth*, ed. F. Falkner and J. M. Tanner, vol. 1, pp. 379–95. New York: Plenum.

Soyre, J. de (1863). De la primiparité à terme. *Gazette des Hôpitaux civils et militaires*, **111**, 441–2.

Spence, J. C., Walton, W. S., Miller, F. J. W. and Court, S. D. M. (1954). *A Thousand Families in Newcastle-upon-Tyne.* London: Oxford University Press.

Spencer, W. G. (trans.) (1938). *Celsus: De medicina.* Loeb Classical Library. London: Heinemann.

Spielrein, I. (1916). Über Kindermessungen in Rostow am Don. *Zeitschrift für Schulgesundheitspflege*, **29**, 451–61, 503–13, 548–60.

Spier, L. (1918). The growth of boys: dentition and stature. *American Anthropologist*, **20**, 37–48.

Spigelius, A. (1632). *De humana corporis fabrica, libri decem.* Frankfurt: Matthew Merian.

Stahl, G. E. (1737). *Theoria medica vera, physiologiam et pathologiam.* Halle: Orphanotropheum. (1st edn, 1707.)

Stahl, W. H. (trans.). *Macrobius' Commentary on the Dream of Scipio.* New York: Columbia University.

Stampfer, P. C. (1879). Dr Guarinoni's Wahlfahrt nach Rom 1613. *Zeitschrift des Ferdinandmuseums für Tirol und Vorarlberg*, **23**, 57–94.

Stanway, S. (1833). See Parliamentary Papers (1833).

Statistical Society (1885). *Jubilee Volume of the Statistical Society (Founded 1834).* London: Stanford.

Steckel, R. H. (1979). Slave height profiles from coastwise manifests. *Explorations in Economic History*, **16**, 363–80.

Steet, G. C. (1874–6). Notes on the development and growth of boys between 13 and 20 years of age. *St George's Hospital Reports*, **8**, 49–56.

Stein, Z., Susser, M., Saenger, G. and Marolla, F. (1975). *Famine and Human Development: the Dutch Hunger Winter of 1944–5.* New York: Oxford University Press.

Stewart, A. M. and Russell, W. T. (1952). Interim report on the Oxford Child Health Survey. *Medical Officer*, **88**, 5–8.

Steyneger, L. (1936). *Georg Wilhelm Steller. The Pioneer of Alaskan Natural History*, Cambridge, Mass.: Harvard University Press.

Stieda, L. (1882–3). Über die Anwendung der Wahrscheinlichkeitsrechnung in der anthropologischen Statistik. *Archiv für Anthropologie*, **14**, 167–82.

Stoeller, J. A. (1726). *De vene sectione secunda in morbis quibusdam cronicis vere secunda*. Halle: Hendel.

Stoeller, J. A. (1729). *Historisch-medizinische Untersuchung des Wachsthums des Menschen in die Länge, so wohl, was Medici, als alle Wachsende und grosse Leute van dessen natürlichen Eigenschaften und Umständen, wie auch von denen mit dem Wachstum des Leibes in die Länge sich zutragenden Kranckheiten, ingleichen, was Soldaten von ihren gemeinsten Maladien und derem Curen zu wissen nöthig haben* ... Magdeberg: Seidel and Scheidhauer.

Stoeller, J. A. (1747). Zuverlässige Nachricht von dem merkwürdigen Leben und Reisen Hern George Wilhelm Stöllers, der russisch Kaiserlich Akademie der Wissenschaften Adiunti und Mitglieds. In: *Ergetzungen der vernünftigen Seele aus der Sittenlehre und der Gelehrsamkeit überhaupt*, ed. J. H. G. von Justi, vol. 5, pp. 362–84. (Cited in Steynegar (1936); not seen.)

Stolz, H. R. and Stolz, L. M. (1944). Adolescent problems related to somatic variations. *Yearbook of the National Society for the Study of Education*, **43**, part 1, 80–99.

Stolz, H. R. and Stolz, L. M. (1951). *Somatic Development of Adolescent Boys. A Study of the Growth of Boys during the Second Decade of Life*. New York: Macmillan.

Stratz, C. H. (1899). *Die Schönheit des weiblichen Körpers*. Stuttgart: Enke.

Stratz, C. H. (1901*a*). *Die Körperpflege der Frau. Physiologische und aesthetische Diätetik für das weibliche Geschlecht*. Stuttgart: Enke.

Stratz, C. H. (1901*b*). *Die Rassenshönheit des Weibes*. Stuttgart: Enke.

Stratz, C. H. (1902). *Die Körperformen in Kunst und Leben der Japaner*. Stuttgart: Enke.

Stratz, C. H. (1903). *Der Körper des Kindes, für Eltern, Erzieher, Ärzte und Künstler*. Stuttgart: Enke. 250pp.

Stratz, C. H. (1908). Menarche und Tokarche. *Verhandlungen der deutschen Gesellschaft für Gynäkologie*, **12**, 777–80.

Stratz, C. H. (1909*a*). Wachstum und Proportionen des Menschen vor und nach der Geburt. *Archiv für Anthropologie*, **8**, 287–97.

Stratz, C. H. (1909*b*). *Der Körper des Kindes und seine Pflege, für Eltern, Erzieher, Ärzte und Künstler*, 3rd ed. 386pp. Stuttgart : Enke.

Stratz, C. H. (1910). Wachstum und Proportionen des Fötus. *Zeitschrift für Geburtshülfe und Gynäkologie*, **65**, 36–51.

Stratz, C. H. (1914*a*). *Die Darstellung des menschlichen Körpers in der Kunst*. Berlin: Springer.

Stratz, C. H. (1914*b*). Gestalt und Wachstum des Kindes. In: *Die Gesundheitspflege des Kindes*, ed. W. Kruse and P. Selter, pp. 7–28. Stuttgart: Enke.

Stratz, C. H. (1915). Betrachtungen über das Wachstum des Menschen. *Archiv für Anthropologie*, **14**, 81–8.

Stuart, H. C. (1933). *Healthy Childhood: Guidance for Physical Care*. New York: Appleton-Century.

Stuart, H. C. (1939). The Center for Research in Child Health and Development, School of Public Health, Harvard University. I. The Center, the group under observation, sources of information and studies in progress. *Monographs of the Society for Research in Child Development*, **4**(1), 261pp.

Stuart, H. C. and Dwinell, P. H. (1942). The growth of bone, muscle and overlying tissues in children six to 10 years of age as revealed by studies of roentgenograms of the leg area. *Child Development*, **13**, 195–213.

Stuart, H. C., Hill, P. and Shaw, C. (1940). Studies from the Centre for Research in Child Health and Development, School of Public Health, Harvard University. III. The growth of bone, muscle and overlying tissues as revealed by studies of roentgenograms of the leg area. *Monographs of the Society for Research in Child Development*, **5**(3), 218pp.

Stuart, H. C. and Meredith, H. V. (1946). The use of body measurements in the school health program. I. General considerations and the selection of measurements. II. Methods to be followed in taking and interpreting measurements, and norms to be used. *American Journal of Public Health*, **36**, 1365–86.

Stuart, H. C. and Reed, R. B. (1959). Longitudinal studies of child health and development, Series II: description of project; bibliography. *Pediatrics*, **24**, 875–85, 972–4.

Stuart, H. C. and Sobel, E. H. (1946). The thickness of the skin and subcutaneous tissue by age and sex in childhood. *Journal of Pediatrics*, **28**, 637–47.

Stuart, H. C. and Stevenson, S. S. (1950). Physical growth and development. In: *Mitchell-Nelson's Textbook of Pediatrics*, 5th edn, ed. W. E. Nelson, pp. 14–73. Philadelphia: W. B. Saunders.

Swieten, G. van (1744–73). *Commentaries upon the Aphorisms of Dr Hermann Boerhaave concerning the Knowledge and Cure of Diseases* (18 vols.: vol. 12, 1765; vol. 13, 1765). London: Horsfield and Longman.

Tacitus. *The 'Agricola' and the 'Germania'*, trans. with an introduction by H. Mattingley and S. A. Handford. Harmondsworth: Penguin Books (1948).

Tacitus. *Germania*, trans. M. Hutton and E. H. Warmington. Loeb Classical Library. Cambridge, Mass.: Harvard University Press (1970).

Tagore, A. N. (1921). *Art et anatomie hindous.*

Talbot, C. H. (1967). *Medicine in Medieval England.* London: Oldbourne.

Tanner, J. M. (1948). A guide to American growth studies. *Yearbook of Physical Anthropology*, **3**, 28–33.

Tanner, J. M. (1951). Some notes on the reporting of growth data. *Human Biology*, **23**, 93–159.

Tanner, J. M. (1952). The assessment of growth and development in children. *Archives of Disease in Childhood*, **27**, 10–33.

Tanner, J. M. (1955). *Growth at Adolescence*, 1st edn. Oxford: Blackwell Scientific Publications.

Tanner, J. M. (1958*a*). *Physical Maturing and Behaviour at Adolescence* (The Convocation Lecture of the National Children's Home). London: National Children's Home.

Tanner, J. M. (1958*b*). The place of human biology in medical education. Suggestions for a revised pre-clinical curriculum in the light of the General Medical Council 1957 recommendations. *Lancet*, **i**, 1185–8.

Tanner, J. M., (1958*c*). The evaluation of physical growth and development. In: *Modern Trends in Paediatrics*, 2nd series, ed. A. Holzell and J. P. M. Tizard, pp. 325–44. London: Butterworth.

Tanner, J. M. (1959). Boas' contributions to knowledge of human growth and form. In: *The Anthropology of Franz Boas*. ed. W. Goldschmidt, pp. 76–111. San Francisco: Chandler. (Also in *American Anthropologist*, **61**, 76–111.)

Tanner, J. M. (ed.) (1960). *Human Growth.* Symposium of the Society for the Study of Human Biology No. 3. Oxford: Pergamon Press.

Tanner, J. M. (1962). *Growth at Adolescence*, 2nd edn. Oxford: Blackwell Scientific Publications.

Tanner, J. M. (1963*a*). Regulation of growth in size in mammals. *Nature* (London), **199**, 845–50.

Tanner, J. M. (1963*b*). The regulation of human growth. *Child Development*, **34**, 817–47.

Tanner, J. M. (1964). Human biology in general university education. *Lancet*, **ii**, 215–18.

Tanner, J. M. (1965). Radiographic studies of body composition. In: *Body Composition*, ed. J. Brozek, Symposium of the Society for the Study of Human Biology No 7, pp. 211–38. Oxford: Pergamon Press.

Tanner, J. M. (1966). Galtonian eugenics and the study of growth. The relation of body size, intelligence test score, and social circumstances in children and adults. *Eugenics Review*, **58**, 122–35. (Reprinted in *Trends and Issues in Developmental Psychology*, ed. P. H. Mussen, J. Largen and M. Covington, New York: Holt (1969).)

Tanner, J. M. (1969). Growth and endocrinology of the adolescent. In: *Endocrine and Genetic Diseases of Childhood*, 1st edn, ed. L. Gardner, pp. 19–60. Philadelphia: W. B. Saunders.

Tanner, J. M. (1973). Trend towards earlier menarche in London, Oslo, Copenhagen, the Netherlands and Hungary. *Nature* (London), **243**, 95–6.

Tanner, J. M. (1975). Growth and endocrinology of the adolescent. In: *Endocrine and Genetic Diseases of Childhood*, 2nd edn, ed. L. Gardner, pp. 14–63. Philadelphia: W. B. Saunders.

Tanner, J. M. (1978*a*). *Fetus into Man*. London/Cambridge, Mass.: Open Books/Harvard University Press.

Tanner, J. M. (1978*b*). Human growth standards: construction and use. In: *Auxology: Human Growth in Health and Disorder*, ed. L. Gedda and P. Parisi, pp. 109–22. London: Academic Press.

Tanner, J. M. (1979). A note on the history of catch-up growth. *Bulletins et Mémoires de la Société d'Anthropologie de Paris*, **6**, 399–405.

Tanner, J. M. (1981). The endocrinology of normal puberty. In: *Clinical Paediatric Endocrinology*, ed. C. G. D. Brook. Oxford: Blackwell Scientific Publications (in press).

Tanner, J. M. and Eveleth, P. B. (1976). Urbanization and growth, In: *Man in Urban Environments*, ed. G. A. Harrison and J. B. Gibson, pp. 144–66. Oxford University Press.

Tanner, J. M., Goldstein, H. and Whitehouse, R. H. (1970). Standards for children's height at ages 2 to 9 years, allowing for height of parents. *Archives of Disease in Childhood*, **45**, 755–62.

Tanner, J. M. and Gupta, D. (1968). A longitudinal study of the excretion of individual steroids in children from 8 to 12 years old. *Journal of Endocrinology*, **41**, 139–56.

Tanner, J. M., Healy, M. J. R., Lockhart, R. D., MacKenzie, J. D. and Whitehouse, R. H. (1956). Aberdeen Growth Study. I. The prediction of adult body measurements from measurements taken each year from birth to five years. *Archives of Disease in Childhood*, **31**, 382–481.

Tanner, J. M. and Inhelder, B. (eds.) (1956–60). *Discussions on Child Development*, vols. 1–4. *A Consideration of the Biological, Psychological and Cultural Approaches to the Understanding of Human Development and Behaviour*. London: Tavistock.

Tanner, J. M. and Israelsohn, W. J. (1963). Parent–child correlations for body measurements of children between the age of one month and seven years. *Annals of Human Genetics* (London), **26**, 245–59.

Tanner, J. M., Lejarraga, H. and Cameron, N. (1975). The natural history of the Silver–Russell Syndrome: a longitudinal study of thirty-nine cases. *Pediatric Research*, **9**, 611–23.

Tanner, J. M. and Thomson, A. (1970). Standards for birthweight at gestation periods from 32 to 42 weeks, allowing for maternal height and weight. *Archives of Disease in Childhood*, **45**, 566–9.

Tanner, J. M. and Whitehouse, R. H. (1973). Height and weight charts from birth to five years allowing for length of gestation: for use in Infant Welfare Clinics. *Archives of Disease in Childhood*, **48**, 786–9.

Tanner, J. M. and Whitehouse, R. H. (1976). Clinical longitudinal standards for height, weight, height velocity and weight velocity and the stages of puberty. *Archives of Disease in Childhood*, **51**, 170–9.

Tanner, J. M., Whitehouse, R. H., Hughes, P. C. R. and Carter, B. S. (1976). Relative importance of growth hormone and sex steroids for the growth at puberty of trunk length, limb length and muscle width in growth hormone deficient children. *Journal of Pediatrics*, **89**, 1000–8.

Tanner, J. M., Whitehouse, R. H., Hughes, P. C. R. and Vince, F. P. (1971). The effect of human growth hormone treatment from one to seven years on the growth of 100 children with growth hormone deficiency, low birthweight, inherited smallness, Turner's syndrome and other complaints. *Archives of Disease in Childhood*, **46**, 745–82.

Tanner, J. M., Whitehouse, R. H., Marshall, W. A., Healy, M. J. R. and Goldstein, H. (1975). *Assessment of Skeletal Maturity and Prediction of Adult Height.* London: Academic Press.

Tanner, J. M., Whitehouse, R. H., Marubini, E. and Resele, L. (1976). The adolescent growth spurt of boys and girls of the Harpenden Growth Study. *Annals of Human Biology,* **3**, 109–26.

Tanner, J. M., Whitehouse, R. H. and Takaishi, M. (1966). Standards from birth to maturity for height, weight, height velocity and weight velocity: British children, 1965. *Archives of Disease in Childhood,* **41**, 454–71, 613–35.

Taranger, J. (1976). The somatic development of children in a Swedish urban community: II. Evaluation of biological maturation by means of maturity criteria. *Acta paediatrica Scandinavica,* **258** (*Supplement*), 77–82.

Taranger, J., Brunning, B., Claesson, I., Karlberg, P., Landström, T. and Lindström, B. (1976*a*). The somatic development of children in a Swedish urban community. IV. Skeletal development from birth to seven years. *Acta paediatrica Scandinavica,* **258** (*Supplement*), 98–108.

Taranger, J., Brunning, B., Claesson, I., Karlberg, P., Landström, T. and Lindström, B. (1976*b*). The somatic development of children in a Swedish urban community. V. A new method for the assessment of skeletal maturity – the MAT method (mean appearance time of bone stages). *Acta paediatrica Scandinavica,* **258** (*Supplement*), 109–20.

Taranger, J., Engström, I, Lichtenstein, H. and Svennberg-Redegren, I. (1976*c*). The somatic development of children in a Swedish urban community. VI. Somatic pubertal development. *Acta paediatrica Scandinavica,* **258** *(Supplement),* 121–35.

Taranger, J., Lichtenstein, H. and Svennberg-Redegren, I. (1976*d*). The somatic development of children in a Swedish urban community. III. Dental development from birth to 16 years. *Acta paediatrica Scandinavica,* **258** (*Supplement*), 83–97.

Tarin, P. (1750). *L'anthropotomie, ou l'art de disséquer.* Paris: Briasson.

Tarn, W. and Griffith, G. T. (1966). *Hellenistic Civilisation,* 3rd edn. London: Methuen.

Temkin, O. (1956). *Soranus' Gynecology.* Baltimore: Johns Hopkins Press.

Theopold, W. (1964). *Schiller: sein Leben und die Medizin im 18 Jahrhundert.* Stuttgart: Fischer.

Theopold, W. (1967). *Der Herzog und die Heilkunst: die Medizin an der Hohen Carlschule zu Stuttgart.* Cologne: Deutsche Ärzteverlag.

Theopold, W., Hövels, O., Hartmann, W. and Uhland, R. (1972). Beobachtung über das Langenwachstum in der zweite Hälfte des 18 Jahrhunderts. *Deutsches Ärzteblatt: Ärztliche Mitteilungen,* **11**, 611–17.

Thissen, D., Bock, R. D., Wainer, H. and Roche, A. F. (1976). Individual growth in stature: a comparison of four growth studies in the USA. *Annals of Human Biology,* **3**, 529–42.

t'Hof, M. A. van, Roede, M. J. and Kowalski, C. J. (1976). Estimation of growth velocities from individual longitudinal data. *Growth,* **40**, 217–40.

t'Hof, M. A. van, Roede, M. J. and Kowalski, C. J. (1977). A mixed longitudinal data analysis model. *Human Biology,* **49**, 165–79.

Thompson, D'A. W. (trans.) (1910). *Aristotle: Historia animalium.* Oxford: Clarendon Press.

Thompson, D'A. W. (1913). *On Aristotle as a Biologist.* Oxford: Clarendon Press.

Thompson, D'A. W. (1917). *On Growth and Form.* Cambridge University Press.

Thompson, D'A. W. (1942). *On Growth and Form,* revised edn. Cambridge University Press.

Thompson, G. W., Popovich, F. and Anderson, D. L. (1976). Maximum growth changes in mandibular length, stature and weight. *Human Biology,* **48**, 285–93.

Thompson, R. D. (1958). *D'Arcy Wentworth Thompson: The Scholar-Naturalist, 1860–1948.* London: Oxford University Press.

Thorndike, L. (1965). *Michael Scot.* London: Nelson.

Thorne, L. T. (1904). The physical development of the London schoolboy: 1890 examinations. *British Medical Journal,* **i**, 829–31.

Thwaites, E. J. (1950). The social background of infancy: the domestic environment of 471 Oxford babies. *Archives of Disease in Childhood,* **25**, 193–203.

Tiisala, R., Kantero, R. l-L. Backström, L. and Hallman, N. (1969). A mixed longitudinal study on skeletal maturation in healthy Finnish children aged one to five years. *Human Biology*, **41**, 560–70.

Tiisala, R., Kantero, R.-L. and Tammien, T. (1971). A mixed longitudinal study on skeletal maturation in healthy Finnish children aged five to 10 years. *Human Biology*, **43**, 224–36.

Todd, R. B. (1976). *Alexander of Aphrodisias on Stoic Physics*. Leiden: Brill.

Todd, T. W. (1930). The roentgenographic appraisement of skeletal differentiation. *Child Development*, **1**, 298–310.

Todd, T. W. (1937). *Atlas of Skeletal Maturation*. London: Kimpton.

Tracy, W. and Savara, B. (1966). Norms of size and annual increments of five anatomical measures of the mandible in girls from three to 16 years of age. *Archives of Oral Biology*, **11**, 587–98.

Trevorrow, V. E. (1957). Longitudinal study of plasma fibrinogen in children. *Human Biology*, **29**, 354–65.

Trevorrow, V. E. (1967). Intra-individual gamma globulin changes with age from four to 20 years. *Human Biology*, **39**, 1–4.

Trevorrow, V. E., Kaser, M., Patterson, J. P. and Hill, R. M. (1942). Plasma albumin, globulin and fibrinogen in healthy individuals from birth to adulthood. II. 'Normal' values. *Journal of Laboratory and Clinical Medicine*, **27**, 471–6.

Trussell, J. and Steckel, R. (1978). The age of slaves at menarche and their first birth. *Journal of Interdisciplinary History*, **8**, 477–505.

Tuddenham, R. D. and Snyder, M. M. (1954). Physical growth of California boys and girls from birth to 18 years. *University of California Publications in Child Development*, **1**, 183–364.

Tuxford, A. W. and Glegg, R. A. (1911). The average height and weight of English school-children. *British Medical Journal*, **i**, 1423–4.

Udjus, L. G. (1964). *Anthropometrical Changes in Norwegian Men in the 20th Century*. Oslo: Universitätsforlaget.

Uhland, R. (1953). *Geschichte der Hohen Karlschule in Stuttgart*. Stuttgart: Kohlhammer.

Uhlitsch, [R.] (1891). Anthropometrische Messungen und deren praktischer Wert. *Allgemeines Statistisches Archiv*, **2**, 419–51.

University of Iowa (1933). *Pioneering in Child Welfare: A History of the Iowa Child Welfare Research Station 1917–1933*. Iowa City: State University of Iowa.

University of Iowa Institute of Child Behavior and Development (1967). *Fifty Years of Research 1917–1967*. Iowa City: State University of Iowa.

Valadian, I., Stuart, H. C. and Reed, R. B. (1959). Longitudinal studies of child health and development, series II: patterns of illness experiences. *Pediatrics*, **24**, 941–71.

Valadian, I., Stuart, H. C. and Reed, R. B. (1961*a*). Contribution of respiratory infections to the total illness experiences of healthy children from birth to 18 years. *American Journal of Public Health*, **51**, 1320–8.

Valadian, I., Stuart, H. C. and Reed, R. B. (1961*b*). Studies of illness of children followed from birth of 18 years. *Monographs of the Society for Research in Child Development*, **26** (3), 125pp.

Valšik, J. A., Stukovsky, R. and Bernátová, L. (1963). Quelques facteurs géographiques et sociaux ayant une influence sur l'âge de la puberté. *Biotypologie*, **24**, 109–23.

Vandenberg, S. J. and Falkner, F. (1965). Hereditary factors in human growth. *Human Biology*, **37**, 357–65.

Variot, G. and Chaumet, —. (1906). Tables de croissance des enfants parisiens de 1 an à 16 ans, dressées en 1905. *Bulletin de la Société de Pédiatrie de Paris*, **8**, 49–58.

Variot, G. and Fliniaux, —. (1914). Tables des croissances comparées des nourissons éléves au sein et au bibéron durant la première année de la vie. *Comptes rendus de l'Académie des Sciences* (Paris), **158**, 1361–4.

Veit, G. (1855). Beiträge zur geburtshülflichen Statistik. *Monatsschrift für Geburtskunde und Frauenkrankheiten*, **5**, 344–81; **6**, 101–32.

Venette, N. (1696). *De la génération de l'homme ou tableau de l'amour conjugal* . . ., 7th edn. Cologne: Joly. (Earlier editions issued under pseudonyn Salocini Venetien.)

Venette, N. (1712). *The Mysteries of Conjugal Love Reveal'd, written in French by Nicholas de Venette . . . the eighth edition done into English by a Gentleman*, 3rd edn, corrected. London: [no publisher given].

Venrooij-Ysseldmuiden, M. E. van and Ipenburg, A. van (1978). Mixed longitudinal data on skeletal age from a group of Dutch children living in Utrecht and surroundings. *Annals of Human Biology*, **5**, 359–80.

Vierordt, H. (1906). *Anatomische, physiologische und physikalische Daten und Tabellen*, 3rd edn. Jena: Fisher. (1st edn, 1888; 2nd edn, 1893.)

Vierordt, K. (1861). *Grundriss der Physiologie des Menschen*. Tübingen: Laupp.

Vierordt, K. (1871). *Grundriss der Physiologie des Menschen*, 2nd edn. Tübingen: Laupp.

Vierordt, K. (1877). Physiologie des Kindesalters. In: *Handbuch der Kinderkrankheiten*, ed. C. Gerhardt, vol. **1**, pp. 59–91. Tübingen: Laupp.

Vierordt, K. (1881). Physiologie des Kindesalters. In: *Handbuch der Kinderkrankheiten*, 2nd edn., ed. C. Gerhardt, vol. 1, pp. 219–91.

Vigeon, E. (1977). Clogs or wooden soled shoes. *Journal of the Costume Society* (Reprinted by Salford Museum, Salford, Lancs.)

Villermé, L. R. (1828). Mémoire sur la mortalité en France dans la classe aisée et dans la classe indigente. *Mémoires de l'Academie de Médecine*, **1**, 51–98.

Villermé, L. R. (1829). Mémoire sur la taille de l'homme en France. *Annales d'hygiène publique*, **1**, 551–99.

Villermé, L. R. (1840). *Tableau de l'état physique et moral des ouvriers employés dans les manufactures de coton, de laine et de soie* (2 vols.). Paris: Renouard.

Vincent of Beauvais, *see* Vincentius Bellovacensis.

Vincentius Bellovacensis. *Speculum quadruplex* (4 vols.), reprinted from 1591 edn. Graz: Akademische Druck (1964–5).

Viola, G. (1932). *La costituzione individuale* (2 vols.) Bologna: Cappelli.

Virey, J. J. (1815). Enfance. *Dictionnaire des sciences médicales*, **12**, 217–58.

Virey, J. J. (1816). Géant. *Dictionnaire des sciences médicales*, **17**, 546–68.

Vitruvius, P. M. *De architectura*, with commentary by Cesariano. Como (1521).

Vitruvius, P. M. *On Architecture*, edited from the Harleian MS 2767 and trans. F. Granger (2 vols.). London: Heinemann (1934).

Völker, A. (1979). Unpublished MS supplied by courtesy of Professor W. Kaiser of the Martin-Luther University of Halle-Wittenburg.

Waddington, C. H. (1957). *The Strategy of the Genes*. London: Allen and Unwin.

Wagner, H. (1856–8). *Geschichte der Hohen Carlschule* (3 vols.). Würzburg: Etlinger.

Walisko, A. and Jedlinska, W. (1976). Wroclaw Growth Study. II. Males. *Studies in Physical Anthropology*, **3**, 27–48.

Walker, H. M. (1929). *Studies in the History of Statistical Method*. Baltimore: Williams and Wilkins.

Wall, R. (1973). *Comparative Statistics in the 19th Century*. Gregg International Publishers.

Wallin, J. E. W. (1938). A tribute to G. Stanley Hall. *Journal of Genetic Psychology*, **113**, 149–53.

Wallis, R. S. (1931). How children grow. An anthropometric study of private school children from two to eight years of age. *University of Iowa Studies in Child Welfare*, **5** (1), 137pp.

Wampen, H. (1864). *Anthropometry, or Geometry of the Human Figure*. London: Boone.

Washburn, A. H. (1950). Growth: its significance in medicine viewed as human biology. *Pediatrics*, **5**, 765–70.

Washburn, A. H. (1953). Medicine as human biology. *Journal of Medical Education*, **28** (12), 9–16.

Washburn, A. H. (1955). Human growth, development and adaptation. *American Journal of Diseases of Children*, **90**, 2–5.

Washburn, A. H. (1957). The child as a person developing. I. A philosophy and program of research. II. More questions raised than answered. *American Journal of Diseases of Children*, **94**, 46–53, 54–63.

Wasse, J. (1724). Concerning the difference in the height of a human body between the morning and night. *Philosophical Transactions of the Royal Society of London*, **33**, 87–8.

Weber, F. (1883). Über die Menstrualverhältnisse der Frauen in St. Petersburg. *St Petersburger Medizinische Wochenschrift*, **8**, 329–32.

Webster, C. (1975). *The Great Instauration: Science, Medicine and Reform 1626–1660*. London: Duckworth.

Wechtler, J. C. (1659). *Homo oriens et occidens*. Frankfurt-am-Main: Caspar Wechtler.

Weir, J. B. de V. (1952). The assessment of the growth of schoolchildren with special reference to secular changes. *British Journal of Nutrition*, **6**, 19–23.

Weise, M. (1726). *De proceritate corporis*. Halle: Hilliger.

Weissenberg, S. (1895). Die sudrussischer Juden: eine anthropometrisches Studie. *Archiv für Anthropologie*, **23**, 347–424, 531–79.

Weissenberg, S. (1909). Menarche und Menopause bei Jüdinnen und Russinen in Südrussland. *Zentralblatt für Gynäkologie*, **33**, 383–5.

Weissenberg, S. (1911). *Das Wachstum des Menschen nach Alter, Geschlecht und Rasse*. Studien und Forschungen zur Menschen und Volkerkunde, vol. 8, ed. G. Buschan. Stuttgart: Strecker and Schröder.

Welon, Z. and Bielicki, T. (1971). Further investigations of parent–child similarity in stature, as assessed from longitudinal data. *Human Biology*, **43**, 517–25.

Welon, Z. and Bielicki, T. (1979). The timing of adolescent growth spurts of 8 body dimensions in boys and girls of the Wroclaw Growth Study. *Studies in Physical Anthropology*, **5**, 75–9.

West, G. M. (1893). Worcester school children: growth of head, body and face. *Science*, **21**, 2–4.

West, G. M. (1894). Anthropometrische Untersuchungen über die Schulkinder in Worcester, Mass. *Archiv für Anthropologie*, **22**, 13–48.

Westerink, L. G. (1954). *Proclus Diadochus: Commentary on the First Alcibiades of Plato*. Amsterdam: North-Holland.

Wetzel, N. C. (1941). Physical fitness in terms of physique development and basal metabolism. *Journal of the American Medical Association*, **116**, 1187–95.

Wetzel, N. C. (1943). Assessing the physical condition of children. *Journal of Pediatrics*, **22**, 82–110, 208–25, 329–61.

White, J. (1712). *New and Exact Observations on Fevers*. London: Brown.

Whitehead, J. (1847). *On the Causes and Treatment of Abortion and Sterility*. London: Churchill.

White House Conference on Child Health and Protection (1932–3). *Growth and Development of the Child* (4 vols.). New York: Century.

Widdowson, E. M. (1947). *A Study of Individual Children's Diets*. Medical Research Council Special Report Series No. 257. London: HMSO. 196pp.

Wiener, C. (1890). Über das Wachstum des menschlichen Körpers. *Verhandlungen des Naturwissenchaftlichen Vereins in Karlsruhe*, **11**, *Abhandlungen*, 22–42.

Wiener, O. (1927). Christian Wiener zum hundertsten Geburtstag. *Naturwissenschaften*, **15**, 81–4.

Wieringen, J. C. van (1972). *Seculaire Groeiverschuiving*: vol. 1, *Lengte en Gewicht Surveys 1964–1966 in Nederland in historisch Perspectief*; vol. 2, *Samenvatting in het Engels Tabellen en Figuren*. Leiden: Netherlands Institute of Preventive Medicine.

Wieringen, J. C. van (1978). Secular growth changes. In: *Human Growth*, vol. 1, ed. F. Falkner and J. M. Tanner, pp. 445–74. New York: Plenum.

Wieringen, J. C. van, Wafelbakker, F., Verbrugge, H. P. and de Haas, J. H. (1971). *Growth Diagrams, 1965, Netherlands.* Leiden: Netherlands Institute of Preventive Medicine.

Wijn, J. F. de and Haas, J. H. de (1960) Groeidiagrammen van 1–25 jarigen in Nederland *Verhandelingen Nederlandshe Instituut voor Preventieve Geneeskunde*, **49**, 30pp.

Williams, T. H. (1926). *Early Greek Elegy*, p. 132. The elegaic fragments of Callinus... Solon... and others... Edited with an introduction, text... and commentary by T. H. Williams. London: Humphrey Milford.

Williamson, G. S. and Pearse, I. H. (1938). *Biologists in Search of Material: An Interim Report on the Work of the Pioneer Health Centre, Peckham.* London: Faber and Faber.

Willis, T. (1664). *Cerebri anatome cui accessit nervorum descriptio et usus.* London: J. Flesher.

Willis, T. (1965). *The Anatomy of the Brain and Nerves*, tercentenary edition 1664–1964, ed. W. Feindel. Montreal: McGill University Press.

Wilmer, H. A. (1952). Richard Everingham Scammon 1883–1952. *Journal–Lancet*, **83**, 506–9.

Wilson, D. C. and Sutherland, A. (1950). Further observations on the age of menarche. *British Medical Journal*, **ii**, 862–6.

Wilson, E. B. (1935). Heights and weights of 275 public school girls for consecutive ages 7 to 16 years inclusive. *Proceedings of the National Academy of Sciences, USA*, **21**, 633–4.

Wilson, E. B. (1942). Norms of growth. *Science*, **95**, 112–13.

Wilson, R. S. (1974). Growth standards for twins from birth to four years. *Annals of Human Biology*, **1**, 175–88.

Wilson, R. S. (1976). Concordance in physical growth for monozygotic and dizygotic twins. *Annals of Human Biology*, **3**, 1–10.

Wilson, R. S. (1979). Twin growth: initial deficit, recovery and trends in concordance from birth to nine years. *Annals of Human Biology*, **6**, 205–20.

Winckelmann, J. J. (1755). *Gedenken über Nachahmung der griechischen Wercke in der Mahleren und Bildhauer-Kunst.* Friedrichstadt. (Reprinted (1885) in *Deutsche Literatur-Denkmale*, ed. B. Seuffert, vol. 20. English translation (1765): *Reflections on the Painting and Sculpture of the Greeks*, trans. H. Fusseli. London: Millar.)

Winckelmann, J. J. (1764). *Geschichte der Kunst des Altertums* (2 parts). Dresden: Walterischen Hofbuchhandlung.

Wingerd, J. (1970). The relation of growth from birth to two years to sex, parental size, and other factors, using Rao's method of the transformed time scale. *Human Biology*, **42**, 105–31.

Wingerd, J., Peritz, E. and Sproul, A. (1974). Race and stature differences in the skeletal maturation of the hand and wrist. *Annals of Human Biology*, **1**, 201–9.

Wingerd, J. and Schoen, E. J. (1974). Factors influencing length at birth and height at five years. *Pediatrics*, **53**, 734–41.

Wingerd, J., Solomon, I. L. and Schoen, E. J. (1973). Parent-specific height standards for pre-adolescent children of three racial groups, with method for rapid determination. *Pediatrics*, **52**, 555–64.

Winter, J. S. D. (1978). Prepubertal and pubertal endocrinology. In: *Human Growth*, ed. F. Falkner and J. M. Tanner, vol. 2, pp. 183–213. New York: Plenum.

Winterberg, C. (1903). Über die Proportionsgesetze des menschlichen Körpers auf Grund von Dürer's Proportionslehre. *Repertorium für Kunstwissenschaft*, **26**, 1–19, 100–16, 204–18, 296–317, 411–24.

Wittkower, R. (1949). *Architectural Principles in the Age of Humanism.* London: Warburg Institute.

Wolanski, N. (1979). Parent–offspring similarity in body size and proportions. *Studies in Human Ecology*, **3**, 7–26.

Woodbury, R. M. (1921). *Statures and weights of children under 6 years of age* Washington: US Department of Labor.

Würst, F. (1964). Untersuchungen zur Akzeleration auf dem Lande. *Öffentliche Gesundheitsdienst*, **26**, 179–86.

Würst, F., Wassertheurer, H. and Kimeswenger, K. (1961). *Entwicklung und Umwelt des Landkindes.* Vienna: Östereichischer Bundesverlag.

Yamasaki, M. (1909). Über den Beginn der Menstruation bei Japanerinnen mit einem Anhang über die Menarche bei den Chinesinnen, den Riukiu- und Aino-frauen in Japan. *Zentralblatt für Gynäkologie*, **33**, 1296–305.

Yates, F. (1949). *Sampling Methods in Censuses and Surveys.* London: Griffin.

Yates, F. A. (1964). *Giordano Bruno and the Hermetic Tradition* London: Routledge and Kegan Paul.

Yates, F. A. (1969). *Theatre of the World.* London: Routledge and Kegan Paul.

Yates, F. A. (1972). *The Rosicrucian Enlightenment.* London: Routledge and Kegan Paul.

Yates, F. A. (1975). *Shakespeare's Last Plays: A New Approach.* London: Routledge and Kegan Paul.

Young, G. M. and Hoardcock, W. D. (1956). *English Historical Documents, 1833–1874.* London: Eyre and Spottiswoode.

Zacharias, L., Rand, W. M. and Wurtman, R. J. (1976). A prospective study of sexual development and growth in American girls: the statistics of menarche. *Obstetrical and Gynecological Survey*, **31**, 325–37.

Zachmann, M., Prader, A., Kind, H. P., Häfliger, H. and Budliger, H. (1974). Testicular volume throughout adolescence: cross-sectional and longitudinal studies. *Helvetica paediatrica acta*, **29**, 61–72.

Zarnack, A. (1824). *Geschichte des Königslichen Potsdamschen Militärwaisenhaus von seiner Entstehung bis auf die jetzige Zeit.* Berlin and Posen: Mittler.

Zeeman, J. (1861). Rapport van de commissie voor statistiek over de lotelingen van de provincie Groningen van 1836–1861. *Nederlands Tijdschrift voor Geneeskunde*, **5**, 691–723.

Zeising, A. (1854). *Neue Lehre von der Proportionen des menschlichen Körpers.* Leipzig: Weigel.

Zeising, A. (1858). Ueber die Metamorphosen in den Verhältnissen der menschlichen Gestalt von der Geburt bis zur Vollendung des Längenwachsthums. *Academia Caesarea Leopoldino Carolina naturae curiosorum nova acta*, **26**, 783–84.

Zeller, W. (1936). *Der erste Gestaltwandel des Kindes.* Leipzig: Barth, 42pp.

Zeller, W. (1938). Die entwicklungsbiologische Diagnostic. In *Handbuch der Jugendärzlichen Arbeitsmethoden*, ed. W. Zeller. Leipzig: Barth 92pp.

Zeller, W. (1952). *Konstitution und Entwicklung.* Göttingen: Psychologische Rundschau.

Zuckermann, S. (ed.) (1950). A discussion on the measurement of growth and form. *Proceedings of the Royal Society of London, Series B*, **137**, 433–523.

NOTES

Chapter 1

1.1 The definitive Greek text is given in Linforth (1919) and in Campbell (1967, p. 35). Linforth, and also Freeman (1926), give English translations; Masaracchia (1958, pp. 332–7) gives a longer commentary and an Italian translation. I have drawn on all three translations but none is entirely satisfactory. Mansfeld's comments (1971, p. 171 especially) are very helpful.

First hebdomad
In the first hebdomad the word *nêpios* (translated 'as an infant') literally means 'unable to speak', as does the Latin *infans*. In both Greek and Latin the words were extended to cover early childhood. In English the confusion still exists; paediatricians consider infancy to extend from birth to about 1 year (the original meaning) whereas schoolteachers (in the United Kingdom) call 'infants' children aged between 5 and 7 years.

Second hebdomad
In describing the second hebdomad, covering ages 7 to 14, the Greek text reads '*hêbês ekphainei sêmata gignomenês*'. The phrase *sêmata phainein* 'is used in Homer always of portents and wonders' (Williams, 1926, p. 131). The phrase translates literally as 'he shows forth the signs [implied: 'and wonders'] of coming-to-be *hêbê*'. The noun *genesis* and its verb *gignesthai* are frequently used in Aristotle's *De generatione* and Peck, the latest and best translator, criticizes the indiscriminate use made of 'coming-to-be' by earlier translators, preferring 'formation' and 'to be formed' (p. lxi). Thus the phrase *hêbês gignomenês* is perhaps best rendered in a modern context by 'initially-appearing *hêbê*' or 'signs of the initial appearance of *hêbê*'.

Hêbê, however, is a difficult word, and puberty (used in the standard versions) is a poor translation of it. *Hêbê* clearly implied the adolescent or immature external genitalia or something closely connected with them. Henderson (1975, p. 173), discussing another similarly early poem, translates '*ekphaine neon hêbês epêlusin chroa*' as 'she uncovered her young flesh, harbinger of youthful womanhood'. 'This *hêbê*', he continues, 'is the antithesis of Neobules' (an older, rejected woman) . . . Here we are at the beginning of a girl's sexual promise; *hêbês* is double in function connoting as an objective genitive the physical seat of sexuality, the speaker's goal'. *Hêbê* is used of both sexes (Henderson, 1975, p. 115). The word seems certainly to have included designation of the pubic hair, and it would be perfectly possible, if crude, to render the passage cited by the words 'new bush' instead of 'young flesh'. Aristotle (*Historia animalium*, 544b; see also 493b) uses *hêbê* to indicate pubic hair alone when he says *'tois d'allois zôois hêbê men ou geinetai'*: 'although in other animals the *hêbê* does not appear'. Neither 'puberty' nor 'genitalia' will at all do in this context: man is peculiar in just this respect of having specifically pubic hair. However, in the same passage (544b) Aristotle uses *hêbê* to refer to the area which becomes covered by pubic hair (our 'pubes', an imprecise word, but with just about this connotation). When discussing the age for sexual intercourse Aristotle writes (Peck's translation) 'in human beings this stage is marked by a change in the voice, and by a change both in the size and in the appearance of the sexual organs, and similarly with respect to the breasts and above all by the growth of the pubic hair'. This last phrase reads *'malista de tê trichôsei tês hêbês'*, literally 'in the becoming hairy of the *hêbê*'. In *De generatione animalium* also (784a, 8–11, Peck's translation) Aristotle uses *hêbê* to refer to what we call the pubes. 'Eunuchs, too, do not go bald, because of their transition into the female state, and the hair (*trichas*) that comes at a later stage they fail to grow at all, or if they already have it, they lose it except for the pubic hair (*plên tês hêbês*; literally, 'of the pubes'). Similarly, women do not have the later hair, though they do grow the pubic hair (*tas d'epi tê hêbê*; literally, 'around the pubes'). Greek usage seems often to have been as imprecise as our own, but where it is specific then 'pubic hair' seems to be meant. Solon's poem indicates the tradition, carried into Roman law and European custom, that the first appearance of the pubic hair (supposedly at 14) signals the beginning of puberty.

The pubic bones were regularly referred to as 'the bones of the *hêbê*' (*tês hêbês ostôn*) by both Soranus *Gynaeciorum IV*, p. 134 1.9 and p. 141 1.4 and *De signis fracturarum*, p. 157 1.25) and Galen (*De usu partium*, 14, 13, 15, pp. 1, 2 and 8; 16, p. 9). The phrase from Soranus (*Gynaeciorum I*, p. 20) which Temkin translates as 'Menstruation in most cases first appears around the fourteenth year, at the time of puberty and swelling of the breasts' is *'hote kai to hêbân kai to diogkousthai tous mastous'* in the original. A more correct, and literal, translation would be 'at the time of the appearance of the pubic hair and of swelling of the breasts'. The same phraseology appears in Rufus of Ephesus, Soranus' contemporary. In Ricci's (1950) translation of Aetios, quoting Rufus, the phrase goes 'About the fourteenth year, the menses begin to appear . . . and at about the same time these girls arrive at the age of puberty and the breasts enlarge' (p. 18): more correct would be 'at about the same time the pubic hair appears and the breasts enlarge'. Our word puberty has become progressively vaguer as time – and perhaps Bowdler – has passed. In most contexts (and see below) it would now simply include the implication of breast growth in girls.

There was, incidentally, a different word, *ephêbaion*, for pubic hair in later Greek. Rufus of Ephesus uses this word several times in his treatise *De partibus corporis humani* (1726 edn, pp. 31, 51, 69). He actually writes 'which some call *hêbê*, others *ephêbaion*' (p. 31). This is translated into the Latin as *'ad pudenda pars pecten atque pubes dicitur, alii ephebaeon vocant.' Pecten* is used by Juvenal and Pliny with the meaning pubic hair; Stephen Batman calls it 'the nether beard' (see Chapter 2). Celsus, however, seems to use *pecten* to mean pubic bone. In a thoroughly obscure (and not satisfactorily translated) passage he describes what must, anatomically, be the pubic bone as the *pecten* but then adds *'idque super intestina sub pube transversum ventrem firmat'* (Book 8, 23). *Sub pube* could most easily mean, as in modern usage, 'below the pubic symphysis', though just what happens there seems unclear (Celsus, Spencer's translation, vol. 3, p. 488). *Pecten* is the name of a particular sort of comb used in combing wool.

Nowadays, it is the enlargement of the testes which is taken in the male as the designative sign of puberty, if any one single designator is chosen. This is because, unlike the advent of the pubic hair, testicular enlargement occurs in all mammals and thus can be used for comparative physiological purposes. But this event was *not* designated *hêbê* by the Greeks. Both Hippocrates (*Epidemics,* 6.4.21) and Galen (*De usu partium,* 14.7) use the word *tragos* to refer to this (or else, or additionally, to the breaking of the voice, the contexts not being entirely clear). In May's translation of *De use partium* (pp. 636–7) *tragos* is translated simply as 'puberty'. Galen, quoting Hippocrates, writes (14.7) '*tragos, hokoteros an phanê exo orchis, dexios arren, aristeros, thêlu*', which May renders as 'At puberty whichever testis appears on the outside . . . . Galen continues '*hopotan gar prôton exairêtai ta gennêtika moria kai metaballê pôs epi to trakuteron te kai baruteron hê phonê – touto gar to tragân esti*' and Mays translates this as 'That is to say, when the generative parts first swell out and the voice becomes rougher and deeper – for this is what puberty is'. The first meaning of *tragos* is 'he-goat', but the word seems to have been used to imply 'smelling like a he-goat' as in *tragomaskalos,* 'arm-pits smelling like a he-goat'. The same word *tragidzein,* 'to bleat like a goat', was apparently used by the Greeks to refer to what we call the 'breaking' of the voice at puberty *(Aristotle, De generatione animalium,* 788a).

Third hebdomad

In the description of the third hebdomad (age 14–21) the translators are again in difficulty. The beard appears; and then follows a phrase all three translators render as 'while the limbs are [still] growing'. The problem is that this is auxologically incorrect; the trunk continues to grow during late adolescence, but the limbs do not. The Greek reads '*tê tritatê de geneiôn auxomenôn eti* [or *epi*] *guiôn laknoutai, chroiês anthos ameibomenês*'. The Greek word *guiôn* is usually used in the plural (as here) and does indeed mean, literally, 'limbs'. But in Homeric times, as Snell (1953, pp. 6–8) has pointed out, there was no word for the body as a whole. Homeric Greeks saw the body as articulated limbs with large muscles, and early Greek art represents a man not as a trunk with limbs attached, but as hugely muscled legs joined to a triangular torso with no intervening abdomen (the classical Sheldonian mesomorph). 'Homer speaks again and again of fleet legs, of knees in speedy motion, of sinewy arms; it is in these limbs . . . that he locates the secret of life' (Snell, p. 8). The implication is that Solon's *guiôn* should be rendered by our word 'body', with the rider that Solon is probably thinking primarily of muscles, for the phrase is a tag attached by Hesiod (*Theogony,* 1.492) to the young Zeus. Such a rendering makes perfect auxological sense (and represents the first written description of the adolescent growth spurt). It is notable that Solon specifically refers to this as an event taking place *during the course* of the third hebdomad (age 14–21), in contrast to his earlier statement about the initiation of the signs of puberty, which he says takes place 'when the god *brings to an end*' the second hebdomad (age 14). This is indeed the auxologically correct timing of events. The fifteenth-century translator of Philo clearly felt the Greek phrase *geneiôn auxomenôn* was obscure for he simply omitted it in his Latin translation, saying merely '*Annis ter septem prima lanugine malas/Vestiet aetatis robore conspicuus*'. The French translator Pierre Bellier (1598) also ignored the limbs (though hardly, one would imagine, because of his auxological knowledge) and rather charmingly renders the sentence '*Et l'an vingt et unisme il n'a si tost atteint/Que la barbe se meste aux roses de son teint*'

Fourth hebdomad

The phrase in the poem indicating that in the fourth hebdomad strength is taken as the criterion of a man's worth is not surprising in the context of seventh-century Greece. Ehrenberg (1973, p. 20) quotes Jacob Burckhardt as saying this was the age of the 'agonal' man; agonistic is the word used to describe the world of athletic competitors and of fairly conducted competition generally, including that of war.

I am most grateful to Dr R. W. Sharples of the Department of Greek, University College London, for help with the Greek syntax and for guidance to many of the bibliographic sources quoted above.

1.2 There has been much discussion about the date of *Peri hebdomadôn.* Though Sarton (1927–48, vol. 1 (1928), p. 97) thought it pre-Hippocratic and of the fifth century B.C., Mansfeld's (1971) recent and exhaustive analysis of the cosmological and philosophical content of the book and of the doxography of the Greek fragments now discovered suggests a data around 100 B.C. The book is in two parts, which Littré thought quite separate, attributing their juxtaposition solely to their being, perhaps, translated by the same man. He called the second part *Des chairs*; it deals with fevers. Mansfeld, however, thinks the second part, though later than the first, was written as an extension of it. He considers that the first part reflects primarily the views of the Stoics, and the second primarily those of the Pneumatists.

Mansfeld (1971) gives a brief and scholarly history of the hebdomadal ages in Hellenistic times (pp. 161–79). He quotes evidence that Diocles (a contemporary of Aristotle) and Strato (d. 269 B.C.) followed Solon's doctrine, and transmitted it to Nichomachus of Gerasa (fl. A.D. 100) and Macrobius (fl. A.D. 400). Nichomachus wrote *Introduction to Arithmetic* and *Divine Properties of Numbers (Arithmetika Theologumena* in its original version) and it is in the latter that the hebdomadal ages appear. Macrobius, a late neo-Platonist writing in Latin and probably coming from North Africa, gives his version in his commentary on Cicero's *Dream of Scipio* (see Stahl, 1952). Both these writers follow Solon in having ten hebdomads. Age 49 is the perfect age, being 7 times 7; 70 signifies the normal completion of the life-span. Both versions have the distinction that growth in length is said to cease at the end of the third hebdomad and growth in breadth at the end of the fourth (Stahl, p. 114, Nicomachus, p. 66). Hellenistic writers had the same problem understanding Solon in this respect as do modern translators. Auxologically, the Diocles–Strato version has something to be said for it; though the adolescent spurt is a spurt really in all dimensions, growth in breadth, particularly of shoulders and chest, does indeed continue somewhat later than growth in stature (see Tanner, 1962). The Roman scholar Varro (116–27 B.C.) a near-contemporary of Philo, also quoted a paraphrased Solon's poem, as reported in fragments given in Gellius' (A.D. *c.* 130–180) *Noctes atticae* and in Censorinus' (fl. A.D. 250) *De die natali.* In several of these versions there is a further elaboration of the events of the first hebdomad; the first teeth are said to appear during the first seven months, and are then shed during the first seven years.

1.3 Marsilio Ficino translated Proclus into Latin, along with works ascribed to Iamblichus (A.D. *c.* 250–325), a neo-Platonist who much influenced Proclus; Iamblichus founded a school in Syria and was largely responsible for introducing the notion of theurgic union into neo-Platonic thought (see Dodds, 1963). Ficino's book, entitled *Iamblichus de mysteriis Aegyptorum, Chaldaerum, Assyriorum, Proclus in Platonicum Alcibiadem de anima atque demone* . . . was first published by Aldus in Venice in 1497. Many editions followed, including one published in London in 1627. Westerink (1954) gives the definitive Greek version and O'Neill (1965) an English translation.

The section headed *Aetates septem planetis septem congruae* is at p. 231 in the London edition. A translation, with an addition to the fourth age from the Greek original given in round brackets, runs as follows: '*Seven Ages Congruent with Seven Planets.* The order of ages follows the order of the Universe. The first age follows the lunar power, for then we live by virtue of the nutritive and vegetal functions. Second, under Mercury, when we practice letters, music [lyre-playing], wrestling and similar games. Third, Venus, when the seminal parts enlarge, and we are incited to procreation. Fourth, the Sun, when the perfection of age is attained and vigour flourishes (since it is the mid-point between coming-to-be and passing-away, for such is the position attached to man's prime). Fifth, Mars, in which we strive for power and victory. Sixth, Jove, in which we seek wisdom [literally, legal knowledge] and a life of action in citizenship. Seventh, Saturn, when nature consents to desist from procreation, to lay aside the body, and to be carried across to that other, incorporeal life.'

Rea (1925; 1928–9) has argued that this passage is the origin of Jacques' speech in *As You Like It* and also of a similar passage in Sir Walter Raleigh's *History of the World.* Shakespeare omits the planetary references but Raleigh keeps them. Raleigh's version is much closer to Proclus than is Shakespeare's; indeed he only seriously departs from Ficino's Latin when he writes of the last age of all. Raleigh's book was written while he was a disgraced prisoner in the Tower of London under long-standing (and eventually executed) sentence of death. It was written for and dedicated to Prince Henry, but published in 1614, the year after Henry's fatal illness. Proclus, the mystic, wrote of the last age: 'To Saturn, the seventh, in which nature consents to desist from procreation, to lay aside the body and to be carried across to that other, incorporeal life'. Raleigh, reflecting perhaps, on his past glory, cannot but look back: 'The last and seventh to Saturne', he wrote, 'wherein our days are sad and overcast, and in which we find by deare and lamentable experience and by the losses which can never be repaired that of all our vaine passions and affections past, the sorrow only abideth'.

Shakespeare's transmutations are naturally greater; indeed the soldier appears in age four instead of five and the civic man (a justice in Shakespeare) in age five instead of six. Shakespeare's sixth age (the pantaloon), not in Proclus, has echoes of Solon's penultimate stage, when 'he is too feeble in mind and speech for the greatest excellence'. Rea bolsters his case by pointing out that a few pages further on in Ficino's book there is a passage on the harmony of immortal souls very like the 'Sit, Jessica' passage in the *Merchant of Venice.* Though Shakespeare's and Raleigh's immediate sources are, of course, not known, Ficino's works must have been in several English libraries. In particular they were in John Dee's great library at Mortlake. Indeed Dee also had a copy of Proclus (French, 1972, p. 47) and a collection of pseudo-Dionysius (Yates, 1969, p. 12), the Christian apologist who a little after Proclus' death copied his book but with Christian emendations. Dee was well known to Walter Raleigh, whose half-brothers were frequent visitors to Mortlake, and it is very possible that Raleigh obtained Ficino's Proclus from that source. There is a sentence in the Greek Proclus which does not appear in Ficino; after Sol, Proclus adds that this age is the perfection 'since it has attained the mid-point between coming-to-be and passing-away for such is the position of man's prime' (O'Neill's translation). Raleigh does not have this, which confirms its Latin source. Dee died in disfavour, oblivion and poverty at just the time Raleigh was writing, and in his last years his great library began to be dispersed. Being confined in the Tower, Raleigh depended on friends for materials for his book and one of these was the mathematician Thomas Hariot who was a friend of Dee's (see French, p. 172). Another possible source of Proclus might have been the library of Henry Percy, Earl of Northumberland (the 'wizard Earl') at Syon House; though Northumberland also was in the Tower, others of his circle were free to help.

The link with Shakespeare is rather more problematical. Though Shakespeare departed from Proclus' characterization of the ages, he kept the number of ages as seven, and this is significant, for after Isidore (see p. 14) six ages were usually reckoned (*infantia, pueritia, adolescentia, juventus, gravitas, senectus*). However, Isidore added an older-than-seventy period, called *senium,* not reckoned one of the stages and in which everything decays. Isidore was much quoted and his account would certainly have been available in its essentials to Shakespeare, especially in *Batman uppon Bartholome,* Stephen Batman's translation of Bartholomew's *De proprietatis rerum.*

Batman uppon Bartholome was published in London in 1582; *As You Like It* was probably written in the late 1590s. Certainly Shakespeare's first and last ages correspond quite closely in description with those of Isidore–Bartholomew and his sixth age, the slippered pantaloon, agrees reasonably with Bartholomew's *senectus*. Yet, as Rea says, Shakespeare's second and fifth ages echo Ficino–Proclus (i.e. his second and sixth): in particular Ficino–Proclus has as his sixth age a person skilled in legal knowledge, and Shakespeare, as his fifth, a justice. Finally the title, so to speak, the *Seven Ages*, is Ficino's and neither Isidore's nor Bartholomew's (nor indeed, Proclus'). It seems likely, as well as natural, that Shakespeare was conflating a number of accounts, amongst them that of Proclus, perhaps at second hand.

I am again indebted to Dr Sharples for the reference to Rea's articles.

1.4 Aristotle's thought as expressed by a later Peripatetic philosopher, Alexander of Aphrodisias (fl. A.D. 200) was as follows:

> When we describe flesh as fluid and in a state of continual dispersal and assimilation, we say that the flesh is affected in this way with respect to its matter. Again, when we describe flesh as remaining the same, we base this attribution on the form itself, and the flesh *qua* form ... for the being of flesh does not lie in the extent of its magnitude – which does not remain the same on account of the fluidity of the matter – but in the sort of form it has, which remains the same as long as any flesh is preserved ... (*De mixtus*, 235.21–33: from Todd, 1976).

A 'uniform part' (roughly equivalent to a 'tissue'), says Alexander, is a compound of matter and form. When growth occurs the form remains unchanged, but the matter changes, with deteriorating matter being quantitatively less than nutritive matter.

Aristotle distinguished between two grades of nourishment (see Peck's Introduction to the *Historia animalium*). Nourishment was concocted, or purified, first in the stomach then in the heart, making blood, and finally on its arrival in each organ. First-grade nourishment provided the animal with 'being', that is the energy of maintenance, and second-grade nourishment caused increase in bulk. After concoction residues remained and the residues of first-grade nourishment differed from those of second-grade nourishment. In the embryo, first-grade residue gave bones and tendons, second-grade nail, horns and hair. Aristotle relegated hair and nails to this class because they went on growing permanently, whereas bones stopped growing. Thus bones were concerned with form and so must have some share of the *pneuma* (see below); this is not so with hair and nails. Accordingly man, the most spiritful of animals, has the least of such base residue, that is, the least hair and smallest nails amongst animals. Aristotle knew that hair goes on growing for a time after death (*De generation animalium*, 745a). Some of the residues are important in other respects than growth. Semen and menstrual fluid are two such: semen is the residue of the greatest degree of concoction in the body (whence, perhaps, the belief that production of semen takes much energy and is exhausting). Menstrual fluid is the product of a lower degree of concoction because the female has less natural heat than the male. The concoction is carried out by *pneuma*, which charges substance with the requisite 'motions'; *pneuma* is a vehicle of the soul. Semen, being more highly charged with what we might now call 'information', imposes shape on the egg to constitute the embryo (see discussion in Peck's Introduction to *Generation of Animals*, p. lxii).

1.5 Humoural pathology and the theory of temperaments goes back beyond Galen to Aristotle and Hippocrates, and originates at a still earlier date (see Sigerist, 1961, chapt. 4). The notion that a proper harmonic relation between the elements that constitute the body is the condition of health was Ionian in origin (if indeed not earlier and more Eastern). It travelled with Pythagoras (580–489 B.C.) from Samos to Croton in southern Italy. The whole-number relations between the lengths of lyre-strings and the harmonics (as for example the octave) that fell easily on the then-unsophisticated ear deeply impressed the Pythagoreans. A well-tuned instrument exemplified the doctrine of perfect numbers. Alcmaeon of Croton (fl. 500 B.C.) taught that health came from perfect harmony amongst the substances comprising the body, and Philolaus of Tarentum (fl. 450 B.C.), a contemporary of Socrates, that the same laws of harmony applied both to the universe and to the individual. The heavenly spheres were separated by musical intervals, and a similar relation existed between the elemental substances of which the body was composed. The Theban Simmias, a pupil of Philolaus, gives his master's views in *Phaedo*; the body he says, is comparable to a lyre, 'strung and held together by heat, cold, moisture, dryness and the like, and the soul is a mixture and a harmony of these same elements, when they are well and properly mixed . . . when the body is too much relaxed or is too tightly strung by diseases or other ills, the soul must of necessity perish (Fowler's translation, p. 299). These views remained unaltered down to the end of the Renaissance and, like the Seven Ages, find their echo in Shakespeare: 'Such harmony is in immortal souls, but whilst this muddy vesture of decay doth grossly close it in, we cannot hear it'.

Peck gives an excellent account of Aristotle's view in his Introduction to the *Historia animalium*. There were four fundamental substances in Aristotle's physical theory: fluid substance (*to hygron*), solid substance (*to xêron*), hot substance (*to thermon*) and cold substance (*to psychron*). Peck remarks that 'fluid' and 'solid' are more in conformity with Aristotle's meanings than the often used 'moist' and 'dry'. The fluid and solid substances are 'passive', the hot and cold substances 'active' or creative. The two passive substances thus serve as 'matter' and the two active substances as instruments of 'form' and 'motion'. Each of the four 'elements' – fire, air, earth and water – consists of a pair of the four fundamental substances. Fire combines the hot and solid (dry), air combines the hot and fluid (wet), earth combines the cold and solid (dry) and water combines the cold and fluid (wet).

The Hippocratic writers expounded the theory that humans, body and soul alike, are compounded of fire and water; ultimately, therefore, of all four of the fundamental substances, hot, cold, solid and fluid. Different invididuals (and different species of animals) had different blends (*krâsis*) of fire and water, and upon the blend health and personal characteristics depended. *Krâsis*, says Peck, is a special case of *symmetria*, 'a condition when every part, ingredient or other factor concerned is at the right strength, or of the right size, or present in the right amount in relation to all the others. In modern English "symmetry" has come to be applied to a much narrower field' (p. lxxxvii). But Blake used it still in the Greek sense.

In the Latin form this is the theory of temperaments (*temperare* means to blend). A sword is tempered by being taken from hot dry fire and plunged into cold wet water, and a well-tempered sword in Ancient and mediaeval times was often the instrument of survival. Each organ has its characteristic temper: the heart, the seat of *pneuma*, was the hottest and driest, and the brain, the source of the residue *phlegma*, the coldest and most fluid. The brain in children was colder and more fluid than in adults; it took a long time to reduce this amount of coldness and fluidity, and this is why the bones around the anterior fontanelle (the bregma) are the last to be completed (*De generation animalium*, 744a). Aristotle also thought that the right side of the body had more hot substance than the left, though his empirical observation led him to disagree with those who said that embryos growing in the right side of the uterus became male, those in the left side female.

Chapter 2

2.1 Not only this. In popular but still professional medicine Proclus' planets had become established, together with the signs of the zodiac into the bargain. This was part of the development of the theme of the microcosm of man representing ('modelling', in contemporary jargon) the macrocosm of the universe. This idea is no longer a self-obvious one. It appears in Plato's *Timaeus* though its origins are certainly older. In its simplest form it says, to quote Sarton (1957, p. 227): 'The microcosm of our body is a much reduced copy of the macrocosm of the universe: our bones are like the rocky skeleton of the earth; there is in us a lake of blood comparable to the seas; the rhythm of our breathing and of the pulse is like that of the tides, the "circulation" of blood [i.e. to-and-fro movement] is like that of water and so on.'

A man's physical constitution arose from the predominance of one of the humours, and the humours were differentially affected by the planets (Saturn corresponding to cold and dry, Mars to hot and dry, etc.), especially by their conjunctions in particular signs of the zodiac. Hence phlebotomy would have by no means the same effect when carried out at one time of the month as it would at another (see also note 2.2). The Chaucerian doctor (see Curry, 1925) had not only to know a good deal of astronomy; he had to be able to calculate what conjuction was occurring at the time when, for example, a relapsing fever started. Thus he carried with him, in the form of seven-by-two-inch strips of parchment, mathematical tables giving him over a number of months or years the dates of eclipses, the phases of the moon and the positions of the relevant astral bodies. 'When the system reached its highest perfection during the late fourteenth and early fifteenth centuries, exact calculations by precise instruments, similar to those of astronomers and navigators, were employed, and many of these physicians' manuscripts contained a volvella with adjustable parts to enable him to work out these calculations with extraordinary accuracy' (Talbot, 1967, p. 128). It is against this background that we have to view the emerging knowledge of the physiology of growth.

2.2 Sherrington's (abridged) translation of a passage of Arnald of Villanova cannot be bettered (1946, p. 25): 'There, you see, near the bottom of the pipkin is the black bile, the melancholy, dark and earthy. Do not, however, suppose it wholly evil. It is part of our fourfold cardinal nature. It is to nourish the melancholy organs, which are solid and cool, the bones and even the spleen. It thickens and fortifies the blood. Excess of it is a disease. Yet there are those whom we regard rightly as healthy, who have, all their life, some excess of the black bile. They make a type which the physicians must recognise: disease in them takes character from that modicum of excess. They are sallow and dark and spare. They are solitary and taciturn, suspicious and sad. The "melancholic" sums them up. On them certain of our medical herbs act with especial power. The melancholy humour remains through all its fermentations an enemy to joy, an ally of age and death. Those who have excess of it are supersensitive to certain phases of the moon, to certain conjuctions of the planets, to Saturn and to certain houses of the Zodiac. The rich have it even as the poor. The planet which acts most on the four qualities of man is the moon, in its first quarter warm and moist; in its second, dry and warm; in its third, cold and dry; and in its fourth, moist and cold. With each moon, therefore, for the first 7 days the blood dominates; then for 7 days the yellow bile, the choler; then for 7 days the black bile, the melancholy; and then the phlegm. From the moon's phases the crisis of an illness can be foretold. The moon's phase can enhance or interfere with a remedy. So it comes that measures can be taken which otherwise would avail nothing. One phase is good for blood-letting, another not.'

2.3 In the following century the great Boerhaave still believed this. At the time of puberty, he wrote, 'a greater acrimony of the humours' occurs as indicated by a yellower and more acrid urine and foetid sweating in the armpits. Nose-bleeds are nature's salutary way of easing the plethora and acrimony of adolescence (van Swieten, 1765, vol. 12, p. 28).

2.4 Arnald of Villanova (c. 1240–1311) was a Catalan, born probably at Valencia. He studied at Montpelier, became physician to a succession of Kings of Aragon and returned to Montpelier around 1290 as professor of medicine in the newly established *studium generale* (i.e. university). He translated a number of Arabic works into Latin, including some of Avicenna and Galen, and himself wrote extensively. His views represent a conjunction of the then-known parts of Hippocrates and Galen with those of Arabic writers and the Salernitans. He occupied himself equally with theology as with medicine, but with less success, escaping imprisonment as a heretic only by virtue of his medical distinction. The last collected edition of his works was published in 1585, but a modern Italian translation of *De conservanda juventute* was issued by Mancini and Fravega in 1963. I have not located an exact source for Clever's sentence – which sounds like a quotation, with its list of fruits – but passages with a similar implication are found in the *De conservanda juventute* (pp. 29–34 in the Italian translation) and in the *De regimine sanitatis* (1585 edn, pp. 667–9). Arnald lays great emphasis on the conservation of innate heat and moisture. Anything which diminishes this brings on old age. Frequent intercourse, for example, is very injurious because it diminishes natural heat; and this is why horses have a shorter life than mules. As to the characteristics of the ages, adolescence is hot and moist, *juventus* hot and dry, and *senectus* increasingly cold and dry.

2.5 The title of this vast tome is somewhat enigmatic. The word *Greul* (*Grewel* is an older spelling) means roughly a horror (presumably in origin something which turns the hair grey – *grauen* – or, as we say, the face ashen). Guarinoni, in each of his seven sections, gives the normal or proper course of events such as eating or sleeping, and then lists a series of *Greuls*. In all he lists some 100 separate *Greuls*. They range from the *Greul* of anger to the *Greul* of an ill-ventilated room and include the *Greul* of taking exercise after eating, the tyrannical *Greul* of many schoolteachers against children, and the dangerous *liebgreul* of irrational youth. At one time Guarinoni seems to be describing a catalogue derived from the Seven Deadly Sins, with *Greul* meaning vice, but at other times it simply means 'fault' or 'outrage'.

This, however, is not all. The title is *Die Grewel der Verwüstung*, and this phrase has a particular and clearly specific connotation, for it was used by Luther in his translation of the Bible to refer to what is rendered in the English Authorized Version as the Abomination of Desolation. The reference is to Daniel's dream as described in Matthew xxiv, 15 and Mark xiii, 14. Christ is foretelling the destruction of the temple, the end of the world and his second coming: 'Many shall come in my name, saying, I am Christ, and shall deceive many . . . there shall be famines, and pestilences and earthquakes, in divers places . . . And many false prophets shall rise, and shall deceive many . . . When ye therefore shall see the abomination of desolation, spoken of by Daniel the prophet, stand in the holy place (who so readeth let him understand) Then let them which be in Judaea flee into the mountains . . . And then shall appear the sign of the Son of man in heaven'. This passage is one of the most politically seminal in the Gospels, for it stoked the fires of populist rebellions throughout mediaeval and Renaissance times and even, arguably, into the nineteenth century (see Cohn, 1957). Christ, or the Evangelist, is referring to Daniel's dream, a passage in the Old Testament which exerted a profound hold on popular imagination for more than a thousand years. Daniel (Daniel vii) dreamed of four great beasts that came up from the sea, each unlike the other. The fourth was the most terrible; he 'made war with the saints, and prevailed against them; Until the Ancient of Days came and judgment was given to the saints of the most High, and the time came that the saints possessed the kingdom'. It was this beast that was called the Abomination of Desolation by Matthew and Mark; and the Anti-Christ by mediaeval and subsequent writers.

That Guarinoni means his title indeed to refer to this beast and hence read *The*

Abomination of Desolation of Humankind is made absolutely plain by the rather amateurish woodcut which occupies the lowest third of the title page. In the central panel is the traditional seven-headed Beast, devouring with one of its heads a woman who is in a bath with a drink in her hand and a piece of food, perhaps fruit, on a tray laid over the bath. Above, in a linked subsidiary panel, is a luxuriously semiclothed woman reclining on a soft bed. Top left is a lady with a sword and scourge in her left hand and a book in her right; bottom left another with a cannon and a clock, bottom right a woman, pierced by an arrow, lying dead over a wine cask; and top right is a very recognizable portrait of Guarinoni flanked by books, pointing downward at the beast. Over the picture is written '*Matthaei 24 Wann jr den Grewel der Verwüstung sehen werdet*' (when ye therefore shall see the Abomination of Desolation).

The Anti-Christ was a key figure in all the millenarian fantasies and revolts of the Middle Ages and later. If the Anti-Christ could be identified, then the longed-for Kingdom of Heaven was at hand, for the rule of the Anti-Christ immediately preceded it. But different people made vastly different identifications. Luther and many others at various times regarded the Pope as Anti-Christ. Guarinoni was clearly not of this persuasion. He seems to be identifying the Beast with a personification of luxury, an identification not far removed from that of his contemporary religious reformers of both Protestant and Catholic faiths. Guarinoni's adolescence was spent with his father in Prague, and Prague in Rudolf's time was by no means unfriendly towards reformers, even heretical ones. In 1585 it housed John Dee and Michael Kelly, as well as Francesco Pucci (1543–1597), a strong millenarian or chiliast (and also an Italian-speaker like Guarinoni) who though reconciled to the Church at one time, ended as a victim of the Inquisition (Evans, 1973, pp. 84–115). Evans, endeavouring to read *Die Grewel der Verwüstung* with eyes focussed on the history of ideas rather than the history of hygiene, describes the book as: 'full of fantasies, spirits and miracles; its central message is a chiliastic appeal for the moral regeneration of mankind' (p. 204n.). A study of the relations of the Guarinonis, father and son, to the theological currents of thought in Prague in the 1580s and 1590s would be of considerable interest.

Chapter 3

3.1 The solids were apparently by a different, and unacknowledged author, Francischi. The Golden Section had a great vogue, and Kepler, in *Mysterium cosmographicum de admirable proportione orbicum* (1596), described the geometrical construction which gives rise to the irrational number of the Golden Section $\frac{1}{2}\sqrt{5}+1)$ as one of the two treasures of geometry, the other being the theorem of Pythagoras. The sequence of numbers represented by successive application of the Golden Section is known as the Fibonacci series, $u_{n+2} = u_{n+1} + u_n$, after Leonardo of Pisa (*c.* 1175–1250) also called Fibonacci (son of Bonaccio). It was Fibonacci who introduced the Hindu system of arithmetic to Europe, in 1202 in a book entitled *Liber abaci.* The Hindu system used a zero, which the Roman did not, and it was so superior in practical reckoning that by the end of the century the abacus and the counting-board had disappeared from Europe. Astonishingly, Fibonacci is also credited with being the first European to recognise that a negative number appearing in a problem concerning money represented a loss instead of a gain (Bell, 1945). Most mathematicians regarded negatives as absurd and their use in arithmetic was not widely understood till several centuries later.

3.2 The senior member of the Hohenzollern family at this time (of which the Brandenburg house were members) was the Elector Frederick William (1620–1688), known as the Great Elector on account of his military activities. His son Frederick I was the first King of Prussia and his grandson was the Frederick-William I who founded the

Militär-Waisenhaus and built up the celebrated regiment of giant Grenadier Guards (see p. 98). Military measuring began some time in the eighteenth century, and one wonders if Elsholtz's anthropometer served as a prototype of the sophisticated stadiometers such as that illustrated in Goethe's drawing of 1779 (see p. 100).

3.3 J. J. Winckelmann (1717–1768) was an outstanding figure of the eighteenth century. Born at Stendal, he studied natural history and physiology with Hamberger at Halle (see p. 87) and Hamberger's influence remained with him always (Schulz, 1962). After graduation he went to the court of August the Strong, Elector of Saxony, as curator of pictures at Dresden and Nöthnitz. Through the good offices of August's grand-daughter, Maria Amalia, who became Queen of Naples, he went to Italy in 1758, 1762, 1764 and 1767 and was associated with the study of the material excavated at Pompeii, Herculaneum and Paestum. In 1755 his first and best-regarded work, *Gedenken über Nachahmung der griechischen Wercke in der Mahleren und Bildhauer-Kunst*, was published in Friedrichstadt. In 1764 his *Geschichte der Kunst der Altertums* (*History of Ancient Art*) appeared in Dresden. He was by then acknowledged throughout Europe as the leading authority on Greek art. The next year he was appointed a custodian at the Vatican Museum. Three years later he was dead, murdered in Trieste by a young travelling companion named Archangeli. The world of German art and literature was profoundly shocked, especially as Winckelmann was about to return in triumph to Germany after a long absence abroad. It seems likely that there was a homosexual background to the murder, but only one contemporary alluded to this and even then in a very guarded fashion (Hatfield, 1943, p. 101).

Winckelmann exerted a profound effect on classical German literature : Goethe, for example, called his books 'epoch-making'. He discovered for Europeans the idealized classical Greek; the physically beautiful, athletic, cultured, free and joyful youth, epitomized in Greek sculpture and Greek painting. The path to the knowledge of perfect beauty, Winckelmann thought, lay in imitation of the spirit of Greek art more than in imitation of nature. In particular this was true of representations of the human body, and in such representations lay the closest approach to beauty that the artist could attain. Thus the proportions of the classical Greek statues, though not to be followed slavishly, were of fundamental importance (see Hatfield, 1943). Winckelmann's influence was widespread. He was a friend of the Pietist Francke, founder of the Francke'sche Stiftung (see Chapter 5), although he eventually became converted to Catholicism. He had a special liking for the Swiss, and was a close friend of Caspar Füssli (1707–1782) whose son, John Henry Fuseli, as he was known in England, became professor at the Royal Academy of Art in London and a highly influential figure in the development of the English art schools' teachings about human proportion and the varieties of human figures. Winckelmann forms, for us, a link, still possible in the eighteenth century, between the science of Hamberger and the art doctrines of Schadow.

3.4 Schadow says the book is by Claude and not Gérard Audran. No author's name is given on the title page nor anywhere else, and the book is always listed under the name of Gérard Audran the engraver. However it seems likely that Schadow, if anyone, knew who the author was. He even explains further that there were five artists called Claude Audran active at about that time but this one was the man who worked with the famous sculptor Lebrun. A second edition was published in Paris in 1801 which was said to be *par Gérard Audran, graveur*. There was a London edition published by Bowles about 1785 (no date is given).

Chapter 4

4.1 Sanctorius' *Medical Statics* was published in many editions and several languages. It consists of several hundred aphorisms, each deduced from what are clearly most careful measurements of the weight of Sanctorius' own body at different times, together with the weights of his food and his excreta. There is no detail given of the methods used, but he does give a picture, now rather well known, illustrating the method of weighing. He sat in a chair suspended from the short end of a steel yard, thus being weighed exactly as a butcher weighs a carcass. No more accurate way was invented until the recent introduction of transducers.

Sanctorius had his followers. White in 1712 advocated relating the amount of blood taken at phlebotomy to the pre-illness weight of the patient, which should be known. Keill (1708, 1738), emulated Sanctorius by weighing himself every day between the ages of 30 and 40, and weighing his night urine each night for months. Keill was one of the first to realize that the calculations he made led to the conclusion that 'in less than a year the whole body must be changed . . . The vulgar opinion gives seven years [again!] to complete a thorough Change of the Body; but for any reason alleged', he writes, 'it might have been 70. Now it appears that the change is almost yearly' (1738 edn. p. 15).

4.2 Martin Weise was one of a family of doctors of this name in Berlin. The older Martin Weise, probably the father, was born in 1605, graduated in 1629 and from then for sixty-two years, almost till his death in 1693, was court physician to successive Electors of Brandenburg, from George-William to Frederick I of Prussia. He was thus a colleague, older in years but perhaps junior in status, to Elsholtz, who arrived to be personal physician to the Great Elector Frederick-William in 1656 but died, in 1688, earlier than the older Weise. Martin Weise the younger presented his thesis under the auspices of Friedrich Hoffman the younger (1660–1742). Hoffman's father, also Friedrich (1626–1675), graduated in Jena and worked in Magdeburg and Halle. He would have known the older Weise; whence the younger Hoffman in graduating the younger Weise refers to the *'Weisianae gentis, vere medicae'* amongst whom the new doctor will flourish as a master amongst masters.

4.3 Spigelius, or Adrian van der Spieghl, born in Brussels, was a student of Fabricius Aquapendente in Padua and from 1616 himself professor of anatomy and surgery there. It seems doubtful if he really was describing catch-up growth. In the passage cited by Weise (Spigelius, 1632, p. 20) he writes 'Illnesses alter the magnitude of the body as may be seen in those who a quartan fever has raised to gigantic size. But I do not believe . . . there is anyone who so far lived in our times who clearly showed smallness and crippling and escaped from disease to gigantism (*ex morbo in maximum gigantem evasit*). Those people who through prodigious growth came early to death the Greeks called *ektrapeloi* [monsters] . . . But the man who goes slowly, with a tardy pace with puberty late-coming (*homo incessu tardus, sensu hebes puber postea factus*), in him suddenly his shortness disappears, as Pliny says (*subit membrorum contractione absumptus est*)'. Like Weise, Spigelius seems to confound the adolescent spurt and catch-up growth and adds the further confusion of gigantism, from, presumably, pituitary tumours or Sotos' disease.

Spigelius, incidently, also remarks (p. 21) that though Pliny said children were often smaller than their parents 'that is the opposite of our experience, which shows the contrary is frequently true, that offspring often depart from their grandfathers and great-grandfathers in length and sons are often taller than their parents'.

4.4 Puberty continued to be thought of as curative of certain diseases – though not necessarily by virtue of its effect on stature. Polinière (1790–1856), a physician of Lyons, writing in the *Dictionnaire des sciences médicales* (1820) says that when puberty takes place naturally and with every event correctly followed, 'it often by natural means dissipates children's diseases which have remained unmoved by therapeutic effort' (p. 53). He cites in particular epilepsy, incontinence and scrofula.

Brière de Boismont (1841 : see p. 296) attributed a specially important part to the beginning of menstruation in addition to the general effect of puberty. 'The appearance of the menses', he wrote (p. 182), 'has the best possible influence on the diseases of this time. It is often the signal for their disappearance. A mass of young persons, delicate, complaining, suffering, a prey to more or less chronic disorders, sometimes grave, seem to be reborn to life after the appearance of the periods ... One sees glandular swellings disperse ... ears cease to discharge' and lots more.

Marina (1896) quotes with approval Regnier (1860) who says that the diseases which tend to increase height are eruptive and typhoid fevers, meningitis, pertussis and 'eclampsia'. The same diseases which increase stature in children and youths have a tendency to cause fatness in adults, in each case by excess activity of the nutritive faculty which, in the case of children, succeeds the slowing of the period of the illness *(succeduta al rallentamento del periodo della malattia)*.

Quite the opposite view was taken by Caspar Joachim Quabeck of Tremona, whose thesis *De insolito corporis augmento frequenti morborum futuorum signo* (Unusual growth of the body a frequent sign of future disease) was presented at Halle under Büchner in 1752. (There is a copy in the U.S. National Medical Library) Quabeck argued that unusually rapid growth indicated a lack of ability of the body's vascular fibres to resist the pounding of the fluids within them, and such lack of resistance during childhood predicated a similar lack later, whence a formidable list of diseases to which such unfortunates were susceptible. The thesis is very much a scholastic exercise, without any actual data on cases; yet one cannot resist the notion that Quabeck may have seen a patient with Marfan's syndrome, to which his explanation would, a bit modernized, apply.

4.5 Hermann Conring (1606–1681) was an astonishing polymath. Born in Friesland, he studied philosophy in Helmstaedt, then, simultaneously, medicine and theology at Leiden from 1625 to 1632. He returned to Helmstaedt as professor of philosophy and later of medicine. At this time he was called in consultation all over Europe and Queen Christina of Sweden invited him, unsuccessfully, to become her personal physician. Changing chairs once again, this time to law, he became an arbitrator on disputes between the royal houses of Europe and was held in honour by one and all. His *De habitus corporum germanicorum ac novi causis* was thrice published in Helmstaedt (1646, 1652 and 1666) and once later in Frankfurt (1727). In it, quoting incessantly from Tacitus, Pliny, Strabo and Caesar, Conring asserted that the Germans of Roman times were 6½ to 7 feet tall with fair skin, blue eyes and red-blond straight hair. In phraseology almost identical to that of Guarinoni (whom he does not quote) Conring wrote:

> It happens nowadays that puberty (*pubertas*) occurs in many earlier than formerly, and by no means a few girls invite men, by the show of their swelling breasts, before the time at which they should be inseminated ... to what other cause can we ascribe this but to the manner of living which brings on heat and dryness more precociously than in earlier times? Just so a gardener brings on precocious fruit by much manuring and liming, or by warming the air artificially. Thus the hot way of life brings on puberty earlier and impels to Venus and indeed even to coition almost before a girl is marriageable. It is easily observable that amongst those luxuriously brought up (*in mollitur educatis*) the boys are clothed (*vesticipes*, presumably meaning pubic hair) much before poor boys or those living in the country; in the same way the girls in their tender years are already itching for pleasure (p. 120).

We see here an early example, if not the acutal birth, of the Tacitan myth that was to continue to Virey (p. 120) in 1816 and beyond. Echoes of it are still with us today, particularly in the German literature on growth, just as echoes of the tall blond Aryan reverberate down the centuries. It was Tacitus (*c.* 55–120) in the *Germania* who described the German tribes as tall, red-blond and blue-eyed, and it was he, together with Julius Caesar, who originated (at least for later Europeans) the notion that late maturing leads to health and strength. Tacitus devoted only two sentences directly to this topic. The first is '*Sera juvenum venus, eoque inexhausta pubertas*', translated in the Loeb edition as 'Late comes love to the young men, and their first manhood is not enfeebled' (1970, pp. 162–3). This is followed by '*Nec virgines festinantur; eadem juventa, similis proceritas: pares validaeque miscentur, ac robora parentum liberi referunt*', translated as 'Nor for the girls is there any hot-house forcing; they pass their youth in the same way as the boys: their stature is as tall; they are equals in age and strength when they are mated, and the children reproduce the vigour of their parents'.

Not only is the translation, made in 1914, unconscionably florid; it is also of dubious accuracy. To suggest that young men and women are the same height is extraordinary; to attribute to them the same strength is ridiculous. The more modern rendering of H. Mattingley and S. A. Handford in the Penguin edition (1948) reads: 'The young men are slow to mate, and thus they reach manhood with vigour unimpaired. The girls, too, are not hurried into marriage. As old and full-grown as the men, they match their mates in age and strength, and the children inherit the robustness of their parents' (p. 118). The two words at issue are *proceritas* and *validae*. The former usually means tallness and it is indeed stretching a point to translate it as 'full-grown'. Tacitus may genuinely have believed the young men and women of that time were of similar height; he certainly induced Conring and Stöller and others to believe so. In fact it would have been impossible, for no human group is known where this, or anything approaching it, obtains. *Validus*, on the other hand, that both translations give as 'strong' (in the muscular sense), can certainly mean 'vigorous' or 'able' (in our sense of 'able-bodied'). This is indeed more likely in the context, for the phrase in which it occurs immediately precedes that describing the handing on of robustness to the children; and at that time and for centuries later it was believed that both the sex and the health and strength of the child were determined by the respective vigour and health of the parents at the moment of conception. (The *Dictionnaire des sciences médicales* of 1812, under the title 'Enfance', says 'Differences in age, in sexual ardour, in manner of life, and in the season in which conception occurs contribute to the inborn differences between children'.) A translation of Tacitus that gives this sense might therefore read: 'The sexual activity of the young men begins late, and so puberty is passed without their powers being exhausted ('being plundered', as used of a city, gives the sense). The girls, too, are unhurried. Alike in the flower of their youth, alike in their full development and in the vigour of their desires, the young men and women unite to produce children whose strength matches their own.'

The origins of this view certainly go back far beyond the times of Tacitus and Julius Caesar: Aristotle wrote that 'the physique of men is also supposed to be stunted in its growth when intercourse is begun before the seed has finished its growth (*Politics*, 8, 383) and Soranus 'men who remain chaste are stronger and bigger than the others' (*Gynecology*, p. 27: see p. 10). But the context of the whole description of the Germans in Tacitus has to be appreciated. Written in A.D. 98, Tacitus' book was strongly moralistic in tone and contrasted the pure, strength-giving life of those free and rural people with the enervating luxury of Rome, where children did *not* reproduce the robustness of their parents, in his view. In Rome sexual promiscuity was considered smart but amongst the Germans there were no corrupting spectacles nor exciting banquets, and chastity was well-nigh universal. Nor is Tacitus only concerned with sexual behaviour. 'In every house', he writes, 'the children go about naked and dirty and grow up with that strength of limb and body that excites our admiration' (Loeb edn, section 20). This also was accepted as part of the picture of the strong and upright barbarian by Guarinoni, Stöller and others; clothes were a part of the vigour-sapping luxury, and should be reduced to the minimum.

Tacitus had no actual first-hand knowledge of the Germans. Apart presumably from the stories of acquaintances, he relied on three sources: the lost histories of the Greek Posidonius, some lost works of Livy and of Pliny the Elder, and the still-available book of Julius Caesar on the Gallic War. Caesar wrote in 52 B.C., just 150 years before Tacitus. The relevant passage, in Handford's translation, reads as follows: 'The Germans spend all their lives in hunting and war-like pursuits, and inure themselves from childhood to toil and hardship. Those who preserve their chastity longest are most highly commended by their friends; for they think that continence makes young men taller, stronger, and more muscular. To have had intercourse with a woman before the age of twenty is considered perfectly scandalous. They attempt no concealment, however, of the facts of sex: men and women bathe together in the rivers, and they wear nothing but hides or short garments of hairy skin, which leave most of the body bare' (Penguin edn, 1951).

4.6. George Frederick Rall (d. 1670) was born in Dam, in Pomerania, and practised in Stettin. His book was written as part of the controversy over Harvey's views on reproduction. The relevant passage reads:

> *Nullum est dubium quin juvencularum etiam uterus non raro ob familiarem, saepiusque institutam cum maribus conversationem, accedentibus in primis blandis lascibisque de rebus venereis confabulationibus, mutuis amplexibus et osculis (cujus rei causam erudite exponit ingeniossisimus Dr Thomas Willis in sua Cerebri et Nervorum Anatomicae XXII p. 176) aliisque amoris, praefertium lascivientis, illecebris, ad maturitatem disponatur citius etiam, quam alias fieri consuetum est, dum nempe his et similibus de causis Spiritus, alias sopiti, commoventur adque partes genitalis copiosus ruentes, earum temperiem sensim immutant, et ad maturitatem debitam easdem perducunt. Idemque maribus etiam adolescentulis simili modo evenire, nec a ratione nec ab experientia alienum est. Ob hanc causam, praestitutum alias a natura ordinarium maturitatis terminum non raro ab utroque sexa anticipari, non est quod dubitemus assevere.*

Thomas Willis (1621–1675) was professor at Oxford, and the founder of modern neurology. The book Rall quotes is most easily consulted in the beautiful facsimile English edition published as a tercentennial commemoration by McGill University Press in 1965. In fact Willis goes by no means so far as Rall, for he talks only of an immediate connection between kissing and sexual arousal, and says absolutely nothing about children or accelerating development. In the English version (pp. 142–3: p. 247 in the Latin version of 1664 published by Flesher) he writes simply 'for as much as many shoots and fibres of the same maxillar Nerve, derived from the fifth pair, interweave themselves with the flesh and skin of the Lips, hence the reason is plain, why these parts are so very sensible, and besides, why the mutual kisses of Lovers impressed on the Lips, so easily irritate love and lust by affecting both the *Precordia* and Genitals; to wit, because the lower branching of the same fifth pair activates the parts constituted in the middle and lower belly, and draws them into the affection with the Lips. The same holds of Love presently admitted by the eyes.'

Evidently Rall was using Willis' by then well-known name to bolster his idea, as Stöller used the name of Spigelius.

4.7 Maygrier (1819) wrote nearly 150 years after Rall, and strongly supported the view that contact with the society of men hastened menarche. So did reading romantic fiction, looking at plays, dancing and *l'habitude des plaisirs*, not to mention masturbation. He blamed mothers for showing off their daughters at society functions and balls and thus bringing on menstruation 'before the time fixed by nature' (p. 384).

Raciborski (1868), writing when the idea that puberty and menstruation were produced by the development of the ovary was just being introduced, said that encounters between boys and girls were 'an excellent occasion for developing the *sens genital*, of making the microscopic ovisacs pass into the Graafian follicles visible to the naked eye (p. 213). Music, he wrote, had an incontestable effect (and then rather spoiled his case by quoting as evidence the celebrated pair of juvenile elephants at the Jardin des Plantes who were the subject of a bizarre experiment on the fourth Prairial, An VI. Music, solo and orchestral, was played to them from a hidden pit. Though both were sexually immature, each made what the observers took to be (unreciprocated) sexual overtures to each other (under the influence, not, regrettably, of a harmonious symphony played to them, but in the case of the female a solo bassoon playing 'Oh ma tendre musette' and in the case of the male a solo clarinet playing 'Nina')).

Raciborski (p. 218) quotes Jean-Jacques Rousseau as a strong protagonist of the maturing effect of music. Rousseau was sceptical about the nutritional explanation for earlier maturing in towns. 'There are areas', he said, 'where the peasant is well nourished and eats tremendously, such as in the Valais or certain mountain cantons of Italy, like the Tyrol. Despite this the age of puberty in both sexes is later than in the towns where, to satisfy vanity, some people eat with extreme parsimony and where the majority 'carry the velvet dress and the hollow belly'. One is astonished to see big lads, as strong as men (*fort comme les hommes*), with a high-pitched voice and beardless, and big girls, already well-rounded (*très formées*) but without any of the periodic signs of their sex. This difference [from town youths] seems to me to come uniquely from the fact that because of the simplicity of their customs their imagination is for longer peaceful and calm, and so makes their blood ferment later and their temperament to be less precocious.' Raciborski adds that he wishes just to substitute 'ripen the ovules' in place of 'ferment the blood'. A modern (or Ancient) auxologist may add that Rousseau was right to be astonished by his strong beardless peasant boys; evidently he did not recognize the connection between muscle growth and the phenomena of puberty, which he thought of as specifically sexual.

4.8 The *Natural History of Birds*, in nine volumes, appeared in 1770–83 and the *Natural History of Minerals*, in five volumes, in 1783–8. There were two posthumous volumes on *Snakes* (1788–9), five on *Fishes* (1798–1803) and one on *Cetaceans* (1804), all based on Buffon's manuscripts. Some 200 later editions of Buffon have been issued; that edited by Cuvier is considered authoritative (1826–8). The first English translation, by Kendrick, was published in six volumes in 1775 and is now a rare work. Very soon after (in 1781) a translation by the noted naturalist and antiquary W. Smellie (1740–95) appeared, but it too is rare, and not even in the British Library. A second edition of Smellie's translation appeared in 1785, but the popularity of the work was really established by *Barr's Buffon* (Barr being a publisher), which appeared in ten volumes in editions of 1792, 1797 and 1810.

4.9 The Parisian *pied*, or foot, divided into 12 *pouces*, or inches, each divided into 12 *lignes*, was longer than the English foot. Isaac Newton, who used the Paris measure on occasion, found 1 *pied* equal to 12.785 inches, but the later official conversion, on the introduction of the metre, gave it as 12.7789 inches. The *pouce*, then, equals 2.71 cm whereas the English inch equals 2.54 cm (Berriman, 1953). The metric system was introduced in France in 1795 but until 1837 the old system and the new were both permitted. After that date only the metric system was legal. The Würtemberg foot, according to Theopold *et al.* (1972), was 28.6 cm, with a *Zoll* or inch equal to 2.38 cm and a *Stritch* equal to approximately 0.6 cm.

4.10 Jean-Joseph Sue (1710–1792), called Sue of the Charité, was an anatomist and surgeon who injected demonstration specimens of the vascular system in cadavers and wrote a book called *L'anthropotomie ou l'art d'injecter* (Paris, 1749).

4.11 The Friedrichshospital was originally a hospital, orphanage and workhouse combined (somewhat like Christ's Hospital, set up in Elizabethan times in London). In 1695 the Elector Frederick III of Brandenburg, who in 1701 became the first King of

Prussia (as Frederick I, or Frederick the Ostentatious) promulgated an edict forbidding begging. A few years later, in the period 1697–1702, the Friedrichshospital was founded in Berlin to look after the poor, sick and orphaned there.

Frederick I died in 1713, but his son Frederick-William I continued to support and enlarge the Friedrichshospital. In 1726–7 the Charité Hospital and the Insane Asylum in Berlin were also enlarged and the sick from the Friedrichshospital were transferred to them. Thus the Friedrichshospital took on more exactly the character of an orphanage or Waisenhaus. In the history of the Militärwaisenhaus in Potsdam (see note 5.1) it is said that in 1741, just after the accession of Frederick the Great, Frederick-William's son, there were in the Friedrichshospital (in addition to civilians) '130 soldiers' daughters (*Soldatenmädchen*) and 87 children of the Second Battalion of the Kings Regiment in receipt of monthly finanical support' (Potsdamsches Waisenhaus, 1824). It is not clear whether these children were boarders in the institution or lived outside under its aegis. The latter is more likely, for in 1784, according to a contemporary account, the Friedrichshospital only actually housed 65 boys and 35 girls together with 19 old men and 13 old women. In addition there were 112 boys and 97 girls looked after as out-boarders, for whom the Institution paid all costs (Nicolai, 1786). Presumably many of these out-boarders were young children, since Nicolai says that only children 8 years old and over were admitted. 'Smaller children who had still to be given the breast, or who were not yet 8 years old . . . were given to knowledgeable persons outside the Institution (*Hause*) in consideration of a monthly stipend' (Nicolai, p. 628).

These then were the children that Jampert measured in 1752 or 1753. Presumably the youngest were out-boarders and the older ones were living in the house. Most such institutions stopped being responsible for children when they reached 15 years old, which leaves the source of Jampert's subjects of 15 years and over very much in doubt.

In 1834 the Friedrichshospital housed 300–400 orphans and supported about 650 out-boarders (Ersch and Graber, 1834). It appears to have been demolished in 1859 (Goldschmidt, 1910, p. 197).

The influence of events in Halle upon the founding of the Berlin Friedrichshospital and also the Potsdam Militärwaisenhaus (see note 5.1) is perhaps worth noting. In 1693 the Elector Frederick III transformed the Academy for Young Noblemen in Halle into a University, with the eminent Pietist pastor August-Hermann Francke as professor of theology and Greek, and the equally famous Friedrich Hoffman as professor of medicine (see p. 71). Halle was a small town just north of Leipzig, owned by Brandenburg-Prussia, but entirely surrounded by Saxon territory. But the University grew and Francke set up the Francke'sche Stiftung, a huge school and orphanage of some 2,000 pupils (Francke, 1787; Halle-Waisenhaus, 1799) upon which many of the subsequent foundations were modelled. In medicine, and especially in the study of growth, Halle in the eighteenth century came to rival or surpass Jena and Göttingen.

4.12 It may be that Mauriceau fell victim to the prevalent uncertainty about measuring units. He wrote that 'a baby born at nine months . . . weighs ordinarily 10 or 12 pounds of 16 ounces each pound'. But his successor, Levret, commenting on this in his textbook of 1761 says that undoubtedly Mauriceau meant to reckon 12 ounces not 16 to each pound. If the weighing was done in ounces, this would mean that instead of '10 or 12 pounds' Mauriceau should have written '$7\frac{1}{2}$ or 9 pounds'.

Not only did Weise in 1726 and Stöller in 1729 know the average weight at birth was about 6 pounds (see p. 77); Haller himself in 1752 said the same thing. But so great, presumably, was Mauriceau's prestige that Jampert, in his thesis done in Halle in 1754 quoted 12 pounds as the newborn weight. Roederer was evidently inveighing against the ignorance of the textbook writers in circumstances in which they should perfectly well have known better: which explains his use of the phrase, applied to Mauriceau, that he was 'hallucinating'.

Chapter 5

5.1 Popular histories have long insisted that Frederick-William specially recruited tall girls as marriage-partners for the Grenadiers, so as to ensure a more reliable supply of giants for the future. Even Robert Ergang, the professional historian and author of the best English-language biography of Frederick-William, writes: 'Having an eye for the future, the King was not satisfied with merely recruiting men of large stature; he also attempted to perpetuate them in Prussia by marrying his giants to women of large stature. Little, however, appears to have come from his experiments' (Ergang, 1941, p. 100). Ergang gives no source for this statement, however. His primary sources are chiefly Fassman and Morgenstern, both members of Frederick-William's Tobacco-Cabinet, Beneckendorf (1787–9) and the life edited by Förster (1834–5). Neither Morgenstern (1793) nor Fassman (1735) has anything to say about girls, marriages or orphanages. Fassman (pp. 719–36) considers the Grenadiers the most beautiful troops in all the world, and says that the tallest man was Jonas, a wood-cutter born in Norway (Udjus' blacksmith, see p. 98; unfortunately he gives no measurement). Ergang's references relate to tales of isolated incidents in which Frederick-William endeavoured to marry off tall girls to particular Grenadiers rather than to any act of policy. At least one of these stories is so amusing and Grimm-like as to be probably apocryphal. (A tall peasant girl was seen by the King and given a note to carry to the Grenadier barracks which read 'marry this woman at once to Grenadier so-and-so': either suspicious or unaccountably literate, the girl gave the note to an old hag she met in the street, who duly married the dashing giant.) All accounts agree that Frederick-William often gave cottages to his Grenadiers when they married, and that he liked to act as godfather to their children. It is quite possible that on his frequent visits to his great Foundation, the Militärwaisenhaus (see below), this or that tall girl amongst the pupils would catch his eye and he would bear her in mind as a possible future wife for one of his Grenadiers. But there is no evidence of any more systematic selection programme.

The Potsdam Militärwaisenhaus, despite its name (meaning Military Orphanage), was primarily a great school for the children of Frederick-William's soldiers, both officers and men. As Frederick-William built up his large and professional army the numbers of soldiers became so great, especially in Berlin and Potsdam, that the civil institutions such as the Friedrichshospital (see note 4.11) were insufficient to cope with them. In 1722, therefore, Frederick-William founded this new institution, clearly modelled on the great Francke'sche Stiftung Waisenhaus in Halle (see note 4.11) which also functioned even more as a school than an orphanage. Frederick-William knew well the younger Francke, being attracted to his Pietist doctrines, which offered, he thought, more hope of salvation than the Calvinist doctrine of the chosen in which he had been brought up (Ergang, 1941, *passim*). The boys' building of the Militärwaisenhaus was opened in 1724, and a separate building for girls, the Mädchenwaisenhaus, in 1725. There was also a separate building for officers' daughters and a hospital (and a mulberry-tree plantation, presumably for silkworms).

All soldiers' children had a right to free education at the Militärwaisenhaus whether or not their fathers were still living or had died in army service. Only children over 6 years of age were admitted, but the Foundation supported financially the upbringing of children of the Guards regiments and certain others right from birth, by means of monthly subventions to the parents. Only legitimate and healthy children were able to enter the Waisenhaus, others being boarded out (at the Foundation's expense) with foster-parents or farmers or in civilian orphanages. When Frederick the Great inherited the school from his father in 1740 there were 1,946 children (about 1,600 being boys) with three priests and eighteen teachers (Potsdamsches Waisenhaus, 1824; Schmidt, 1825; Mertins, 1968). In 1786, according to Nicolai (1786, p. 129), there were 704 boys and 666 girls living in, with a further 640 boys boarding out with teachers and others. In addition the Foundation supported nearly 3,000 younger children, of whom a little over 2,000 were those belonging to the specially designated regiments.

The Militärwaisenhaus was one of the longest-lived, as well as one of the biggest, of all such institutions. It functioned right up to the end of World War II, when it was dissolved as an institution and the buildings put to other use. Measurements of height, weight and chest circumference were started in 1869 and probably were made continuously thereafter. Only very brief accounts of them have been published however (Militärwaisenhaus, 1874; Potsdamsches Grosses Waisenhaus, 1924, 1930) and the original documents have now, I suppose, been lost.

Duffy (1974, p. 60), incidentally, states that members of the First Battalion of the Guards were permitted to set themselves up in private quarters in the Potsdam Militärwaisenhaus 'with the girl of their choice', on obtaining the permission of their company commander. Any children of these, often temporary, liaisons were alleged to be consigned to the orphanage. Duffy gives no authority for this statement, and I can find none. Under Frederick-William soldiers could not get a marriage certificate without the permission of their commanding officer, and when Frederick the Great came to the throne he found the system was being abused in the sense that officers were charging the soldiers for the certificates. He therefore strictly forbade any such charges to be made, adding that they discouraged marriage at a time when the population needed replenishing (Büsch, 1962, pp. 39, 53). It is unlikely that the Militärwaisenhaus had anything to do with this; children under 6 were not admitted and illegitimate children were excluded totally.

I am most grateful to Herr Ewald Mertins of Berlin and Frau Liesel Mewes of Hanover for much helpful and detailed correspondence concerning the Militärwaisenhaus and other matters discussed above.

5.2 In Finland, at that time a province of Sweden, records of recruits survive from 1768 (Kajava, 1927), and in the canton of St Marie Vesubie in Liguria from 1792 (Kiil, 1939). In Prussia the cantonal administration begun by Frederick-William in the 1730s was fully developed for recruiting purposes by Frederick the Great, and by the end of the Seven Years' War, in 1762, 'every duty-bound male was entered on a roll at birth . . . and regimental chiefs kept themselves up to date on the details of the age, height and appearance of all such young disposable men . . . Every year before the spring review a regimental officer and a civilian official jointly selected such recruits as were needed' (Dugg, 1974, p. 54). No analysis of these records has been made, so far as I know. In France, records go back to about 1800 (see Hargenvilliers, 1817, reprinted 1937; Villermé, 1829; Quetelet, 1830). In Holland records began in about 1820 in Groningen and Nijmegen, about 1830 in Rotterdam and in 1851 in the country as a whole (Zeeman, 1861; Oppers, 1966; van Wieringen, 1978). Württemberg records are analysable from 1829 (Riedle, 1834) and Baden records from 1840 (see Ammon, 1894). In Denmark measurements began on conscripts in 1774 but have been analysed only from 1815 (MacKeprang, 1907–11). Swedish statistics begin in 1840 (Hultkräntz, 1927), Belgian ones in 1842 (Quetelet, 1869*a*, vol. 2, p. 64). Livi (1896, 1905) gives Italian figures for the years 1879–83, and Scheiber (1881) figures for Hungarian recruits in 1865–8.

5.3 Pierre Tarin, in the article on 'Accroissement' in the first edition of Diderot's *Encyclopédie*, published in 1751, and entirely derived from Buffon's *Natural History*, often verbatim, says that an Englishman has shown that height in the morning is greater than in the evening, by 6 or 7 *lignes* (i.e. about $\frac{1}{2}$–$\frac{3}{4}$ inch). This finding, he says, annoyed the Abbé de Fontenu, apparently because he also had made a similar observation, but been beaten to the publication by his fellow-clergyman. The Abbé reported the finding in a memoir sent to the Académie Royale des Sciences in 1725 (Fontenu, 1727, an abstract only). He claimed to have discovered that height increased about 6 *lignes* after meals also, and gradually then went down again. This was due to the absorbed nutrient juices stretching the blood vessels before they were sufficiently attenuated to escape by transpiration in the manner demonstrated by Sanctorius. Buffon must have known of Wasse's finding, but perhaps did not quote him because the style of the *Supplément* article is quite different from that of the *Natural History* itself and includes no quotations of the works of others.

5.4 Carl Eugen was born and brought up in the Brussels palace of his mother, Princess of Thurn and Taxis, in the cultured atmosphere of the French Enlightenment. He was 8 years old when he first went to Stuttgart, his capital. His father, Duke Karl Alexander, a general in the Austrian army and field-marshal under Prince Eugene of Savoy died in 1737 when Carl Eugen was 9. In 1741 Carl Eugen (aged 13) and his brother were sent to complete their education at the Prussian court in Berlin, where Frederick the Great had just come to the throne. The Grenadier Guards – the battalion of giants that Frederick's father, Frederick-William, had created – must still have been fresh in people's minds and the guardsmen themselves must still have been wandering the streets of Potsdam. Perhaps still more important an influence on the young Carl Eugen was the huge Potsdam Military Orphanage. (see note 5.1). The rules and customs of the Carlschule were very similar to those in the Militärwaisenhaus as described by Zarnack (1824), and almost certainly derived from them. Carl Eugen's custom of measuring the Carlschule boys was all his own invention, however; perhaps his later obsession with height was born during his three years as major-general in Frederick's army. When he was 16 Carl Eugen was declared of age and returned home to rule Württemberg.

5.5 The first full statement of the regulations for admission and recruitment to the Society appeared in 1798. The persons qualifying for the Charity were:

1. Boys sent by various Magistrates of the Cities of London and Westminster from their Courts, Offices and Prisons.
2. Orphans lurking about the streets; many of them recommended by Magistrates, some by Citizens of London, and others occasionally by Gentlemen who have been witnesses to their complicated Distresses.
3. Apprentices, their Masters consenting to the Indentures being cancelled before Magistrates, under Complaint of Petty Offences, Disinclination, or total Inability to follow their Trades, and requesting the Society to fit them out for Sea Service.
4. Vagabonds, the major part of them overwhelmed with Filth and Rags, in danger of perishing through Cold, Hunger, Nakedness or Disease, several being cured of the Itch, Scald Head, etc.
5. Sons of poor Widows, and other worthy poor labouring People with numerous Families, being Boys of a hardy bold disposition, wanting employment, and applying to be fitted out, their parents consenting, *these being the greatest Number* [my italics].
6. Recommended by Magistrates from the Country as dangerous in civil Society (only in wartime).
7. From the Country, when Town Boys did not offer in sufficient numbers (only in wartime).

In 1809, these categories were restated, and became :

1. Such boys as are literally in a vagrant state, of whom some are recommended by magistrates, either as found wandering, or as guilty of some petty offence. (fewest of these offer)
2. Those who live chiefly by begging, or seldom do any work, but appear in filth and rags, and sometimes half naked. (more than of the former)
3. Some who have occasionally earned their bread by going on errands, or in markets, brick kilns, glass-houses, feeding hackney coach horses, draw-boys and such like. (often naked and unemployed and apply to this office) (if exception is taken to this, note that the boys leave vacancies for younger or less hardy boys of the same class)
4. The sons of poor people who have numerous families, and, upon enquiry, are in too great a state of indigence to provide clothing for the sea. Such boys, what ever their inclinations may be for a sea life, are not likely to be accepted by any master, but by means of the society. (*majority of these*) [my italics]
5. Boys whose parts have been wrong cast, being so contrary to their genius, that they are more inclined to hazard their necks, than to live a sedentary life. (*No inconsiderable number of these*) [my italics]

It was also noted that boys from London were accepted first.

I am indebted to Professor Floud for copies of these regulations.

5.6 J. J. Virey (1775–1846) was a leading French pharmacist in the period of the wars of the Revolution and the First Empire. He qualified additionally as a doctor in 1814, and was secretary of the pharmacy section of the Academy of Medicine from 1823 onwards. He was also president of the Société de Pharmacie in 1830 (see Berman, 1965). His wide-ranging interests centred on natural history and the impact of social customs and conditions on disease, and he opposed the mechanicians of his day. His chief general work, amongst very many writings, was entitled *Hygiène philosphique appliquée à la politique et à la morale* (Paris, 1828; 2nd edn, 1831). In the introduction he wrote 'Customs and habits, created by laws and giving rise to institutions, whether monarchical, republican or mixed, whether theocratic or feudal, ought to be studied as to their influence on health'.

Chapter 6

6.1 Quetelet's person and his data – in all probability his data on growth – were crucial in the founding of the Statistical Society in 1834. The events have been well documented. On 27 February 1833 the Reverend Richard Jones, just elected to the chair of political economy at King's College, London, expressed the hope, in his inaugural address, that 'a Statistical Society will be added to the number of those which are advancing the scientific knowledge of England' (Royal Statistical Society, 1934, p. 8). In June the British Association for the Advancement of Science held its third annual meeting in Cambridge and Quetelet was sent as delegate by the newly inaugurated Belgian government. Quetelet read papers concerning falling stars and magnetism at the top and bottom of mountains; but also, to quote Charles Babbage, 'he had brought over with him some highly interesting statistical documents which unfortunately could find a reception in none [of the British Association's current Sections]. Under these circumstances a gentleman [Richard Jones] who fully understood their value invited a few of his private friends most interested in the subject [they were Babbage, himself and Malthus] to meet M. Quetelet . . . in College for the purpose of talking over this valuable budget' (Royal Statistical Society, pp. 7–8). Jones, Babbage and Malthus then formed themselves, with others, into a Statistical

Section of the British Association, quite irregularly, and the following morning asked the president, Adam Sedgwick, to ratify what they had done and to add, officially, a sixth Section to the then-existing five. Sedgwick was a little hesitant (British Association for the Advancement of Science, 1834, p. xxvii), pointing out that the Section could not be legitimate till it had been recognized by the General Committee and ratified by the governing body. He saw no obstacle to recognition provided – and only provided – that there was 'some limitation perhaps of the specific objects of inquiry . . . to facts, relating to communities of men, which are capable of being expressed by numbers and which promise when sufficiently multiplied to indicate general laws'. Sedgwick, and evidently a body of Association members, were worried that 'these inquiries were most intimately connected with moral phenomena and economical speculations – they touched the mainsprings of passion and feeling – they blended themselves with the generalizations of political science'. If this should happen the meetings would be 'distracted by the worst human passions . . . [and] the daemon of Discord would find its way into their Eden of Philosophy'.

It seems most unlikely these remarks were aimed at Quetelet, whose sentiments exactly echoed Sedgwick's. Perhaps Sedgwick's use of the words 'political economy' implies a certain distrust of Richard Jones (who succeeded Malthus at the East Indian College at Haileybury in 1835; he never read a paper to the Statistical Society and died in 1855). At all events, the General Committee duly approved the formation of the Section, but laid down the rule that it should restrict its inquiries to classes of facts expressed by numbers which promise . . . to indicate general laws' (British Association, 1834, p. 483). Babbage was the first chairman and Drinkwater the secretary.

The British Association only met once a year, however, and as Babbage relates (Babbage, 1851, p. 17; Royal Statistical Society, 1934, p. 9): 'At the concluding meeting of the Statistical Section at Cambridge, it was resolved that a more permanent body was necessary to carry out the views and wishes of the Section and it was agreed to establish a Statistical Society in London'. Babbage himself carried out the necessary arrangements, which culminated in a public inaugural meeting held in March 1834. The committee remarked that 'though the want of such a Society has been long felt and acknowledged, the successful establishment of it, after every previous attempt had failed, has been due altogether to the impulse given by the last Meeting of the Association. The distinguished foreigner (M. Quetelet) who contributed so materially to the formation of the Statistical Section, was attracted to England principally with a view to attending that meeting' (British Association, 1834, p. 484). Though F. J. Mouat, in his account of the Society's history in the jubilee volume of 1885 (Statistical Society, 1885, pp. 14–71), may be read as implying that some division of opinion may have taken place in the Statistical Section leading to the setting up of the Society, there seems little reason to suppose this was so (Mouat is unreliable elsewhere in that he gives the wrong man – Whewell instead of Sedgwick – as president at the 1833 meeting). Babbage was the chairman of one and the organizer and member of committee of the other. Furthermore, in the 'Prospects of the Objects and Plan of the Statistical Society of London', Sedgwick's proviso was carefully and precisely followed: 'An essential rule of its conduct [is] to exclude carefully all *opinions* from its transactions and publications, to confine its attention rigorously to facts and as far as it may be found possible, to facts which can be stated numerically and arranged in tables' (British Association, 1834, p. 492). The first president of the Society was the Marquis of Lansdowne, with the historian Henry Hallam as treasurer. The committee included Babbage, Drinkwater, Richard Jones, Malthus, Porter, Nassau Senior, Thomas Tooke, Vardon and Whewell. The meeting of 15 March ended with the passage of a motion proposed by Babbage and Jones that 'In the consideration of the distinguished character of Monsieur August [sic] Quetelet of Brussels and the part taken by him in the formation of the Statistical Section of the British Association meeting at Cambridge in June 1833, to which the Society owes its establishment, he be now chosen the first Foreign Member of the Society' (Royal Statistical Society, 1934, p. 13).

6.2 Quetelet himself left a record of his meeting with Goethe, published in Biedermann's *Goethes Gespräche* (see Pfeiffer-Belli, 1950, pp. 623–34).

25 August 1829
I was honoured by a most kindly welcome from this illustrious old man ... after enquiring about the reason for my journey he desired to see the apparatus which I used to observe magnetic intensity; then he offered me his garden, near the park in Weimar, in which to make my experiments ... I accepted, with gratitude, as much, I admit, because of my natural feelings of curiosity and veneration as from any scientific end ... My observations, as one might imagine, were not made with all the necessary calm; I had to go back to the garden and then take the measures a third time in an isolated part of the park ... When Goethe knew that I also was interested in optical experiments, he showed me with great pleasure what he had done in this fascinating part of physics: he even had the kindness to give me several lenses for experiments on polarization, and a work in which he had set forth his ideas on various phenomena which depended on it, and on the theory of colours.

25–31 August 1829
Goethe had the custom of receiving graciously the numerous foreigners who stopped at Weimar to see him ... which sometimes gave rise to serious inconvenience. However, most of the distinguished men then at Weimar had the desire to listen to him; but they knew his repugnance to talk, as it were publicly, although in the midst of his own salon. Several of them asked me to help them ... my youth and my desire to oblige them perhaps made me forget a little what I owed to our illustrious Maecenas. I talked to him about various of his journeys and works ... and when I came to talk of Venice, his youth and his first compositions, I realized that I had touched on a subject that interested him ... Goethe, whose noble appearance inspired the deepest respect, and whose language was both brilliant and penetrating, talked eloquently of the first memories of his youth ...

Apart from these large soirées Goethe held evening meetings to which just a few special persons were admitted. In these little committees the illustrious poet wanted to discuss further with me the science of optics and the research which he himself had undertaken ... At the end of an evening he would happily say: tomorrow we shall consider such-and-such a scientific point. I came to Weimar to spend a day there and I found myself already there more than a week before I could bring myself to say to him that I had to leave to attend a scientific congress in Heidelberg, which was about to open ... When he saw I really had to part he asked me to go for a moment with him into another room. 'Well' he said, smiling, when we were alone, 'I will tell you the whole truth; if I tried to hide it you are perspicacious enough to see it. As a poet, my reputation is built and I can regard it with assurance; but as a physicist, it is not the same, and opinions on my research vary greatly'. Then, after a moment of silence: 'So you are going to Heidelberg to take part in this big scientific bazaar: everyone will go to spread out his merchandise, to praise it to the skies and perhaps to run down that of his neighbour. Well I also am one of the neighbours, and I admit that I would be very curious to know what they think of my merchandise and whether they like it a bit. You must promise to tell me the truth'.

We re-entered the salon. 'Before you go', he said, 'I will give you proof of a talent that they certainly don't think I possess' . . . He put some objects for me, also for my wife, into an envelope, which I have kept since with the greatest care as one of the most precious presents I ever had. He included a copy of his work *Zür Naturwissenschaft überhaupt, besonders zür Morphologie* and on the title page wrote these words of friendship. 'To Director Quetelet as a gentle remembrance of 28th Aug 1829' [Goethe's eightieth birthday] . . . Between the leaves of the book he placed six coloured glasses, seven centimetres long and four wide, for experiments on light and to demonstrate his ideas on polarisation . . . also four lines of verse in French for my wife. They were written in his own hand, with firm characters looking as if etched with a chisel rather than traced by the pen of an octogenarian.

'Chaque jour est un bien que du ciel je reçois
Profitons aujourd'hui de celui qu'il nous donne :
Il n'appartient pas plus aux jeunes gens qu'à moi,
Et celui de demain n'appartient à personne'.

When, towards the end of the same evening, it was time to leave and to make my final farewells, words failed me; the good old man perceived it and embraced me with the tenderness of a father.

6.3 François Chaussier (1746–1828) was the leading physiologist of his time in France, as well as the leading medico-legal expert *(Biographie médicale* (1821), **3**, 232; Pariset, 1836). Born in Dijon of poor parents, he studied briefly in Paris and then returned to Dijon to practice. With a few others he built up a school of instruction in anatomy and physiology at the Dijon Academy which was said to be second only to that in Paris. Then, aged 50, he was called to Paris to be professor, not of legal medicine, as he probably wished, but of anatomy and physiology in the faculty of medicine. When the Ecole Polytechnique was founded in 1804 he was appointed lecturer in chemistry and medicine there, and in the same year he joined the staff of the Maternité Hospital, looking after pre- and post-natal care, and the care of the newborn. (The conduct of labour was in the hands first of Baudelocque, then Dubois.)

One of Chaussier's first appointments in Dijon had been to the two prisons in the town; and presumably from this grew his interest in accusations of infanticide, on which he wrote a thesis. Probably he devised ways of measuring the fetus primarily to see if it was at the stage of viability. His *mécomètre*, according to Murat (1816), consisted of a square wooden rod, marked in decimetres, centimetres and millimetres. There were two ends made of copper, one fixed and the other movable. The fetus or child was measured by placing its head against the fixed end and moving the other end until it came in contact with the feet, much as in the modern neonatometer. Murat says that Chaussier invented a way of ageing a fetus by its body proportions. He showed that at term the half-way point between head and heels was the umbilicus; at 8 months 'several *lignes* higher'; at 7 months nearer the bottom of the sternum than the umbilicus; and at 6 months at the bottom of the sternum. Murat also says Chaussier showed that the fetus grew fastest from the middle of the fourth month to the sixth month and that growth in the last month was very slow (see Murat, 1816, p. 54). He gives no details as to how this was determined (presumably on abortuses), and whether he is talking of length or weight or some combination of the two is unclear. Chaussier was dismissed from his chair in 1822 as the result of political intrigue and never really recovered from the blow. But he was still in Paris, and attending functions at the Maternité, when Quetelet arrived in 1823, and must have been well known to Villermé. He died in 1828.

6.4 Bowditch thought there were only ten children of each sex at each age (1877, pp. 14–15). However this is almost certainly a misunderstanding. Bowditch quotes the text of the later work *Anthropométrie* (1870) and not Quetelet's 1835 book. In the studies made for *Anthropométrie,* which included many more measurements than two, Quetelet did indeed only use ten subjects of each sex at each year; and specially selected subjects at that, as Bowditch points out. But there is no evidence that he did a similar thing in the height survey. Pagliani (1879, p. 362) makes the same error, or copies it from Bowditch, and later Carlier (1892) repeats it, as does Camerer (1893), in trying to account for the peculiarities of Quetelet's curves. Camerer's son continued the error in his chapter in Pfaundler and Schlossman's textbook of paediatrics (Camerer, 1908: see Chapter 11, p. 280) and Baldwin (1921, p. 220) repeated it. Variot and Chaumet (1906, p. 56) and also Godin (1913) thought that Quetelet made longitudinal studies, but there is no evidence that he did any apart from those on his own and his friends' children.

6.5 Polinière (1790–1856) in the article on puberty in the *Dictionnaire des sciences médicales,* published (1820) ten years before Quetelet's study, gave currency to the notion that girls were generally always shorter than boys. Young men, he wrote, gained 4, 5, 6 or even 7 inches in one year without their health being affected. Young women also 'grew more or less rapidly', but 'kept in general a stature inferior to that of men' (p. 57).

6.6 Quetelet made no study of age of menarche himself, but he wished to include a short section on it in the second edition of *Physique sociale* issued in 1869. Accordingly, he wrote to Professor Adolphe Hannover in Copenhagen, who had already published on the subject (Hannover, 1865), requesting a brief review. Hannover's contribution arrived too late to be put into *Physique sociale* but Quetelet presented it to the Belgian Royal Academy of Science, Letters and Fine Arts at a meeting in 1869 (Quetelet, 1869*b*). Hannover's title was 'Les rapports de la menstruation en Danemark et l'époque, en général, de la première menstruation chez les differents peuples'. He gives the average menarcheal age of 2,129 women in Denmark, from recollection data, as 17.03 years in rural areas and 16.76 in Copenhagen. A table containing the statistics available up to that time is presented, with bibliographic sources; but Hannover is very sceptical about the value of most of the series, because most come from small and select samples. Being himself medical officer to the four administrators of the territory of Greenland, he feels that he has sufficiently certain information to conclude that the Eskimo there have an early (mostly 13+ or 14+) menarche 'almost as early as the Hindus'. He thus rejects the prevalent notion that climate is of major importance, and thinks racial differences are what chiefly matter. He corresponded with the well-known Dr Finlay of Havana, Cuba, who told him that little difference in menarcheal age could be discerned between Blacks and Whites in Havana. Another Cuban doctor, in the interior, however, alleged that Blacks matured earlier unless living in very poor circumstances. Hannover cites (at second hand from Raciborski) a Dr Dropsy who made enquiries in 'the little town of Zaclaw' (probably Zaslaw in East Poland, now part of the USSR) and found average ages of menarche of 14.32 years in 100 Jewish women, 15.86 in peasants of Slavic origin, and 13.96 in 70 nobly-born ladies 'living in very easy circumstances'.

Chapter 7

7.1 Earlier censuses were made by the Romans, by the Tuscans of the fifteenth century and also in 1624 in Virginia and from 1666 for some years in parts of French Canada. Particularly comprehensive censuses of the modern type were carried out from 1742 onwards in several German states.

7.2 De Moivre presumably worked amongst those Huguenot weavers who made up the majority of the Spitalfields Mathematical Society. This flourished from 1717 to 1846. According to Professor John Cassels (*The Times,* 18 November 1978, p. 3) the Spitalfields Society originally had sixty-four members, later reduced to forty-nine.

The members were craftsmen who assembled after work for quiet intellectual recreation and mutual instruction in mathematics. Anyone who spoke during the evening's hour of silence was fined, and anyone introducing points of divinity or politics in his mathematical lecture was fined quite seriously. In 1846 the Society was absorbed into the Royal Astronomical Society and its last president became a founder of the London Mathematical Society.

7.3 I have not been able to trace the original of this quotation, or to identify Gibbins. The source that Pinchbeck and Hewitt give (Knowles, *The Industrial and Commercial Revolutions in Great Britain during the Nineteenth Century*, 3rd edn) is incorrect.

7.4 Chadwick's entry into the world of social and administrative reform owed a good deal to Nassau Senior. It was Senior who proposed him as a member of the powerful Political Economy Club (in 1834). In 1832 (the year of Bentham's death) Chadwick was one of the sixteen Assistant Poor Law Commissioners; Senior was the Principal Commissioner. Senior was in contact with Quetelet at just this time: Quetelet's biographer Levy (1970) quotes a letter to Quetelet which enclosed Senior's 1832 *Instructions to Assistant Poor Law Commissioners* (p. 307). Perhaps these instructions served as a model for Chadwick's instructions to Assistant Factory Commissioners. T. C. Cowell, incidentally, was a friend of Senior's and another of the Assistant Poor Law Commissioners in 1832 (Levy, 1970, pp. 83, 84). He must have switched over to the Factory Commission in the following year.

7.5 Quetelet says the shoe was $\frac{1}{2}$ to $\frac{1}{3}$ inches thick in boys and $\frac{1}{6}$ to $\frac{1}{8}$ inches thick in girls. He makes no mention of heels. His translator, Knox, wonders whether Manchester children did not wear clogs, even to Sunday school. Miss T. M. Swann, Keeper of the Shoe Collection at Northampton Central Museum, has kindly assured me that Quetelet was perfectly correct in his allowances. The normal wear for women and girls at that period was a flat shoe without a heel; the boys usually had a stacked heel, but only $\frac{1}{4}$ to $\frac{1}{2}$ inch high. Miss Swann thinks that clogs were worn as overshoes at that time in the Manchester region, but Miss E. Vigeon (1977) of the City of Salford Art Gallery and Museum is of a contrary opinion: she feels sure that if clogs had been worn, they would have been kept on in the school. However she adds that clogs followed the fashion in shoes and it is probable that in the 1830s even clogs would have been thin-soled and with only a minimum heel. I have used Quetelet's corrections of 0.5 cm for girls and 1.0 cm for boys in reducing Cowell's values for plotting.

7.6 The present-day growth curves which most resemble those of the factory children come from children with the clinical diagnosis of Silver–Russell syndrome. These children suffer some unidentified pathology of embryo or placenta and are born with a low birthweight. They remain always small, but are not retarded in growth (see Tanner, Lejarraga and Cameron, 1975).

Chapter 8

8.1 John Henry Bridges (1832–1906) combined a medical-administrative career with high-level philosophic activity, something perhaps a little less unusual in his day than now. His entry in the *Dictionary of National Biography* (Second Supplement, vol. 1, p. 222) starts simply 'Positivist philosopher . . .'. He was born in Suffolk and took a degree in Classics at Oxford. He then studied medicine at St George's Hospital London, where he was a fellow-student with Charles Roberts, and in Paris. Bridges and Roberts qualified in the same year, 1859, Bridges with an Oxford degree in medicine, and Roberts with a Membership of the Royal College of Surgeons. Bridges married at once and emigrated to Melbourne, but his wife died soon after their arrival and he returned to England and settled in Bradford, Yorkshire where he was appointed physician to the Bradford Infirmary in 1861, and Factory Inspector for

North Riding in 1869. It seems very likely that it was Roberts' presence in Yorkshire which influenced Bridges to go to Bradford, and subsequently they worked together in the public health sphere in the North Riding, Bridges as factory inspector and Roberts as prison surgeon. In 1870 Bridges removed to London to take up the post of medical inspector to the Local Government Board of the Metropolitan Region, and it was in this capacity that, with Holmes, he directed the 1872–3 survey. Roberts followed Bridges to London in 1871 and Bridges appointed him as one of the survey doctors. Bridges retired from his Local Government Board appointment in 1898.

At Oxford Bridges came strongly under the influence of Richard Congreve and throughout his life he devoted himself to promulgating in England the views of Auguste Comte. He translated Comte's *Politique positive I* (1865 and 1875), contributed large numbers of articles to the *Positivist Review* (1893–1906) and edited Roger Bacon's *Opus majus.* He was much concerned with practical social reform, especially in its relation to health, and published lectures entitled *The Influence of Civilisation on Health* (1869), *A Catechism of Health for Primary Schools* (1870) and *Moral and Social Aspects of Health* (1877).

8.2 The present writer's earlier statement that Charles Roberts was a general practitioner of Uxbridge, Middlesex (Tanner, 1959) is wrong, and arose from a confusion about Roberts' identity.

8.3 The British Association for the Advancement of Science, at its Bristol meeting in 1875, set up an Anthropometric Committee, instructed to 'continue the collection of observations on the systematic examination of heights, weights etc. of human beings in the British Empire, and the publication of photographs of the typical races of the Empire'. William Farr was the chairman and Lane Fox (Pitt-Rivers) secretary. Galton was a member from the beginning and took over the chairmanship in 1881. John Beddoe, the author of the celebrated *Races of Britain* (1885), was a member, as was also Sir Rawson Rawson. Surprisingly, Charles Roberts was not a member till 1879. Perhaps this was because the early emphasis was on race differences in adults, with much stress on photographs (including Galton's famous superimpositions). Roberts, according to the Committee report for 1879, brought a collection of 50,000 observations of weight and height to join the 'nearly 12,000' assembled through the activities of the Committee (British Association Anthropometric Committee, 1879). In this report some interesting frequency distributions are given. 'Mr Whiteley's shopmen', a presumably relatively unselected group, averaged 5 feet 6½ inches at ages 20–40 (211 persons). Nearly 2,000 observations on boys at Christ's Hospital School, London (perhaps on as many boys, but more likely a mixed longitudinal series) aged 9 + to 16 + gave a maximum annual increment of height of 6.9 cm/yr centred at age 15.0.

In the 1880 report of the Committee, Marlborough College statistics for the years 1874–78 are given (that is data collected subsequent to Rodwell and Fergus' report of 1874). These give a maximum yearly height increment of 6.1 cm/yr centred as early as 14.0 years, but with increments spread rather evenly at centres 14.0, 15.0 and 16.0. Maximum yearly weight increment was at 16.0. This report also gives (pp. 142–3) the raw data, and individual growth centres, of Bowditch's thirteen girls and twelve boys followed longitudinally, that were described in his 1872 summary.

In 1881 a lengthy report was issued, almost entirely contributed, in separate sections, by Roberts and by Galton. Roberts divides his subjects now into four social classes, and Galton writes of interquartile and interdecile (10th–90th centile) ranges.

In 1883 a Final Report was issued. As regards growth only two new points were made. There is a table of birth lengths and weights, taken from Queen Charlotte's Lying-In Hospital in London and the Edinburgh Royal Maternity Charity, on 451 boys and 466 girls. The mean lengths were 19.52 inches (49.6 cm) and 19.32 inches (49.1 cm), and naked weights 7.12 lb (3.2 kg) and 6.94 lb (3.15 kg) (see Chapter 11).

Secondly a table is given comparing heights of boys aged 9–17 at the 'York Friends' School' from 1853 (the earliest school heights in Britain probably!) to 1879. There was no discernible change over this time.

This was the end of the first Anthropometric Committee of the British Association, but a second was convened in 1892, with Sir Douglas Galton (a cousin of Francis) as chairman and Dr F. Warner as secretary, to investigate physical deviations from normal in children at elementary schools; they found 19 per cent of boys and 16 per cent of girls had such deviations. A third committee was set up at the Nottingham meeting of the Association in 1893 to organize measurements of girls, but it only resulted in a rather small sample. A fourth committee was convened in Glasgow in 1901 to study pigmentation and a fifth at the same meeting, this time under the auspices of the Education Section of the British Association rather than the Anthropology Section, to report on conditions of health essential to work in schools (see Brabrook, 1904–5; British Association Anthropometric Committee, 1878, 1880, 1881, 1882, 1883).

8.4 Livy (1885) a few years later examined the teeth of 2,000 schoolchildren under 10 years old and 2,000 boys and girls over 10 working in mills and factories to see if he could propose a surer system. All these had, he says, valid birth certificates. Girls were more advanced in dentition than boys. For each tooth Livy gives the number of boys and girls who have it 'as their oldest tooth' at a given chronological age. He appears to ignore differences in order of eruption. If the lateral incisor, for example, is the oldest erupted tooth in a boy, then the boy is unlikely (from Livy's tables) to be more than age 12, and virtually certain to be not more than 14. The way Livy presented his results makes their use confusing and probably inefficient.

R. B. Bean, professor of anatomy at Tulane University, Louisiana, revived the subject with a massive survey of the numbers of erupted or erupting teeth in children in Ann Arbor, Michigan, in 1906–7 (Bean , 1914). Bean gave the ages at which each tooth had erupted in 50 per cent of the children. Girls were a half to one and a half years ahead of boys, depending on the tooth considered. Bean thought that periods of particularly active eruption initiated or preceded periods of rapid growth in stature. (He accepted Stratz's view of the rhythm of growth: see Chapter 9.) Children below the average grade for their age in school, Bean found, had 0.9 tooth less than those in the average grade; and 1.7 teeth less than those in above-average grades: indeed at age 10 these differences increased to 1.5 and 3.3 teeth respectively. He concluded that 'eruption of the teeth is a better criterion than age as a standard for both physical and mental development' (1914, p. 154).

8.5 At Bootham School heights and weights were taken six times a year, at the beginning and end of each school term, at least in the 1890s. Two hundred and fifty such records were analysed by A. H. Gray in 1909, though unfortunately with very little useful result. Gray gives a number of references to the belief, still current in the nineteenth century, that fevers, at least in some exceptional cases, cause a vast increase of growth (something Stöller was concerned to deny in the century before).

Gray obtained the records from the Anthropometric Register of the school. Sadly, the present headmaster tells me this appears now to be irretrievably lost.

8.6 B. A. Gould (1824–1896), a distinguished Bostonian astronomer who took his PhD with Gauss in Göttingen in 1848, and was author of *Investigations in the Military and Anthropological Statistics of American Soldiers* (1869), had already realized this. Henry Bowditch consulted him over the working up of his Boston data, and in a letter dated 4 August 1877 Gould wrote: 'The jumps – *schüsse* as the Germans call them – which appear in the growth of children near the times of second dentition and puberty . . . are liable to be obliterated in curves deduced from measures of individuals of different classes, since small differences in the normal epochs for these classes cause the effects of these phenomena to be superposed [sic]. But I feel sure that by a careful assessment of these cases these jumps can be made clearly manifest and their laws determined'. (Gould, 1877). Bowditch, rather understandably, seems not to have been much enlightened.

In fact, Gould had expressed the same thing with much greater clarity in his book (1869, pp. 116–17):

> Our discussion would be incomplete did it omit to recognize and illustrate the truth, that inferences drawn from the mean of all the men at each year of age may not always represent the facts for the average man with perfect correctness. This is well illustrated by Lehmann, in an able and ingenious memoir, in which he treats of the possibility of applying to individual cases the laws which have been deduced for the average man. That these laws may fail to indicate phenomena even of a strikingly marked character, occurring in every individual, and get so masked in the averages as actually to escape notice, will be manifest when we consider the so-called 'shoot' or sudden increase of growth, which occurs at or just preceding the chief epoch of physical development. The rate of increase in stature seems to diminish, regularly or nearly so, from birth, until the time at which the shoot takes place; it is then suddenly augmented by a very considerable amount, after which it diminishes again . . . in as much as the epoch of the shoot is extremely variable, fluctuating between the eleventh and the nineteenth year of age, the tokens of its occurrence disappear from the corresponding curve of mean growth. This latter manifests a nearly even progression during the ages in question, and rises at the average age for the shoot scarcely more rapidly than at adjacent ages, since the sudden accession of growth does not in the majority of cases occur at the average age. All indications of a sudden change in the rate of growth are wanting in the curve of mean stature, so that the investigator who studies the Law of Growth solely by the mean results from many individuals, might easily allow one of the most salient and unfailing phenomena connected with this law to remain unnoticed.

Lehmann's papers (1841, 1843*b*: see Chapter 6) were published a few years before Gould's stay in Göttingen.

The single reference to Galton appeared in Backman's (1934) scholarly monograph. Backman criticizes Manouvrier for writing that the curve of the average 'can be exactly compared with curves of each individual' and says 'the opinion of the English Anthropometric Committee of 1881 was quite the opposite', then followed the quotation (Backman, 1934, pp. 37, 72).

I am grateful to Mr Richard Wolffe, Curator of Manuscripts and Rare Books at the Countway Library, Boston, Mass., for drawing my attention both to Gould's letter and to the other aspects of Bowditch's correspondence discussed in this chapter.

8.7 The letter continues: 'and I hope this winter to get the result of observations on some 17,000 children in that city'. It is not clear whether such a survey did actually take place in Leipzig at that time. I have not found a record of one. (The Saxon Minister of Culture in an order of 1873 gave permissible limits for the height of Saxon boys aged 6–8, 8–10, 10–12, 12–14 (see Geissler and Uhlitzsch, 1888) but perhaps that is not quite the same thing.) It was only in 1876, the year after Bowditch's letter, that Arthur Geissler (1832–1902) joined the Royal Saxon Statistical Bureau as medical assistant. Geissler was a physician who graduated at Leipzig, served for a while as a factory doctor and ophthalmologist, and then turned to statistics, being one of the first physicians in continental Europe to do so. In the 1880s he made an excellent analysis of the survey of the growth of children in Freiberg (modern Freyburg), a town close to Leipzig, done by the school-director Lose (Geissler and Uhlitzsch, 1888); but the first survey in Leipzig itself seems to have been a few years later than this, in 1889 (see Uhlitzsch, 1891; Geissler, 1892). In 1898 Geissler was made director of the Statistical Bureau. He died in 1902. The head of the Bureau at the time of Bowditch's letter was Viktor Böhmert.

Uhlitzsch, Geissler's collaborator, in 1891 wrote an excellent review of anthropometry and its usefulness. Clearly addressed to statisticians and administrators, it summarized the major cross-sectional studies of schoolchildren, and left longitudinal studies aside. Uhlitzsch was particularly struck by the work of Galton, whose *Natural Inheritance* (1889) he had evidently just read. After first pointing out the difference in growth between rich and poor, he deals with hereditary family influences, being one of the first to do so. He mentions the usefulness of anthropometry in the design of school desks and in the monitoring of health in school.

Chapter 9

9.1 One of the few complete copies of Key's work is in the library of the Karolinska Institute in Stockholm. All aspects of school hygiene are dealt with, from the incidence of signs of disease to the length of hours worked and the time demanded for homework. There is much concern about the problem of 'over-pressure' caused by schoolwork beyond the supposed powers of the pupil's bodily development to support it; a chapter on history shows that this idea, still present in Porter's mind in the 1890s, goes back at least as far as 1836 (Key, 1885, p. 16), and was very generally held in the German states. In the chapter on bodily growth Key cites Bowditch, Pagliani and Kotelmann as well as Quetelet, but he evidently does not know Roberts' work at first-hand.

I am much indebted to Dr Gunilla Lindgren, of the School of Education in Stockholm, for help in locating this book.

9.2 A useful review of all American and many non-American studies up to 1898 is given by Frederic Burk (1898), who worked in Stanley Hall's department (see Chapter 10) in Clark University, and published in the journal that Stanley Hall edited. There are tables of mean height and weight each year for all studies reported, also of the *enfant moyen americain* (mean values of all American studies, comprising about 45,000 boys and 42,000 girls (peak height increments at 15.0 and 13.0 years)). The frequency distributions of heights at each year of age are also given for these data. The raw data of Christian Wiener's four sons are reprinted, with their increments, and a graph of the velocities. The bibliography lists 109 items, many annotated.

Chapter 11

11.1 Michel Friedlander (Friedlaender) was yet another graduate of Halle. Born in Königsberg, he completed his thesis in 1791 and then studied further in Berlin. About 1800 he went to Paris where he remained for the rest of his life. He edited a German-language journal designed to bring the new French findings in physics, chemistry, physiology and biology to German readers (*Neueste Entdeckungen der französischen Gelehrten . . .*, Hamburg, 1802–15). Most of his *De l'éducation physique de l'homme* originally appeared in sections in the new journal *Annales de l'éducation* first issued in 1811. His interest in education and popular enlightenment extended to social reform in general and his last work was a *Bibliographie méthodique des ouvrages publiés en Allemagne sur les pauvres* (1822: *Biographische Lexicon herrorrä gender Aerzte* (1930), vol. 2, 622).

Friedlander was a younger contemporary of Johann Pestalozzi (1746–1827), the Swiss educational reformer whose revolutionary ideas and didactic novels exerted a profound effect on educational thought at the beginning of the nineteenth century. Pestalozzi had a famous school in Yverdon from about 1800 to 1820 and it seems likely that M. Schwartz was one of its many trainees. The great Friedrich Froebel (1782–1852) was another, but Schwartz, if there at all, would probably have been before Froebel's time.

11.2 Quetelet's data came from newborns at the St Pierre Maternity Hospital in Brussels. Length was measured with Chaussier's *mécomètre* (as it had been in his earlier series of lengths and heights only) and weight 'using an ordinary balance . . . taking into account the clothes in which the child was swaddled' (Quetelet, 1833). Quetelet gives frequency distributions of weight but not length of sixty-three boys and fifty-six girls. The lowest weights were 2.34 kg in boys and 1.12 kg in girls, so presumably some pre-term infants were included. His own means (3.20 kg and 2.91 kg) seem to be incorrectly calculated, for the means calculated from his frequency distributions come to 3.29 kg and 3.05 kg in the two sexes. Quetelet gives means of length as 49.6 cm and 48.3 cm. (Roberts, incidently, remarked in his *Manual of Anthropometry* that Quetelet's observations were made on 'dead' infants, but 'carried to term and normally proportioned' (Quetelet's words translated) and naturally criticizes this source of data. Roberts is quoting *Anthropométrie*, however, in which Quetelet details large numbers of measurements, such as would indeed be very hard to take on a living newborn infant. The infants of 1833 who only had lengths and weights measured were evidently living. Bowditch made a similar confusion between the *Anthropométrie* series and the 1830s survey (see note 6.4).)

11.3 The use of fetal length as a measure of fetal age became quite widespread. One famous – it should have been infamous – rule was that of Haase (1875). Haase reported the birthlength statistics for the Berlin Charité for 1875, giving modal lengths of 50 cm for both sexes. In a footnote he added what was intended clearly as an approximate mnemonic for rough clinical use: if length is greater than 25 cm, divide by 5 to get the month of gestation of the stillborn; if less than 25 cm, take the square root (9 cm gives 3 months, 16 cm 4 months, etc.). Figures derived from Haase's rule – 1, 4, 9, 16, 25, 30, 35, 40, 45, 50 cm – turn up quite frequently in later literature, for example in Stratz's *Der Körper des Kindes* (1903).

11.4 It may be that these were the data, said to be from His (1874), that D'Arcy Thompson used for drawing his curve of fetal growth in *On Growth and Form* (1942, p. 111). Actually His (1874) contains no overall length measurements at all, so D'Arcy for once has slipped up in his references. A search of other articles by His at around this time reveals nothing. Stratz's smoothed curve, however, does give approximately the values that D'Arcy lists, if D'Arcy's value of 27.5 cm at five months is a misprint for 25.7 cm, and the second of the two final values he gives, 49.0, is ignored. Perhaps there was some intermediate source between Stratz and D'Arcy.

11.5 Crêches had been started in 1844 in France and 1850 in England (Hewitt, 1958), but they produced no weight measurements, perhaps because they were not run by doctors and perhaps because they only took in children over six months old.

11.6 Lehmann in the 1840s (see Chapter 6, p. 138) was the first to realize the importance of the difference between longitudinal and cross-sectional data. B. A. Gould quoted him in 1869. Though lost in the United States, it seems that an understanding of this issue lingered on in Germany, for Karl Vierordt, in 1877, wrote with understanding of the 'individualizing' and 'generalizing' methods, though without actually quoting Lehmann. Understanding was reintroduced in America by Boas, and Boas, of course, had his undergraduate training in Germany.

11.7 So far as I can ascertain, the word 'menarche' was first introduced by Stratz in

1908 in the proceedings of the German Gynaecological Society. In the same article he introduced also the word 'tokarche' for the beginning of the fertile period (which he believed was some years after menarche). The latter word never caught on.

11.8 Women whose dates of menarche were actually recorded at the time (because they were subjects in the longitudinal studies described in Chapters 12 and 13) have been interrogated at age 30 or thereabouts. Despite having been made acutely conscious in their youth of everything to do with their growth and development, they were frequently several months, and sometimes a year in error (Damon, Damon, Reed and Valadian, 1969; Damon and Bajema, 1974: see also Livson and McNeill, 1962, and Bergsten-Brucefors, 1976, for similar results).

11.9 The calculation of mean ages of menarche in the nineteenth century is attended by considerable confusion. For the most part authors give frequency distributions of numbers of women having their first period 'at age 14, age 15' etc. The problem is whether these age classes represent age-at-last-birthday, age-at-nearest-birthday, or, even, perish the thought, age-at-next-birthday. Evidently this depends on the precise nature of the query made to the women in question, and this is practically never stated.

Nowadays it is customary to use age-at-last-birthday; thus when a women is asked 'How old were you when you first menstruated?' and she replies 'Fourteen', one assumes she belongs to a group of women who have a mean age of menarche of 14.5 years. But there is no guarantee this was common usage a hundred years ago. Even nowadays when patients are asked the age of one of their children they frequently say 'Fourteen' when the child is perhaps 13.8 or 13.9, though not when he is 13.5 or even 13.6 (see also Lenner, 1944, p. 114). Further, if the questioner persists in requiring an answer to the month, then the reply 'Fourteen' will be taken to mean 14.0, just as 'Fourteen and four months' will be taken to mean 14.3. In this case there will probably be a great piling-up of values at 14.0 (and probably at 14.5). Such a piling-up might well induce a grouping around 14, called 'age 14', and centred at least approximately on 14.0.

We do not know what was the question that Roberton actually asked. However, in one instance, that of Roberton's contemporary Guy (1845), the question is explicitly stated. Guy says clearly that he used age-at-nearest-birthday. Another contemporary, Whitehead (1847), curiously but carefully states that those entered, in his main table, under 'age 15' were girls less than eight months above 15.0, but eight months and more above 14.0; fortunately he also gives the total sums of the ages of his groups in months, making it clear that what was actually asked was age in years and months. Weissenberg (1909) in Russia, who specified exactly what he did, also used age-at-nearest-birthday. On the whole then it seems that we do better to assume that during most of the nineteenth century age-at-nearest-birthday was used rather than age-at-last-birthday. Brudevoll, Liestøl and Walløe (1979) reached a similar conclusion in a careful study of the Norwegian archives of the same period. They were able to find the original records, and enough of these gave ages such as '14½ years', '14 years 3 months' and suchlike to make the designation '14 years' clearly imply '14.0 years as the nearest I can remember'. Thus, throughout this section on age at menarche I have assumed that prior to 1890 (unless the author seems to imply to the contrary) the class labelled 'age 14 years' has its centre at age 14.0 years, and means have been calculated accordingly. Such means will err on the low side, if they err at all, and may perhaps underestimate by 0.1 or 0.2 years.

All Roberton's papers, except the first, used the age classes '14 years' etc. and give means calculated on the basis of 14.0 being the centre of the class. In the 1830 paper the age classes are labelled 'in the eleventh year, in the twelfth year', etc. Now the 'eleventh year' should be the year *ending* at 11.0. The centre of such an age class would therefore be 10.5 years. This leads to a result out of line with Roberton's subsequent work; and he himself says, in his 1851 book (p. 29), that the classes in 1830 were wrongly labelled and 'in the eleventh year' should actually read 'age 11 years'. German usage in the late nineteenth century frequently makes the same error: it is clear, for example, that when Krieger (1869) uses the words *'im 15 ten Lebensjahr'* he means the year 15 +, for one particular case menstruated *in ihrem 9 Jahr* at 9 years some months old. Robertson did exactly the same, writing in English in 1908. Weissenberg (1909), on the other hand, clearly specifies that in one class go all those who had menarche between 14½ and 15½ years old: he then calls this class the *15 Lebensjahr.*

Chapter 12

12.1 An earnest of this concern was the series of meetings held under the auspices of the World Health Organization in Geneva in 1953, 1954, 1955 and 1956. Instigated by Ronald Hargreaves, chief of Mental Health at WHO and later professor of psychiatry at Leeds University, with the help of the Macey Foundation, the meetings brought together in discussion a quite remarkable galaxy of diverse talent. At the opening meeting were Margaret Mead, Jean Piaget, Konrad Lorenz, W. Grey Walter, John Bowlby and René Zazzo; at subsequent ones the same plus Erik Erikson, Julian Huxley and L. Von Bertalanffy. J. M. Tanner and Bärbel Inhelder (Piaget's chief assistant), also there, reduced the transcripts of the meetings to four volumes of talk which, though still spontaneous, is surprisingly grammatical and consecutive (Tanner and Inhelder, 1956–60). A very wide range of subjects pertinent to child development was discussed. Understanding between the participants can be seen to be growing in each successive year. In the final meeting Piaget gives an account of the basic model for development, endeavouring in it to find a structure common to the physical, intellectual and behavioural development of the child. The four volumes exerted a considerable influence amongst educationists in the 1960s.

12.2 Lawrence Frank appears to be in error when he writes (1962, p. 209) that the Station 'was initiated and financed by a grant from the Women's Christian Temperance Union'. In fact the money was voted by the State Legislature. The Temperance Union was one of the numerous groups in Iowa listed as helping in the campaign to get the Bill passed. It also provided, in 1919, a grant of $10,000 a year for five years to support work in eugenics (University of Iowa, 1933, p. 48). A major source of later support was the Laura Spelman Memorial Fund, by way of Lawrence Frank himself. The following notice (Baldwin, 1921, p. 33) was sent to medical officers and others:

> The University of Iowa, through its Child Welfare Research Station and Extension Division, has formulated a cooperative plan to assist school officers and parents in the State in recording and evaluating the semi-annual physical measurements of the growth of their boys and girls, between the ages of 5½ and 18 years. These measurements, which are few and simple, offer the best indices of growth, health and nutrition. The measurements are to be taken in the school by the medical inspector or the teacher of mathematics or science who should be especially accurate. The cards are forwarded in June to the Extension Division and the Research Station records the results, compares them with standards for Iowa children and returns the cards and tabulated results with critical comments through the Extension Division.
>
> The essential principle of the plan lies in the cooperation between the University specialists and the school or parents, affording continuous observations on the same children for periods from one to twelve years. In no homes or schools in the

United States have an appreciable number of children been measured consecutively. Iowa is the first state to organize, as one phase of its child welfare work, a standardized method for repeated measurements on the same boys and girls for long periods of time, resulting in *individual history* curves of scientific value. An extension of the plan will be issued subsequently, to include a similar method for the measurement of infants from birth to six years of age.

Arrangements can be made by the larger school systems of the state to have a limited number of the staff trained by the Research Station at the University or through the institutes, in accurate methods of making and recording data on the measurements to be taken.

The individual growth cards of the child whose measurements are recorded, follow one child from grade to grade in school, or from one portion of the state to another, if a change of locality becomes necessary. No expense is involved except for the cost of the individual record card, which may be used for twelve years, and the postage on the part of the school or parent that sends the cards to the University. For details of this work see Extension Bulletin No. 59.

12.4 This priceless book is hard to come by. The writer's own copy was the extraordinarily generous gift of Howard Meredith to a young and totally unknown lecturer from the University of Oxford travelling through the United States in 1948 prior to setting up his own longitudinal growth study in Harpenden, England (Tanner, 1948). I spent a week in August learning measurement techniques from Meredith and his colleagues; the original techniques in use at the Station had been set up personally by the legendary Ales Hrdlicka in 1918 and thirty years later they remained untarnished by any trace of slackness or over-familiarity.

12.5 It is impossible to comment in detail on this vast assembly of papers. A few points are of interest, however. Montbeillard is mentioned, indeed, as 'probably the first instance of successive measurement of a child', but very cursorily and with a reference to an edition of Buffon published in 1837. Pagliani's 1879 paper is listed, but is one of the few major papers given without any annotation, suggesting that Baldwin only read it in outline; his summaries of Pagliani's other papers are not precisely accurate so perhaps he had trouble translating Italian.

12.6 Rotch took radiographs of the hands of 200 'normal' children attending the Children's Hospital in Boston and classified the degree of development of the bones using an alphabetic rating. He later took the average age of all those with similar radiological appearances to be the 'anatomic age' of a child with this particular appearance, in a method that was the prototype of Todd's atlas, Rotch expressed forcefully, if somewhat inelegantly, the reasons for this departure. 'The people must be educated', he wrote, 'up to the plane of intelligently seeing that because an individual has been born three or four years, this does not necessarily mean that such chronological age should be rigidly adopted for entering a kindergarten; that because it is six or seven years of chronological age it should necessarily be in the usual grade of school corresponding to that age; that because it is 10 or 12 chronological years of age, it should necessarily be grouped in athletics with boys or girls of that chronological age; or because it is 14 or 15 or 16 years of age it should be allowed to work beyond what its anatomic development shows it can do without physical harm, as for instance in the mills. Anatomic age, like the physiological age recently proposed by Boas and Crampton, would be a much better criterion' (1909, p. 619). Rotch was still concerned with the problem of over-pressure. The radiographic grading, he believed, 'represents what an individual child or adolescent is able to take without mental or physical overstrain' (1910*a*, p. 399).

It is interesting to read F. K. Shuttleworth's opinion in 1938 of this development. A series of papers by Rotch, beginning in 1908 [he wrote] attributed large educational significance to skeletal development and profoundly altered the course of later research in America. A pressing problem at that time was the discovery of some criterion other than mere chronological age for the educational classification of school children. The problem was acute since psychologists beginning with Cattell in 1896 had failed to find significant relationships between educational progress and such tests as sensory acuity, weight discrimination, etc. Rotch advocated school grading by anatomical age determined from X-rays of the carpal bones. Antedating Rotch by a year, Crampton had attacked the same problem; advocating, instead, educational classification by physiological age, as indicated by pubic hair. The true key was already available in 1905 but in an almost unnoticed article by Binet and Simon ... Crampton has been undeservedly neglected these 30 years, whereas Rotch has survived to plague scores of students with a laborious and expensive and fruitless search for educational correlations with carpal development (Shuttleworth, 1938*a*, p. 6).

12.7 Howard Meredith is one of the few people to have had an obituary notice published while he was still alive. The perpetrator was the present writer; and it occurred in what was ostensibly a review of Meredith's collected orthodontic papers (Meredith, Knott and collaborators, 1973, reviewed in *Annals of Human Biology*, vol. 2, p. 413). Some months previously I had been approached by the editors of an encyclopaedia who told me that Meredith, on his sick-bed, had requested that I should be asked to see the proofs of his article on 'Growth' through the press, changing anything I thought necessary and adding my own name as co-author if I wished. The letter was couched, as I remember it, in terms which suggested Meredith, who had had severe arthritis for many years, was not in a position to recover. Fortunately I had the pleasure of sending him a copy of the review-obituary a year or two later.

12.8 The selection of persons whose help Shuttleworth acknowledges in the preface to his first monograph is not without interest. There was Lawrence Frank 'for continued emphasis on developmental problems'; Mark May (his head of department) 'for the first impetus towards the serious study of the statistical problems involved in the analysis of longitudinal data'; and, very significantly, Oscar W. Richards and Carroll E. Palmer 'for a critical reading of the entire manuscript and for a great many constructive and valuable suggestions'. Oscar Richards was a zoologist, and joint author of one of the most penetrating papers ever written on the analysis of growth data (Richards and Kavanagh, 1945): it appeared in D'Arcy Thompson's Festschrift.

12.9 Horace Gray's work on growth in private-school children began in Boston towards the end of World War I (Gray and Gray, 1917). In 1921 he published a paper with W. J. Jacomb, the physical educationist at Groton School, on the heights and weights of the pupils there, pointing out they were above those of the average American boy, though with a lower weight-for-height (at given age). He concluded that Wood's Life Extension Institute table of weight-for-height, at that time much used, and disseminated by the Bureau of Education, was quite inappropriate for well-off children, and proposed some weight-for-height charts of his own (Gray and Jacomb, 1921; Gray, 1921). He was one of the first to point out (in 1927) the occurrence of the secular trend in American children. Gray was also one of the few who raised a sceptical voice in the 1930s regarding the possibility of studying the 'whole child'. There is a widening tendency', he wrote, 'in some circles at least, to flourish phrases about integration and correlation, as if these mathematical terms in themselves lent security to the statements affirmed' (Gray and Ayres, 1931, p. 240). In 1948 he published a paper in which various methods of predicting adult height were compared.

12.10 Davenport, in the same paper, used the phrase 'peak year of male growth', which combined with D'Arcy Thompson's use of 'velocity' seems likely to be the origin of the expression 'peak height velocity' popularized by Tanner. Both Davenport and Shuttleworth (whose equivalent phrase was 'maximal growth age') used 'growth' as synonymous with 'growth velocity': Tanner felt this to be insufficiently explicit.

The introduction of the famous (perhaps infamous) terminology 'longitudinal' and 'cross-sectional' came later, but the details are not clear. Bayley and Davis in 1935 did not use the term 'longitudinal' but 'seriatim', a word used in 1927 by Scammon in talking of Montbeillard. But in 1937 Shuttleworth used both 'longitudinal' and 'cross-sectional' without feeling much explanation of their meaning was necessary and in 1938 MacFarlane did the same. Todd also used 'longitudinal' at one point in the commentary to his 1937 atlas. It seems likely the terms were introduced via developmental psychology in 1935–6. But possibly Shuttleworth himself was responsible.

12.11 The writer has good reason to remember Carroll Palmer with affection. In 1948, visiting the USA for a Viking Fund Anthropology Conference and subsequently touring the growth studies, I called on Palmer, then in charge of the tuberculosis service of the US Public Health Service, located in Washington. In the course of conversation it emerged that I could only get as far as Ohio in my tour because of lack of funds; Denver and California were denied me. Palmer was aghast: Frank Shuttleworth was spending a sabbatical year in Berkeley and would put me up, he was sure; Nancy Bayley was there and on no account to be missed; in Denver Washburn was pre-eminent and his large team involved in work not duplicated elsewhere. Within ten minutes a cheque on the USPHS petty cash account had been endorsed and I was on my way to California.

Chapter 13

13.1 This small and inexperienced team fell into a trap well known to all journal editors. Recognizing the scientific goldmine they had wandered into, they labelled their first batch of ore 'Aberdeen Growth Study I'. But as so often in such situations, events overtook them and the planned Aberdeen Growth Study II, III, ... never appeared at all.

13.2 Tanner had a school background of mathematics and engineering but changed plans and entered medical school in London. Soon after, World War II started and each year a number of British medical students were sent for training in the United States in a scheme run by the Rockefeller Foundation. Tanner was one of these and spent two and a half years at the University of Pennsylvania Medical School. He had always intended a career in research and specifically in an area of research where, as he naively put it in the United States selection interview, 'biology, psychology and sociology meet'. At Pennsylvania he began by working on human cardiovascular physiology under Isaac Starr, using his fellow students as subjects, and was soon forced to realize the importance of body size and shape in assessing his results. Accordingly, he trained briefly in somatotyping with W. H. Sheldon and his associates, before finishing his medical studies at Johns Hopkins and returning to London. When he got to Oxford in 1946 the professor, Le Gros Clark, knowing his interest, required him to give a course, not only in differences in adult physique, but in how they came about during the growing period. Clark had an ulterior motive; as an anatomist he had been plagued by his neighbour Ryle who complained that the infants he was studying had a habit of diminishing in length when they should be growing, and demanded an explanation (which a glance at the technique of measuring employed soon provided). Tanner's interest in growth was aroused and Bransby's invitation fell on receptive ears.

13.3 The circumstances were as follows. At the annual meeting of the British Paediatric

Society in 1952 Debré gave the Windermere Lecture, an address traditionally delivered by a distinguished foreign paediatrician. At the same session Falkner presented a short paper describing the longitudinal research just started at the Institute of Child Health. Debré thereupon approached Moncrieff with the idea of starting an exactly similar study and requested Falkner's aid in setting it up. Falkner spent 1953 in Paris doing this, sponsored partly by the Hôpital des Enfants Malades and partly by the Centre International de l'Enfance. In the following year Natalie Masse joined the Staff of the Centre and took over the general direction of the French study and also the coordination of the other studies as they entered in 1955 and 1956.

13.4 On 6 February 1958 seven of the persons attending the symposium and engaged in research in what they thought of as human biology met at Tanner's laboratory at the Institute of Child Health, and resolved to form the Society. Besides Tanner himself, there was J. S. Weiner and D. R. Roberts, the editors of the symposium, which appeared eventually as volume 1 in the Society symposia series, G. A. Harrison of the University of Oxford, A. E. Mourant, E. Ashton from Birmingham, J. C. Trevor from the University of Cambridge and K. Oakley from the Natural History Museum. The Society was formally inaugurated on 7 May 1958 at the Natural History Museum, with J. Z. Young as its first chairman. In the symposium volume Young argued forcefully for human biology as a liberal and scientific discipline: 'I believe', he wrote, 'that the objective study of man by biological techniques such as an anthropologist can use provides as firm a foundation as any other for a liberal education'. Weiner outlined the content of a proposed course in human biology, and Tanner, considering medicine as essentially the practice of applied human biology, proposed redesigning the pre-clinical part of the medical curriculum as a course on the life of man, dropping entirely the traditional divisions into anatomy, physiology and so on.

Chapter 14

14.1 The early history of United States growth standards is described in Benedict and Talbot (1921). It seems that the first large-scale source of values from 6 months to 4 years was provided by Better-Baby Contests inaugurated by the *Women's Home Companion* about 1913 and soon taken over as Baby Health Conferences by the American Medical Association. The contests (or conferences) were held in twenty-three states. The values for weight and length of 5,602 boys and 4,821 girls were reported by Crum (1916), who gave means for each month of age from 6 months to 4 years. More than 100 children of each sex were represented at each month up to age 3.0. In addition to length and weight, circumferences of head, chest and abdomen were taken, two diameters of the chest, and the length of the arms and legs.

For schoolchildren Boas' 1898 standards were used, but as Gray pointed out in 1917 they were too low for private-school boys (partly because of the social class difference and partly because of the secular trend affecting children in general). In addition the weights for given height given by Wood's Life Extension Institute tables were too high for private-school children. At this point Baldwin entered the field and produced with Wood tables of weight-for-height for each year of age based on a large number of schoolchildren, many in private schools (Baldwin and Wood, 1923). Gray replied by producing his famous private-school measurements, giving means and standard deviations for a large number of measurements simply in terms of age (Gray and Ayres, 1931). Many subsequent authors used these as standards.

As the Baldwin–Wood table of 1923 began to get out of date because of the secular trend, the new growth-data being accumulated at Iowa were pressed into service. They first appeared as a practical standard only in 1946, under the joint authorship of Harold Stuart and Howard Meredith. Centiles were used, and the 10th, 25th, 50th, 75th and 90th were charted from age 5 to 18 for height, weight, hip width and chest circumference, which Stuart felt were the most important measurements in judging healthy progress (Stuart and Meredith, 1946). Several years later the gap from birth to 5 was filled from the Harvard School of Public Health Longitudinal Studies data and the whole 0–18 range appeared as two tables in the fifth edition of *Mitchell-Nelson's Textbook of Pediatrics* (Stuart and Stevenson, 1950). For the ages 0–5 the 3rd and 97th centiles were given in addition to the others; this seems to have been the beginning of the custom followed by most later makers of centile standards, who give the 3rd and 97th as their outer centiles. (In the fourth edition of *Mitchell-Nelson* standard deviations were given: see Palmer and Ciocco, 1946.)

These weight and height standards were used (usually under the name of Harvard–Iowa standards) for thirty years, yet the numbers on which they were based, as well as the provenance of the subjects, rendered them far from suitable as a national (still less an international) reference. At age 5 there were 235 boys and 210 girls; at subsequent ages a decreasing number, since the Iowa study was mixed longitudinal. At age 12 the numbers were 95 boys and 112 girls; then new adolescents entered the study and the numbers increased a little, to revert to 86 and 87 at age 18. The effective numbers were of course considerably lower, since most children were present on four or five occasions. This statistical disadvantage of longitudinal over cross-sectional data had been pointed out by Pagliani in 1879; and Shuttleworth, that great exponent of longitudinal studies, ignoring for once the shape-of-curve problem, had bluntly written 'Longitudinal data have no advantage whatever over cross-sectional data for the purpose of determining the average size of children in general at any age' (Shuttleworth, 1937, p. 180). Furthermore the children from 5 to 18 were drawn almost exclusively from the well-off class of Iowa society – from families of businessmen, proprietors, managers and university teachers – while the values from birth to 5 were based on lower middle or working class Boston children. It is astonishing that this statistically improbable chimaera should have been used as the growth standard for thirty years in the United States and indeed in many a developing country. Such is the effect of professional prestige and political power.

14.2 This paper became one of the most widely quoted in the paediatric literature of the 1970s. *Current Contents* (1979, vol. 7, p. 12) listed it as a 'citation classic', cited over 300 times in articles in major journals between 1966 and 1979.

INDEX

Note: page numbers in *italic* typre refer to figures.